Handbook of Plants with Pest-Control Properties

Handbook of Plants with Pest-Control Properties

Michael Grainge and Saleem Ahmed

with the special assistance of

Ponciano Epino
Eric Mercadejas
Boniface Peiris
Xenia Wolff

computer programming by

Hong-chun "June" Yang
D. Patricia Goddard

Resource Systems Institute
East-West Center
Honolulu, Hawaii

WILEY

A Wiley-Interscience Publication

JOHN WILEY & SONS

New York   Chichester   Brisbane   Toronto   Singapore

Library of Congress Cataloging-in-Publication Data

Grainge, Michael.
    Handbook of plants with pest-control properties/Michael Grainge
and Saleem Ahmed, with the special assistance of Ponciano Epino . . .
[et al.]; computer programming by Hong-chun "June" Yang, Patricia
Goddard.
      p. cm.
    "A Wiley-Interscience publication."
    Includes index.
    Bibliography: p.
    ISBN 0-471-63257-0
    1. Pesticidal plants—Handbooks, manuals, etc. I. Ahmed, Saleem.
II. Title.
SB292.A2G73 1988
632'.96—dc19   87-28572
                                                    CIP

Printed in the United States of America

10 9 8 7 6 5 4 3 2 1

Foreword

Plants are nature's "chemical factories," providing the richest source of organic chemicals on Earth. Although only about 10,000 secondary plant metabolites have been chemically defined, it is estimated that the total number of plant chemicals may amount to 400,000 or more. During their evolution from the Devonian period, many plant taxa have evolved highly sophisticated defense systems, largely a complex array of defense chemicals produced by the plants themselves, against insects, mites, pathogens, and even weeds. They provide a rich source of botanical pesticides. Many of the oldest and most common pesticides, such as nicotine, pyrethrins, and rotenone, were derived from plants. The chemical or pesticidal approach in agriculture had its beginning in the use of botanical materials.

The current problem of pesticide resistance and detrimental effects on nontarget organisms, including humans, associated with the use of synthetic broad-spectrum pesticides has revived interest in exploiting the unlimited pest control potential of plants. It is in this context that the *Handbook of Plants with Pest-Control Properties* by Michael Grainge and Saleem Ahmed, which catalogues 2,400 plant species, is an important milestone. The handbook provides useful clues on plant characteristics, bioactive materials and their properties, methods of preparation and extraction, methods of application, cautions in use, and other complementary uses. This compendium is obviously the result of the authors' sustained search through highly scattered and sometimes inaccessible data collected over several years from different corners of the globe. This encyclopedia of information will motivate plant protection researchers to study the use of the pest control potential of the plant species it describes. The authors deserve our gratitude for this valuable book.

M.S. SWAMINATHAN
DIRECTOR GENERAL
INTERNATIONAL RICE RESEARCH INSTITUTE

October 1987

v

About the Authors

This *Handbook* is the result of the collaboration of Michael Grainge, a graduate student at the University of Hawaii working under an East-West Center scholarship, and his East-West Center advisor, Saleem Ahmed, at the Resource Systems Institute of the East-West Center in Honolulu, Hawaii.

Michael Grainge is a Peace Corps volunteer in the Cameroons and supports the work of other, non-governmental, organizations concerned with bettering living conditions of the poor in developing countries. Born in England and raised in Canada, Michael Grainge is now a U.S. national. He speaks Tagalok, Vietnamese, and French. Mr. Grainge earned his Masters of Science degree in Plant Pathology at the University of Hawaii in 1987. He also holds an MAgr degree in Soil Science/ Horticulture (1979) from the University of the Phillipines at Los Banos, an MEd in Special Education (1972) and a BA in Asian Studies (1970), both from the University of Hawaii. He is also the co-author of *Plant Species Having Pest-control Properties: An EWC/UH Database,* published by the East-West Center in 1985.

Saleem Ahmed is Research Associate, and Leader, Botanical Pest Control Project at the Resource Systems Institute of the East-West Center in Honolulu, where he has worked since 1973. Dr. Ahmed is concerned with rural development and is Convenor of Americans for Impartiality in the Middle East (AIME). Born in India and raised in Pakistan, Mr. Ahmed is now a U.S. national. He speaks Urdu, Hindi, and Persian. Dr. Ahmed was a Senior Fulbright Scholar in India in 1985, and has worked under grants from American, Asian, and international agencies for botanical pest control multidisciplinary research. He has served as consultant to national and international agencies on projects concerning crop production, fertilizer marketing, and rural development, and has edited the proceedings of several international conferences on fertilizer use and plant protection. Mr. Ahmed received his doctorate in Soil Science from the University of Hawaii in 1965, while studying under an East-West Center scholarship, and he earned an MSc in Geology (1961) and a BSc (1960) in Geology and Chemistry from the University of Karachi, Pakistan.

Preface

Approximately one-third of global agricultural production, valued at more than U.S. $1 billion, is reportedly destroyed each year by more than 20,000 species of insects, mites, and nematodes, as well as plant diseases, weeds, rodents, and other plant pests. Losses are higher in developing countries, and may exceed 40 percent on crops such as rice, sugarcane, and cotton. Synthetic pesticides used for control are not without problems: toxicity to nontarget organisms, development of pest resistance, and environmental degradation often result from their continued and injudicious use. Alternative pest control strategies are needed, and selected plant materials may provide an answer under certain conditions. Although many plants have been used as medicines or as a source of medicinal compounds, few have gained importance as pesticides, perhaps due more to a lack of scientific attention than to a lack of pest-control properties.

This database, the fruit of more than six years' labor, provides information on approximately 2,400 plant species reportedly possessing pest-control properties and it lists an additional 1,000 plants that may have pest-control properties because of the poisonous nature of some of their constituents or because of their use in controlling human and animal diseases. The species included represent an assortment of plant types, from aquatic weeds to giant trees, from tropical evergreens to desert succulents, and from highly poisonous to completely edible.

The pest control properties of plants may be utilized in two ways: One approach is to use plant tissue or crude derivate, such as an aqueous or organic extract, directly. The second approach is to isolate, identify, and process the active compound and then, if possible, to produce it or its active analogs through industrial processes. The first, more traditional, approach is practiced in some farming communities in developing countries; the second, more scientific, approach is used to produce plant-based products such as pyrethrum and chemical products such as pyrethrins. These approaches need not be mutually exclusive, a point probably best illustrated by the potential offered by South Asia's neem tree (*Azadirachta indica*).

In the process of compiling information we were struck by the inadequacy of knowledge in some areas. For example, few of the catalogued plant species have been subjected to needed toxicological assessment. Information on the extent of current use of plant materials for pest control and reasons for nonuse was also found lacking. Our preliminary surveys in South Asia, undertaken in collaboration with local institutions, indicate that some farmers have discontinued the traditional pest-control use of neem tree materials for fear of being labeled "backward." This stems in part from well-intentioned government efforts to increase food production through the use of "modern" inputs. Follow-up interdisciplinary research in this and other areas is suggested. Human problems transcend disciplinary boundaries; and although these boundaries are essential for in-depth scholarly analyses of their disparate segments, comprehensive deliberations are necessary to help policymakers arrive at solutions.

We were also struck by the fact that the listed 2,400 plant species have been found effective in collectively controlling only 800 pest species out of the more than 20,000 that reportedly cause damage to food and agriculture. We suspect this stems in part from the fact that laboratory testing is usually conducted with pests that are easier to rear and work with. *Attagenus piceus* (black carpet beetle), for example, is controlled by 240 plant species; on the other hand, few plants have been found thus far to control some major insect pests such as *Tryporyza incertulas* (yellow rice stem borer) and *Nilaparvata lugens* (brown planthopper). There is a need, therefore, to focus on important pest species in various areas, and to expand the testing program beyond insects to include other pests also.

Finally, we were struck by the observation that technical and social scientists in various countries are, at times, unaware of the knowledge possessed by the area's traditional farmers regarding pest-control properties of some local plants. This underscores the need for increased interaction, not only among scientists, but also between them and local farmers. This may facilitate greatly the task of developing viable pest-control technologies based on indigenous renewable resources which are in harmony with nature.

The Resource Systems Institute of the East-West Center addresses emerging issues in resources and development as countries of the Asia-Pacific region strive to enhance quality of life. Issues pertaining to the development of safe and effective pest-control strategies are among the ones currently being actively debated in international circles. On the one hand, there cannot be adequate agricultural production without pest control; on the other, the environment may be harmed by the continued injudicious use of synthetic pesticides. We hope this database encourages efforts to develop alternative pest control strategies that will help alleviate this dilemma. Many plant materials inspire hope that solutions can be found; these may serve as models for the next generation of safe and effective pest-control (as opposed to pesticidal) chemicals.

To facilitate the use of this document, we urge you to read the section "How to Use This Database Effectively." It provides an explanation of the approach we have followed and also defines the codes we have used to describe various characteristics of the listed plant species and of their pest-control materials.

SALEEM AHMED

Honolulu, Hawaii
August 1987

Acknowledgments

The information provided in this document is based upon literature search, backed up by survey responses we received from relevant national, regional, and international agencies located in Bangladesh, China, Costa Rica, France, Fiji, India, Malaysia, Mauritius, Mexico, New Zealand, Pakistan, the Philippines, Sri Lanka, Switzerland, Thailand, the United Kingdom, the United States, Vietnam, and West Germany. We gratefully acknowledge this support.

The task of putting together this database would not have been possible without the enthusiastic and dedicated help of the following individuals, University of Hawaii graduate students on East-West Center scholarships: Ponciano "Boy" Epino (Philippines), Boniface Peiris (Sri Lanka), and Xenia Wolff (USA)—each of whom took a specific portion of the task and completed it diligently; research interns Hong-chun "June" Yang (China) and D. Patricia Goddard (USA)—who entered data into the computer; and Eric Mercadejas, a Filipino student who helped transfer data onto cards while Michael Grainge was on field study at the International Rice Research Institute (IRRI) and the University of the Philippines at Los Baños (UPLB) during 1985–1986. These members of our in-house Botanical Pest Control Project team put in considerable time and energy—much beyond the call of duty.

We also warmly thank University of Hawaii professors W.C. Mitchell (entomology) and J.W. Hylin (agricultural biochemistry), who had co-authored with us an earlier "homegrown" version of this database and who provided valuable guidance and support as this project moved ahead, and Ashrafuddin Chaudhry (Bangladesh), an EWC alumnus, who had played a leading role in computer programming for the earlier version and who helped us in an advisory capacity this time also. Support for compiling the earlier version had been provided in part by the joint University of Hawaii–East-West Center Collaborative Research Program, and support for its publication had been provided by the University of Hawaii Foundation. We gratefully acknowledge these.

Our thanks are also due to Dr. M.S. Swaminathan, IRRI Director-General, for his encouragement in our task, and Dr. J.A. Listinger, IRRI Entomologist, who first suggested six years ago that we examine the feasibility of using plant materials for pest control.

Finally, at the Resource Systems Institute, we thank Dr. Seiji Naya, Director, for the institutional support provided, Ms. Lynn Garrett, Institute Editor, for her editorial guidance and contract negotiations with the publishers, and Dr. Bruce Koppel, Rural Transformations Program Leader, for his valuable contribution to an examination of how micro-macro socioeconomic linkages have a bearing on the use of plant materials for pest control as well as for his sustained support throughout the six-year period of this project. It was particularly appropriate that this project was carried out under the support of the East-West Center, whose mandate called for problem-oriented research undertakings by multidisciplinary and multicultural staff–student teams.

SALEEM AHMED
MICHAEL GRAINGE

How to Use This Database: Codes Used for Characterizing Plant Species

Information in this document is presented in three sections. Section I catalogues about 2,400 plants having pest-control properties; Section II is a listing of about 800 pests and the plants that reportedly control them; and Section III lists another 1,000 plants that are either poisonous in nature or reportedly control diseases and nematodes of humans and animals. The latter are candidate plants for screening for activity against crop pests. Comments on each section follow.

SECTION I LIST OF PEST-CONTROL PLANT SPECIES

On page 1 we have listed 41 plant species having broad-spectrum pest-control properties; these might be looked upon as the "cream of the crop," potentially the best candidates for large-scale utilization for pest control in developing countries. Before such use, however, the cautions listed under Code L should be carefully reviewed and appropriate toxicological studies conducted.

The alphabetical listing of pest-control plant species begins on page 2. Information about these plant species and pests they reportedly control appears in the following order (subject to availability) in tabular form.

I. Plant Name

Plant species are listed according to their scientific names, followed by their common names in English (in parentheses), whenever known, and their family names. The scientific and common names were checked with Bailey's *Hortus Third*, and family names follow the Cronquist's classification system wherever possible. Because our computer does not underline words, scientific names of plants and pests appear without the usual underling. *The scientific names of plants and pests may have been changed to conform to the current approved usage; this should be kept in mind when looking up references.*

II. Description of the Plant and Its Pest-Control Ingredients

Some important characteristics of the plants and of their pest-control properties are indicated in a horizontal format under the headings A–O. Not all information was available in many cases; corresponding headings are then excluded. Conversely, multiple listings of certain headings denote multiple characteristics. Thus, for example, the first plant, *Abelmoschus esculentus*, does not have headings E, F, G, K, and L, but has four listings under M, 3 under N, and 2 under O. The following codes are used under the headings A–O.

Plant Characteristics

A Plant Life Cycle
1 = perennial   2 = biennial   3 = annual   4 = other

B Type of Plant
1 = tree   5 = creeper   9 = aquatic/semiaquatic plant
2 = shrub, subshrub   6 = herb   10 = parasitic/epiphytic plant
3 = woody climber   7 = woody herb
4 = herbaceous vine   8 = cactus

C Plant Classification
1 = fungus   4 = moss   7 = cycad
2 = alga   5 = early vascular   8 = conifer
3 = liverwort   6 = fern   9 = flowering plant

D Climate in Which Plant Is Commonly Found
1 = tropical   5 = arid   9 = hotlands
2 = subtropical   6 = Mediterranean   10 = cosmopolitan
3 = temperate   7 = arctic/subarctic   (1 + 2 + 3 categories)
4 = semiarid   8 = alpine

E Problem Soils to Which Plant Is Adaptable
1 = acid sulfate soils   5 = calcerous soils   9 = rocky areas
2 = other acid soils   6 = other alkaline soils 10 = waterlogged areas
3 = sodic soils   7 = sandy soils   11 = peat/bog/marshy areas
4 = saline soils   8 = heavy clays   12 = infertile soils

Description of the Active Materials

F Effective Life
   Consideration
1 = material breaks down in sunlight   2 = plant has seasonal activity

   Duration for Which Activity Is Maintained in the Field
3 = >2 months   5 = 2 weeks   7 = 2–3 days
4 = 1 month   6 = 1 week   8 = 1 day or less

   Duration for Which Activity Is Maintained in Storage
 9 = >1 year   12 = 2–7 weeks   15 = <24 hours
10 = 6–11 months   13 = 4–13 days
11 = 2–5 months   14 = 1–3 days

G Maximum Dilution of the Active Principle for Effectiveness
1 = 1:2   4 = 1:100   7 = 1:100,000
2 = 1:5   5 = 1:1,000
3 = 1:10   6 = 1:10,000

H Type of Pest-Control Activity Observed

We have retained descriptors used by the authors, although sometimes these are vague (for example, antivermin). The "anti" category (for example, anti-insect, antimite) has also been used when the mode of action was not defined. An X after a descriptor (for example, No. 12 below) indicates that the pest controlled is an animal pest. No. 20 includes antimicrobial action when the microbes controlled are not defined.

1 = anti-insect	12 = antimite-X	25 = rodenticidal
2 = insecticidal	13 = antitick	25a = rodent repellent
2a = contact poison	14 = antifungal	26 = antifertility
2b = stomach poison	15 = antifungal-X	(rodents)
3 = growth inhibitor	16 = antinematode	27 = antivermin
4 = antifeedant	17 = antinematode-X	28 = antisnail/leech
5 = repellent	18 = antibacterial	29 = pest-free
6 = attractant	19 = antibacterial-X	30 = synergistic
6a = trap crop	20 = antibiotic/anti-	31 = adjuvant
7 = chemosterilant	septic/antimicrobial	32 = fish poison
8 = termite resistant	21 = antiviral	33 = poisonous
9 = insectivorus	22 = antiviral-X	34 = anaesthetic/sedative
10 = sticky trap	23 = herbicidal	35 = narcotic
11 = antimite	24 = allelopathic	

I Plant Part(s) Used/Responsible for the Pest-Control Activity

1 = whole plant	8 = flowers	15 = shoots/buds/tops
2 = roots/tubers/rhizomes	9 = fruit/fruiting body	16 = aerial parts
3 = bulbs/corms	10 = seeds/nuts/spores	17 = trichomes
4 = bark	11 = pods	18 = oilcake/residue
5 = wood/pulp	12 = oil	19 = tissue culture
6 = stem/branches	13 = gum/resin	20 = crop residue
7 = leaves	14 = sap/latex/juice	

Method of Preparation/Extraction of the Pest-Control Material

J a Nonchemical Preparation

1 = no preparation needed	4 = powdering of the plant part
2 = drying of the plant part	5 = tapping for sap/latex
3 = aqueous extraction	6 = pressing/distilling for oil
3a = crude extraction, using village facilities	

J b Chemical Extraction; Solvent Used

7 = ether	11 = petroleum ether	15 = kerosene
8 = alcohol	12 = methanol	16 = chloroform
9 = ethanol	13 = benzene	17 = ethyl acetate
10 = acetone	14 = ethyl ether	

Method of Application of the Pest-Control Material

K Method of Use

1 = merely planting	6 = spraying the preparation
2 = mixing with bait	7 = fumigating/burning the plant part
3 = surface spreading	8 = rubbing the material on the plant
4 = using as mulch	9 = mixing with stored produce
5 = dusting on the crop	

Environmental Conditions in Use

L Cautions in Use
1 = material is toxic to honeybees
2 = material is toxic to grazing cattle
3 = material is an oral poison
4 = material is a contact poison
5 = the plant is a potential weed

Additional Economic Value of the Plant

M Other Plant Uses/Sources

1 = as food/drink for humans
1a = is edible after cooking
2 = as animal food substitute
3 = provides fiber
4 = provides materials to make tools
5 = provides medicine/drugs
6 = is a source of fuel/light
7 = as a wind break
8 = as a sand-binder
9 = for erosion control as a cover crop
10 = as a fertilizer
11 = for soil reclamation
12 = fixes N_2
13 = is an ornamental plant
14 = used as a spice/flavoring
15 = for soap making/as a soap
16 = is a source of dye/ink
17 = is a source of perfume/incense
18 = is a source of honeybee nectar
19 = for wood carving/in carpentry
20 = in paints/varnish
21 = is a source of tannin
22 = is a source of paper
23 = is a source of beads for jewelry
24 = used in weaving
25 = is a source of wood preservative
26 = is a source of rubber
27 = is a source of sulfur
28 = is a source of cooking oil/fat

N Plant Parts Used for Food by Humans (same codes as I above)

O Plant Parts Used for Medicine/Drugs (same codes as I above)

III. Organisms Controlled (OC)

Under the heading OC we have listed alphabetically pests that are reportedly controlled by the plant. Because of their specific nature, the type of pest-control activity (code H) and plant part(s) responsible for the pest-control action (code I) are listed against each pest here rather than being listed in the horizontal format described above. Reference(s) for each entry are then listed in parentheses.

IV. Other References (OR)

Listed here (in parentheses) are references that do not name any specific pest but describe some pest-control properties of the plant under consideration (under codes H and I).

V. Active Principles (AP)

The following codes describe the active principles found in the plants:

Alk = alkaloids
Cou = coumarinds
Fla = flavanoids
Sap = saponins
Sfr = sulfur
Str = steroids
Tan = tannins
Tri = triterpenoids

The plant part in which these compounds are found is indicated using Code I, followed by references (in parentheses).

SECTION II   PESTS AND PLANTS THAT CONTROL THEM

This listing is alphabetized first by the type of pest (such as bacteria, fungi, and insects) and then, within each pest type, by pest species. In some cases, however, authors have not indicated the pest species controlled, but have only indicated the general category of pest in question (such as ants, cockroaches). These are also included in this listing.

Plant species that reportedly control the pest in question are then indicated alphabetically for each pest species. Thirty five insect species that are also controlled by rotenone, a plant-derived compound, are listed on page 293, at the end of Section I.

SECTION III   POISONOUS PLANTS AND PLANTS CONTROLLING HUMAN AND ANIMAL PESTS

This section has been included to provide a list of candidate plants for scientists to screen for effectiveness against crop pests. Codes used are the same as in Section I; numbers in brackets are references.

Contents

Section I. **List of Pest-Control Plant Species**
A. Plants with Broad-Spectrum Pest Control   1
B. Alphabetical Listing of Pest-Control Plant Species   2

Section II. **Pests and Plants That Control Them**
Bacteria   294
Fungi   297
Insects   311
Leeches and Mollusks (including snails)   365
Mites and Ticks   366
Nematodes   369
Rodents   374
Viruses   375
Weeds   377

Section III. **Poisonous Plants and Plants which Control Non-Insect**   **380**
Animal Parasites and Diseases

References   **406**

Handbook of Plants
with Pest-Control
Properties

SECTION I. LIST OF PEST-CONTROL PLANT SPECIES

A. PLANTS WITH BROAD-SPECTRUM PEST CONTROL

The following are plants that reportedly possess broad-spectrum pest-control properties; these may be considered the "cream of the crop." A listing of all pest-control plant species begins on the next page.

Aconitum ferox   (Indian aconite)   Ranunculaceae
Acorus calamus   (Sweetflag)   Araceae
Ageratum conyzoides   (Goatweed)   Asteraceae
Aleurites fordii   (Tung tree)   Euphorbiaceae
Annona reticulata   (Custard apple)   Annonaceae
Annona squamosa   (Sugar apple)   Annonaceae
Arachis hypogaea   (Peanut)   Fabaceae
Artabotrys hexapetalus   (Ylang-ylang)   Annonaceae
Azadirachta indica   (Neem tree)   Meliaceae
Chrysanthemum cinerariifolium   (Pyrethrum)   Asteraceae
Croton tiglium   (Purging cotton)   Euphorbiaceae
Datura metel   (Angel trumpet)   Solanaceae
Datura stramonium   (Jimsonweed)   Solanaceae
Derris elliptica   (Derris)   Fabaceae
Haplophyton cimicidum   (Cockroach plant)   Apocynaceae
Justicia adhatoda   (Malabar nut tree)   Acanthaceae
Madhuca indica   (Mowra)   Sapotaceae
Mammea americana   (Mammee apple tree)   Clusiaceae
Melia azedarach   (Chinaberry)   Meliaceae
Mundulea suberosa   (Sweetcane)   Fabaceae
Nicotiana rustica   (Wild tobacco)   Solanaceae
Nicotiana tabacum   (Tobacco)   Solanaceae
Pachyrhizus erosus   (Chinese yam bean)   Fabaceae
Piper nigrum   (Black pepper)   Piperaceae
Pogostemon patchouli   (Patchouli)   Lamiaceae
Pongamia pinnata   (Poonga oil tree)   Fabaceae
Quassia amara   (West Indian quassia)   Simaroubaceae
Ricinus communis   (Castor bean)   Euphorbiaceae
Ryania speciosa   (Not known)   Flacourtiaceae
Schoenocaulon officinale   (Sabadilla)   Liliaceae
Tagetes erecta   (African marigold)   Asteraceae
Tagetes patula   (French marigold)   Asteraceae
Tephrosia virginiana   (Devil's shoestring)   Fabaceae
Tephrosia vogelii   (Vogel tephrosia)   Fabaceae
Tripterygium forrestii   (Three-winged nut)   Celastraceae
Tripterygium wilfordii   (Thunder-god vine)   Celastraceae
Veratrum album   (European white hellebore)   Liliaceae
Veratrum viride   (American flase hellebore)   Liliaceae
Vitex negundo   (Indian privet)   Verbenaceae
Zanthoxylum clava-herculis   (Southern prickly ash)   Rutaceae
Zingiber officinale   (Ginger)   Zingiberaceae

1

B. ALPHABETICAL LISTING OF PEST-CONTROL PLANT SPECIES

Abelmoschus esculentus   (Okra)   Malvaceae
   A   B   C   D   J   M   M   M   M   N   N   N   O   O
  03 06 09 10 03 01 03 05 22 09 10 11 07 09
OC: Anthonomus grandis H-6a, I-? (1276)   Fusarium nivale H-14, I-1, 14 (1048)
    Drechslera oryzae H-14, I-2, 6, 7 (113, 432)

Abelmoschus moschatus   (Musk mallow)   Malvaceae
   A   B   C   D   M   M   M   M   M   N   N   O
  03 06 09 10 01 03 05 17 24 07 15 10
OC: Culex quinquefasciatus H-2, I-10 (463)
OR: H-5, I-? (80)

Abies balsamea   (Balsam fir)   Pinaceae
   A   B   C   D   D   M   M   M   O
  01 01 08 03 08 04 05 20 05
OC: Pyrrhocoris apterus H-3, I-5 (684, 721)

Abrus precatorius   (Red-bead vine)   Fabaceae
   A   B   C   D   F   J   J   J   J   L   M   M   M   M   M   N   N   O   O   O   O
  01 03 09 01 08 02 03 04 10 03 01 03 05 12 14 02 07 02 07 10 12
                                                     23 24
OC: Mosquitoes H-2, I-9 (98)   Poecilocera picta H-3, I-10 (87, 782)
OR: H-33, I-10 (507, 968, 983)
AP: **Alk**=I-6 (1361, 1379); **Alk**=I-10 (abrine)(1392); **Fla**=I-? (1361)

Abutilon striatum   (Indian mallow)   Malvaceae
   A   B   C   D   J   J   M
  01 02 09 01 03 09 03
OC: Tobacco mosaic virus H-21, I-7, 12 (1310)

Abutilon theophrasti   (China jute)   Malvaceae
   A   B   B   C   D   J   J   L   M
  03 02 06 09 01 03 11 05 03
OC: Attagenus piceus H-4, I-15 (48)
OR: H-24, I-7, 10 (1187, 1192)
AP: **Alk**=I-10 (1391)

Acacia albida   (White acacia)   Mimosaceae
   A   B   C   D   M   M   M
  01 01 09 01 02 12 21
OC: Locusta migratoria H-4, I-7 (998)
AP: **Tan**=I-4 (504)

Acacia catechu   (Catechu tree)   Mimosaceae
   A   B   C   D   D   J   M   M   M   M   M   M
  01 01 09 01 04 09 04 05 06 12 16 21
OC: Colletotrichum falcatum H-14, I-5, 9 (1047) Pyricularia oryzae H-14, I-5, 9 (1047)
OR: H-19, I-? (411); H-22, I-6 (881)
AP: **Fla**=I-4 (1361); **Tan**=I-5, 9 (504, 1047, 1352)

Acacia chiapensis   (Not known)   Mimosaceae
    A   C   D   M
    01   09   01   12
OC: Spodoptera eridania H-1, I-? (619)

Acacia concinna   (Soap nut)   Mimosaceae
    A   B   C   D   D   J   K   M   M
    01   02   09   05   10   04   05   12   15
OC: Chirida bipunctata H-1, I-? (966)   Sitotroga cerealella H-2, I-? (88)
    Rhizopertha dominica H-4, I-? (124, 132)   Tribolium castaneum H-5, I-? (517)
    Sitophilus oryzae H-5, I-? (517)
OR: H-2, I-8, 11 (88, 1345)
AP: Alk=I-4 (1392)

Acacia confusa   (Not known)   Mimosaceae
    A   B   C   D   M
    01   01   09   01   12
OR: H-24, I-? (1247)
AP: Alk=I-7, 8 (1392)

Acacia dealbata   (Silver wattle)   Mimosaceae
    A   B   C   D   M   M   M   M
    01   01   09   10   12   13   17   21
OR: H-29, I-1 (87)
AP: Alk=I-7 (1392); Tan=I-4 (504)

Acacia farnesiana   (Sponge tree)   Mimosaceae
    A   B   C   D   D   M   M   M   M   M   M   M   M   O   O   O
    02   02   09   01   02   04   05   12   13   16   17   19   21   02   04   07
OC: Spodoptera eridania H-4, I-7 (619)
OR: H-2, I-8 (87)
AP: Alk=I-6, 7 (1392); Alk=I-10 (1391); Tan=I-4, 11 (1352)

Acacia hockii   (Not known)   Mimosaceae
    A   C   D   M
    01   09   01   12
OC: Locusta migratoria H-4, 5, I-7 (596)

Acacia loculata   (Not known)   Mimosaceae
    A   C   D   J   M
    01   09   01   03   12
OC: Alternaria tenuis H-14, I-7, 14 (480)   Helminthosporium sp. H-14, I-7 (480)
    Curvularia penniseti H-14, I-7 (480)

Acacia longifolia   (Sydney wattle)   Mimosaceae
    A   B   B   C   D   M   M   N
    01   01   02   09   02   01   12   10
OC: Popillia japonica H-5, I-? (1243)
AP: Alk=I-6, 7 (tryptamine)(1392); Alk=I-6, 7, 8 (phenethylamine)(1392)

Acacia nilotica   (Egyptian mimosa)   Mimosaceae
    A   B   C   D   M   M   M   M   M   M   M   M   M   M   M   M   N   O
    01   01   09   01   01   02   03   04   05   06   12   14   16   19   20   21   22   24   13   11
OC: Colletotrichum falcatum H-14, I-4 (1047)   Termites H-1, I-5 (1352)
    Pyricularia oryzae H-14, I-4 (1047)

OR: H-8, I-5 (504)
AP: Alk=I-9 (1392); Tan=I-4, 11 (1047, 1352, 1353)

Acacia pinnata   (Aila)   Mimosaceae
   A   B   C   D   D   M   M   M   M   O   O   O
   01 03 09 04 10 03 05 12 21 01 07 20
OC: Maggots H-1, I-? (87)
OR: H-32, I-4, 9 (73, 459, 1352)
AP: Alk, Tan=I-11 (1360); Tan=I-4, 9 (1352)

Acacia tortilis   (Not known)   Mimosaceae
   A   B   C   D   D   E   M   M   M   M   O
   01 02 09 05 09 07 02 05 06 08 12 04
OC: Locusta migratoria H-4, I-7 (998)
OR: H-20, I-4 (1353)

Acalypha indica   (Indian nettle)   Euphorbiaceae
   A   B   C   D   J   J   L   M   M   O   O
   03 06 09 01 04 08 05 05 21 02 07
OC: Caterpillars H-1, I-4 (1029)   Pericallia ricini H-1, I-4 (105)
    Euproctis fraterna H-1, I-4 (105)   Plutella xylostella H-1, I-4 (105)
    Maggots H-1, I-7 (105)
AP: Alk=I-? (acalyphine, triacetonomine)(1392); Tan=I-? (504)

Acanthospermum hispidum   (Not known)   Asteraceae
   B   C   D   M   M
   06 09 01 05 12
OC: Aspergillus oryzae H-14, I-12 (649)   Pseudomonas solanacearum H-18, I-7 (650)
    Fusarium oxysporum H-14, I-12 (649)
AP: Alk=I-6, 7 (1392)

Acer carpinifolium   (Hornbean maple)   Aceraceae
   A   B   C   D   J   M
   01 01 09 03 03 19
OC: Drosophila hydei H-2, I-6, 7 (101)

Acer circinatum   (Vine maple)   Aceraceae
   A   B   B   C   D   M
   01 01 02 09 03 04
OR: H-24, I-7 (1180)

Acer negundo   (Box elder)   Aceraceae
   A   B   C   D   D   M   M   M   N
   01 01 09 02 03 01 04 19 14
OC: Melanoplus femurrubrum H-2, I-7 (87)

Acer platanoides   (Norway maple)   Aceraceae
   A   B   C   D   J   M
   01 01 09 03 03 04
OC: Rodents H-25, I-10 (945)
OR: H-19, I-7 (984, 1042); H-19, 22, I-? (1052); H-19, 22, I-9 (536, 537)

Acer pseudoplatanus   (Sycamore)   Aceraceae
   A   B   C   D   J   M
   01 01 09 03 03 19

OC: Venturia inaequalis H-14, I-? (16)
OR: H-19, I-7 (1101)

Acer rubrum   (Red maple)   Aceraceae
   A   B   C   D
   01 01 09 03
OC: Malacosoma disstria H-4, I-7 (87)
OR: H-24, I-2 (1251)

Acer saccharium   (Sugar maple)   Aceraceae
   A   B   C   D   M   M   N
   01 01 09 03 01 19 14
OC: Leptinotarsa decemlineata H-4, I-7 (548)   Panonychus citri H-11, I-7 (669)
    Manduca sexta H-4, I-7 (669)
AP: Alk=I-6, 7 (1392)

Achillea micrantha   (Yarrow)   Asteraceae
   A   B   C   D   M
   01 06 09 03 17
OR: H-2, I-8 (101)

Achillea millefolium   (Common yarrow)   Asteraceae
   A   B   C   D   M   M   M   N   O   O
   01 06 09 03 01 05 17 07 07 08
OC: Dermacentor marginatus H-13, I-8 (87, 1133) Musca domestica H-2, I-? (87)
    Haemaphysalis punctata H-13, I-8 (87, 1133) Rhipicephalus rossicus H-13, I-8 (87, 1133)
    Ixodes redikorzevi H-13, I-8 (87, 1133)
OR: H-19, I-? (592); H-20, I-7 (504)
AP: Alk=I-6, 7 (achiceine, achilleine, moschatine)(1392); Alk=I-10 (1391)

Achillea millefolium var. lanulosa   (Nose bleed)   Asteraceae
   A   B   C   D   J   M   O
   01 06 09 03 11 05 02
OC: Aedes aegypti H-2, I-2 (165)
OR: H-34, I-2 (504)

Achillea nobilis   (Camphor yarrow)   Asteraceae
   A   B   C   M
   01 06 09 17
OR: H-2, I-8 (104, 105)

Achillea sibirica var. ptarmicoides   (Yarrow genus)   Asteraceae
   A   B   C   D   J   M
   01 06 09 03 03 05
OC: Drosophila sp. H-2, I-6, 7, 8 (101)

Achyranthes aspera   (Prickly chaff flower)   Amaranthaceae
   A   B   C   D   J   M   M   M   M   N   O   O
   01 02 09 01 03 01 05 15 16 07 02 07
OR: H-19, I-7 (982); H-24, I-2, 10, 15 (1184)
AP: Alk=I-2, 6, 7, 10 (1360); Alk=I-16 (1392)

Achyranthes bidentata   (Not known)   Amaranthaceae
   B   C   D   J
   06 09 01 12
OR: H-3, I-2 (853)

Achyranthes fauriei   (Not known)   Amaranthaceae
   B   C   D   J
   06 09 01 12
OR: H-3, I-2 (853)

Achyranthes japonica   (Not known)   Amaranthaceae
   B   C   D   J
   06 09 01 12
OR: H-3, I-2 (853)

Achyranthes longifolia   (Not known)   Amaranthaceae
   B   C   D   J
   06 09 01 12
OR: H-3, I-2 (853)

Achyranthes mollicula   (Not known)   Amaranthaceae
   B   C   D   J
   06 09 01 12
OR: H-3, I-2 (853)

Achyranthes obtusifolia   (Not known)   Amaranthaceae
   B   C   D   J
   06 09 01 12
OR: H-3, I-2 (853)

Achyranthes ogatai   (Not known)   Amaranthaceae
   B   C   D   J
   06 09 01 12
OR: H-3, I-2 (853)

Achyranthes radix   (Not known)   Amaranthaceae
   B   C   D
   06 09 01
OC: Chilo suppressalis H-1, I-? (1003)   Pieris rapae H-1, I-? (1003)
    Culex pipiens molestus H-1, I-? (1003)   Plutella xylostella H-1, I-? (1003)
    Musca domestica H-1, I-? (1003)
OR: H-3, I-? (1055)

Achyranthes rubrofusca   (Not known)   Amaranthaceae
   B   C   D   J
   06 09 01 12
OR: H-3, I-2 (853)

Acokanthera oblongifolia   (Winter sweet)   Apocynaceae
   A   B   B   C   D   J   L
   01 01 02 09 01 11 03
OC: Spodoptera littoralis H-4, I-7 (964)
OR: H-33, I-1 (507)

Aconitum anthora   (Monkshood genus)   Ranunculaceae
   A   B   C   D   L   M   O   O
   01 06 09 03 03 05 02 07
OC: "Bugs" H-2, I-8, 9 (101)   Lice H-2, I-8, 9 (101)
    Flies H-2, I-8, 9 (101)   Mosquitoes H-2, I-8, 9 (101)

OR: H-17, 33, I-2 (504)
AP: Alk=I-2 (anthorine, atisine)(1392)

Aconitum baicalense   (Monkshood genus)   Ranunculaceae
   A B C D L
   01 06 09 03 03
OC: "Bugs" H-2, I-8, 9 (101)   Lice H-2, I-8, 9 (101)
   Flies H-2, I-8, 9 (101)   Mosquitoes H-2, I-8, 9 (101)
OR: H-33, I-2 (504)

Aconitum barbatum   (Monkshood genus)   Ranunculaceae
   A B C D J L
   01 06 09 03 03 03
OC: Anopheles quadrimaculatus H-2, I-8, 9 (101) Flies H-2, I-8, 9 (101)
   "Bugs" H-2, I-8, 9 (101)   Lice H-2, I-8, 9 (101)
   Cockroaches H-2, I-8, 9 (101)
AP: Alk=I-? (aconitine)(1392)

Aconitum carmichaelii   (Monkshood genus)   Ranunculaceae
   A B C D
   01 06 09 03
OC: Caterpillars H-1, I-2, 6 (1241)

Aconitum chinense   (Monkshood genus)   Ranunculaceae
   A B B C D J K L
   01 05 06 09 03 04 05 03
OC: Epilachna varivestis H-2, I-2 (29, 101)
OR: H-33, I-2, 10 (504, 507)
AP: Alk=I-? (aconitine)(1392)

Aconitum excelsum   (Monkshood genus)   Ranunculaceae
   A B C D L
   01 06 09 03 03
OC: Flies H-2, I-8, 9 (101)   Mosquitoes H-2, I-8, 9 (101)
   Lice H-2, I-8, 9 (101)
OR: H-33, I-2 (504)
AP: Alk=I-? (aconitine, hypaconitine)(1392); Alk=I-2 (acsinatine, acsine, lappaconitine, mesaconitine)(1392)

Aconitum ferox   (Indian aconite)   Ranunculaceae
   A B C D J J J K L M
   01 06 09 03 02 03 11 05 03 05
OC: Aphis maidis H-2, I-2 (87, 502)   Lipaphis erysimi H-2, I-2 (87, 91, 502)
   Athalia proxima H-2, I-2 (87, 502)   Siphocoryne indobrassicae H-2, I-2
   Aulacophora foveicollis H-2, I-2 (87, 502)   (87, 502)
   Hieroglyphus nigrorepletus H-2, I-2 (87, 502)
OR: H-2a, I-1 (129); H-33, I-2 (504)
AP: Alk=I-2 (aconitine)(1392)

Aconitum japonicum   (Japanese aconite)   Ranunculaceae
   A B C D J J L
   01 06 09 03 03 10 03
OC: Culex pipiens H-1, I-? (101)   Drosophila hydei H-2, I-7, 8 (101)
OR: H-33, I-2 (504, 839)
AP: Alk=I-16 (aconitine, ignavine, isohypognavine, mesaconitine)(1392)

7

Aconitum kusnezoffii   (Monkshood genus)   Ranunculaceae
   A   B   C   D   J
   01 06 09 03 03
OC: Caterpillars H-1, I-2, 6 (1241)   Puccinia graminis tritici H-14, I-2, 6
    Colaphellus bowringi H-1, I-2, 6 (1241)   (1241)
    Grasshoppers H-1, I-2, 6 (1241)   Puccinia rubigavera H-14, I-2, 6 (1241)
    Maggots H-1, I-2, 6 (1241)   Rice borers H-1, I-2, 6 (1241)
    Phytophthora infestans H-14, I-2, 6 (1241) Spodoptera litura H-1, I-2, 6 (1241)
    Pieris rapae H-1, I-2, 6 (1241)

Aconitum lycoctonum   (Yellow wolfsbane)   Ranunculaceae
   A   B   C   D   L
   01 06 09 03 03
OC: Flies H-2, I-8, 9 (101)   Mosquitoes H-2, I-8, 9 (101)
    Lice H-2, I-8, 9 (101)
OR: H-33, I-2 (504); H-35, I-2 (504)
AP: Alk=I-? (aconitine)(1392); Alk=I-2 (lycaconitine, nycotonine)(1392)

Aconitum napellus   (Garden wolfsbane)   Ranunculaceae
   A   B   C   D   J   L   M   O   O
   01 06 09 03 03 03 05 02 07
OC: Blow flies H-2a, I-? (105)   Nilaparvata lugens H-1, I-? (143)
    Lucanus cervus H-1, I-? (104, 105, 1025)
OR: H-33, I-2 (504)
AP: Alk=I-2 (aconine, aconitine, benzaconine, ephedrine, hypaconitine, mesaconitine,
    napelline, neoline, neopelline, sparteine)(1392); Alk=I-16 (nepellonine)(1392)

Aconitum subrosulatum   (Monkshood genus)   Ranunculaceae
   A   B   C   D
   01 06 09 03
OC: Maggots H-1, I-2 (1241)

Aconitum villosum   (Monkshood genus)   Ranunculaceae
   A   B   C   D   J   K   L
   01 06 09 03 04 05 03
OC: Epilachna varivestis H-2, I-1 (29, 101)
OR: H-33, I-2 (504)

Aconitum volubile   (Not known)   Ranunculaceae
   A   B   B   C   D   L
   01 04 06 09 03 03
OC: Flies H-2, I-8, 9 (101)   Mosquitoes H-2, I-8, 9 (101)
    Lice H-2, I-8, 9 (101)
OR: H-33, I-2 (504)

Acorus calamus   (Sweetflag)   Araceae
   A   B   B   B   C   D   D   E   F   F   G   J   J   J   J   K   K   L   M   M   M   O
   01 05 06 09 09 06 10 11 03 04 03 02 03 04 10 03 05 03 05 13 17 02
                                     11 14 15   06
OC: Aedes aegypti H-5, I-2, 12 (825)   Aulacophora foveicollis H-1, I-? (600)
    Ants H-1, I-? (105)   Bagrada picta H-1, I-? (600)
    Athalia proxima H-1, I-7 (438); H-2, Bombyx mori H-1, I-2 (603)
      I-2 (148, 1011); H-4, I-2 (607)   Callosobruchus analis H-1, I-4 (179)

Callosobruchus chinensis H-1, I-2
   (189, 1124); H-2, I-12 (965); H-3,
   I-2 (707); H-5, I-? (599)
Ceratitis capitata H-6, I-2, 12 (840)
Cimex lectularius H-4, I-2 (609)
Clothes moths H-1, I-? (105, 802);
   H-4, I-2 (609)
Culex fatigans H-2, I-2 (87)
Dactynotus carthani H-2, I-2 (91)
Dacus cucurbitae H-6, I-2, 12 (840)
Dacus dorsalis H-6, I-2, 12 (840)
Dermestres maculatus H-1, I-6 (544)
Dysdercus cingulatus H-1, I-6
   (88, 124, 544); H-2, I-2 (435);
   H-3, I-2 (441)
Dysdercus koenigii H-2, I-2 (91); H-3,
   I-2, 12 (543, 707, 710, 759, 787, 788)
Empoasca devastans H-1, I-? (600)
Fleas H-1, I-? (1197); H-5, I-2 (87)
Flies H-1, I-? (802)
Fowl lice H-1, I-? (802); H-4, I-2 (609)
Graphosoma italicum H-1, I-? (544)
Graphosoma mellonella H-1, I-? (544)
Latheticus oryzae H-1, I-2 (602)
Lipaphis erysimi H-2, I-2 (84, 91)

Mites H-11, I-2 (1320)
Mosquitoes H-1, I-? (1197); H-2, I-1 (101)
Moths H-5, I-2 (87)
Musca domestica H-1, I-2 (84, 1197); H-2,
   I-2 (101); H-3, I-2 (856)
Musca nebulo H-1, I-2 (603); H-2, I-2 (87)
Oryctes rhinoceros H-1, I-? (802)
Pericallia ricini H-2, I-2, 6 (124, 435)
Pieris brassicae H-1, I-? (600)
Pyrrhocoris apterus H-3, I-2 (544, 707)
Rats H-26, I-2 (1325)
Rhizopertha dominica H-1, I-?
   (93, 126, 189)
Sitophilus oryzae H-1, I-2, 12 (602, 1124);
   H-3, I-2 (707)
Sitotroga cerealella H-1, I-2 (121, 613)
Spodoptera litura H-2, I-6 (124, 435);
   H-4, I-7 (1064)
Stored grain pests H-1, I-? (91, 802)
Thermobia domestica H-3, I-2, 12
   (707, 714)
Tribolium castaneum H-1, I-2 (189, 602);
   H-1, I-4 (179); H-3, I-2, 12 (264, 707)
Trogoderma granarium H-3, I-2, 12
   (189, 462, 688, 707)

OR: H-2, I-? (73); H-4, I-? (100); H-19, I-2 (429); H-24, I-2, 7 (1325); H-33, I-2 (220)
AP: Sap, Tan=I-2 (1320)

Acorus gramineus   (Grass-leaved sweetflag)   Araceae
   A   B   B   C   D   M
   01 06 09 09 03 13
OR: H-2, I-2 (105, 220); H-2, 5, I-2 (1345)

Acorus tatarinowii   (Sweetflag genus)   Araceae
   A   B   C   D   E   J
   01 06 09 03 11 03
OC: Aphids H-1, I-2, 6 (1241)   Phytophthora infestans H-14, I-2 (1241)

Acrostichum aureum   (Leather fern)   Polypodiaceae
   A   B   C   D   E   J   M   M   M   M   N   O   O
   01 09 06 01 11 09 01 02 04 05 07 02 07
OC: Rats H-26, I-16 (1393)
AP: Str, Tri=I-? (1376)

Actaea arguta   (Baneberry)   Ranunculaceae
   A   B   C   D   J   J   L   M
   01 06 09 03 03 11 03 13
OC: Attagenus piceus H-4, I-2 (48)
OR: H-33, I-9 (507)

Actaea spicata   (Bugbane)   Ranunculaceae
   A   B   C   D   L   M   M
   01 06 09 03 03 05 13
OR: H-2, 33, I-? (87); H-33, I-9 (220, 507)

Actinidia chinensis   (Chinese gooseberry)   Actinidiaceae
   B   C   D   J   M   N
   03 09 02 03 01 09
OC: Cnaphalocrocis medinalis H-1, I-2, 7 (1241) Oulema oryzae H-1, I-2, 7 (1241)
   Colaphellus bowringi H-1, I-2, 7 (1241)   Pieris rapae H-1, I-2, 7 (1241)
   Euproctis pseudoconspersa H-1, I-2, 7   Rice borers H-1, I-2, 7 (1241)
    (1241)

Adansonia digitata   (Baobab)   Bombacaceae
   A   B   C   D   K   M   M   M   M   M   M   N   N   O
   01 01 09 01 07 01 03 04 05 14 15 22 09 10 07
OC: Flies H-1, I-9, 20 (504, 1345)
OR: H-5, I-9 (1352)
AP: Alk, Sap=I-9 (1360); Sap=I-10 (1360); Str=I-4 (B-sitosterol)(1385)

Adenostoma fasciculatum   (Greasewood)   Rosaceae
   A   B   C   D   M
   01 02 09 02 09
OC: Bromus rigidus H-24, I-6 (597)
OR: H-19, I-9 (1262); H-24, I-7 (988, 1246)

Adiantum capillus-veneris   (Venus's-hair fern)   Polypodiaceae
   A   C   D   M
   01 06 10 05
OC: Erwinia carotovora H-18, I-7 (854)
   Xanthomonas campestris pv. phaseoli-sojensis H-18, I-7 (854)

Adiantum trapeziforme   (Giant maidenhair fern)   Polypodiaceae
   A   C   D
   01 06 01
OC: Pseudomonas solanacearum H-18, I-7 (854)
   Xanthomonas campestris pv. phaseoli-sojensis H-18, I-7 (854)

Adina cordifolia   (Karam)   Rubiaceae
   A   B   B   C   D   D   F   J   J   M
   01 01 02 09 01 02 10 3a 08 04
OC: Beetles H-1, I-7 (87)   Maggots H-2, I-7 (220, 504)
OR: H-2, I-14 (73, 94, 220); H-20, I-4 (105)
AP: **Alk**=I-5 (coumarin, isocoumarin)(1383); **Alk**=I-6, 9 (1360); **Sap**=I-6 (1360); **Tan**=I-6,
   8, 9 (1360)

Adonis vernalis   (Spring adonis)   Ranunculaceae
   A   B   C   D   J   J   M   M   O
   01 06 09 03 03 11 05 13 01
OC: Attagenus piceus H-4, I-1 (48)

Aegle glutinosa   (Tabon)   Rutaceae
   A   B   C   D   D
   01 01 09 01 02
OC: Rice field insects H-5, I-6, 7 (137)

Aegopodium podagraria   (Bishop's weed)   Apiaceae
   A   B   C   D   J   J   L   M   M   M   N   O
   01 06 09 03 04 10 05 01 05 09 07 07

OC: Alternaria brassicae H-14, I-2, 6 (1115)   Fusarium culmorum H-14, I-2, 6 (1115)
    Colletotrichum lagenarium H-14, I-2, 6   Glomerella cingulata H-14, I-2, 6 (1115)
    (1115)   Septoria nodorum H-14, I-2, 6 (1115)

Aeonium arboreum   (Not known)   Crassulaceae
  A   B   C   D   D   G   J
  01   02   09   02   06   05   3a
OC: Potato virus Y H-21, I-7 (184)   Tobacco mosaic virus H-21, I-7 (184)

Aeonium balsamiferum   (Not known)   Crassulaceae
  A   B   C   D   J
  01   02   09   06   3a
OC: Tobacco mosaic virus H-21, I-7 (184)

Aeonium haworthii   (Pinwheel)   Crassulaceae
  A   B   C   D   J
  01   02   09   02   3a
OC: Potato virus Y H-21, I-7 (184)   Tobacco mosaic virus H-21, I-7 (184)

Aerva javanica   (Not known)   Amaranthaceae
  A   B   C   D   L
  03   06   09   10   09
OC: Musca domestica H-1, I-1 (785); H-2, I-? (100)

Aerva lantana   (Apug-apugan)   Amaranthaceae
  A   B   C   D   J   M   M   M   O
  03   02   09   01   03   01   05   07   01
OC: Drechslera oryzae H-14, I-7 (113)

Aeschynomene indica   (Kuhilia)   Fabaceae
  A   B   C   D   D   M   M   M
  03   02   09   01   02   04   06   12
OC: Rats H-26, I-16 (1393)

Aeschynomene sensitiva   (Swamp grass)   Fabaceae
  C   D   D   M
  09   01   02   12
OC: Diaphania hyalinata H-2, I-10, 11 (101)   Musca domestica H-1, I-6 (78)
    Mosquitoes H-1, I-6 (78)
OR: H-2, I-? (1112)

Aesculus californica   (California horse chestnut)   Hippocastanaceae
  A   B   C   D   J   J   L   M   N
  01   01   09   03   02   04   01   1a   10
OC: Epilachna varivestis H-2, I-8 (96); H-2b, I-10 (101, 105)

Aesculus hippocastanum   (Common horse chestnut)   Hippocastanaceae
  A   B   C   D   M   M   M   M   N   O
  01   01   09   03   1a   04   05   06   10   10
OC: Reticulitermes flavipes H-2, I-9 (87)   Venturia inaequalis H-14, I-7 (16)
    Rodents H-25, I-9 (945)
OR: H-17, I-10 (504)
AP: Tan=I-10 (504)

Aesculus pavia   (Dwarf buckeye)   Hippocastanaceae

A	B	C	D	J	J	K	L
01	02	09	03	03	04	06	03

OC: Livestock pests H-2, I-2 (101)   Popillia japonica H-2, 6, I-8 (101)
OR: H-5, I-5 (1345); H-32, I-10 (504)

Afrormosia laxiflora   (Not known)   Fabaceae

A	B	C	D	J	M	M	M	M	O
01	01	09	01	11	04	05	12	19	04

OC: Attagenus piceus H-4, I-6, 7 (48)   Termites H-8, I-5 (1353)
    Stored grain pests H-5, I-7 (533)
OR: H-33, I-2 (1353)

Aganope gabonica   (Not known)   Fabaceae

B	C	D	M
04	09	01	12

OC: Aphis fabae H-2, I-2, 6 (105)

Agapanthus africanus   (Lily-of-the-Nile)   Amaryllidaceae

A	B	C	D	J
01	06	09	01	3a

OC: Tobacco mosaic virus H-21, I-7 (184)

Agauria salicifolia   (Not known)   Ericaceae

A	B	C	D	K	M	O
01	01	09	01	02	05	07

OC: Beetles H-2b, I-7 (87)

Agave americana   (Century plant)   Agavaceae

A	B	C	D	D	E	F	F	G	J	J	L	M	M	M	M	M	N	N
01	02	09	01	04	04	03	05	05	3a	07	02	01	03	05	09	13	07	14

OC: Potato virus Y H-21, I-7 (184)   Termites H-5, I-7 (220, 1352)
    Sitophilus oryzae H-1, I-? (116)   Tobacco mosaic virus H-21, I-7 (184)
    Stored grain pests H-1, I-? (89)
OR: H-2, I-7 (101, 104, 105, 116, 118, 127); H-32, 33, I-7 (220)

Agelaea pentagyna   (Not known)   Connaraceae

A	B	C	D	J
01	01	09	01	11

OC: Attagenus piceus H-4, I-9 (48)

Ageratum conyzoides   (Goatweed)   Asteraceae

A	B	C	D	F	G	J	J	J	K	L	L	M	M	M	M	M	O	O	O	O
03	06	09	01	05	06	03	07	10	06	04	05	02	05	08	13	27	02	06	07	12

OC: Drosophila melanogaster H-2, I-? (134)   Meloidogyne javanica H-16, I-7 (100, 616)
    Dysdercus cingulatus H-2, I-7   Musca domestica H-1, I-1 (785); H-2, I-?
      (119, 125, 134)   (134)
    Dysdercus flavidus H-2, I-12, 16 (618);   Panstrongylus megistus H-2, I-? (144)
      H-3, I-? (164)   Rhodnius prolixus H-2, I-? (144)
    Locusta migratoria H-2, I-? (545)   Sitophilus zeamais H-2, I-7 (125, 134)
    Meloidogyne incognita H-16, I-7 (100, 616)   Tribolium castaneum H-2, I-1, 7 (125, 785)
OR: H-1, I-? (100); H-2, I-2, 7 (127); H-19, 24, I-7 (132)
AP: Alk=I-7 (1390, 1392); Sap, Sfr, Tan=I-7 (1321)

Ageratum houstonianum   (Flossflower)   Asteraceae

A	B	C	D	D	E	F	G	J	K	M
03	06	09	01	02	02	05	05	07	06	05

OC: Drosophila melanogaster H-3, 7, I-? (703)   Pediculus humanus capitis H-1, I-? (914)
    Locusta migratoria H-3, I-? (700, 743)   Rhodnius prolixus H-1, I-? (144, 913)
    Oncopeltus fasciatus H-3, 7, I-1   Schistocerca gregaria H-3, I-? (701)
      (56, 702, 705, 743)   Tristoma infestans H-3, 4, 5, I-1 (144)
    Panstrongylus megistus H-3, I-? (144, 915)
OR: H-3, I-? (558)
AP: Alk=I-10 (1391)

Aglaomorpha meyenianum   (Bear's paw fern)   Polypodiaceae

A	B	C	D	M
01	10	06	01	13

OC: Xanthomonas campestris pv. phaseoli-sojensis H-18, I-7 (854)

Agrimonia pilosa   (Agrimony genus)   Rosaceae

A	B	C	D	J	J	K
01	06	09	04	03	04	05

OC: Aphids H-1, I-1 (1241)   Puccinia rubigavera H-14, I-1 (1241)
    Puccinia graminis tritici H-14, I-1 (1241)

Agropyron repens   (Quack grass)   Poaceae

A	B	C	D	J	L	M	M	N	O
01	06	09	03	03	05	01	05	02	02

OR: H-24, I-1, 2, 7 (1193, 1194, 1198, 1255, 1259, 1260, 1261, 1263)

Agrostemma githago   (Corn cockle)   Caryophyllaceae

A	A	B	C	D	J	L	L	M	N
01	02	06	09	06	09	03	05	01	07

OC: Rodents H-25, I-1 (946)   Tobacco mosaic virus H-21, I-19 (1125)
OR: H-33, I-10 (507)

Ailanthus altissima   (Tree-of-heaven)   Simaroubaceae

A	B	C	D	J	L	M	M	M	M	N	O	O
01	01	09	01	09	03	01	04	05	13	07	02	04

OC: Leptinotarsa decemlineata H-4, I-7 (531)
OR: H-17, I-4 (105, 504); H-24, I-5, 6, 7 (587, 1207); H-33, I-7 (220)
AP: Alk=I-? (1391); Alk=I-2, 6, 7 (1392); Alk=I-10 (1374, 1391)

Ailanthus excelsa   (Tree-of-heaven)   Simaroubaceae

A	B	C	D	J	J	J	K	M	M	M	M	O	O
01	01	09	03	03	04	11	05	04	05	13	22	06	07

OC: Meloidogyne incognita H-16, I-7 (616)   Tribolium castaneum H-1, I-? (179)
    Meloidogyne javanica H-16, I-7 (616)
OR: H-29, I-1 (507)
AP: Alk=I-9 (1374)

Ailanthus giralolii   (Varnish tree genus)   Simaroubaceae

A	B	C	D	J	J
01	01	09	01	03	04

OC: Aphids H-1, I-2, 7 (1241)   Maggots H-1, I-2, 7 (1241)
    Ascotis selenaria H-1, I-2, 7 (1241)   Pectinophora gossypiella H-1, I-2, 7 (1241)

Pieris rapae H-1, I-2, 7 (1241)   Puccinia graminis tritici H-14, I-2, 7
Pseudoperonospora cubensis H-14, I-2, 7   (1241)
   (1241)   Rice leafhoppers H-1, I-2, 7 (1241)

Ajuga bracteosa   (Bugleweed genus)   Lamiaceae
   B   C   D   M
   06 09 03 09
OC: Lice H-2, 5, I-? (105, 1276)

Ajuga remota   (Bugleweed genus)   Lamiaceae
   B   C   D   F   J   J   K   M
   06 09 03 06 03 12 06 09
OC: Bombyx mori H-3, I-2, 7 (369, 726)   Spodoptera exempta H-3, I-2, 7 (169, 369)
    Pectinophora gossypiella H-3, I-2, 7   Spodoptera frugiperda H-3, I-2, 7 (369, 726)
    (369, 726)   Spodoptera littoralis H-1, I-? (169)
OR: H-2b, I-1, 7 (107); H-3, I-? (132)

Akebia quinata   (Five-leaf akebia)   Lardizabalaceae
   A   B   C   D   J   M   M   N
   01 02 09 03 08 01 17 09
OC: Attagenus piceus H-4, I-6, 7 (48)

Akebia trifoliata   (Three-leaf akebia)   Lardizabalaceae
   A   B   C   D   J   M   N
   01 02 09 03 03 01 09
OC: Aphis gossypii H-1, I-6, 7 (1241)   Phytophthora infestans H-14, I-6, 7 (1241)

Alangium platanifolium   (Not known)   Alangiaceae
   A   B   C   D   J
   01 02 09 03 03
OC: Aphids H-1, I-2 (1241)

Alangium salviifolium   (Akola)   Alangiaceae
   A   B   C   D   D   J   M   M   M   M   O
   01 01 09 05 09 14 04 05 17 19 04
OC: Ceratitis capitata H-1, I-4 (956)   Dacus cucurbitae H-1, I-4 (956)
AP: Alk=I-9 (1382)

Albizia glaberima   (Not known)   Mimosaceae
   A   C   D   M
   01 09 01 12
OC: Locusta migratoria H-4, 5, I-? (596)

Albizia julibrissin   (Silk tree)   Mimosaceae
   A   B   C   D   M   M   M   N
   01 01 09 10 01 12 13 07
OR: H-1, I-4 (1241)
AP: Alk=I-6, 7 (phenethylamine)(1392)

Albizia lebbek   (Woman's tongue tree)   Mimosaceae
   A   B   C   D   D   M   M   M   M   M   M   M   M   O   O   O
   01 01 09 01 02 04 05 06 12 15 16 19 21 04 07 10
OC: Mosquitoes H-1, I-? (101)   Musca domestica H-1, I-? (101)
AP: Alk=I-4, 6, 7, 8 (1368); Tan=I-4 (1352)

Albizia lucida   (Not known)   Mimosaceae
  A C D D M M M N
  01 09 01 02 01 12 13 10
OR: H-2, I-? (100)
AP: Alk=I-6, 7 (1392)

Albizia odoratissima   (Ceylon rosewood)   Mimosaceae
  A B C D M M M M
  01 01 09 01 02 12 16 19
OC: Rats H-26, I-4, 6 (1393)   Termites H-8, I-5 (1352)

Albizia procera   (Tall albizia)   Mimosaceae
  A B C D D J J M M M M M
  01 01 02 01 02 03 09 04 12 13 16 19 21
OR: H-1, I-? (1276); H-2, I-4 (73, 105, 871); H-2, I-7 (1345); H-22, I-16 (790); H-32, I-4
(220, 459, 968, 983, 1352)
AP: Tan=I-4 (1352); Tan=I-7, 11 (1360)

Albizia saponaria   (Not known)   Mimosaceae
  A B C D M M
  01 01 09 01 12 15
OR: H-32, I-4 (408, 864, 968, 983); H-33, I-4 (968)
AP: Alk=I-4 (1392); Sap=I-4 (968)

Albizia stipulata   (Not known)   Mimosaceae
  A B C D D J M M
  01 01 09 01 02 03 12 13
OC: Diaphania hyalinata H-1, I-10 (1108)   Dysdercus sanguinarius H-1, I-10 (1108)
OR: H-2, I-? (105, 1109, 1276); H-5, I-? (105); H-32, I-? (73, 1109, 1276)
AP: Alk=I-10 (1373)

Alchemilla procumbens   (Lady's mantle genus)   Rosaceae
  B C D J J
  06 09 01 02 03
OC: Spodoptera frugiperda H-1, I-1 (176)

Alchornea triplinervia   (Not known)   Euphorbiaceae
  A C D J
  01 09 01 09
OC: Anthonomus grandis H-4, I-7 (1356)   Heliothis virescens H-4, I-7 (1356)

Aleurites fordii   (Tung tree)   Euphorbiaceae
  A B C D D F G J J K M M M
  01 01 09 02 03 04 04 03 06 06 04 16 20
OC: Acalymma vittata H-3, 4, I-? (469, 1139)   Diabrotica undecimpunctata H-4, I-? (469)
   Aleurothrixus floccosus H-4, I-? (469)   Laspeyresia pomonella H-4, I-? (469, 1139)
   Anthonomus grandis H-4, I-? (1139); H-4,   Oregma lanigera H-2, I-12 (101); H-4, I-?
    I-10 (319); H-4, I-12 (394)   (469)
   Aonidiella aurantii H-4, I-? (469)   Panonychus citri H-4, 11, I-? (469)
   Argyrotaenia velutinana H-4, I-?   Peronospora tabacina H-14, I-? (469, 482)
    (469, 1139)   Planococcus citri H-4, I-? (469)
   Conotrachelus nenuphar H-4, I-? (469, 1139)
OR: H-15, I-9 (651)

15

Aleurites saponaria   (Bagilumbang)   Euphorbiaceae
    A   B   C   D   D   M
    01 01 09 01 02 20
OR: H-2, I-10, 12 (163)

Alisma orientale   (Water plantain genus)   Alismataceae
    A   B   B   C   D   J
    01 06 09 09 03 13
OC: Spodoptera litura H-4, I-7 (1064)

Allamanda cathartica   (Golden trumpet)   Apocynaceae
    A   B   C   D   J   L   M   M   M   O   O
    01 03 09 01 03 03 05 13 21 02 07
OC: Alternaria tenuis H-14, I-8 (1295)   Fusarium nivale H-14, I-7 (181, 1295)
    Curvularia lunata H-14, I-8 (1295)   Ustilago hordei H-14, I-7 (181)
    Drechslera graminea H-14, I-8 (1295)   Ustilago tritici H-14, I-7 (181)
AP: Alk=I-6, 7 (1363, 1369); Alk=I-9 (1363); Glu, Tan=I-6, 7 (1318); Tri=I-7 (1390)

Allamanda neriifolia   (Bush allamanda)   Apocynaceae
    A   B   C   D
    01 02 09 01
OC: Manduca sexta H-4, 5, I-7 (669)   Panonychus citri H-11, I-7 (669)
AP: Alk=I-6, 7 (1392)

Allionia incarnata   (Not known)   Nyctaginaceae
    B   C   D   J   J
    06 09 03 03 11
OC: Attagenus piceus H-4, I-1 (48)   Tineola bisselliella H-4, I-1 (48)

Allium ampeloprasum var. porrum   (Leek)   Amaryllidaceae
    A   B   C   D   J   M   M   N   O
    01 06 09 06 03 01 05 03 01
OC: Flies H-1, I-? (105)

Allium cepa   (Onion)   Amaryllidaceae
    A   B   C   D   J   J   K   M   M   M   N   O
    01 06 09 03 3a 04 05 01 05 21 03 12
OC: Alternaria tenuis H-14, I-? (426); H-14,   Dermacentor marginatus H-13, I-14
    I-7 (182, 480); H-14, I-12 (185)   (87, 1133)
    Aspergillus niger H-14, I-? (385)   Diplodia maydis H-14, I-7 (480)
    Botrytis allii H-14, I-? (426, 448)   Drechslera graminea H-14, I-7 (182)
    Callosobruchus analis H-5, I-10 (179)   Fusarium culmorum H-14, I-? (385)
    Ceratocystis ulmi H-14, I-12 (185, 426)   Fusarium graminearum H-14, I-14 (935)
    Claviceps purpurea H-14, I-12 (185, 426)   Fusarium moniliforme H-14, I-? (385)
    Colletotrichum circinans H-14, I-14   Fusarium nivale H-14, I-7 (182)
       (491, 935)   Fusarium oxysporum
    Colletotrichum lindemuthianum H-14, I-14   f. sp. conglutinans H-14, I-? (426)
       (935)   f. sp. lycopersici H-14, I-? (426)
    Colletotrichum trifolii H-14, I-14 (935)   Fusarium poae H-14, I-? (385)
    Corynebacterium michiganense H-18, I-?   Fusarium sp. H-14, I-? (1166)
       (1081)   Gibberella fujikuroi H-14, I-? (426)
    Curvularia lunata H-14, I-7 (182, 385)   Haemaphysalis punctata H-13, I-14
    Curvularia penniseti H-14, I-7 (480)   (87, 1133)

Helminthosporium sp. H-14, I-7 (480)   Schistocerca gregaria H-4, I-? (615)
Ixodes redikorzevi H-13, I-14 (87, 1133)   Ticks H-13, I-12 (101)
Myrothecium verricaria H-14, I-12 (185)   Tribolium castaneum H-5, I-10 (179)
Phyllobius oblongus H-4, I-7 (480)   Ustilago avenae H-14, I-12 (185, 426)
Phytodecta fornicata H-4, I-7 (480)   Venturia inaequalis H-14, I-12 (185)
Pieris napi H-5, I-? (1002)   Verticillium albo-atrum H-14, I-12
Pieris rapae H-5, I-? (1002)   (185, 426)
Rhipicephalus rossicus H-13, I-14   (87, 1133)

OR: H-19, I-3, 7 (497, 672, 704, 879, 880, 1166); H-20, I-12, 14 (504, 530); H-24, I-14 (569)
AP: Sfr=I-3 (1116); Tan=I-3 (1321)

Allium cernuum   (Wild onion)   Amaryllidaceae
   B   C   D   J   M   N
   06 09 03 3a 01 03
OC: Agrobacterium tumefaciens H-18, I-7 (585)   Erwinia carotovora H-18, I-7 (585)

Allium fistulosum   (Spanish onion)   Amaryllidaceae
   A   B   C   D   M   N
   01 06 09 03 01 07
OC: Aphids H-1, I-1 (1241)
OR: H-19, I-? (172)

Allium nipponicum   (Not known)   Amaryllidaceae
   B   C   D   J   M   N
   06 09 03 03 01 01
OC: Drosophila hydei H-2, I-2, 4, 7, 8 (101)

Allium oleraceum   (Field garlic)   Amaryllidaceae
   B   C   D   M
   06 09 03 14
OC: Locusta oleraceae H-4, 5, I-7 (596)

Allium sativum   (Garlic)   Amaryllidaceae
   A   B   C   D   F   G   J   J   J   J   J   K   K   M   M   N   N
   01 06 09 10 14 03 03 04 06 09 12 05 06 01 05 03 07
OC: Aedes aegypti H-1, I-? (325)   Colletotrichum capsici H-14, I-? (783)
    Aedes nigromaculis H-1, I-? (325)   Colletotrichum circinans H-14, I-14 (935)
    Aedes sierrensis H-1, I-1 (325)   Colletotrichum lindemuthianum H-14, I-12
    Aedes triseriatus H-1, I-? (325)   (425, 935)
    Agrobacterium tumefaciens H-18, I-3 (1357)   Colletotrichum trifolii H-14, I-14 (935)
    Alternaria tenuis H-14, I-7   Corynebacterium flaccumfaciens H-18, I-3
        (182, 426, 480, 783, 927)   (1357)
    Aspergillus niger H-14, I-? (385, 783)   Corynebacterium michiganense H-18, I-3
    Botrytis allii H-14, I-3, 12 (185, 426)   (1081, 1357)
    Callosobruchus chinensis H-2b, I-3 (116);   Culex peus H-1, I-? (325)
        H-5, I-? (599)   Culex quinquefasciatus H-1, I-? (1046)
    Cephalosporium sacchari H-14, I-? (927)   Culex tarsalis H-1, I-? (325)
    Ceratocystis ulmi H-14, I-? (426)   Curvularia lunata H-14, I-7
    Cercospora cruenta H-14, I-3 (1122)   (182, 385, 927)
    Cladosporium cucumerinum H-14, I-? (425)   Curvularia penniseti H-14, I-7 (480)
    Cladosporium fulvum H-14, I-12, 14   Dermacentor marginatus H-13, I-? (1133)
        (185, 426)   Diplodia maydis H-14, I-? (426)
    Claviceps purpurea H-14, I-12, 14   Drechslera graminea H-14, I-7
        (185, 426)   (182, 778, 927)

Drechslera oryzae H-14, I-3, 7, 8
  (113, 432, 783)
Dysdercus cingulatus H-1, I-?
  (88, 124, 435)
Erwinia aroideae H-18, I-? (1082)
Erwinia carotovora H-18, I-3 (1357)
Fusarium culmorum H-14, I-? (385)
Fusarium graminearum H-14, I-12 (935)
Fusarium moniliforme H-14, I-? (385, 783)
Fusarium nivale H-14, I-? (182, 927)
Fusarium oxysporum H-14, I-? (783)
Fusarium oxysporum
    f. sp. conglutinans H-14, I-? (426)
    f. sp. lycopersici H-14, I-? (426)
    f. sp. udum H-14, I-? (783)
Fusarium poae H-14, I-? (385)
Fusarium sp. H-14, I-? (1166)
Gibberella fujikuroi H-14, I-? (426)
Glomerella cingulata H-14, I-? (783)
Haemaphysalis punctata H-13, I-? (87, 1133)
Helminthosporium sp. H-14, I-7 (480)
Ixodes redikorzevi H-13, I-? (87, 1133)
Lentinus lepideus H-14, I-? (428)
Lenzites trabea H-14, I-? (428)
Meloidogyne incognita H-16, I-3 (100, 616)
Meloidogyne javanica H-16, I-3 (100, 616)
Monilinia fructicola H-14, I-? (425)
Mosquitoes H-2, I-? (55)
Musca domestica H-1, 5, I-1 (123, 889, 890)

Myrothecium verricaria H-14, I-12 (185)
Pericallia ricini H-1, I-? (435)
Pestalotia sp. H-14, I-? (927)
Phomopsis sp. H-14, I-? (927)
Polyporus versicolor H-14, I-3 (386, 428)
Pseudomonas lachrymans H-18, I-? (425)
Pseudomonas phaseolicola H-18, I-3
  (425, 1357)
Pseudomonas solanacearum H-18, I-3 (1357)
Pseudoperonospora cubensis H-14, I-? (425)
Pyricularia oryzae H-14, I-3, 7
  (114, 432, 783, 1133)
Rhipicephalus rossicus H-13, I-? (87, 1133)
Sphaceloma ampelinum H-14, I-12 (1342)
Spodoptera littoralis H-1, I-? (783)
Spodoptera litura H-1, I-?
  (88, 124, 435, 436)
Thrips H-1, I-? (783)
Trichoderma viride H-14, I-? (927)
Trogoderma granarium H-1, I-? (889, 890)
Ustilago avenae H-14, I-? (426)
Ustilago hordei H-14, I-? (181)
Ustilago tritici H-14, I-? (181)
Verticillium albo-atrum H-14, I-? (426)
Xanthomonas campestris
    pv. campestris H-18, I-3 (1357)
    pv. oryzae H-18, I-3 (1298, 1357)
    pv. vesicatoria H-18, I-3 (1357)
Xanthomonas sp. H-18, I-14 (148)

OR: H-3, I-? (441); H-17, I-12 (166, 220); H-19, I-3, 10, 12, 14 (186, 399, 497, 659, 660, 672, 704, 755, 761, 980, 982, 1080, 1166, 1167, 1168, 1291); H-20, I-? (530)
AP: **Alk, Sap, Tan=I-3** (1321)

Allium schoenoprasum   (Chive)   Amaryllidaceae
    A   B   C   D   J   M   N
    01   06   09   03   03   01   07
OC: Mosquitoes H-2, I-1 (98, 105)
OR: H-19, I-? (448)
AP: **Alk=I-10** (1369); **Sfr=I-10** (1116)

Allium tricoccum   (Wild leek)   Amaryllidaceae
    B   C   D   M   N
    06   09   03   01   03
OC: Agrobacterium tumefaciens H-18, I-7 (585)   Erwinia carotovora H-18, I-7 (585)
AP: **Alk=I-2** (1392)

Allium tuberosum   (Chinese chive)   Amaryllidaceae
    A   B   C   D   M   M   N   O
    01   06   09   01   01   05   07   10
OC: Aphids H-1, I-1 (1241)   Spider mites H-11, I-1 (1241)
    Phytophthora infestans H-14, I-1 (1241)

Alnus firma   (Alder genus)   Betulaceae
    A   B   B   C   D   D   J   M
    01   01   02   09   02   03   03   13
OC: Drosophila hydei H-2, I-2 (101)

18

Alnus hirsuta   (Manchurian alder)   Betulaceae

    A   B   C   D   J   M

    01 01 09 03 03 16

OC: Drosophila hydei H-1, I-6, 7 (101)

Alnus rugosa   (Hazel alder)   Betulaceae

    A   B   B   C   D

    01 01 02 09 03

OC: Lymantria dispar H-5, I-? (340)

Alocasia macrorrhiza   (Taro)   Araceae

    A   B   C   D   J   J   K   L   M   N

    01 06 09 01 03 04 03 03 01 03

OC: Curvularia lunata H-14, I-2 (1295)   Rice field insects H-5, I-1 (137)

OR: H-32, I-5, 7 (983); H-33, I-14 (968)

AP: Alk=I-2 (1392)

Aloe barbadensis   (Medicinal aloe)   Liliaceae

    A   B   C   D   D   D   J   J   M   M   O   O

    01 06 09 01 05 06 03 04 05 13 07 14

OC: Athalia proxima H-4, I-7 (438, 607, 1011)   Popillia japonica H-5, I-7 (105, 1243)

   Meloidogyne incognita H-16, I-7 (542)

OR: H-2, I-? (104); H-5, I-? (100); H-17, I-14 (504)

Aloe ferox   (Aloe genus)   Liliaceae

    A   B   C   D   M

    01 02 09 02 05

OC: Flies H-5, I-14 (1345)

Aloe striata   (Coral aloe)   Liliaceae

    A   B   C   D

    01 06 09 05

OC: Fleas H-6, I-16 (105)

Aloe succotrina   (Aloe genus)   Liliaceae

    A   B   C   M   O

    01 06 09 05 14

OC: Scale insects H-1, I-? (105)

Aloysia triphylla   (Lemon verbena)   Verbenaceae

    A   B   B   B   C   D   D   K   M   M   N

    01 02 05 06 09 01 02 06 01 17 07

OC: Aphis gossypii H-2, I-12 (87, 105)

   Tetranychus cinnabarinus H-11, I-12 (87, 105)

Alpinia afficinarum   (Ginger lily)   Zingiberaceae

    A   B   C   D   M   M   O

    01 06 09 10 05 13 02

OC: Musca domestica H-2, I-2, 12 (87)

OR: H-15, 19, I-2 (411)

Alpinia galanga   (Galanga ginger lily)   Zingiberaceae

    A   B   C   D   J   M   M   M   M   M   N

    01 06 09 10 06 01 02 13 14 17 08

OR: H-2, I-? (80); H-19, I-2, 12 (450, 897)

Alysicarpus vaginalis   (Alyce clover genus)   Fabaceae

```
A   B   C   D   E   J   M   M   M
03  06  09  10  02  03  02  09  11  12
```

OR: H-24, I-1 (1212)
AP: Alk=I-10 (1391)

Amanita mappa   (Not known)   Agaricaceae

```
A   C
04  01
```

OR: H-4, I-? (132)
AP: Alk=I-9 (bufotenine)(1392)

Amanita muscaria   (Fly agaric)   Agaricaceae

```
A   C   D   J   L   M
04  01  10  03  03  05
```

OC: Aphis persicae-niger H-2, I-9 (105)   Flies H-2, I-9 (101, 105)
    Attagenus piceus H-4, I-9 (48)   Musca domestica H-1, I-9 (742)
OR: H-22, I-9 (969); H-33, 35, I-9 (504)
AP: Alk=I-9 (bufotenine, hercynine, muscarine, myketosine)(1392)

Amanita pantherina   (Not known)   Agaricaceae

```
A   C   D   J   L
04  01  03  10  03
```

OC: Culex pipiens H-2, I-9 (101)   Flies H-2, I-9 (104)
OR: H-33, I-9 (504)
AP: Alk=I-9 (bufotenine, muscarine)(1392)

Amaranthus ascendens   (Amaranth genus)   Amaranthaceae

```
A   B   C   D
03  06  09  10
```

OC: Leptinotarsa decemlineata H-4, I-7 (531)

Amaranthus hybridus   (Pigweed)   Amaranthaceae

```
A   B   C   D   L   M   N   N
03  06  09  10  05  01  07  10
```

OC: Botrytis cinerea H-14, I-8, 16 (945)

Amaranthus retroflexus   (Redroot)   Amaranthaceae

```
A   B   C   D   L   M   N   N
03  06  09  03  05  01  07  10
```

OC: Athalia rosae H-4, I-7 (531)   Phytodecta fornicata H-4, I-7 (531)
    Hyphantria cunea H-4, I-7 (531)   Pieris brassicae H-4, I-7 (531)
    Leptinotarsa decemlineata H-4, I-7 (531)

Amaranthus spinosus   (Spiny amaranth)   Amaranthaceae

```
A   B   C   D   J   L   M   M   M   N   O   O
03  06  09  01  03  05  01  02  05  07  02  07
```

OC: Drechslera oryzae H-14, I-7 (113)

Amaranthus viridis   (Green amaranth)   Amaranthaceae

```
A   B   C   D   J   M   M   M   M   N   N   O
03  06  09  01  03  01  02  05  21  06  07  07
```

OC: Alternaria tenuis H-14, I-7 (480)   Helminthosporium sp. H-14, I-7 (480)
    Curvularia penniseti H-14, I-7 (480)
AP: Tan=I-7 (1353)

Ambrosia cumanensis   (Ragweed genus)   Asteraceae
      B   C   D   L
      06   09   10   05
OR: H-24, I-? (1221)

Ambrosia psilostachya   (Western ragweed)   Asteraceae
      A   B   C   D   J   L   M
      01   06   09   10   03   05   05
OR: H-24, I-2, 15 (1233)

Ambrosia trifida   (Giant ragweed)   Asteraceae
      B   C   D   J   L
      06   09   10   03   05
OR: H-24, I-7 (1215); H-24, I-16 (1254)
AP: Alk=I-10 (1391)

Amianthium muscitoxicum   (Fly poison)   Liliaceae
      A   B   C   D   J   J   J   J   K   K   K   L
      01   06   09   03   03   04   08   11   02   05   06   03
OC: Aedes aegypti H-2a, I-7 (48)   Flies H-2, 2b, I-3 (104, 105)
    Aphids H-2, I-3 (105)   Grasshoppers H-2b, I-3 (104, 105)
    Attagenus piceus H-4, I-3, 7 (48)   Leptinotarsa decemlineata H-2, I-3, 7
    Bombyx mori H-2, I-3, 7 (105)   (96, 105)
    Caterpillars H-2, I-3, 7 (104, 105)   Ostrinia nubilalis H-2, I-3, 7 (101)
    Cockroaches H-2, I-3, 7 (104, 105)
OR: H-33, I-3 (507)
AP: Alk=I-2, 7 (amianthine, jervine)(1392)

Ammania baccifera   (Not known)   Lythraceae
      A   B   C   D   J   L
      03   06   09   01   03   03
OC: Exserohilum turcicum H-14, I-15 (827)
OR: H-33, I-? (220)

Ammania multiflora   (Not known)   Lythraceae
      B   C   D   J
      06   09   10   03
OC: Exserohilum turcicum H-14, I-15 (827)

Amomum melegueta   (Melegueta pepper)   Zingiberaceae
      A   B   C   D   J   M
      01   06   09   01   03   17
OC: Glossina sp. H-5, I-? (105)

Amoreuxia wrightii   (Not known)   Cochlospermaceae
      B   C   D   J
      06   09   03   11
OC: Attagenus piceus H-4, I-2 (48)

Amorpha fruticosa   (False indigo)   Fabaceae
      A   B   C   D   F   J   J   J   J   J   K   M   M   M
      01   02   09   05   01   03   09   10   11   16   03   07   09   12
OC: Acalymma vittata H-5, I-9 (299)   Aedes aegypti H-2, I-2, 9 (484); H-3, I-?
    Acyrthosiphum pisum H-2, I-10, 11 (484)   (156, 165)

Aphis fabae H-2, I-10 (1104)
Aphis gossypii H-2, I-2, 9 (78, 101); H-2,
   I-10, 11 (484)
Blissus leucopterus H-2, I-2, 9, 10, 11
   (78, 101, 484); H-5, I-9 (299)
Cerotoma trifurcata H-2, I-2, 9, 10, 11
   (78, 101, 484)
Culex tarsalis H-3, I-? (156, 165)
Diabrotica duodecimpunctata H-2, I-10, 11
   (101, 484)

Empoasca fabae H-2, I-10, 11 (78, 101)
Epicauta lemniscata H-2, I-10, 11 (484)
Epilachna varivestis H-1, I-10 (284)
Leptinotarsa decemlineata H-2, I-10, 11
   (484)
Livestock pests H-1, I-9 (101, 447)
Lygus lineolaris H-2, I-? (101)
Macrobasis unicolor H-2, I-10, 11 (484)
Meloidae sp. H-2, I-10, 11 (101, 484)
Tribolium confusum H-2, I-10, 11 (484)

OR: H-20, I-15 (832); H-30, I-10 (815); H-32, I-1 (192)
AP: Alk=I-10 (1371)

Amorpha glabra   (Not known)   Fabaceae
   B   C   D   J   J   M
   02 09 03 03 11 12
OC: Attagenus piceus H-4, I-2 (48)   Periplaneta americana H-1, I-2 (78)

Amorphophallus campanulatus   (Whitespot giant arum)   Araceae
   A   B   C   D   J   L   M   N   N
   01 06 09 01 03 03 1a 03 07
OC: Drechslera oryzae H-14, I-6, 7 (113, 432)   Pyricularia oryzae H-14, I-7 (114)
   Flies H-6, I-8 (105)   Rice field insects H-5, I-6, 7 (137)
OR: H-19, I-7 (449); H-33, I-7 (968, 982, 983, 1348)
AP: Alk=I-2 (1392)

Ampelopsis japonica   (Not known)   Vitaceae
   A   B   C   D
   01 03 09 03
OC: Pieris rapae H-1, I-1 (1241)

Amphiroa fragilissima   (Red algae)   Corallinaceae
   A   B   C   J
   04 09 02 03
OC: Mucor racemosus H-14, I-1 (1024)   Rhizopus oryzae H-14, I-1 (1024)

Amsonia elliptica   (Bluestar genus)   Apocynaceae
   A   B   C   D   J   M
   01 06 09 03 03 13
OC: Drosophila hydei H-2, I-7 (101)
AP: Alk=I-2 (amsonine)(1392)

Anabasis aphylla   (Anabasine)   Chenopodiaceae
   A   B   C   D   J   J
   01 02 09 03 03 04
OC: Acyrthosiphum pisum H-2, I-? (97)   Laspeyresia pomonella H-2, I-? (297)
   Anasa tristis H-1, I-? (239)   Macrosiphum rosae H-2, I-? (489)
   Aphids H-1, I-6 (30, 81, 97, 1033)   Mites H-11, I-? (97)
   Aphis fabae H-2, I-? (231, 320, 489)   Oncopeltus fasciatus H-1, I-? (239)
   Aphis pomi H-2, I-? (253, 489)   Rhopalosiphum rufomaculata H-2, I-? (489)
   Culex quinquefasciatus H-1, I-? (217)   Scirtothrips citri H-1, I-? (277)
   Eriosoma lanigerum H-1, I-? (253)
OR: H-2, I-6 (80, 101)
AP: Alk=I-6 (anabasine)(217, 231); Alk=I-16 (anabasine, aphyllidine, aphylline, base
   V-lupinine, N-methylanabasine, oxyaphyllidine, oxyaphylline, supinine)(1392)

Anacardium occidentale   (Cashew)   Anacardiaceae

A	B	C	D	D	F	J	J	L	M	M	M	M	M	M	N	N	O	O
01	01	09	01	04	03	03	06	03	01	05	08	16	20	21	09	10	04	07

OC: Ahasverus advena H-1, I-? (1124)   Oryzaephilus surinamensis H-1, I-? (1124)
    Culex fatigans H-1, I-10, 12 (1049)   Powderpost beetles H-1, I-10, 12 (87)
    Meloidogyne incognita H-16, I-7 (100, 616)   Sitophilus granarius H-1, I-12 (87, 90)
    Meloidogyne javanica H-16, I-7 (100, 616)   Termites H-1, I-10, 12 (73, 87, 90); H-5,
    Mosquitoes H-1, I-10, 12 (87, 90)   I-10, 12 (105, 504)
    Moths H-1, I-12 (87, 90)

OR: H-5, I-13 (1353); H-17, I-10 (1318); H-24, I-7 (1353); H-33, I-10, 12 (220, 968)
AP: Sap=I-6 (1318); Tan=I-4, 6, 7 (1318, 1353)

Anacyclus pyrethrum   (Pellitory)   Asteraceae

B	C	D	J	M	O
06	09	06	02	05	12

OC: Acyrthosiphum pisum H-2, I-2 (1252)   Tenebrio molitor H-2, I-2 (87)
OR: H-2, I-2 (61, 803)
AP: Alk=I-2 (anacyclin, pellitorine)(803, 1392)

Anagallis arvensis   (Scarlet pimpernel)   Primulaceae

A	B	C	D	D	G	J	J	J	L	M	M	N
03	06	09	03	06	07	03	04	07	05	01	05	07

OC: Anguina tritici H-16, I-15 (920)   Leeches H-28, I-? (73, 220)
    Athalia rosae H-4, I-7 (531)   Pythium aphanidermatum H-14, I-2, 6, 7, 8
    Bipolaris maydis H-14, I-1 (338)   (855)
    Colletotrichum falcatum H-14, I-2, 6,   Rhizopus nigricans H-14, I-2, 6, 7, 8 (855)
      7, 8 (855)   Ustilago hordei H-14, 1-7 (181)
    Colletotrichum papayae H-14, I-1 (388, 855) Ustilago tritici H-14, I-7 (181)
    Exserohilum turcicum H-14, I-2, 6, 7, 8, 15 (827, 855)

OR: H-2, I-? (105); H-5, I-16 (87); H-5, 32, I-? (1276); H-22, I-1 (790); H-32, I-? (73, 220)
AP: Alk=I-2, 6, 7, 9 (1368)

Ananas sativus   (Pineapple)   Bromeliaceae

A	B	C	D	J	J	M	M	N
01	06	09	01	03	08	01	03	09

OC: Blatta orientalis H-6, I-12 (105)
OR: H-17, I-7 (1135, 1320)

Ancistrocladus barterii   (Not known)   Ancistrocladaceae

C	D	J	J
09	01	03	08

OC: Attagenus piceus H-4, I-2 (48)

Andrachne cordifolia   (Not known)   Euphorbiaceae

A	B	C	D	L
01	02	09	03	03

OR: H-2, I-2, 4, 7 (73, 105, 220); H-33, I-? (220)

Andrachne ovalis   (Not known)   Euphorbiaceae

A	B	C	D	J
01	02	09	03	04

OC: Flies H-2, I-2 (87)

Andrographis paniculata   (Kariyat)   Acanthaceae
    B   C   D   M   O   O
    06 09 01 05 02 07
OC: Grasshoppers H-2, I-? (95)   Nematodes H-16, I-? (100)
    Hydrellia philippia H-2, I-? (95)
OR: H-19, I-? (411)

Andropogon schoenanthus   (Lemongrass)   Poaceae
    A   B   C   D   D   M   M   N
    01 06 09 01 03 01 14 07
OC: Myrothecium verricaria H-14, I-12 (185)

Andropogon virginicus   (Virginia beard grass)   Poaceae
    A   B   C   D   D   M
    01 06 09 01 05 08
OR: H-24, I-? (1188); H-24, I-7 (1204)

Anemia mexicana   (Flowering fern)   Schizaeaceae
    A   C   D   J   J
    01 06 01 08 11
OC: Attagenus piceus H-4, I-1 (48)

Anemia pinquis   (Not known)   Schizaeaceae
    A   C   D
    01 06 01
OR: H-4, I-? (132)

Anemone altaica   (Windflower)   Ranunculaceae
    A   B   C   D   J
    01 06 09 03 03
OC: Flies H-2, I-? (101)   Mosquitoes H-2, I-? (101)
    Lice H-2, I-? (101)

Anemone chinensis   (Pulsatilla chinensis)   Ranunculaceae
    A   B   C   D   J
    01 06 09 03 03
OC: Agrotis sp. H-1, I-1, 2, 6 (1241)   Puccinia graminis tritici H-14, I-1, 2, 6
    Aphids H-1, I-1, 2, 6 (1241)   (1241)
    Caterpillars H-1, I-1, 2, 6 (1241)   Puccinia rubigavera H-14, I-1, 2, 6 (1241)
    Phytophthora infestans H-14, I-1, 2, 6 (1241)

Anemone coronaria   (Anemone genus)   Ranunculaceae
    A   B   D   J   M
    01 06 03 03 13
OC: Drosophila hydei H-2, I-2 (101)

Anemone hupehensis   (Japanese anemone)   Ranunculaceae
    A   B   C   D   J   M
    01 06 09 03 03 13
OC: Aphis gossypii H-1, I-1 (1241)   Pieris rapae H-1, I-1 (1241)
    Maggots H-1, I-1 (1241)   Puccinia glumarum H-14, I-1 (1241)
    Oulema oryzae H-1, I-1 (1241)   Spider mites H-11, I-1 (1241)
    Phytophthora infestans H-14, I-1 (1241)

Anemone nemorosa   (European wood anemone)   Ranunculaceae
    A   B   C   D   F   J   M
    01 06 09 03 03 02 05
OC: Venturia inaequalis H-14, I-? (16)

Anemone nuttalliana   (Lion's beard)   Ranunculaceae
    A   B   C   D   J   J   J   M
    01 06 09 03 03 08 11 13
OC: Attagenus piceus H-4, I-1 (48)

Anemone raddeana   (Windflower)   Ranunculaceae
    A   B   C   D   J   M
    01 06 09 03 03 13
OC: Drosophila hydei H-2, I-6, 7 (101)

Anethum graveolens   (Dill)   Apiaceae
    A   B   C   D   J   J   M
    03 06 09 03 02 04 14
OC: Alternaria tenuissima H-14, I-? (383)   Fusarium poae H-14, I-? (383)
    Curvularia lunata H-14, I-? (383)   Haemaphysalis punctata H-13, I-? (87, 1133)
    Dermacentor marginatus H-13, I-? (87, 1133)   Ixodes redikorzevi H-13, I-? (87, 1133)
    Drosophilia melanogaster H-1, I-10 (930)   Rhipicephalus rossicus H-13, I-? (87, 1133)
    Fusarium chlamydosporum H-14, I-? (383)   Rhizoctonia solani H-14, I-? (383)

Anethum sowa   (Sowa)   Apiaceae
    B   C   D   M
    06 09 01 12
OC: Schistocerca gregaria H-4, I-12 (1008)   Tribolium castaneum H-4, I-12 (375, 1008)
OR: H-30, I-? (132)

Angelica archangelica   (Wild parsnip)   Apiaceae
    A   A   B   C   D   M   M   N   O
    01 02 06 09 03 01 05 07 02
OC: Ceratitis capitata H-6, I-10 (87)
OR: H-19, I-? (429)

Angelica dahurica   (Angelica genus)   Apiaceae
    A   B   C   D   J
    01 06 09 03 03
OC: Aphids H-1, I-? (1241)   Puccinia graminis tritici H-14, I-? (1241)
    Pieris rapae H-1, I-? (1241)

Angelica glauca   (Angelica genus)   Apiaceae
    A   B   C   D   G   J   J   K   M   O
    01 06 09 03 04 04 11 05 05 02
OC: Tribolium castaneum H-1, 5, I-2 (179)

Angelica japonica   (Japanese angelica)   Apiaceae
    A   B   C   D
    01 06 09 03
OC: Spodoptera litura H-4, I-7 (1067)

Angelica pubescens   (Angelica genus)   Apiaceae
    A   B   C   D
    01 06 09 03
OC: Aphelencoides besseyi H-16, I-2 (52, 187)

Angelica sylvestris   (Wild angelica)   Apiaceae
    A   B   C   D   J   M   N   N
    01 06 09 03 04 01 06 07 15
OC: Dermestres maculatus H-3, 4, I-? (544)   Pyrrhocoris apterus H-3, 4, I-? (544)
    Dysdercus cingulatus H-3, 4, I-? (544)   Sitophilus granarius H-4, I-? (1398)
    Graphosoma italicum H-3, 4, I-? (544)   Tribolium confusum H-4, I-? (1398)
    Graphosoma mellonella H-3, 4, I-? (544)   Trogoderma granarium H-4, I-? (1398)
    Pediculus humanus capitis H-2, I-9 (504)
OR: H-2, I-2 (87)

Angiopteris evecta   (Turnip fern genus)   Marattiaceae
    A   C   D   M   N
    01 06 01 01 05
OC: Pseudomonas solanacearum H-18, I-7 (854)
    Xanthomonas campestris pv. phaseoli-sojensis H-18, I-7 (854)

Anisomeles malabarica   (Chodhava)   Lamiaceae
    B   C   D   M   O
    06 09 01 05 07
OR: H-24, I-15 (1184)
AP: Alk=I-? (1392); Str=I-1 (B-sitosterol)(1381)

Annona cherimola   (Cherimoya)   Annonaceae
    A   B   C   D   D   J   J   J   L   M   N
    01 01 09 01 02 03 08 11 05 01 09
OC: Attagenus piceus H-4, I-? (48)   Parasites (human) H-1, I-10 (87)
    Manduca sexta H-4, I-7 (669)   Tineola bisselliella H-4, I-? (48)
    Panonychus citri H-4, 11, I-7 (669)
OR: H-1, I-10 (104, 105); H-19, I-7 (186); H-26, I-7 (118); H-32, I-10 (105, 192)
AP: Alk=I-10 (caffeine)(1392)

Annona glabra   (Pond apple)   Annonaceae
    A   B   C   D   D   J   J   J   J   M   N
    01 01 09 01 02 03 04 08 11 01 09
OC: Aedes aegypti H-4, I-10 (48)   Oncopeltus fasciatus H-4, I-10 (48)
    Attagenus piceus H-4, I-10 (48)   Tineola bisselliella H-4, I-10 (48)
    Macrosiphoniella sanborni H-2, I-7 (105)
OR: H-2, I-10 (104, 105)
AP: Alk=I-6, 7, 9 (1392)

Annona montana   (Mountain soursop)   Annonaceae
    A   B   C   D   J   M   N
    01 01 09 01 03 01 09
OC: Oncopeltus fasciatus H-2, I-10 (48)

Annona muricata   (Soursop)   Annonaceae
    A   B   C   D   D   J   J   J   J   M   M   N   O
    01 01 09 01 02 03 04 08 11 01 05 09 07

OC: Acyrthosiphum pisium H-2, I-10 (101, 1252)   Pediculus humanus capitis H-2, I-2, 7 (87)
    Aedes aegypti H-4, I-10, (48)   Pseudaletia unipuncta H-2, I-10 (221, 1252)
    Aphids H-2, I-? (22)   Spodoptera eridania H-2, I-10 (101)
    Attagenus piceus H-4, I-10 (48)
    Macrosiphoniella sanborni H-2, I-10 (22, 674)
AP: Alk=I-? (anonaine, anoniine)(1392); Alk=I-4 (muricine, muricinine)(1392); Alk, Sap=I-6, 7
    (1318)

Annona nolitu   (Ritnaj)   Annonaceae
   A   B   C   D   M   N
   01 01 09 01 01 09
OR: H-1, I-4, 10 (127)

Annona palustris   (Alligator apple)   Annonaceae
   A   B   C   D   M   N
   01 01 09 01 01 09
OC: Aphids H-2, I-? (22)   Oncopeltus fasciatus H-2, I-10 (48, 130)
    Macrosiphoniella sanborni H-2, I-7 (22, 674)

Annona reticulata   (Custard apple)   Annonaceae
   A   B   C   D   D   F   J   J   J   J   K   L   M   M   N   O   O
   01 01 09 01 02 03 02 03 04 08 06 05 01 05 09 02 04
OC: Achaea janata H-2, I-4, 7, 10 (105)   Idiocerus sp. H-2, I-4, 10 (105)
    Aphis fabae H-2a, I-10 (15)   Leanium sp. H-1, I-10 (87)
    Callosobruchus chinensis H-2, I-10 (105)   Macrosiphoniella sanborni H-2, I-2, 4, 6,
    Callosobruchus maculatus H-5, I-10   7, 10 (90, 105, 674); H-2a, I-10 (15, 87)
      (460, 1124)   Macrosiphum solanifolii H-2a, I-10 (15)
    Coccus viridis H-1, I-4, 7, 10 (1029)   Pediculus humanus capitis H-2, I-10 (87, 90)
    Crocidolomia binotalis H-2, I-4, 10 (105)   Plutella xylostella H-2, I-4, 7, 10 (105);
    Dysdercus cingulatus H-3, I-? (712)   H-2b, I-10 (15)
    Epacromia tamulus H-2, I-4, 7, 10 (105)   Rice field insects H-5, I-6, 7 (137)
    Euproctis fraterna H-2, I-4, 10 (105)   Spodoptera litura H-2, I-4, 7, 10 (105)
    Hypsa ficus H-2, I-4, 10 (105)   Tribolium castaneum H-1, I-16 (785)
OR: H-2, I-10 (67, 73, 92, 104, 105, 220); H-17, I-7, 9 (1319)
AP: Alk=I-2 (linocline)(1381); Alk=I-4 (anonaine)(1392); Alk=I-6, 7 (1319, 1392)

Annona senegalensis   (Wild custard apple)   Annonaceae
   A   B   C   D   M   M   N   O   O
   01 01 09 01 01 05 09 02 07
OC: Oncopeltus fasciatus H-2, I-6 (958)   Stored grain pests H-5, I-7 (533)
OR: H-2, I-2 (1353); H-15, I-9 (1353); H-17, I-7, 9 (1351, 1353)

Annona spinescens   (Not known)   Annonaceae
   A   B   C   D   J   J   M   N
   01 01 09 01 03 04 01 09
OR: H-2, 14, I-10 (104)

Annona squamosa   (Sugar apple)   Annonaceae
   A   B   C   D   D   J   J   J   J   J   K   M   N
   01 01 09 01 02 03 04 07 08 11 06 01 09
OC: Aedes aegypti H-2a, I-10 (48)   Aulacophora foveicollis H-1, I-7 (87)
    Aphids H-1, I-? (78); H-2b, I-2, 10 (101)   Aulacophora hilaris H-2b, I-10 (87, 101)
    Aphis fabae H-2a, I-10 (15)   Bombyx mori H-2b, I-10 (87)
    Attagenus piceus H-4, I-10 (48)   Brevicoryne brassicae H-2b, I-10 (87, 101)

27

Callobruchus chinensis H-1, I-10
 (95, 120, 475); H-2, I-10, 12 (605, 784);
 H-3, 5, I-? (1327)
Caterpillars H-1, I-? (84)
Coccus viridis H-1, I-10 (120)
Crocidolomia binotalis H-1, I-10 (605)
Dysdercus koenigii H-2, I-10 (91)
Epacromia tamulus H-1, I-? (605)
Epilachna viginti-octopunctata H-1, I-?
 (605)
Euproctis fraterna H-1, I-10 (120, 605)
Lice H-1, I-10 (87)
Lipaphis erysimi H-2, I-10 (91)
Macrosiphoniella sanborni H-2a, I-7, 10
 (15, 674)
Macrosiphum solanifolii H-2a, I-10 (15)
Meloidogyne incognita H-16, I-7 (542)
Musca domestica H-1, I-7, 9, 10, 16
 (87, 785)
Musca nebulo H-1, I-10 (87)

Nephotettix virescens H-1, I-10, 12
 (140, 942); H-4, I-10, 12 (141)
Nilaparvata lugens H-2a, I-10, 12 (141)
Oncopeltus fasciatus H-2a, I-? (48)
Oryzaephilus surinamensis H-1, I-? (78)
Pericallia ricini H-1, I-? (605)
Phymatocera aterrima H-2b, I-2, 10 (101)
Plutella xylostella H-1, I-10 (78, 605);
 H-2b, I-2, 10 (15, 101)
Rhizopertha dominica H-3, 5, I-? (1327)
Sitophilus oryzae H-3, I-10 (888); H-4,
 I-10 (517)
Sogatella furcifera H-4, I-10, 12 (141)
Spodoptera litura H-2, I-10 (120, 605)
Stegobium paniceum H-2, I-10, 12 (784)
Tineola bisselliella H-4, I-10 (48)
Tribolium castaneum H-1, I-10, 16
 (87, 785); H-4, I-10 (517)
Urentius echinus H-1, I-10, 12 (87)
Vermin H-27, I-2, 7, 9, 10 (87, 1318)

OR: H-2, I-7, 9 (15, 80, 100, 105); H-19, I-4 (982); H-30, I-? (520); H-32, I-7, 9 (87)
AP: Alk=I-5, 6 (1375); Alk=I-7, 10 (annonaine)(1392); Alk, Tan=I-6, 7 (1318)

Anodendron affine   (Not known)   Apocynaceae
    A   B   C   J
    01 03 09 03
OC: Drosophila hydei H-2, I-6, 7 (101)

Anogeissus leiocarpus   (Not known)   Combretaceae
    A   B   C   D   M   M   N   O
    01 01 09 01 01 05 16 13 07
OC: Pseudomonas solanacearum H-18, I-4 (945)

Anomospermum japurense   (Not known)   Menispermaceae
    · C
    09
OR: H-5, 14, I-? (1276)

Anthemis arvensis   (Corn chamomile)   Asteraceae
    B   C   D   J
    06 09 06 03
OC: Mice H-25a, I-7 (104)   Musca domestica H-2a, I-8 (87)
OR: H-1, I-20 (104)

Anthemis cotula   (Mayweed)   Asteraceae
    A   B   C   D   J   J   J   K   L   M   O
    03 06 09 03 02 03 04 05 05 05 07
OC: Cimex lectularius H-2, I-8 (104)   Flies H-2, I-8 (104)
    Fleas H-2, I-8 (104)
OR: H-2, 5, I-7 (104, 1345)

Anthemis tinctoria   (Golden chamomile)   Asteraceae
    A   B   C   D   M
    02 06 09 03 16
OC: Musca domestica H-2a, I-7, 8 (87)

Anthocephalus cadamba   (Kadam tree)   Rubiaceae
    A   B   C   D   J   M   M   M   M   N   O
    01 01 09 01 10 01 04 05 19 09 04
OC: Dysdercus cingulatus H-3, I-6 (471, 690)
AP: Alk=I-? (1392); Alk=I-4 (1368); Tri=I-2 (quinovic acid)(1384)

Anthriscus vulgaris   (European chervil)   Apiaceae
    A   B   C   D   M   N
    03 06 09 03 01 07
OC: Ants H-5, I-7 (105)

Antiaris toxicaria   (Upas tree)   Moraceae
    A   B   C   D   J   J   J   L   M   M
    01 01 09 01 02 3a 05 03 03 22
OC: Mice H-25, I-14 (102)   Rats H-25, I-14 (102)
OR: H-24, I-1 (504); H-33, I-2, 7, 14 (139, 220, 431, 453, 504, 507, 637, 849, 863, 983,
    1348, 1352)
AP: Alk=I-14 (antiarin)(1348, 1352); Glu=I-? (antiarin)(1348); Tri=I-7 (1390)

Antidesma pentandrum   (Binayoyong pugo)   Euphorbiaceae
    A   B   C   D   J
    01 01 09 01 03
OC: Drechslera oryzae H-14, I-7 (113)   Pyricularia oryzae H-14, I-7 (114)

Antirrhinum majus   (Snapdragon)   Scrophulariaceae
    A   B   C   D
    01 06 09 03
OC: Leptinotarsa decemlineata H-4, I-7 (531)

Aphanamixis polystachya   (Pithraj)   Meliaceae
    A   B   C   D   J   J   K   M   M   M
    01 01 09 01 02 06 04 04 05 06
OC: Rice field insects H-2, I-9 (155)   Sitotroga cerealella H-3, I-7 (161)

Apium graveolens   (Celery)   Apiaceae
    A   B   C   D   J   J   M   M   N   N   O   O
    02 06 09 10 03 11 01 05 02 06 02 10
OC: Attagenus piceus H-4, I-10 (48)
OR: H-24, I-10 (569)

Apocynum androsaemifolium   (Common dogbane)   Apocynaceae
    B   C   D   J   M   M   O
    06 09 03 03 05 26 02
OC: Aedes aegypti H-1, I-2, 6, 7, 8, 10 (77, 165, 843)

Apocynum cannabinum   (Indian hemp)   Apocynaceae
    B   C   D   J   J   M   M   M   O
    06 09 03 03 08 03 05 26 02
OC: Attagenus piceus H-4, I-2 (48)   Leptinotarsa decemlineata H-1, I-? (548)
OR: H-19, I-7 (1171)
AP: Alk=I-10 (1391)

Apocynum subiricum   (Dogbane genus)   Apocynaceae
      B   C   D   J   J   M
      06 09 03 03 11 03
OC: Attagenus piceus H-4, I-2, 6, 7 (48)   Drosophila melanogaster H-3, I-6 (165)

Aquilaria agallocha   (Eaglewood)   Thymelaeaceae
      A   B   C   D   D   J   M   M
      01 01 09 01 02 04 17 19
OC: Fleas H-1, I-5 (105)   Lice H-1, I-5 (105)

Arachis hypogaea   (Peanut)   Fabaceae
      A   B   C   D   D   D   E   M   M   M   M   M   M   N
      03 06 09 01 02 06 07 01 02 06 10 12 30 09
OC: Acromyrmex octospinosus H-6, I-12 (470)   Leaf-cutting ants H-6, I-12 (87)
    Aphelenchus avenae H-16, I-18 (100, 617)   Locusta migratoria H-4, 5, I-7 (596)
    Cacoecia argyrospila H-1, I-? (1086)   Mealy bugs H-1, I-6, 12 (78)
    Callosobruchus maculatus H-1, I-10 (1124)   Meloidogyne incognita H-16, I-18
    Coccids H-1, I-6, 12 (78)   (100, 539, 541, 617, 923)
    Colletotrichum atramentarium H-14, I-18   Meloidogyne javanica H-16, I-18
      (923)   (925, 962, 992)
    Ditylenchus cypei H-16, I-18 (100, 617)   Peronospora tabacina H-14, I-12 (482)
    Fusarium oxysporum   Phenacoccus gossypii H-1, I-? (1086)
      f. sp. lini H-14, I-18 (957)   Pratylenchus penetrans H-16, I-12 (1054)
      f. sp. lycopersici H-14, I-18 (541, 923)   Pythium aphanidermatum H-14, I-18 (875)
      f. sp. udum H-14, I-18 (957)   Rhizoctonia solani H-14, I-14 (541, 923)
      f. sp. vasinfectum H-14, I-18 (957)   Rotylenchulus reniformis H-16, I-18
    Fusarium sp. H-14, I-18 (923)   (539, 923)
    Helicotylenchus erythrinae H-16, I-18   Sphaerothica humuli H-14, I-12 (496)
      (541, 923)   Tylenchorhynchus brassicae H-16, I-18
    Hoplolaimus indicus H-16, I-18   (100, 539, 541)
      (100, 539, 541, 612, 617)   Urentius echinus H-2, I-12 (87)
OR: H-2, 5, I-? (105)
AP: Alk=I-10 (arachine)(1392)

Aralia elata   (Not known)   Araliaceae
      A   B   C   J   J   J
      01 02 09 03 08 11
OC: Attagenus piceus H-4, I-6, 7 (48)

Aralia humilis   (Hercules' club)   Araliaceae
      A   B   C   J
      01 06 09 11
OC: Attagenus piceus H-4, I-6 (48)

Aralia spinosa   (Devil's walking stick)   Araliaceae
      A   B   C   D   J   M   O
      01 01 09 01 11 05 04
OC: Attagenus piceus H-4, I-4 (48)

Araujia sericifera   (Not known)   Asclepiadaceae
      A   B   B   C   D   M
      01 03 06 09 01 03
OC: Caenurgin crassiuscula H-6, I-? (341)   Heliothis zea H-6, I-? (341)
    Caenurgin erechtea H-6, I-? (341)
AP: Alk=I-6, 7 (1368)

Arbutus menziesii   (Madrona)   Ericaceae
  A B C D M M M
  01 01 09 03 06 19 21
OR: H-24, I-7 (1180)

Arbutus unedo   (Strawberry tree)   Ericaceae
  A B C D
  01 01 09 03
OR: H-4, I-? (132)

Arctium minus   (Burdock)   Asteraceae
  B C D J J M O
  06 09 03 03 09 05 02
OC: Bipolaris sorokiniana H-14, I-1 (944)   Pieris brassicae H-4, I-7 (531)
    Leptinotarsa decemlineata H-4, I-7 (531)
OR: H-19, I-6, 7, 9, 14 (704, 765, 940, 1086)
AP: Alk=I-6, 7, 8 (1392); Alk=I-9, 10 (1391)

Arctium tomentosum   (Beggar's-buttons genus)   Asteraceae
  B C D L
  06 09 03 05
OC: Athalia rosae H-4, I-7, (531)   Phytodecta fornicata H-4, I-7 (531)
    Leptinotarsa decemlineata H-4, I-7 (531)

Arctostaphylos glandulosa   (Eastwood manzanita)   Ericaceae
  B C D
  02 09 03
OR: H-24, I-7 (988)

Arctostaphylos glauca   (Big-berry manzanita)   Ericaceae
  A B C D M N
  01 02 09 03 01 09
OR: H-24, I-7 (988)

Arctostaphylos pungens   (Mexican manzanita)   Ericaceae
  A B C D J M N
  01 02 09 02 03 01 09
OC: Spodoptera frugiperda H-3, I-1 (176)

Arctostaphylos uva-ursi   (Bearberry)   Ericaceae
  A B C D M M M N
  01 02 09 03 01 09 21 09
OC: Popillia japonica H-5, I-? (105, 1243)
AP: Tan=I-7 (504, 507)

Ardisia crispa var. dielsii   (Mata ajam)   Myrsinaceae
  A B C D J K M N
  01 02 09 10 04 05 01 09
OC: Epilachna varivestis H-1, I-2 (29)

Ardisia escallonioides   (Marlberry)   Myrsinaceae
  A B C D D J
  01 01 09 01 02 11
OC: Attagenus piceus H-4, I-6 (48)

Ardisia neriifolia   (Not known)   Myrsinaceae
   A   C   D   J
   01 09 02 09
OC: Rats H-26, I-16 (1393)

Ardisia picardae   (Not known)   Myrsinaceae
   A   B   C   J
   01 02 09 08
OC: Attagenus piceus H-4, I-4 (48)   Tineola bisselliella H-4, I-4 (48)

Areca catechu   (Betel palm)   Arecaceae
   A   B   C   D   J   J   M   M   M   M   M   M   M   N   O
   01 01 09 01 03 08 01 04 05 13 19 22 23 24 07 10
OR: H-17, 20, I-10 (1318); H-19, I-7, 10 (982); H-33, I-? (968); H-35, I-10 (504)
AP: Alk=I-? (nororecaidine, nororecoline)(1392); Alk=I-6, 7 (1392); Alk=I-7 (1318); Alk=I-10
   (arecaidine, arecaine, arecolidine, arecoline, guvacine, guvacoline, isoguacine)(1392);
   Tan=I-7, 10 (1318)

Arenaria leptoclados   (Sandwort)   Caryophyllaceae
   B   C   C   D   J
   06 03 09 03 03
OC: Drosophila hydei H-1, I-7 (101)

Arenaria peploides   (Sea chickweed)   Caryophyllaceae
   B   C   D   D   J   M   N
   06 09 03 07 11 01 07
OC: Attagenus piceus H-4, I-1 (48)

Argemone fruticosa   (Prickly poppy)   Papaveraceae
   B   C   D
   06 09 03
OR: H-2, I-2 (105)

Argemone mexicana   (Mexican prickly poppy)   Papaveraceae
   A   B   C   D   D   G   J   J   J   J   J   K   L   M   M   M   M   O
   03 06 09 01 02 04 03 09 11 12 15 06 05 05 06 10 15 02
OC: Alternaria tenuis H-14, I-6 (523)   Meloidogyne javanica H-16, I-2, 7 (100, 616)
   Bagrada cruciferarum H-2, I-7, 10 (608)   Pieris brassicae H-1, I-? (600)
   Bagrada picta H-1, I-? (600)   Sitophilus oryzae H-2, I-8 (87, 400, 1124)
   Dysdercus koenigii H-2, I-10 (91)   Spodoptera litura H-1, I-? (88, 124)
   Helminthosporium sp. H-14, I-? (523)   Termites H-1, I-10, 12 (1352, 1353)
   Lipaphis erysimi H-2, I-10 (91, 600)
   Meloidogyne incognita H-16, I-2, 7 (100, 616)
OR: H-2a, I-1, 7 (124, 127, 148); H-19, 22, I-1 (127, 881); H-33, 35, I-10 (220, 1352)
AP: Alk=I-? (1391); Alk=I-? (argenomine, codeine, morphine)(1392); Alk=I-1 (dihydro-
   sanguinarine, protopine)(1392); Alk=I-2 (allocrytopine, berberine, chelerythrine,
   coptisine, dihydrochelerythrine, sanguinarine)(1392); Alk=I-2, 6, 7 (1360); Alk=I-6, 7
   (berberine)(1392); Alk=I-6, 7, 8, 9 (1371); Alk=I-9 (norargemonine)(1392); Alk=I-10
   (sanguinarine)(1392)

Arisaema consanguineum   (Not known)   Araceae
   B   C   M
   06 09 13
OC: Epilachna varivestis H-2, I-? (29)

Arisaema dracontium   (Dragon root)   Araceae
    B   C   D   M
   06 09 03 13
OR: H-2, I-3 (104)

Arisaema erubescens   (Not known)   Araceae
    B   C   J   M
   06 09 04 13
OC: Epilachna varivestis H-2, I-2 (101)

Arisaema japonicum   (Not known)   Araceae
    B   C   D   M   M   N
   06 09 03 01 13 03
OR: H-2, I-2 (104)

Arisaema purpureogaleatum   (Not known)   Araceae
    B   C   J   K   M
   06 09 04 05 13
OC: Aphis fabae H-2, I-2 (29, 101)   Epilachna varivestis H-2, I-2 (29, 101)

Arisaema serratum   (Dragon root genus)   Araceae
    B   C   D   J
   06 09 03 03
OC: Aphids H-1, I-2, 6 (1241)   Rhizoctonia solani H-14, I-2, 6 (1241)
    Caterpillars H-1, I-2, 6 (1241)   Spider mites H-11, I-2, 6 (1241)
    Pieris rapae H-1, I-2, 6 (1241)
    Puccinia graminis tritici H-14, I-2, 6 (1241)

Arisaema speciosum   (Cobra lily)   Araceae
    B   C   D   D   L   M   M   O
   06 09 03 08 08 05 13 02
OR: H-2, I-2, 9 (73, 104, 105, 220); H-5, I-2 (105); H-17, I-2 (504); H-33, I-2 (220)

Arisaema thunbergii   (Not known)   Araceae
    A   B   C   D   J
   01 06 09 03 03
OC: Aphids H-1, I-2, 6 (1241)   Puccinia graminis tritici H-14, I-2, 6
    Caterpillars H-1, I-2, 6 (1241)   (1241)
    Maggots H-1, I-2, 6 (1241)   Rhizoctonia solani H-14, I-2, 6 (1241)
    Pieris rapae H-1, I-2, 6 (1241)   Spider mites H-11, I-2, 6 (1241)

Arisaema tortuosum   (Not known)   Araceae
    B   C   D   L   M   O
   06 09 03 03 05 02
OR: H-2, I-2 (220, 1276, 1345); H-5, I-? (1276); H-17, I-2 (504); H-33, I-2, 9 (220)

Arisaema triphyllum   (Dragon root)   Araceae
    B   C   D   J   J   J   M   M   M   N   O
   06 09 03 03 09 12 01 05 13 02 02
OC: Botrytis cinerea H-14, I-1, 8, 9 (946)
OR: H-19, I-9 (504)

Aristolochia bracteata   (Birthwort genus)   Aristolochiaceae
  A   B   C   D   J   J   J   M   O   O
  01 03 09 01 03 08 09 05 02 07
OC: Aedes aegypti H-1, I-? (791)   Maggots H-2, I-7, 14 (105)
    Callosobruchus chinensis H-2, I-2   Mosquitoes H-2, I-6, 7 (101)
      (605, 1029)   Tribolium castaneum H-2, I-? (991)
    Dysdercus koenigii H-1, I-? (791)
OR: H-2, I-14 (73); H-5, I-? (1276); H-15, I-7 (504); H-17, I-2 (504)

Aristolochia brasiliensis   (Brazil Dutchman's pipe)   Aristolochiaceae
  A   B   C   D   K   M   O
  01 03 09 01 01 05 02
OR: H-2, I-? (105)

Aristolochia cornuta   (Not known)   Aristolochiaceae
  A   B   C   D   D   K
  01 03 09 01 02 01
OR: H-2, I-? (104)

Aristolochia elegans   (Calico flower)   Aristolochiaceae
  A   B   C   D
  01 03 09 01
OC: Manduca sexta H-4, I-7 (669)   Panonychus citri H-11, I-7 (669)
OR: H-2, I-? (105); H-20, I-2 (504)
AP: Alk=I-6, 7 (1392)

Aristolochia grandiflora   (Pelican flower)   Aristolochiaceae
  A   B   D   J   J   M   O   O
  01 03 01 03 04 05 02 07
OC: Pieris rapae H-1, I-4, 10 (105)
OR: H-32, I-4, 10 (105)

Aristolochia indica   (Birthwort genus)   Aristolochiaceae
  A   B   B   C   J   J
  01 03 05 09 03 08
OC: Coccus viridis H-1, I-7 (1029)   Mice H-25, I-? (708)
    Euproctis fraterna H-2, I-6, 7, 9 (105)   Spodoptera litura H-2, I-7 (105)
    Idiocerus sp. H-2, I-6, 7, 9 (105)
AP: Alk=I-2 (aristolochine, isoaristolochic acid)(1392)

Aristolochia maxima   (Contracapetano)   Aristolochiaceae
  A   B   B   C   D   D   J   M   O
  01 03 05 09 01 02 03 05 02
OC: Pieris rapae H-1, I-4, 10 (105)

Aristolochia ringens   (Birthwort genus)   Aristolochiaceae
  A   B   C   D
  01 03 09 01
OC: Locusta migratoria H-4, 5, I-7 (596)

Aristolochia rotunda   (Birthwort genus)   Aristolochiaceae
  A   C   D   D   M   O
  01 09 02 06 05 02

OC: Lice H-1, I-2 (105)
OR: H-17, I-2 (504)
AP: Alk=I-2 (aristolochine)(1392)

Aristolochia serpentaria   (Virginia snake root)   Aristolochiaceae
   B   C   D   D   J   M   O
   06 09 02 03 03 05 02
OC: Attagenus piceus H-4, I-2 (48)

Armoracia rusticana   (Horseradish)   Brassicaceae
   A   B   C   D   D   J   J   M   N
   01 06 09 03 06 04 06 01 02
OC: Blatta orientalis H-5, I-12 (105)   Haemaphysalis punctata H-13, I-7 (87, 1133)
   Culex pipiens H-2, I-2 (87)   Ixodes redikorzevi H-13, I-7 (87, 1133)
   Dermacentor marginatus H-13, 1-7 (87, 1133) Rhipicephalus rossicus H-13, I-7 (87, 1133)
OR: H-19, I-2, 12 (1290, 1292); H-24, I-7, 14 (569)

Arnica chamissonis var. incana   (Not known)   Asteraceae
   A   B   C   D
   01 06 09 01
OC: Dacus cucurbitae H-6, I-6, 7, 8, 9 (956)

Artabotrys hexapetalus   (Ylang-ylang)   Annonaceae
   A   B   C   D   F   G   J   J   M   M   N
   01 03 09 01 10 04 03 3a 01 17 08
OC: Agrobacterium tumefaciens H-18, I-7 (1357) Pseudomonas phaseolicola H-18, I-7 (1357)
   Alternaria tenuis H-14, I-7 (181)   Pseudomonas solanacearum H-18, I-7 (1357)
   Corynebacterium flaccumfaciens H-18, I-7   Ustilago hordei H-14, I-7 (181)
     (1357)   Ustilago tritici H-14, I-7 (181)
   Corynebacterium michiganense H-18, I-7   Xanthomonas campestris
     (1357)   pv. campestris H-18, I-7 (1357)
   Curvularia lunata H-14, I-7 (181)   pv. oryzae H-18, I-7 (1298, 1357)
   Drechslera graminea H-14, I-7 (181)   pv. phaseoli H-18, I-7 (1357)
   Erwinia carotovora H-18, I-7 (1357)   pv. vesicatoria H-18, I-7 (1357)
   Fusarium nivale H-14, I-7 (181)   pv. vitians H-18, I-7 (1357)
OR: H-19, I-? (164, 904); H-24, I-7, 14 (1357)

Artemisia abrotanum   (Old man)   Asteraceae
   A   B   C   D   D   M
   01 02 09 03 05 15
OC: Pieris brassicae H-5, I-? (1002)
OR: H-17, 20, I-7 (504)
AP: Alk=I-7 (abrotine)(1392)

Artemisia absinthium   (Wormwood)   Asteraceae
   A   B   C   D   D   D   E   J   J   J   J   J   L   M   M   O
   01 06 09 03 05 06 12 01 02 03 04 06 03 05 17 07
OC: Aspergillus niger H-14, I-7, 12 (898)   Pieris brassicae H-5, I-? (1002)
   Caterpillars H-1, I-? (105)   Rhipicephalus rossicus H-13, I-7 (87, 1133)
   Dermacentor marginatus H-13, I-7 (87, 1133) Sitophilus granarius H-1, I-7 (101)
   Haemaphysalis punctata H-13, I-7 (87, 1133) Sitotroga cerealella H-1, I-7 (101)
   Ixodes redikorzevi H-13, I-7 (87, 1133)   Spodoptera litura H-1, I-? (143)
   Meloidogyne incognita H-16, I-? (1138)   Tinea granella H-1, I-7 (101)
OR: H-1, I-1 (1026); H-2, I-? (1276); H-4, I-? (132); H-5, I-10 (73); H-14, 19, I-7, 12
   (898); H-24, I-12, 17 (559, 1113, 1207); H-33, 35, I-12 (220, 1352)
AP: Alk=I-6, 7, 8 (1372)

35

Artemisia annua   (Sweet wormwood)   Asteraceae
    A   B   C   D
    03   02   09   03
OC: Aphis gossypii H-1, I-1 (1241)   Phytophthora infestans H-14, I-1 (1241)
    Ceratocystis fimbriata H-14, I-1 (1241)   Puccinia glumarum H-14, I-1, (1241)
    Flies H-1, I-1 (1241)   Puccinia graminis tritici H-14, I-1 (1241)
    Heliothis armigera H-1, I-1 (1241)   Rice borers H-1, I-1 (1241)
    Mosquitoes H-1, I-1 (1241)   Spider mites H-11, I-1 (1241)

Artemisia apiacea   (Ching hao)   Asteraceae
    A   B   C   D   M   O
    03   02   09   03   05   07
OC: Agrotis sp. H-1, I-1 (1241)   Puccinia graminis triciti H-14, I-1 (1241)
    Aphis gossypii H-1, I-1 (1241)   Rhizoctonia solani H-14, I-1 (1241)
    Phytophthora infestans H-14, I-1 (1241)   Rice borers H-1, I-1 (1241)
    Pieris rapae H-1, I-1 (1241)

Artemisia argyi   (Sagebrush genus)   Asteraceae
    C
    09
OC: Aphis gossypii H-1, I-? (1241)   Pieris rapae H-1, I-? (1241)
    Caterpillars H-1, I-? (1241)   Spider mites H-11, I-? (1241)
    Maggots H-1, I-? (1241)   Spodoptera litura H-1, I-? (1241)
    Mosquitoes H-1, I-? (1241)

Artemisia californica   (California sagebrush)   Asteraceae
    A   B   C   D   D   E
    01   02   09   02   04   12
OC: Avena fatua H-24, I-7 (578, 733)   Bromus rubens H-24, I-7 (578)
    Bromus mollis H-24, I-7 (578)   Erodium cicutarium H-24, I-7 (578)
    Bromus rigidus H-24, I-7 (578)   Festuca megalura H-24, I-7 (578)

Artemisia campestris   (Sagebrush genus)   Asteraceae
    A   B   C   D   E   J
    01   06   09   05   12   04
OC: Dermacentor marginatus H-13, I-10   Ixodes redikorzevi H-13, I-10 (87, 1133)
      (87, 1133)   Rhipicephalus rossicus H-13, I-10
    Haemaphysalis punctata H-13, I-10 (87, 1133) (87, 1133)

Artemisia capillaris   (Sagebrush genus)   Asteraceae
    B   C   D   D   E   J
    06   09   01   04   12   13
OC: Spodoptera litura H-4, I-7 (1064)
OR: H-15, I-? (170); H-15, I-12 (970)
AP: Str=I-2 (B-sitosterol)(1386)

Artemisia cina   (Levant wormseed)   Asteraceae
    B   C   D   D   E
    06   09   03   04   12
OC: Meloidogyne incognita H-16, I-? (1138)
OR: H-15, I-2, 12 (1157); H-17, I-8 (504)

Artemisia dracunculus (Tarragon) Asteraceae
   A   C   D   D   E   M   M   M   O
   01 09 03 04 12 05 14 17 07
OC: Meloidogyne incognita H-16, I-7 (1138)
OR: H-17, 34, I-7 (504)

Artemisia fasciculata (Sagebrush genus) Asteraceae
   B   C   D   E
   06 09 04 12
OC: Avena fatua H-24, I-7 (578)   Bromus rubens H-24, I-7 (578)
    Bromus mollis H-24, I-7 (578)   Erodium cicutarium H-24, I-7 (578)
    Bromus rigidus H-24, I-7 (578)   Festuca megalura H-24, I-7 (578)

Artemisia japonica (Sagebrush genus) Asteraceae
   A   B   C   D
   01 06 09 03
OC: Aphids H-1, I-1 (1241)   Heliothis armigera H-1, I-1 (1241)

Artemisia ludoviciana (White sage) Asteraceae
   A   B   C   D   D   E   M
   01 06 09 03 04 12 17
OC: Spodoptera eridania H-1, I-? (143)
AP: Alk=I-2, 6, 7 (1392)

Artemisia maritima (Wormseed) Asteraceae
   A   B   C   D   D   D   E   G   J   J   K   L   M   M   M   O
   01 02 09 02 03 05 12 04 04 11 05 03 05 14 17 07
OC: Tribolium castaneum H-5, I-1 (179)
OR: H-33, I-? (220)
AP: Alk=I-? (1392)

Artemisia monosperma (Sagebrush) Asteraceae
   A   B   C   D   E   G   M
   01 06 09 05 12 04 17
OC: Drosophila melanogaster H-1, I-? (101)   Musca domestica H-2a, I-7, 8 (87)

Artemisia rigida (Sagebrush genus) Asteraceae
   B   C   D   E
   06 09 04 12
OC: Leptinotarsa decemlineata H-4, I-7 (1118)

Artemisia roxburghiana (Sagebrush genus) Asteraceae
   B   C   D   E   J   M
   06 09 04 12 09 07
OC: Musca domestica H-1, I-1 (785)   Tribolium castaneum H-1, I-1 (785)

Artemisia tridentata (Common sagebrush) Asteraceae
   A   B   C   D   D   D   E   J   M   M   N   N   O
   01 02 07 02 03 04 12 03 01 05 07 10 07
OC: Agropyron spicatum H-24, I-7 (1174)   Sitanion hystrix H-24, I-7 (1174)
   Leptinotarsa decemlineata H-4, I-7 (1118) Stipa thurberiana H-24, I-7 (1174)
OR: H-19, I-7 (625); H-24, I-7 (1172)
AP: Alk=I-7 (1392)

37

Artemisia vulgaris (Mugwort) Asteraceae

A	B	C	D	D	E	F	J	J	M	M	O
01	06	09	03	05	12	05	3a	06	05	17	07

OC: Athalia rosae H-4, I-7 (531)   Mosquitoes H-2, I-? (101)
    Cassida nebulosa H-4, I-7 (531)   Phytodecta fornicata H-4, I-? (531)
    Cockroaches H-2, I-? (101)   Pieris brassicae H-4, I-7 (531)
    Flies H-2, I-? (101)   Stored grain pests H-1, I-? (89)
    Leptinotarsa decemlineata H-4, I-7 (531)   Tanymecus dilaticollis H-4, I-7 (531)
    Lice H-2, I-? (101)
OR: H-15, 19, I-? (898); H-17, I-7 (1318); H-32, I-? (220)
AP: Alk=I-2, 6, 7 (1318, 1372)

Artocarpus altilis (Breadfruit) Moraceae

A	B	C	D	J	J	M	M	N
01	01	09	01	03	11	01	13	09

OC: Attagenus piceus H-4, I-7 (48)
OR: H-19, I-2, 4, 7 (172, 982)
AP: Alk=I-2, 4, 6, 7, 9 (1362)

Artocarpus heterophyllus (Jackfruit) Moraceae

A	B	C	D	J	M	M	N	N	N	O
01	01	09	01	11	01	05	08	09	10	07

OC: Attagenus piceus H-4, I-6, 7 (48)   Rotylenchulus reniformis H-16, I-15 (1330)
    Helicotylenchus indicus H-16, I-15 (1330) Tylenchorhynchus brassicae H-16, I-15 (1330)
    Hoplolaimus indicus H-16, I-15 (1330)   Tylenchus filiformis H-16, I-15 (1330)

Arum maculatum (Adam & Eve) Araceae

A	B	C	D	L	M	M	N	N	O
03	06	09	06	03	1a	05	02	07	02

OC: Locusta migratoria H-4, 5, I-? (596)
OR: H-33, I-1 (504)
AP: Alk=I-? (coniine)(1392)

Arundo donax (Giant reed) Poaceae

A	B	C	D	D	M	M	M	M	M
01	06	06	01	06	04	09	19	22	24

OC: Schistocerca gregaria H-2, I-? (143)
AP: Alk=I-6, 7 (donaxarine, gramine)(1392)

Asclepias curassavica (Bloodflower) Asclepiadaceae

A	B	C	D	J	J	L	L	L	M	M	M	O
03	06	09	01	03	14	02	03	05	05	13	30	02

OC: Dacus cucurbitae H-6, I-7 (956)   Vermin H-5, 27, I-? (104)
    Fleas H-5, I-? (104)
OR: H-15, I-7 (504); H-32, I-? (73, 220, 983); H-33, I-? (1350)
AP: Alk=I-7 (1392)

Asclepias eriocarpa (Milkweed genus) Asclepiadaceae

A	B	C	D	D	J	J	J	M
01	06	09	02	04	03	08	11	03

OC: Attagenus piceus H-4, I-6, 7 (48)

Asclepias incarnata   (Swamp milkweed)   Asclepiadaceae
   A B C D J J J M M M M N N O
   01 06 09 03 02 03 09 01 03 05 13 03 08 02
OC: Aedes aegypti H-1, I-7 (165)
   Drosophila melanogaster H-1, I-2, 6, 7, 9 (165)
OR: H-17, I-2 (504); H-19, I-? (1171); H-22, I-1 (909, 944)
AP: Alk=I-10 (1391)

Asclepias kansana   (Milkweed genus)   Asclepiadaceae
   A B C D J
   01 06 09 03 08
OC: Attagenus piceus H-4, I-2, 9 (48)

Asclepias labriformis   (Milkweed genus)   Asclepiadaceae
   B C J J J J M
   06 09 03 04 08 11 13
OC: Attagenus piceus H-4, I-1 (48)   Ostrinia nubilalis H-2, I-6 (101)

Asclepias longifolia   (Milkweed genus)   Asclepiadaceae
   A B C
   01 06 09
OC: Locusta migratoria H-4, 5, I-7 (596)

Asclepias speciosa   (Showy milkweed)   Asclepiadaceae
   A B C D J J J J M M M N N
   01 06 09 03 02 03 08 11 01 13 14 02 07 15
OC: Aedes aegypti H-2, I-7, 8 (77, 165, 843)   Drosophila melanogaster H-1, I-9 (77)
   Attagenus piceus H-4, I-2 (48)

Asclepias syriaca   (Common milkweed)   Asclepiadaceae
   A B C D J J M M N N
   01 06 09 03 03 11 01 13 11 15
OC: Aedes aegypti H-1, I-? (156)   Culex tarsalis H-1, I-? (156)
   Aphids H-2, I-? (80)   Drosophila melanogaster H-1, I-2, 6 (165)
   Attagenus piceus H-4, I-7 (48)   Mealy bugs H-2, I-? (80)
   Cassida nebulosa H-4, I-7 (531)
OR: H-34, I-2 (504)
AP: Alk=I-2 (nicotine)(1392); Alk=I-6, 7, 8 (1392); Alk=I-10 (1391)

Asclepias tuberosa   (Butterfly weed)   Asclepiadaceae
   A B C D J K M M M N N N O
   01 06 09 03 03 06 01 05 13 02 06 11 02
OC: Blattella germanica H-2, I-2 (48)

Asclepiodora viridis   (Milkweed genus)   Asclepiadaceae
   B C D J J J
   06 09 03 03 08 11
OC: Attagenus piceus H-4, I-4, 7 (48)
AP: Alk=I-6, 7, 8 (1370)

Asparagus cochinchinensis   (Asparagus genus)   Liliaceae
   C
   09
OC: Maggots H-1, I-1, 2 (1241)

Asparagus officinalis   (Asparagus genus)   Liliaceae
  A B   C   D   E   F   G   J   K   M   N
  01 06 09 02 03 03 05 03 01 01 10
OC: Helicotylenchus nannus H-16, I-? (1000)   Meloidogyne incognita H-16, I-2 (147)
    Heterodera glycines H-16, I-? (685)   Meloidogyne sp. H-16, I-2 (525)
    Heterodera rostochiensis H-16, I-? (685)   Nematodes H-16, I-? (100)
    Meloidogyne hapla H-16, I-? (685)   Pratylenchus curvitatus H-16, I-? (685)
    Pratylenchus penetrans H-16, I-?   Trichodorus christiei H-16, I-?
     (685, 1000)   (1000, 1100)
    Rotylenchulus reniformis H-16, I-2 (147) Xiphinema americanum H-16, I-? (1000)
    Rotylenchulus sp. H-16, I-2 (525)
OR: H-19, I-1 (982)
AP: Alk=I-6, 7, 9 (1370, 1373)

Asparagus racemosus   (Asparagus genus)   Liliaceae
  A B   C   D   D   J   M   M   N   O
  01 04 09 01 02 03 01 05 09 02
OC: Meloidogyne arenaria H-16, I-2 (796)   Meloidogyne javanica H-16, I-2 (796)
AP: Alk=I-2 (1360, 1378); Alk=I-6, 7 (1373)

Asphodelus tenuifolius   (Not known)   Liliaceae
  A B   C   D   D
  03 06 09 01 06
OC: Musca domestica H-2, I-1 (785)   Tribolium castaneum H-2, I-1 (785)
OR: H-2, I-? (100)

Asplenium nidus   (Bird's nest fern)   Polypodiaceae
  A B   C   D   M
  01 10 06 01 13
OC: Erwinia carotovora H-18, I-7 (854)
    Xanthomonas campestris pv. phaseoli-sojensis H-18, I-7 (854)

Astelia cunninghamii   (Not known)   Liliaceae
  B C   D   J   J
  06 09 02 03 11
OC: Attagenus piceus H-4, I-9 (48)

Aster lateriflorus   (Aster genus)   Asteraceae
  A B   C   D   J
  01 06 09 03 09
OC: Rodents H-25, I-1 (944)

Aster macrophyllus   (Rough tongues)   Asteraceae
  B C   D   J   M   N
  06 09 03 09 01 07
OC: Ceratocystis ulmi H-14, I-1 (946)   Xanthomonas campestris
    Pseudomonas solanacearum H-18, I-1, 8 (945) pv. phaseoli H-18, I-1 (946)
OR: H-24, I-? (559)

Aster novae-angliae   (New England aster)   Asteraceae
  A B   C   D   J
  01 06 09 03 08
OC: Popillia japonica H-5, I-7 (1243)

Aster sagittifolius   (Aster genus)   Asteraceae
    A   B   C   D
    01 06 09 03
OC: Rodents H-25, I-1, 8 (945)

Astragalus adsurgens var. robustior   (Milk vetch genus)   Fabaceae
    B   C   D   J   J   M
    06 09 03 02 03 12
OC: Aedes aegypti H-2, I-7, 8 (165)

Astragalus canadensis   (Canadian milk vetch)   Fabaceae
    A   B   C   D   J   J   J   M   M   N
    01 06 09 03 02 03 08 01 12 02
OC: Aedes aegypti H-2, I-6, 7, 8, 9 (77, 165, 843)
    Drosophila melanogaster H-3, I-7, 8, 9 (165)

Astragalus flexuosus   (Not known)   Fabaceae
    B   C   J   J   M
    06 09 02 03 12
OC: Aedes aegypti H-2, I-7, 8 (165)

Astragalus pectinatus   (Milk vetch genus)   Fabaceae
    B   C   D   J   J   M
    06 09 03 02 03 12
OC: Aedes aegypti H-2, I-9 (165)   Drosophila melanogaster H-1, I-2 (165)

Astragalus racemosus   (Not known)   Fabaceae
    B   C   J   J   M
    06 09 02 03 12
OC: Aedes aegypti H-2, I-9 (165)

Astragalus tenellus   (Not known)   Fabaceae
    B   C   D   J   J   M
    06 09 03 02 03 12
OC: Aedes aegypti H-2, I-8, 9 (165)
AP: Alk=I-10 (1391)

Astragalus verilliflexus   (Not known)   Fabaceae
    B   C   D   J   J   M
    06 09 03 02 03 12
OC: Aedes aegypti H-2, I-7, 8 (165)

Atalantia monophylla   (Not known)   Rutaceae
    A   B   C   D   G   J   K   M   O   O
    01 01 09 10 02 02 04 05 09 12
OC: Sitophilus oryzae H-5, I-6, 7 (117)   Sitotroga cerealella H-5, I-6, 7 (117)
AP: Cou=I-2, 4 (1385); Str=I-2, 5 (stigmasterol)(1385)

Athyrium pterorachis   (Not known)   Polypodiaceae
    A   C   D   J
    01 06 01 03
OC: Drosophila hydei H-2, I-6, 7 (101)

Atractylis gummifera   (Add-add)   Asteraceae
    B   C   D   D   K
    06 09 04 06 07
OC: Hadena oleracea H-2, I-2 (78, 101)   Phaedon cochleariae H-2, 5, I-2 (78, 101)

Atractylis ovata   (Not known)   Asteraceae
    B   C   D   D   K
    06 09 03 06 07
OC: Stored grain pests H-1, I-? (105)

Atractylodes chinensis   (Not known)   Asteraceae
    C
    09
OC: Puccinia graminis tritici H-14, I-1 (1241) Rhizoctonia solani H-14, I-1 (1241)

Atriplex nummularia   (Saltbrush genus)   Chenopodiaceae
    A   B   C   D   E   M
    01 02 09 06 04 02
OC: Locusta migratoria H-4, 5, I-7 (596)

Atriplex patula   (Saltbrush genus)   Chenopodiaceae
    B   C   D   D   E   F
    06 09 02 03 04 10
OC: Venturia inaequalis H-14, I-? (16)

Atropa accuminata   (Indian belladonna)   Solanaceae
    A   B   C   D   G   J   J   K   M   O
    01 06 09 03 04 04 11 05 05 02
OC: Tribolium castaneum H-5, I-2 (179)
AP: Alk=I-6 (1371)

Atropa belladonna   (Belladonna)   Solanaceae
    A   B   B   C   D   L   M   O   O
    01 06 10 09 10 03 05 02 07
OC: Leptinotarsa decemlineata H-2, I-? (87)
OR: H-2, I-2, 7 (97); H-2, 5, I-? (1276); H-4, I-? (132); H-5, I-10 (118); H-22, I-7 (881);
    H-33, I-1 (220, 504); H-33, 35, I-2, 7 (504)
AP: Alk=I-2, 16 (apoatropine, atropine, belladonnine, bellaradine, cusconygrine, hypocyamine,
    hyposcine, N-methylpyrroine, scopolamine)(1392); **Alk**=I-6, 7 (1371); **Alk**=I-8 (1392)

Aureolaria pedicularia   (False foxglove)   Scrophulariaceae
    A   B   B   C   D   E   J
    03 06 10 09 03 02 08
OC: Popillia japonica H-5, I-7, 8 (105, 1243)

Aureolaria virginica   (Downy false foxglove)   Scrophulariaceae
    A   B   B   C   D   E
    01 06 10 09 03 02
OC: Flies H-5, I-? (105)

Avena fatua   (Wild oats)   Poaceae
    A   B   C   D   J   M   N
    03 06 09 03 03 01 10
OR: H-24, I-2, 7 (1213, 1238)

Avena sativa (Oats) Poaceae
A B D J J M N
03 06 03 10 12 01 10

OC: Alternaria solani H-14, I-2 (1021)
   Bipolaris sorokiniana H-14, I-2 (1021)
   Ceratocystis ulmi H-14, I-2 (1021)
   Colletotrichum pisi H-14, I-2 (1021)
   Culex quinquefasciatus H-2, I-10 (463)
   Fusarium solani
      f. sp. phaseoli H-14, I-16 (362)

Meloidogyne incognita H-16, I-? (354)
Meloidogyne sp. H-16, I-20 (424)
Mosquitoes H-2, I-10 (101)
Ophiobolus graminis H-14, I-2 (1021)
Pythium irregulare H-14, I-2 (1021)
Rhizoctonia solani H-14, I-? (353)
Verticillium albo-atrum H-14, I-2 (1021)

OR: H-19, I-2 (1021); H-24, I-7 (1228)
AP: Alk=I-7 (hordenine)(1392); Alk=I-10 (ergothioneine, trigonelline)(1392)

Averrhoa bilimbi (Cucumber tree) Oxalidaceae
A B C D F G J M M M M N O
01 01 09 01 13 03 03 01 05 14 15 09 07
OC: Drechslera oryzae H-14, I-7 (113)
AP: Tri=I-7 (1390)

Azadirachta indica (Neem tree) Meliaceae
A B C D D E F F G J J J J K K M M M M M O
01 01 09 01 02 02 03 07 07 02 04 07 08 03 04 02 04 05 06 07 07
                              06 09   10 11 15

OC: Acalymma vittata H-4, 5, I-? (168, 469)
   Achaea janata H-4, I-? (168)
   Acrida exatana H-4, I-? (168)
   Agrotis ipsilon H-4, I-? (168)
   Aleurothrixus floccosus H-4, I-? (168);
      H-5, I-? (469)
   Alternaria tenuis H-14, I-? (539)
   Amsacta moorei H-1, I-? (123, 505);
      H-4, I-7 (87)
   Antestiopsis orbitalis bechuana H-1, I-7
      (852); H-3, I-10 (168)
   Antestiopsis sp. H-1, I-? (725)
   Anthrenus flavipes H-4, I-? (168)
   Antigastra catalaunalis H-1, I-? (468)
   Aonidiella aurantii H-4, I-? (168, 1139);
      H-5, I-? (469)
   Aonidiella citrina H-4, I-? (168, 1139);
      H-5, I-? (469)
   Aphelenchus avenae H-16, I-18 (100)
   Aphids H-1, I-? (89)
   Aphis gossypii H-1, I-? (116)
   Aphis mellifera H-4, I-? (63)
   Argyrotaenia velutinana H-4, I-? (63);
      H-5, I-? (469)
   Atherigona soccata H-1, I-? (342)
   Aulacophora foveicollis H-4, I-? (168)
   Boarmia selenaria H-1, I-? (345)
   Callosobruchus chinensis H-3, I-10 (95,
      124, 475, 516, 520, 1279); H-5, I-? (599)
   Callosobruchus maculatus H-4, I-?
      (168, 604, 878)

Carpophilus hemipterus H-5, I-10 (87, 469)
Chirida bipunctata H-1, I-? (124, 966)
Chrotogonus trachypterus H-4, I-? (168)
Chrotoicetes terminifera H-4, I-? (168)
Cnaphalocrocis medinalis H-1, I-? (51);
   H-3, I-10 (134); H-4, I-10 (166)
Cockroaches H-2, I-? (1343)
Colletotrichum atramentarium H-14, I-?
   (923)
Conotrachelus nenuphar H-4, 5, I-?
   (63, 469)
Corcyra cephalonica H-4, I-? (95, 1687)
Crocidolomia binotalis H-1, I-12
   (124, 1333)
Cryptolestes pusillus H-4, I-? (168)
Culex fatigans H-1, I-? (1279); H-2, I-12
   (1267); H-4, I-? (63)
Diabrotica undecimpunctata H-4, 5, I-?
   (63, 469)
Diacrisia obliqua H-4, I-? (168)
Ditylenchus cypei H-16, I-18 (100)
Dysdercus cingulatus H-7, I-? (119, 692)
Dysdercus suturellus H-4, I-? (168)
Earias insulana H-4, I-? (168, 346, 348)
Ephestia cautella H-4, I-? (168)
Epilachna varivestis H-4, I-? (168)
Euproctis fraterna H-1, I-? (124)
Euproctis laniata H-4, I-? (168)
Euproctis lunata H-4, I-? (365)
Fusarium oxysporum
   f. sp. lycopersici H-14, I-18 (539, 541)

43

Fusarium sp. H-14, I-? (923)
Galleria mellonella H-4, I-? (168)
Grasshoppers H-5, I-9 (148)
Helicotylenchus erythrinae H-16, I-18
    (541, 923)
Helicotylenchus indicus H-16, I-4, 7, 8,
    9, 13, 18, (793, 1123, 1335)
Heliothis armigera H-1, I-? (124)
Heliothis virescens H-4, I-? (168)
Hellula rogatalis H-1, I-? (350)
Hirschmanniella oryzae H-16, I-18 (1335)
Holotrichia consanguinea H-6, I-7 (606)
Holotrichia insularis H-6, I-7 (606)
Holotrichia serrata H-6, I-7 (606)
Hoplolaimus indicus H-16, I-4, 7, 8, 9,
    13, 18 (100, 168, 538, 539, 541, 793,
    923, 1123, 1335)
Indarbela quadrinotata H-4, I-? (168)
Lasioderma serricorne H-4, I-? (168)
Laspeyresia pomonella H-4, 5, I-? (63, 469)
Latheticus oryzae H-4, I-? (168)
Leptinotarsa decemlineata H-4, I-? (168)
Leucinodes orbonalis H-1, I-? (93, 126)
Liriomyza sativae H-1, I-10 (1337)
Liriomyza trifolii H-1, I-10, 12 (1334)
Locusta migratoria H-1, I-? (878);
    H-1, I-10 (87); H-4, 5, I-7 (596)
Locusts H-1, I-? (866)
Lymantria dispar H-4, I-? (168)
Meloidogyne arenaria H-16, I-? (616);
    H-16, I-10 (1335)
Meloidogyne incognita H-16, I-2, 7, 18
    (100, 168, 539, 541, 542, 616, 793,
    923, 1349)
Meloidogyne javanica H-16, I-7, 18
    (616, 922, 925, 962, 992, 1265, 1335)
Musca domestica H-1, I-? (520, 1094)
Myllocerus sp. H-1, I-? (93, 126)
Mythimna separata H-3, I-10 (1341)
Nematodes H-16, I-18 (1010, 1335)
Nephantis serinopa H-1, I-? (124)
Nephotettix virescens H-4, I-10, 12
    (140, 141, 942, 1299, 1331, 1339)
Nilaparvata lugens H-1, I-7, 10, 12
    (50, 93, 126, 1289, 1331); H-2a, 4,
    I-10, 12 (141); H-4, I-10, 12 (166, 1338)
Oncopeltus fasciatus H-3, I-? (130)
Ophiomyia reticulipennis H-1, I-? (126)
Orseolia oryzae H-4, I-? (63)
Oryzaephilus surinamensis H-1, I-? (1284)
Panonychus citri H-4, 5, 11, I-?
    (168, 469, 1139)
Paramyelois transitella H-4, I-? (63)
Parasaissetia nigra H-4, I-? (168)
Phyllocnistis citrella H-1, I-? (93, 126)
Phyllotreta downsei H-1, I-? (124)

Pieris brassicae H-3, I-10 (168)
Piesma quadratum H-3, I-? (168)
Planococcus citri H-4, I-? (63, 1139);
    H-5, I-? (469)
Plutella xylostella H-1, I-10 (124, 1336)
Poecilocera picta H-4, I-? (168)
Popillia japonica H-1, I-? (129); H-4,
    I-? (332); H-4, I-10 (1293)
Pratylenchus brachyurus H-16, I-? (168);
    H-16, I-7 (1273, 1335)
Pratylenchus delattrei H-16, I-? (93, 126)
Pratylenchus sp. H-16, I-10 (756)
Rhizoctonia solani H-14, I-? (541, 923);
    H-14, I-18 (539)
Rhizopertha dominica H-1, I-? (124, 126,
    454, 461, 520, 601, 878); H-1, I-12
    (1117, 1124); H-5, I-10 (87)
Rhopalosiphum nympheae H-1, I-? (168)
Rotylenchulus reniformis H-16, I-7, 10, 18
    (168, 538, 539, 793, 923, 1269, 1335)
Saissetia nigra H-4, I-? (63)
Schistocerca gregaria H-1, I-? (167, 369,
    434, 878, 1001, 1277, 1279); H-4, I-7,
    10, 12 (151, 611, 614, 615, 756, 1266,
    1268, 1272); H-5, I-2, 5, 7, 10 (87, 611,
    615, 1270, 1272)
Sclerotium rolfsii H-14, I-10 (1300)
Sitophilus oryzae H-3, 5, I-10 (87, 124,
    126, 517, 601, 878, 888, 1279)
Sitotroga cerealella H-1, I-7, 10
    (88, 121, 124)
Sogatella furcifera H-2a, I-10, 12 (141)
Spodoptera frugiperda H-4, I-? (130, 1095)
Spodoptera littoralis H-1, I-? (348)
Spodoptera litura H-4, I-? (123, 124, 343,
    348, 440, 1095, 1288); H-5, I-10 (347)
Stegobium paniceum H-4, I-? (168)
Stored grain pests H-1, I-? (89); H-5,
    I-7, 9, 10 (148)
Stored rice pests H-1, I-7 (119)
Tribolium castaneum H-4, I-? (168); H-5,
    I-10 (517)
Tribolium confusum H-4, I-? (168)
Trogoderma granarium H-1, I-? (454, 601,
    878); H-4, I-? (168); H-5, I-10 (87, 167)
Tryporyza incertulas H-1, I-? (93)
Tungro virus of rice H-21, I-10, 12
    (140, 141, 942, 1299, 1331, 1339)
Tylenchorhynchus brassicae H-16, I-10, 18
    (100, 168, 538, 539, 617, 793, 923, 1335)
Tylenchorhynchus elegans H-16, I-18 (1335)
Tylenchus filiformis H-16, I-4, 7, 8, 9, 18
    (538, 793, 1123)
Urentius echinus H-4, I-? (63)
Urentius hystricellus H-4, I-? (168)
Utetheisa pulchella H-4, I-? (168)

OR: H-1, I-7, 9, 10 (50, 73, 80, 84, 88, 132, 420, 504, 1326); H-2, 4, I-7, 9 (107, 117, 220);
    H-2a, 5, I-7 (116); H-2b, 4, 5, I-7, 10 (128); H-3, I-7, 10 (122); H-3, 4, I-? (1093);
    H-3, 4, I-7, 8, 10 (132); H-4, I-10 (123, 124); H-5, I-7, 9, 10 (126, 220, 504); H-5,
    16, 20, I-7, 10 (136); H-20, I-7 (420, 504); H-22, I-? (1287)
AP: Tri=I-10 (azadirachtin, meliantriol, salannin)(168)

Azolla sp.   (Water fern)   Salviniaceae
   B   C   D   D   J   M
   09   06   01   02   01   10
OC: Mosquitoes   H-1, I-1 (472)

Baccharis coridifolia   (Not known)   Asteraceae
   A   B   C   J
   01   02   09   03
OC: Attagenus piceus   H-4, I-6, 7, 8 (48)
AP: Alk=I-? (baccharine)(1392)

Baccharis floribunda   (Not known)   Asteraceae
   A   B   C   J
   01   02   09   11
OC: Attagenus piceus   H-4, I-2 (48)   Tineola bisselliella   H-4, I-2 (48)
OR: H-2, I-? (105)

Baccharis glutinosa   (Water willow)   Asteraceae
   A   B   C   J   M
   01   02   09   03   09
OC: Blattella germanica   H-2a, I-7 (48)
OR: H-32, I-6, 7 (401)

Baccharis halimifolia   (Consumption weed)   Asteraceae
   A   B   C   D   J
   01   02   09   03   03
OC: Pseudomonas solanacearum   H-18, I-8, 16 (945)
OR: H-15, 19, I-7 (651)
AP: Alk=I-6, 7 (1392)

Baccharis ramulosa   (Escobilla)   Asteraceae
   A   B   C   D
   01   02   09   01
OC: Spodoptera frugiperda   H-1, I-1 (176)

Backhousia myrtifolia   (Grey myrtle)   Myrtaceae
   A   C   D   M
   01   09   01   04
OC: Aedes sp.   H-2, 5, I-12 (101)   Anopheles sp.   H-2, 5, I-12 (101)
OR: H-30, I-12 (101)

Baeckea frutescens   (Bruere de Tonkin)   Myrtaceae
   A   B   C   D   M   M
   01   02   09   01   15   17
OC: Ants   H-1, I-6, 7 (1241)   Gryllotalpa sp.   H-1, I-6, 7 (1241)
   Cnaphalocrocis medinalis   H-1, I-6, 7 (1241)

Balanites aegyptica (Desert date) Simaroubaceae

A	B	B	C	D	J	J	J	L	M	M	M	M	M	M	M	N	N	N	O	O	O
01	01	02	09	01	03	08	11	03	01	03	04	05	06	15	19	09	10	12	04	10	12

OC: Attagenus piceus H-4, I-2, 9 (48) Snails H-28, I-4, 9 (1353)
Diaphania hyalinata H-2, I-9 (101) Vermin H-27, I-7, 9 (1353)
Plutella xylostella H-2, I-9 (101)
OR: H-32, 33, I-2, 4, 10 (1276, 1353)
AP: Alk=I-7 (1374); Sap=I-2, 4 (1353)

Balanites roxburghii (Ingudi) Simaroubaceae

A	C	D	D	J	J	J	J	M	M	M	N	N
01	09	01	04	02	03	04	08	01	15	30	10	12

OC: Aphids H-2, I-4 (105) Grasshoppers H-2, I-4 (105)
OR: H-32, I-4 (73, 105, 220, 605, 1276)

Balsamodendron playfairii (Not known) Burseraceae

A	B	C	D	D
01	01	09	01	05

OC: Lice H-2, I-13 (105, 129)

Bambusa arundinacea (Spiny bamboo) Poaceae

A	B	C	D	J	J	M	M	M	M	M	N	N	O
01	07	09	01	10	12	01	04	05	13	19	10	15	15

OC: Dysdercus koenigii H-3, I-6 (795) Mosquitoes H-1, I-15 (73)
OR: H-2, I-16 (220); H-24, I-7 (1334)

Bambusa bambos (Bamboo genus) Poaceae

A	B	C	D	D
01	07	09	01	02

OC: Maggots H-2, I-15 (1345) Mosquitoes H-2, I-15 (1345)

Bambusa oldhamii (Oldham bamboo) Poaceae

A	B	C	D	D	M	O
01	07	09	01	02	01	15

OR: H-24, I-? (1247)

Bambusa vulgaris (Bamboo) Poaceae

A	B	C	D	D	M	M	M	M	M	N
01	07	09	01	02	01	03	04	06	22	15

OC: Rice field insects H-5, I-6 (137)

Bandeiraea simplicifolia (Not known) Caesalpiniaceae

C	D	M
09	01	24

OC: Lice H-1, I-7 (105)

Baptisia tinctoria (Yellow wild indigo) Fabaceae

A	B	C	D	J	M	M	M	M	N	O
01	06	09	03	01	01	05	12	16	15	02

OC: Flies H-5, I-16 (104)
AP: Alk=I-2, 4, 7, 8, 9 (1370); Alk=I-2, 10 (cytisine)(1392); Alk=I-6, 7 (1392); Alk=I-6, 7, 8 (1369)

Barbarea aristata   (Winter cress genus)   Brassicaceae
```
B    C    D    J    J    J
06   09   03   03   08   11
```
OC: Lipaphis erysimi H-2, I-9 (600)

Barbarea vulgaris   (Winter cress)   Brassicaceae
```
A    B    C    D    J    L    M    N
01   06   09   03   03   05   01   07
```
OC: Agrobacterium tumefaciens H-18, I-1 (585)   Erwinia carotovora H-18, I-1 (585)
OR: H-24, I-10 (1192)

Barringtonia asiatica   (Potat)   Barringtoniaceae
```
A    B    C    D    J    M    M    M    M    N    O
01   01   09   01   08   1a   05   06   19   11   07
```
OC: Attagenus piceus H-4, I-9 (48)   Vermin H-27, I-10 (1350)
OR: H-32, I-2, 4, 9, 10 (73, 220, 416, 459, 528, 983, 1350)
AP: Sap=I-11 (1350)

Barringtonia racemosa   (Potat)   Barringtoniaceae
```
A    B    C    D    J    J    JL   L    M    M    M    N    N    O    O    O
01   01   09   01   03   08   11   03   01   05   06   07   09   04   07   10
```
OC: Attagenus piceus H-4, I-4 (48)   Toxoptera aurantii H-2, I-4 (87, 681)
OR: H-1, I-10 (73, 105); H-32, I-2, 4, 10 (220, 459, 504, 983); H-33, I-7 (968)
AP: Sap=I-7 (968); Tri=I-7 (1390)

Basella alba   (Malabar spinach)   Basellaceae
```
B    C    D    M    M    N
04   09   01   01   16   07
```
OC: Drechslera oryzae H-14, I-7 (113)

Bauhinia purpurea   (Camel's foot tree)   Caesalpiniaceae
```
A    B    C    D    M    N    N
01   01   09   01   01   08   15
```
OR: H-2, I-? (88); H-2a, I-7 (124); H-24, I-7 (1247)
AP: Alk=I-? (1391)

Begonia pearcei   (Not known)   Begoniaceae
```
A    B    C    D    D
03   06   09   01   02
```
OC: Plutella xylostella H-1, I-16 (87)

Belamcanda chinensis   (Leopard flower)   Iridaceae
```
B    C    D    M    O
06   09   03   05   02
```
OC: Phytophthora infestans H-14, I-1 (1241)   Puccinia rubigavera H-14, I-1 (1241)
    Puccinia graminis tritici H-14, I-1 (1241)

Bellis perennis   (English daisy)   Asteraceae
```
A    B    C    D    F    J    J    L    M    M    N
01   06   09   03   11   02   03   05   01   05   07
```
OC: Attagenus piceus H-4, I-1 (48)   Venturia inaequalis H-14, I-1 (16)
    Locusta migratoria H-4, 5, I-7 (596)

Berberis aristata   (Indian barberry)   Berberidaceae
  A B C D J L M M N
  01 02 09 03 03 03 01 16 09
OC: Mosquitoes H-2, I-6 (101)   Tribolium castaneum H-5, I-2 (179)
OR: H-2, I-? (105); H-2, I-7 (101); H-2, 32, I-4 (1276); H-32, 33, I-? (73, 220)
AP: Alk=I-4 (berberine, palmatine)(1392)

Berberis thunbergii   (Japanese barberry)   Berberidaceae
  A B C D J
  01 02 09 03 3a
OC: Agrobacterium tumefaciens H-18, I-9, 14 (585)
    Erwinia carotovora H-18, I-9, 14 (585)
OR: H-19, I-1, 7, 14 (652, 704)
AP: Alk=I-2 (berbamine, berberine, berlambine, columbamine, jatrorrhizine, lambertine, magno-
    florine, oxyacanthine, oxyberberine, palmatine, shobakunine, tetrahydroshobakunine)(1392);
    Alk=I-6, 7 (1392)

Bergenia cordifolia   (Winter begonia genus)   Saxifragaceae
  A B C D M M M N
  01 06 09 03 01 09 13 07
OC: Tobacco mosiac virus H-21, I-? (1061)

Bersama abyssinica   (Not known)   Melianthaceae
  A B D J
  01 09 01 14
OC: Ceratitis capitata H-6, I-2, 4, 8 (956)   Heloithis zea H-4, I-4 (1354)
    Dacus cucurbitae H-6, I-2, 4, 8 (956)
OR: H-19, I-4 (1354)

Bersama paulinioides   (Not known)   Melianthaceae
  A B C D J M O O
  01 01 09 01 03 05 02 04
OC: Attagenus piceus H-4, I-2 (48)   Vermin H-27, I-16 (1353)
    Tineola bisselliella H-4, I-2 (48)

Beta altissima   (Beet genus)   Chenopodiaceae
  A B C D
  01 06 09 03
OC: Leptinotarsa decemlineata H-4, I-7 (531)

Beta esculenta   (Beet genus)   Chenopodiaceae
  A B C D
  01 06 09 03
OC: Leptinotarsa decemlineata H-4, I-7 (531)   Pieris brassicae H-4, I-7 (531)
    Phytodecta fornicata H-4, I-7 (531)

Beta nana   (Beet genus)   Chenopodiaceae
  B C D J
  07 09 03 03
OC: Tobacco necrotic virus H-21, I-7 (1316)

Beta vulgaris   (Beet)   Chenopodiaceae
  A A B C D
  02 03 06 09 03

OC: Alternaria tenuis H-14, I-? (480)   Fusarium solani H-14, I-? (736)
    Aspergillus oryzae H-14, I-? (736)   Helminthosporium sp. H-14, I-? (480, 523)
    Curvularia penniseti H-14, I-? (523)   Rhizopus nigricans H-14, I-? (736)
    Fusarium oxysporum H-14, I-? (736)
OR: H-24, I-9 (569)

Betula lenta (Sweet birch)   Betulaceae
    A   B   C   D   M   M   M   M   M   M   N   O   O
    01 01 09 03 01 04 05 06 14 19 14 05 12
OC: Alternaria tenuis H-14, I-? (426)   Fusarium oxysporum
    Botrytis allii H-14, I-? (426)   f. sp. conglutinans H-14, I-? (426)
    Ceratocystis ulmi H-14, I-? (426)   f. sp. lycopersici H-14, I-? (426)
    Cladosporium fulvum H-14, I-? (426)   Gibberella fujikuroi H-14, I-? (426)
    Claviceps purpurea H-14, I-? (426)   Ustilago avenae H-14, I-? (426)
    Cochliomyia americana H-1, I-12 (244)   Verticillium albo-atrum H-14, I-? (426)
    Diplodia maydis H-14, I-? (426)

Betula papyrifera (White birch)   Betulaceae
    A   B   C   D   M   N   N
    01 01 09 03 01 04 07
OC: Ceratocystis ulmi H-14, I-12 (185)   Cladosporium fulvum H-14, I-12 (185)

Betula platyphylla (Birch genus)   Betulaceae
    A   B   C   D   E   J   M   M
    01 01 09 03 07 06 13 19
OC: Culex pipiens H-2, I-4 (87, 101)

Betula sollennis (Birch genus)   Betulaceae
    A   B   C   D   J
    01 01 09 03 03
OC: Drosophila hydei H-2, I-7 (101)

Bidens bipinnata (Spanish-needles genus)   Asteraceae
    A   B   C   D   L   M   M   N   O
    03 06 09 01 05 01 05 07 14
OC: Aphids H-1, I-6, 7 (1241)
OR: H-15, I-6, 7 (652)

Bidens cernua (Tickseed)   Asteraceae
    B   C   D   J
    06 09 10 09
OC: Mosquitoes H-1, I-1 (944)

Bidens laevis (Tickseed)   Asteraceae
    B   C   D
    06 09 10
OR: H-24, I-7 (1215)

Bidens pilosa (Spanish-needles genus)   Asteraceae
    A   B   C   D   J   J   J   M   M   N   O
    03 06 09 10 03 08 11 01 05 07 07
OC: Attagenus piceus H-4, I-8, 15 (48)   Periplaneta americana H-2, I-1 (78)
    Oncopeltus fasciatus H-2, I-1 (78)
OR: H-19, I-? (172)
AP: Alk=I-2 (1369); Alk=I-2, 6, 7 (1392)

49

Bixa orellana   (Lipstick tree)   Bixaceae
   A   B   C   D   J   M   M   M   M   O
   01 01 09 01 04 03 05 13 16 09
OC: Mosquitoes H-5, I-9, 10 (101)   Pseudomonas solanacearum H-18, I-2 (945)
OR: H-5, I-9 (1352); H-19, I-? (982)

Blechnum spicant   (Deer fern)   Polypodiaceae
   A   C   D   M
   01 06 06 13
OC: Pseudomonas solanacearum H-18, I-7 (854)
   Xanthomonas campestris pv. phaseoli-sojensis H-18, I-7 (854)

Bletilla striata   (Not known)   Orchidaceae
   B   C   D   M
   06 09 03 05
OC: Plutella xylostella H-1, I-7 (87)

Blumea aurita   (Not known)   Asteraceae
   B   C   D   M   O
   06 09 01 05 07
OR: H-5, I-? (105)

Blumea balsamifera   (Ngai camphor)   Asteraceae
   B   B   C   D   D   J   M   M   M   N   N
   02 06 09 01 08 03 01 17 21 02 07
OC: Drechslera oryzae H-14, I-7 (113)   Snails H-28, I-7 (662)
   Pyricularia oryzae H-14, I-7 (114)
OR: H-17, I-7 (1318); H-32, I-7 (983)
AP: Alk=I-6, 7 (1318, 1392); Alk=I-7 (1390); Tan=I-6, 7 (1318)

Blumea eriantha   (Not known)   Asteraceae
   C
   09
OC: Aspergillus oryzae H-14, I-12 (649)   Pseudomonas solanacearum H-18, I-? (650)
   Fusarium oxysporum H-14, I-12 (649)

Blumea lacera   (Not known)   Asteraceae
   C   D   M   M   N   O
   09 01 01 05 07 07
OC: Fleas H-5, I-? (87, 105)
OR: H-17, I-7 (504); H-30, I-? (132, 687)

Blumea lyrata   (Not known)   Asteraceae
   C   D
   09 01
OC: Mosquitoes H-1, I-16 (101)

Bocconia frutescens   (Not known)   Papaveraceae
   A   B   B   C   D   J   M   M   O
   01 01 02 09 01 3a 05 16 02
OC: Ticks H-13, I-14 (105)
AP: Alk=I-? (allocryptopine, chilerythrine, protopine)(1392); Alk=I-4, 5, 9 (sanguinarine)
   (1392); Alk=I-6, 7, 9 (1369)

Boenninghausenia albiflora   (Not known)   Rutaceae
    A   B   C   D   D   J
    01   06   09   02   03   04
OC: Cimex lectularius H-5, I-7 (87)   Ctenocephalides canis H-5, I-7, 12 (87)
AP: Alk=I-2, 6, 7 (dictamnine)(1392)

Boerhavia diffusa   (Hogweed)   Nyctaginaceae
    C   D   J   M   M   M   N   N   N   O
    09   01   03   01   02   05   02   07   10   02
OC: Gomphrena mosaic virus H-21, I-2   Tobacco mosaic virus H-21, I-2
      (557, 841)   (557, 841, 976)
    Sunnhemp rosette virus H-21, I-2 (590, 841) Tobacco ring spot virus H-21, I-2 (590, 841)
OR: H-17, I-? (1353)
AP: Alk=I-? (punarnavine)(1392); Alk=I-2, 6, 7 (1360); Alk=I-7 (1379)

Borago officinalis   (Talewort)   Boraginaceae
    A   B   C   D   J   M   M   N
    03   06   09   06   09   01   18   07
OC: Botrytis cinerea H-14, I-1 (946)
AP: Alk, Tan=I-9, 10 (1391)

Borreria articularis   (Landrena)   Rubiaceae
    B   C   D   D   J   M   O
    06   09   01   04   03   05   07
OR: H-24, I-1 (1212)
AP: Tan=I-6, 7 (1360)

Boscia angustifolia   (Not known)   Capparidaceae
    A   B   C   D   M   M   M   N   N   N
    01   01   09   05   01   02   05   04   09   10
OC: Locusta migratoria H-4, 5, I-7 (596)

Boswellia carteri   (Frankincense)   Burseraceae
    A   B   C   D   D   K   M
    01   01   09   01   05   07   17
OC: Mosquitoes H-5, I-13 (105)

Boswellia dalzielii   (Frankincense tree)   Burseraceae
    A   B   C   D   D   K
    01   01   09   01   05   08
OC: Flies H-5, I-13 (87)   Termites H-5, I-13 (87)
    Mosquitoes H-5, I-13 (87)

Boswellia serrata   (Frankincense tree)   Burseraceae
    A   B   C   M   M   M   M   M   N   O
    01   01   09   01   05   06   17   20   09   13
OC: Helminthosporium sp. H-14, I-12 (901)   Pythium aphanidermatum H-14, I-12 (901)
OR: H-2, I-? (1276); H-15, I-12 (901)
AP: Tan=I-4, 6 (1360)

Bothriochloa intermedia   (Australian bluestem)   Poaceae
    A   B   C   D   D   E   M   M
    01   06   09   01   04   05   02   09
OC: Grasshoppers H-2, I-? (143); H-4, I-7 (828)

Bothriochloa ischaemum   (Yellow bluestem)   Poaceae

   A   B   C   D   D
   01 06 09 03 06

OC: Leptinotarsa decemlineata H-4, I-7 (531)

Bougainvillea sp.   (Bougainvillea)   Nyctaginaceae

   B   C   J
   03 09 07

OC: Sitophilus oryzae H-2, I-8 (87)

Bougainvillea spectabilis   (Bougainvillea)   Nyctaginaceae

   A   B   C   D
   01 02 09 01

OC: Physalis shoestring mosaic virus H-21,   Tobacco mosaic virus H-21, I-7 (1301)
   I-7 (1301)   Tomato yellow mottle mosaic virus H-21,
   Sunnhemp rosette virus H-21, I-7 (1301)   I-7 (1301)

Brassica cernua   (Not known)   Brassicaceae

   B   C
   06 09

OC: Drosophila hydei H-2, I-6, 7 (101)
OR: H-33, I-16 (220)

Brassica geniculata   (Mustard genus)   Brassicaceae

   B   C
   06 09

OC: Mosquitoes H-10, I-10 (916)

Brassica hirta   (White mustard)   Brassicaceae

   A   B   C   D   D   L   M
   03 06 09 03 06 02 06

OC: Heterodera rostochiensis H-16, I-12 (677)   Locusta migratoria H-4, 5, I-7 (596)
AP: Alk=I-? (sinapine)(1392)

Brassica integrifolia   (Mustard genus)   Brassicaceae

   A   B   C   D   J   M   M   N
   03 06 09 01 03 01 28 07

OC: Drechslera oryzae H-14, I-7, 12 (113, 432)   Pyricularia oryzae H-14, I-7, 12 (113, 432)

Brassica juncea   (Indian mustard)   Brassicaceae

   A   B   C   D   M   M   N   N   O
   03 06 09 03 01 05 07 12 10

OC: Crocidolomia binotalis H-1, I-? (1333)   Musca domestica H-6, I-2 (932)
   Meloidogyne javanica H-16, I-? (992)
OR: H-24, I-10 (1192)
AP: Alk=I-2, 6, 7, 8 (1370)

Brassica kaber   (Charlock)   Brassicaceae

   A   B   C   D   J   J   M   N
   03 06 09 03 03 04 01 07

OC: Leptinotarsa decemlineata H-4, I-7 (531)   Tylenchulus semipenetrans H-16, I-10 (540)

Brassica latifolia   (Mustard genus)   Brassicaceae
    B   C   D   J   J   K
    06 09 03 06 08 06
OC: Callosobruchus chinensis H-3, I-12 (516)   Caterpillars H-2, I-4, 7 (1029)

Brassica napus   (Rape)   Brassicaceae
    A   B   C   D
    03 06 09 03
OC: Acromyrmex octospinosus H-1, I-12 (470)   Sphaerothica humuli H-14, I-12
    Bipolaris sorokiniana H-14, I-12 (593)   (493, 496, 760)
    Peronospora tabacina H-14, I-12 (482)

Brassica napus var. arvensis   (Mustard genus)   Brassicaceae
    A   B   C   D   M
    03 06 09 03 28
OC: Phyllobius oblongus H-4, I-? (531)   Tanymecus dilaticollis H-4, I-? (531)

Brassica nigra   (Black mustard)   Brassicaceae
    A   B   C   D   J   J   M   M   N   N   N   O
    03 06 09 03 03 06 01 05 07 10 12 10
OC: Aphelenchus avenae H-16, I-? (100)   Heterodera rostochiensis H-16, I-12
    Aspergillus alliaceus H-14, I-10 (550)   (675, 677, 682)
    Aspergillus niger H-14, I-10 (550)   Hoplolaimus indicus H-16, I-? (100)
    Botrytis alii H-14, I-10 (550)   Meloidogyne incognita H-16, I-18 (100, 617)
    Callosobruchus maculatus H-3, I-10 (1124)   Mosquitoes H-2, I-10 (504)
    Colletotrichum circinans H-14, I-10 (550)   Papilio polyxenes H-1, I-10 (595)
    Ditylenchus cypei H-16, I-18 (100, 617)   Tylenchorhynchus brassicae H-16, I-18
    Heliothis virescens H-5, I-? (1240)   (100, 617)
OR: H-20, I-19 (1148)
AP: **Alk**=I-2, 6, 7, 8, 9 (1370); **Alk**=I-2, 10 (1368); **Alk**=I-10 (sinapine)(1392)

Brassica oleracea   (Wild cabbage)   Brassicaceae
    A   A   B   C   D   M   N
    01 03 06 09 03 01 07
OC: Drosophila melanogaster H-2, I-2 (87)   Musca domestica H-2, I-2 (87)
    Inopus rubriceps H-2, I-2 (87, 860)   Tobacco mosaic virus H-21, I-? (1307)
    Locusta migratoria H-4, 5, I-7 (596)
AP: **Alk**=I-7 (narcotine)(1392); **Alk**=I-10 (1369)

Brassica oleracea var. acephala   (Kale)   Brassicaceae
    A   B   C   D   M   M   N
    03 06 09 10 01 02 07
OC: Drosophila melanogaster H-2, I-2 (87)   Papilio polyxenes H-1, I-? (54)
    Musca domestica H-2, I-2 (87); H-6, I-2 (932)

Brassica oleracea var. botrytis   (Broccoli and cauliflower)   Brassicaceae
    B   C   D   M   N
    06 09 03 01 08
OC: Cassida nebulosa H-4, I-7 (531)   Musca domestica H-6, I-2 (932)

Brassica oleracea var. capitata   (Cabbage)   Brassicaceae
    B   B   C   J   M   N
    03 06 09 3a 01 07

OC: Cephalosporium sacchari H-14, I-1 (1048)   Leptinotarsa decemlineata H-4, I-? (531)
    Drosophila melanogaster H-2, I-2 (87)   Musca domestica H-2, 6, I-2 (87, 932)
    Fusarium nivale H-14, I-1 (1048)   Pyrausta nubilalis H-1, I-7 (719)
    Galleria mellonella H-1, I-7 (719)
OR: H-15, I-7 (719); H-19, I-? (622); H-24, I-7, 10 (569)

Brassica oleracea var. gemmifera   (Brussel sprouts)   Brassicaceae
    B   C   D   M   N
    06 09 03 01 07
OC: Drosophila melanogaster H-2, I-2 (87)   Musca domestica H-2, 6, I-2 (87, 932)

Brassica oleracea var. gongyloides   (Kohlrabi)   Brassicaceae
    A   B   C   D   M   N   N
    02 06 09 03 01 06 07
OC: Aspergillus oryzae H-14, I-? (736)   Hyphantria cunea H-4, I-? (531)
    Fusarium oxysporum H-14, I-? (736)   Musca domestica H-6, I-2 (932)
    Fusarium solani H-14, I-? (736)   Rhizopus nigricans H-14, I-? (736)

Brassica oleracea var. italica   (Italian broccoli)   Brassicaceae
    B   C   D   M   N   N
    06 09 03 01 06 08
OC: Musca domestica H-2, 6, I-2 (87, 932)
OR: H-24, I-? (951)

Brassica rapa   (Field mustard)   Brassicaceae
    B   C   D   J
    06 09 03 3a
OC: Cephalosporium sacchari H-14, I-? (1048)   Fusarium oxysporum
    Fusarium oxysporum H-14, I-? (1048)   f. sp. udum H-14, I-? (959)
    Fusarium oxysporum   f. sp. vasinfectum H-14, I-? (959)
      f. sp. lini H-14, I-? (959)   Hoplolaimus indicus H-16, I-? (612)

Brassica rapa var. rapifera   (Turnip)   Brassicaceae
    A   B   C   D   M   M   N
    02 06 09 03 01 03 02
OC: Acyrthosiphum pisum H-2, I-2, 7 (87, 387)   Fusarium solani H-14, I-? (735)
    Aspergillus oryzae H-14, I-? (735)   Musca domestica H-2, I-7 (387, 392)
    Blattella germanica H-2, I-7 (387)   Tetranychus atlanticus H-11, I-2, 7
    Drosophila melanogaster H-2, I-7 (387, 392)   (87, 387)
    Epilachna varivestis H-2, I-7 (387)   Tribolium confusum H-2, I-7 (387)
    Fusarium oxysporum H-14, I-? (735)
OR: H-2, I-2 (390)

Brassica sinapistrum   (Field mustard)   Brassicaceae
    B   C   D   F   J   M   N   N
    06 09 03 11 02 01 06 08
OC: Venturia inaequalis H-14, I-? (16)

Brassica sp.   (Mustard genus)   Brassicaceae
    A   B   C   D   J
    03 06 09 03 06
OC: Bipolaris sorokiniana H-14, I-6, 7 (593)   Callosobruchus maculatus H-1, I-? (644)
    Callosobruchus chinensis H-1, I-?   Chrysomya macellaria H-5, I-12 (199)
      (516, 644)   Cochliomyia americana H-2, I-12 (244)

Fusarium oxysporum
    f. sp. lini H-14, I-? (959)
    f. sp. udum H-14, I-? (959)
    f. sp. vasinfectum H-14, I-? (959)
OR: H-15, I-? (399)

Plasmodiophora brassicae H-14, I-? (933)
Sphaerothica humuli H-14, I-? (496)

Brexia madagascariensis   (Not known)   Saxifragaceae
    A   B   C   D
    01 01 09 01
OC: Dacus dorsalis H-6, I-8 (473)

Brongniantia benthamiana   (Not known)   Fabaceae
    C   M
    09 12
OC: Ceratitis capitata H-6, I-6, 7, 9 (956)   Dacus cucurbitae H-6, I-6, 7, 9 (956)

Broussonetia kazinoki   (Paper mulberry genus)   Moraceae
    A   B   C   D
    01 01 09 01
OC: Fusarium roseum H-14, I-15 (917)

Broussonetia papyrifera   (Paper mulberry)   Moraceae
    A   B   C   D   D   M   M   N
    01 01 09 01 03 01 22 09
OC: Fusarium roseum H-14, I-15 (917)

Brucea javanica   (Not known)   Simaroubaceae
    A   B   C   D   J   M   O
    01 02 09 01 03 05 09
OC: Cnaphalocrocis medinalis H-1, I-7, 10   Rice borers H-1, I-7, 10 (1241)
    (1241)   Rice leafhoppers H-1, I-7, 10 (1241)
OR: H-17, I-9 (504)
AP: Alk=I-6, 7, 10 (yatanine)(1392); Alk=I-7 (1390); Alk=I-10 (1369)

Brugmansia arborea   (Marikoa)   Solanaceae
    A   B   B   C   D   D   M
    01 01 02 09 01 02 13
OC: Ants H-5, I-? (101)
OR: H-5, I-? (1276); H-35, I-? (504, 507)
AP: Alk=I-2 (atropine, hyoscyamine)(1392); Alk=I-6 (hyoscyamine)(1392); Alk=I-7 (atropine, hyoscine)(1392); Alk=I-10 (hyoscine)(1392)

Bryonia alba   (Wild hop)   Cucurbitaceae
    B   B   D   J   M   M   N   O
    04 06 03 02 01 05 15 02
OC: Aphids H-1, I-2 (104, 1025)
AP: Alk=I-2 (bryonicine)(1392)

Bucea sumatrana   (Not known)   Simaroubaceae
    A   B   C   D   M   O
    01 01 09 01 05 09
OR: H-5, I-2 (504)

Buddleia lindleyana   (Butterfly bush genus)   Loganiaceae
    A   B   C   D   J   J
    01 02 09 03 02 04
OC: Aphis fabae H-2, I-7 (101)

Buddleia officinalis   (Butterfly bush genus)   Loganiaceae
    A   B   C   D
    01 02 09 03
OC: Rice field insects H-1, I-2, 6, 7 (1241)

Bumelia retusa   (Not known)   Sapotaceae
    A   B   C   D   D
    01 02 09 02 03
OC: Fleas H-5, I-1 (118)   Lice H-5, I-1 (118)

Bursera graveolens   (Not known)   Burseraceae
    A   B   C   D
    01 01 09 01
OR: H-5, I-14 (1276)

Bursera simaruba   (West Indian birch)   Burseraceae
    A   B   C   D
    01 01 09 01
OC: Lymantria dispar H-5, I-? (340)

Butea monosperma   (Flame-of-the-forest)   Fabaceae
    A   B   C   D   M   M
    01 01 09 01 12 13
OC: Rats H-26, I-10 (799)
OR: H-2, I-10 (220)

Butea superba   (Flame-of-the-forest)   Fabaceae
    A   B   C   D   D   D   E   J   M   M   M   M   M   O   O
    01 01 09 01 02 04 04 03 03 05 12 13 16 07 08
OC: Mosquitoes H-2, I-10 (101)
OR: H-1, I-? (73, 105); H-17, I-10 (504)

Butryospermum parkii   (Butter tree)   Sapotaceae
    A   B   C   D   M   M   M   M   M   M   N
    01 01 09 01 01 02 04 06 15 19 10
OC: Stored grain pests H-5, I-7, 9 (533)

Buxus japonica   (Japanese buxus)   Buxaceae
    A   B   B   C   D   M
    01 01 02 09 03 19
OC: Drosophila hydei H-2, I-6, 7 (101)

Buxus sempervirens   (Common buxus)   Buxaceae
    A   B   C   D   D   J
    01 01 09 03 06 03
OC: Plutella xylostella H-4, I-? (87)   Popillia japonica H-5, I-7 (105, 1243)
OR: H-19, I-6, 7 (446)
AP: Alk=I-7 (bebeerine, isochondodendrine)(1392)

Caesalpinia coriaria   (Divi-divi)   Caesalpiniaceae
    A   B   B   C   D   M   M   M   M
    01 01 02 09 01 12 16 19 21
OC: Attagenus piceus H-4, I-6, 7, 9 (48)   Rats H-26, I-9 (1393)
    Popillia japonica H-5, I-? (105, 1066)
AP: Tan=I-11 (507, 1352)

Caesalpinia pulcherrima   (Peacock flower)   Caesalpiniaceae
    A   B   C   D   F   J   J   J   K   M   M   O
    01 02 09 01 15 04 11 15 06 05 12 02
OC: Manduca sexta H-1, I-7 (669)   Sitophilus oryzae H-2a, I-8 (87, 598)
    Panonychus citri H-4, 11, I-7 (669)
OR: H-2, I-2, 4, 7, 8 (127); H-22, I-? (975)

Caladium bicolor   (Mother-in-law plant)   Araceae
    A   B   C   D   J   J   M   M   N   O
    01 06 09 01 03 04 1a 05 03 03
OR: H-2, I-7 (104); H-19, I-7 (982)

Calla palustris   (Water arum)   Araceae
    A   B   B   C   D   J   M   N
    01 06 09 09 03 16 01 02
OC: Fusarium roseum H-14, I-1 (837)
OR: H-19, I-1 (837)

Callicarpa americana   (French mulberry)   Verbenaceae
    A   B   C   D   J   M   M   O
    01 02 09 10 03 05 13 07
OC: Heliothis virescens H-5, I-7 (1240)
OR: H-15, 19, I-7 (499)

Callicarpa candicans   (Beautyberry genus)   Verbenaceae
    A   B   B   C   D   D   J   L   M   M   O   O   O
    01 01 02 09 01 02 03 03 05 13 02 04 07
OC: Drechslera oryzae H-14, I-7 (113)
OR: H-20, I-2, 4, 7 (504); H-32, 33, I-7 (504, 528, 983)

Callicarpa japonica   (Beautyberry genus)   Verbanaceae
    A   B   C   D   D   J   M
    01 02 09 02 03 13 13
OC: Spodoptera litura H-4, I-7 (378, 867)

Callilepis laureola   (Not known)   Asteraceae
    B   D   D   J
    09 01 03 04
OC: Maggots H-1, I-? (87)
OR: H-2, I-2 (104)

Callistemon citrinus   (Crimson bottlebrush)   Myrtaceae
    A   B   B   C   D   J   J
    01 01 02 09 01 03 08
OC: Exserohilum turcicum H-14, I-15 (827)
OR: H-19, I-7, 8 (777, 982)
AP: Alk=I-7 (1392)

Callistephus chinensis (China aster) Asteraceae

   A   B   C   D   J
   03 06 09 03 15

OC: Sitophilus oryzae H-2, I-8 (598)

Calodendrum capense (Cape chestnut) Rutaceae

   A   B   C   D   J   M   M   M
   01 01 09 01 14 04 13 15

OC: Oncopeltus fasciatus H-2, I-4 (958)

Caloncoba glauca (Not known) Flacourtiaceae

   A   B   C   D   J   K
   01 01 09 01 03 02

OC: Rats H-25, I-10 (1345)

Calophyllum inophyllum (Laurelwood) Clusiaceae

   A   B   C   D   E   J   M   M   M   M   M   M   M   M   N   O   O   O
   01 01 09 01 04 03 01 04 05 06 13 17 19 20 09 07 10 12

OC: Meloidogyne incognita H-16, I-8 (100, 616) Mites H-11, I-12 (105)
   Meloidogyne javanica H-16, I-8 (100, 616)
OR: H-2, I-? (73); H-20, I-10, 12 (504, 1349); H-32, I-7 (105, 968); H-33, I-10, 12 (1349)
AP: **Fla**=I-7 (amentoflavone, biflavone)(1383); **Sap**=I-7 (968); **Tan**=I-4 (968)

Calopogonium coeruleum (Jicuma) Fabaceae

   B   C   D   J   M   M
   04 09 01 02 12 14

OC: Diaphania hyalinata H-2, I-10, 11 (101) Spodoptera frugiperda H-2, I-10, 11 (101)
   Plutella xylostella H-2, I-10, 11 (101)

Calopogonium vellutium (Catinga de macaco) Fabaceae

   C   D   J   M
   09 01 08 12

OC: Lice H-2, I-? (105) Ticks H-13, I-? (105)
OR: H-32, I-? (105)

Calotropis gigantea (Crown plant) Asclepiadaceae

   A   B   C   D   D   E   J   J   J   J   K   K   L   L   M   M   M   M   N   O
   01 01 09 01 05 01 03 3a 04 15 05 06 02 03 01 03 05 06 24 08 07

OC: Diacrisia obliqua H-3, I-7 (161) Schistocerca gregaria H-4, I-7
   Flies H-1, I-? (101) (614, 615, 1039)
   Meloidogyne indica H-16, I-? (616) Sitophilus oryaze H-2, I-8 (87, 400)
   Meloidogyne javanica H-16, I-? (616) Termites H-1, I-7, 14 (871)
   Mosquitoes H-1, I-? (101) Vermin H-27, I-4, 12 (1352)
OR: H-20, I-4, 14 (1352); H-33, I-? (220); H-34, I-14 (1352)

Calotropis procera (Swallowwort) Asclepiadaceae

   A   B   C   D   D   J   J   J   L   M   M   M   M   M   O   O   O
   01 02 09 01 04 03 08 11 03 03 05 13 20 24 02 07 14

OC: Attagenus piceus H-4, I-2 (48) Fowl lice H-2, I-6, 7, 8 (105)
   Aulacophora foveicollis H-1, I-7, 8 (120) Locusts H-1, I-? (866)
   Bagrada picta H-1, I-? (600) Pieris brassicae H-1, I-7, 8 (120, 600)
   Empoasca devastans H-1, I-? (600) Schistocerca gregaria H-4, I-7 (1001)
OR: H-2, 5, I-? (1276); H-15, I-7 (1353); H-16, 20, 33, I-2, 14 (1353); H-17, I-14 (973)
AP: **Alk**=I-4 (1392)

Calpurnia aurea   (East African laburnum)   Fabaceae
    B   C   D   D   M
    02   09   01   02   12
OC: Pediculus humanus humanus H-1, I-7 (1143)

Calpurnia intrusa   (Not known)   Fabaceae
    C   D   M   M
    09   02   12   13
OC: Maggots H-1, I-? (87)

Calpurnia subdecandra   (Not known)   Fabaceae
    A   B   B   C   D   M   M
    01   01   02   09   01   12   13
OC: Maggots H-1, I-? (87)

Caltha palustris   (Marsh marigold)   Ranunculaceae
    A   B   C   D   E   J   L   M   N   N   N
    01   06   09   03   11   09   02   01   02   06   07   08
OC: Rats H-26, I-1 (1393)   Venturia inaequalis H-14, I-? (16)
AP: Alk=I-? (berberine)(1392)

Calvatia gigantea   (Giant puffball)   Lycopodiaceae
    A   B   C   K   M   M   N   O
    04   03   01   05   01   05   09   09
OR: H-2, I-10 (105); H-22, I-9 (969, 1016)

Calystegia sepium   (Wild morning glory)   Convolvulaceae
    A   B   C   D   M   N   N
    01   06   09   03   01   02   06   15
OC: Leptinotarsa decemlineata H-4, I-7 (531)

Calystegia soldanella   (Sea bindweed)   Convolvulaceae
    A   B   C   D   E   J   M   N   N
    01   06   09   10   04   14   01   06   15
OC: Dacus cucurbitae H-6, I-6, 7, 8 (956)

Camelina microcarpa   (Not known)   Brassicaceae
    C   J
    09   03
OC: Aedes aegypti H-2, I-2, 7, 10 (165)

Camelina sativa   (False flax)   Brassicaceae
    B   C   D   M   M   N   N
    06   09   02   01   03   10   12
OC: Linum usitatissimum H-24, I-7 (178)

Camellia japonica   (Camellia)   Theaceae
    A   B   C   D   M
    01   01   09   03   13
OC: Culex quinquefasciatus H-2, I-7 (463)

Camellia oleifera   (Tea-oil plant)   Theaceae
    A   B   C
    01   02   09

OC: Ceratocystis fimbriata H-14, I-10, 18
   (1241)
   Glomerella gossypii H-14, I-10, 18 (1241)
   Maggots H-1, I-10, 18 (1241)
   Puccinia graminis tritici H-14, I-10, 18 (1241)

Puccinia rubigavera H-14, I-10, 18 (1241)
Rice borers H-1, I-10, 18 (1241)
Rice field insects H-1, I-10, 18 (1241)

Camellia sinensis   (Tea)   Theaceae
   A   B   B   C   D   D   J   J   K   K   M   M   N
   01 01 02 09 01 02 03 04 05 06 01 13 07
OC: Anasa tristis H-1, I-12 (105)
    Aphis craccivora H-2, I-18 (503)
    Aphis gossypii H-2, I-18 (503)
    Aphis maidis H-2, I-18 (503)
    Aphis tavaresi H-2, I-18 (503)
    Eupterote mollifera H-2, I-18 (503)

Ferrisiana virgata H-2, I-18 (503)
Macrosiphum rosae H-2, I-18 (503)
Oregma lanigera H-2, I-7, 12 (101)
Termites H-4, I-? (87)
Tetranychus cinnabarinus H-11, I-18 (503)

OR: H-2, I-7 (87); H-2, 32, I-10, 18 (1276); H-22, I-? (1056); H-30, I-12 (105)
AP: **Alk**=I-7 (theobromine, theophylline)(1392); **Alk**=I-7, 8, 9 (caffeine)(1392)

Campanula trachelium   (Throatwort)   Campanulaceae
   A   B   C   D   J
   01 06 09 03 09
OC: Botrytis cinerea H-14, I-1 (946)
OR: H-22, I-1 (946)

Camptotheca acuminata   (Not known)   Nyssaceae
   A   B   C   D   D   J   J
   01 01 09 02 03 09 14
OC: Oncopeltus fasciatus H-2, I-4, 5 (958)   Tobacco mosaic virus H-21, I-? (1126)
OR: H-19, I-7 (1262)

Canarium amboinense   (Java almond)   Burseraceae
   A   B   C   D   M   M   N   O
   01 01 09 01 01 05 10 14
OR: H-5, I-14 (504)

Canarium schweinfurthii   (Incense tree)   Burseraceae
   A   B   C   D   M   M   M   N   N   O
   01 01 09 01 01 05 17 09 10 04
OC: Mosquitoes H-1, I-? (1353)
OR: H-2, I-? (87)

Canavalia ensiformis   (Jack bean)   Fabaceae
   A   B   C   D   J   L   M   M   M   M   M   M   N
   01 04 09 01 03 03 01 02 09 10 12 13 11
OC: Attagenus piceus H-4, I-10 (48)
OR: H-33, I-10 (507)
AP: **Alk**=I-10 (1391)

Canavalia maritima   (Not known)   Fabaceae
   B   C   D   D   D   J   L   M   M   M   M   M   N
   04 09 01 02 04 03 03 01 02 10 12 13 10
OC: Drosophila hydei H-2, I-6, 7 (101)

Canella winterana   (Wild cinnamon)   Canellaceae
   A   B   C   D   J   M   M   O
   01 06 09 01 03 05 14 04
OR: H-4, I-? (132)

Canna flaccida   (Not known)   Cannaceae
   A   B   C   D   D   J   M
   01 06 09 01 02 11 13
OC: Attagenus piceus H-4, I-1 (48)

Canna generalis   (Common garden canna)   Cannaceae
   A   B   C   D
   01 06 09 01
OR: H-2, I-6, 7 (1345)

Canna indica   (Indian shot)   Cannaceae
   A   B   C   D   F   J   K   L   M   M   O   O
   01 06 09 01 13 15 06 05 05 21 02 10
OC: Dysdercus cingulatus H-2a, I-2, 7   Phytodecta fornicata H-4, I-7 (531)
   (117, 124)   Schistocerca gregaria H-4, I-? (615)
   Leptinotarsa decemlineata H-4, I-7 (531)   Sitophilus oryzae H-2, I-8 (400)
AP: Tan=I-7 (1321)

Cannabis sativa   (Marijuana)   Cannabaceae
   A   B   C   D   J   J   J   J   K   K   M   M   M   M   M   M   M   N   O   O   O
   03 06 09 02 03 04 08 09 03 06 01 02 03 05 06 20 28 10 08 10 15
OC: Cimex lectularius H-5, I-7 (87)   Mosquitoes H-5, I-? (87)
   Corynebacterium michiganense H-18, I-?   Musca domestica H-5, I-? (87)
   (1081)   Rhipicephalus rossicus H-13, I-7 (87, 1133)
   Dermacentor marginatus H-13, I-7 (87, 1133) Stored grain pests H-1, I-7 (104)
   Fleas H-5, I-7 (87)   Ustilago hordei H-14, I-7 (181)
   Haemaphysalis punctata H-13, I-7 (87, 1133) Ustilago tritici H-14, I-7 (181)
   Ixodes redikorzevi H-13, I-7 (87, 1133) Weevils H-1, I-7 (104)
OR: H-2, 19, I-6, 7, 8 (871, 1262); H-22, I-? (975); H-32, I-16 (220); H-35, I-14 (1352)
AP: **Alk**=I-? (nicotine)(1392); **Alk**=I-10 (trigonelline)(1392)

Canthium euroides   (Not known)   Rubiaceae
   A   B   C   D
   01 01 09 01
OC: Epilachna varivestis H-2, I-? (143)
OR: H-4, I-? (132)

Capparis aphylla   (Caper bush)   Capparaceae
   A   B   C   D   D   M   M   M   N   N
   01 02 09 01 04 01 04 06 09 15
OC: Termites H-8, I-5 (1352)
OR: H-2, I-? (105, 1276)

Capparis cantoniensis   (Not known)   Capparaceae
   A   C   D   J   M
   01 09 02 03 12
OR: H-1, I-7 (1241)

Capparis horride   (Not known)   Capparaceae
    A   B   C   D   F   J   L
    01 02 09 01 05 03 03
OC: Pseudaletia unipuncta H-2, I-9 (117)
OR: H-34, I-2 (504)

Capsella bursa-pastoris   (Shepherd's purse)   Brassicaceae
    B   C   D   D   J   L   M   O
    06 09 02 03 03 05 05 07
OC: Mosquitoes H-10, I-10 (916)   Rats H-26, I-7 (1325)
    Popillia japonica H-5, I-1 (104, 1243)
OR: H-24, I-7 (811)
AP: Alk=I-? (tyramine)(1392); Alk=I-6, 7, 8 (1370)

Capsicum annuum   (Pepper)   Solanaceae
    A   B   C   D   J   M
    03 06 09 10 02 14
OC: Callosobruchus maculatus H-1, I-? (644)   Leptinotarsa decemlineata H-5, I-7 (531)
    Hyphantria cunea H-5, I-7 (531)   Pieris brassicae H-5, I-7 (531)
    Lentinus lepideus H-14, I-? (428)   Polyporus versicolor H-14, I-? (428)
    Lenzites trabea H-14, I-? (428)
OR: H-2, I-? (87); H-5, I-9 (83); H-24, I-9, 14 (569); H-33, I-9 (620)
AP: Alk=I-16 (capsaicine, solonidine)(1392)

Capsicum frutescens   (Red pepper)   Solanaceae
    A   B   C   D   D   J   J   J   J   K   K   M   M   N   O
    01 06 09 01 03 02 03 3a 04 11 05 07 01 05 09 09
OC: Cucumber mosaic virus H-21, I-7, 14   Tobacco etch virus H-21, I-7, 14
      (1070, 1075)   (1070, 1075)
    Culex quinquefasciatus H-2, I-11 (463)   Tobacco mosaic virus H-21, I-14
    Potato virus X H-21, I-14 (1136)   (1070, 1126, 1131)
    Sitophilus oryzae H-1, I-9 (437)   Tobacco ring spot virus H-21, I-14 (1070)
    Stored grain pests H-2, I-20 (105)
OR: H-5, I-9 (101); H-15, I-18 (1319); H-32, I-? (1276)
AP: Alk=I-6, 7 (1319, 1366); Alk=I-7 (1390); Alk=I-7, 9 (1392)

Caraipa fasciculata   (Not known)   Clusiaceae
    A   B   C   D   M   O
    01 01 09 01 05 13
OR: H-2, I-13 (105)

Carapa guianensis   (Crabwood)   Meliaceae
    A   B   C   D   J   J   J   J   J   M   M   M
    01 01 09 01 03 06 07 08 11 04 06 19
OC: Attagenus piceus H-4, I-9 (48)
OR: H-1, I-? (104); H-2, 5, I-10, 12 (105)

Carapa procera   (Kunda oil tree)   Meliaceae
    A   B   C   D   L   M   M   M   M   O   O
    01 01 09 01 03 04 05 15 19 10 12
OR: H-2, I-10, 12 (105); H-12, I-10 (87); H-15, 33, I-10 (87)

Cardaria draba   (Not known)   Brassicaceae
    B   C   D   D   M   M   N   N
    06   09   03   04   01   14   07   10
OC: Aphids H-2, I-10, 12 (105)   Mites H-11, I-10, 12 (105)
OR: H-32, I-10, 12 (73)
AP: Alk=I-2, 6, 7, 8, 9 (1370)

Cardiocrinium cordatum   (Not known)   Liliaceae
    A   B   C   D   J   M   N
    01   06   09   03   03   01   03
OC: Drosophila hydei H-2, I-6, 7 (101)

Carex clivorum   (Not known)   Cyperaceae
    A   B   C   D   J   M   M
    01   06   09   03   03   02   09
OC: Drosophila hydei H-2, I-2 (101)

Carex lacustris   (Sedge genus)   Cyperaceae
    A   B   C   D   J   J   J
    01   06   09   03   03   09   16
OC: Fusarium roseum H-14, I-1 (837)
OR: H-19, I-1 (837)

Carex siderosticata var. glabra   (Not known)   Cyperaceae
    A   B   C   D
    01   06   09   03
OC: Drosophila hydei H-2, I-6, 7 (101)

Careya arborea   (Slow match tree)   Lecythidaceae
    A   B   C   J   M   N
    01   01   09   08   01   09
OC: Attagenus piceus H-4, I-4 (48)   Leeches H-28, I-4 (220)
OR: H-32, I-2, 4, 7 (73, 220)

Carica papaya   (Papaya)   Caricaceae
    A   B   C   D   J   J   J   M   M   N   O
    01   01   09   01   03   3a   08   01   05   09   07
OC: Dacus diversus H-6, I-8 (473)   Meloidogyne incognita H-16, I-7 (115)
    Dacus fergugineus H-6, I-8 (473)   Rotylenchulus reniformis H-16, I-15 (1330)
    Dacus zonatus H-6, I-8 (473)   Tylenchorhynchus brassicae H-16, I-15
    Drechslera oryzae H-14, I-2, 7   (1330)
      (113, 193, 432)   Tylenchus filiformis H-16, I-15 (1330)
    Helicotylenchus indicus H-16, I-15 (1330) Vermin H-27, I-2, 7, 15 (1353)
    Hoplolaimus indicus H-16, I-15 (1330)   Weevils H-1, I-4, 14 (1353)
OR: H-17, I-7 (1318); H-19, I-4, 7 (982); H-24, I-9, 14 (569)
AP: Alk=I-7, 9, 10 (carpaine)(1392)

Carissa carandas   (Bengal currants)   Apocynaceae
    A   B   C   D   M   M   M   M   N
    01   01   09   01   01   06   13   21   09
OC: Flies H-5, I-? (105)
AP: Alk=I-4 (1392)

Carissa opaca   (Natal plum genus)   Apocynaceae
   A C D M
   01 09 01 21
OC: Colletotrichum falcatum H-14, I-7 (1047)   Pyricularia oryzae H-14, I-7 (1047)
AP: Tan=I-7 (1047)

Carlina acaulis   (Carline thistle)   Asteraceae
   A B C D D M O
   01 06 09 03 08 05 02
OC: Dermestres maculatus H-3, 4, I-? (544)   Graphosoma mellonella H-3, 4, I-? (544)
    Dysdercus cingulatus H-3, 4, I-? (544)   Pyrrhocoris apterus H-3, 4, I-? (544)
    Graphosoma italicum H-3, 4, I-? (544)

Carpesium abrotanoides   (Not known)   Asteraceae
   C
   09
OC: Agrotis sp. H-1, I-6, 7, 9 (1241)   Puccinia rubigavera H-14, I-6, 7, 9 (1241)
    Phytophthora infestans H-14, I-6, 7, 9   Verticillium albo-atrum H-14, I-6, 7, 9
     (1241)   (1241)
    Pieris rapae H-1, I-6, 7, 9 (1241)
    Puccinia graminis tritici H-14, I-6, 7, 9 (1241)

Carpinus tschonoskii   (Ironwood genus)   Betulaceae
   A B C D
   01 01 09 03
OC: Erwinia carotovora H-18, I-15 (917)   Pseudomonas syringae H-18, I-15 (917)
    Fusarium oxysporum f. sp. mori H-14, I-15 (917)

Carthamus oxycantha   (Wild safflower)   Asteraceae
   A B D M M
   01 06 05 06 28
OC: Nematodes H-16, I-? (100)

Carthamus tinctorius   (Safflower)   Asteraceae
   A B C D D J M M M M M M N N N
   03 06 09 01 06 13 01 02 06 12 16 20 09 10 12
OC: Aphelencoides besseyi H-16, I-2, 8   Nematodes H-16, I-? (100)
     (52, 187, 377)   Phytophthora drechsleri H-14, I-? (1137)
    Hoplolaimus indicus H-16, I-? (612)
OR: H-15, 19, I-? (411)
AP: Alk=I-9 (1372)

Carum bulbocastanum   (Caraway genus)   Apiaceae
   B B M M M N N N
   03 06 01 06 16 02 09 10
OC: Clothes moths H-5, I-? (105)

Carum carvi   (Caraway genus)   Apiaceae
   A A B D D D J J M M M N O
   02 03 06 02 03 06 09 10 01 05 14 02 12
OC: Cochliomyia americana H-2, I-12 (244)   Pediculus humanus humanus H-2, I-10, 12
    Mosquitoes H-2, I-10 (98, 105)   (105)
OR: H-15, 19, I-10 (411)

Carum copticum   (Caraway genus)   Apiaceae
   A   B   C   D   J   J   K   M   M   O   O
  03 07 09 10 04 11 05 05 14 10 12
OC: Callosobruchus analis H-2, I-10 (179)   Sitotroga cereallela H-2, I-10 (179)
OR: H-15, 19, I-10 (411); H-20, I-10 (504)

Carya glabra   (Hickory)   Juglandaceae
   A   B   C   D   K   M   M   N
  01 01 09 03 08 01 04 10
OC: Flies H-5, I-7 (105)

Carya ovata   (Hickory)   Juglandaceae
   A   B   C   D   J   M   N
  01 01 09 03 13 01 10
OC: Scolytus multistriatus H-4, I-4 (1006)
OR: H-4, I-? (132)

Caryopteris divaricata   (Bluebeard genus)   Verbenaceae
   A   B   C   D   J   M   M   M   M   O   O
  01 02 09 10 13 05 12 13 21 07 10
OC: Spodoptera litura H-4, I-6, 7 (378, 867, 1065)
OR: H-4, I-? (132)
AP: Alk=I-6, 7 (1374)

Cassia absus   (Four-leaf senna)   Caesalpiniaceae
   B   C   D   D   G   J   J   J   K   M   M   M   O   O
  02 09 01 02 04 02 04 11 05 05 12 21 07 10
OC: Tribolium castaneum H-5, I-10 (179)
OR: H-15, I-10 (504, 1351)
AP: Alk=I-10 (chaksine, isochaksine)(1392); Tan=I-7 (1351)

Cassia alata   (Ringworm cassia)   Caesalpiniaceae
   A   B   C   D   D   J   M   M   M   M   O   O   O
  01 02 09 01 02 3a 05 12 13 21 02 07 10
OC: Ants H-5, I-? (87)   Ticks H-13, I-4 (105)
    Mosquitoes H-2, I-7 (78, 101)
OR: H-17, I-10 (504); H-20, I-7 (504); H-32, I-? (220)
AP: Alk=I-7 (1392)

Cassia auriculata   (Cassia)   Caesalpiniaceae
   A   B   C   D   J   J   M   M   M
  01 02 09 01 06 09 12 13 21
OC: Cochliomyia americana H-1, I-12 (244)   Lenzites trabea H-14, I-? (428)
    Fusarium nivale H-14, I-14 (1048)   Polyporus versicolor H-14, I-? (428)
    Lentinus lepideus H-14, I-? (428)   Schistocerca gregaria H-4, 5, I-7 (611)
OR: H-15, I-12 (908); H-19, I-4, 12 (908, 1052); H-22, I-1, 2, 4 (881, 908, 1040, 1052);
    H-24, I-2 (1184)
AP: Tan=I-4 (507); Tan=I-6, 7, 8, 9 (1360)

Cassia didymobotiya   (Not known)   Caesalpiniaceae
   B   C   D   D   J   J   M
  02 09 01 02 03 08 12
OC: Aphids H-2, I-10 (105)   Toxoptera aurantii H-2, I-7, 10 (681)
OR: H-32, I-10 (105)

Cassia fistula (Golden-shower) Caesalpiniaceae
A B C D J J M M M M M N O O
01 01 09 01 02 09 01 05 12 13 21 09 09 11
OC: Callosobruchus chinensis H-1, I-7 (95) Phytophthora parasitica H-14, I-? (911)
Colletotrichum falcatum H-14, I-? (1047) Pyricularia oryzae H-14, I-? (1047)
Dacus dorsalis H-6, I-8 (329, 473) Reticulitermes flavipes H-2, I-5 (87)
Meloidogyne javanica H-16, I-7 (922)
OR: H-15, I-7 (1321); H-22, I-4, 6, 11 (881, 1190, 1287)
AP: Alk, Sap, Tan=I-9 (1360); Sap=I-10 (1360); Str, Tan=I-7 (1359); Tan=I-7 (1321)

Cassia hirsuta (Shower tree genus) Caesalpiniaceae
A B C D J L M
01 02 09 01 03 03 12
OC: Mosquitoes H-2, I-? (78) Musca domestica H-2, I-? (78)
OR: H-2, I-? (101); H-32, I-? (105)

Cassia javanica (Apple blossom cassia) Caesalpiniaceae
A B C D J M M M M O
01 01 09 01 09 04 05 12 21 11
OC: Rats H-26, I-16 (1393)
AP: Tan=I-4 (504)

Cassia laevigata (Smooth senna) Caesalpiniaceae
A C D M M M N O
01 09 01 01 05 12 10 10
OC: Aphids H-2, I-7 (772)
AP: Alk=I-7 (1392)

Cassia multijuga (Shower tree genus) Caesalpiniaceae
A B C D D M
01 01 09 01 02 12
OC: Aphids H-2, I-? (772)

Cassia obtusifolia (Sicklepod) Caesalpiniaceae
A B D J M
01 09 01 03 12
OC: Heliothis virescens H-5, I-7 (1240)

Cassia occidentalis (Coffee senna) Caesalpiniaceae
A B C D J M M M N O
01 06 09 10 09 01 05 12 10 10
OC: Amaranthus spinosus H-24, I-10 (159) Pediculus humanus humanus H-1, I-? (87)
Meloidogyne javanica H-16, I-7 (922, 1184)
OR: H-24, I-? (1184)
AP: Alk=I-6, 7, 8 (1373); Alk=I-10 (1391); Sap=I-2, 6, 7 (1360)

Cassia spectabilis (Shower tree genus) Caesalpiniaceae
A B C D M
01 01 09 01 12
OC: Mosquitoes H-2, I-7 (78)
AP: Alk=I-7, 8, 9 (1392)

Cassia stipulacea   (Shower tree genus)   Caesalpiniaceae
    C M
    09 12
OR: H-2, I-7 (104)

Cassia tora   (Coffee weed)   Caesalpiniaceae
    A B C D J J J K L M M M M N N O
    03 06 09 10 02 03 08 05 05 01 05 12 16 07 10 02
OC: Autographa brassicae H-2, I-6 (78)   Pyricularia oryzae H-14, I-6 (1047)
    Colletotrichum falcatum H-14, I-6 (1047)
OR: H-17, 20, I-1, 2, 7, 10 (1349, 1353); H-19, I-7 (982); H-22, I-1 (881, 1040)
AP: Tan=I-? (1047)

Cassine xylocarpum   (Not known)   Celastraceae
    A B C D D J
    01 01 09 01 02 03
OC: Attagenus piceus H-4, I-4 (48)

Cassytha ciliolata   (Not known)   Lauraceae
    B B C D
    04 10 09 01
OR: H-2, I-? (87)

Cassytha filiformis   (Not known)   Lauraceae
    B B C D J M M N
    04 10 09 01 08 01 16 09
OC: Vermin H-27, I-? (73)
OR: H-2, I-? (105, 220)
AP: Alk=I-? (laurotetanine)(1392); Alk=I-6 (1366); Alk, Fla=I-6 (1377)

Castanea crenata   (Japanese chestnut)   Fagaceae
    A B C D M M M N
    01 01 09 03 01 04 19 10
OC: Fusarium roseum H-14, I-15 (917)
    Fusarium solani f. sp. mori H-14, I-15 (917)

Castanea dentata   (American chestnut)   Fagaceae
    A B C D M N
    01 01 09 03 01 09
OC: Popillia japonica H-1, I-1 (105, 1243)

Castanea mollissima   (Chinese chestnut)   Fagaceae
    A B C D J J J M M N
    01 01 09 03 03 07 08 01 21 10
OC: Endothia parasitica H-14, I-4 (947)
AP: Tan=I-4 (947)

Castanea sativa   (Spanish chestnut)   Fagaceae
    A B C D J M M N N
    01 01 09 06 09 01 19 09 10
OC: Venturia inaequalis H-14, I-? (16)
OR: H-19, I-9, 15 (1101); H-22, I-4, 6 (881, 1040)

Castela texana (Amargoso) Simaroubaceae
```
B   C   D   J   J   M   O
02  09  02  03  08  05  04
```
OC: Attagenus piceus H-4, I-2 (48)

Catalpa ovata (Chinese catawba) Bignoniaceae
```
A   B   C   D
01  01  09  03
```
OC: Leafhoppers H-1, I-4, 7 (1241)   Rice borers H-1, I-4, 7 (1241)

Catalpa speciosa (Western chestnut) Bignoniaceae
```
A   B   C   D   J   M
01  01  09  03  09  19
```
OC: Alternaria sp. H-14, I-5 (638)   Lymantria dispar H-4, I-7 (829)
    Fusarium sp. H-14, I-5 (638)   Rhizoctonia solani H-14, I-5 (638)
    Helminthosporium sp. H-14, I-5 (638)   Rhizopus nigricans H-14, I-5 (638)
AP: Alk=I-6, 7 (1370)

Catharanthus roseus (Rose periwinkle) Apocynaceae
```
A   B   C   D   J   J   L   M   M   M   O
01  06  09  01  03  06  02  05  12  13  07
```
OC: Drechslera oryzae H-14, I-7 (113)   Meloidogyne javanica H-16, I-2, 7, 8
    Dysdercus cingulatus H-2a, I-9 (88, 124)   (100, 616)
    Meloidogyne incognita H-16, I-2, 7, 8   Spodoptera littoralis H-4, I-7 (938)
      (100, 616)   Tryporyza incertulas H-1, I-? (95)
OR: H-17, 34, I-2 (1318)
AP: Alk=I-1 (catharanthine, leurosine, lochnericine, vindoline, vindolinine, virosine)
    (1392); Alk=I-2 (1368); Alk=I-2 (ajmalicine, akuammine, alstonine, lochenerine,
    reserpine, serpentine, tetrahydroalstonine, vincaleucoblastine, vincamine, vinceine,
    yohimbine)(1392); Alk=I-2, 6, 7 (1372, 1392); Alk=I-6, 7 (1318); Alk=I-7 (vinceine)
    (1392); Gly=I-6, 7 (1318)

Caulophyllum thalicitroides (Papooseroot) Berberiaceae
```
B   C   D   J   J   J   M   O
06  09  03  03  08  11  05  02
```
OC: Attagenus piceus H-4, I-2 (48)
AP: Alk=I-2 (caulophylline, N-methylcytisine)(1392)

Cayoponia ficifolia (Not known) Cucurbitaceae
```
B   C   D
04  09  02
```
OC: Attagenus piceus H-4, I-6, 7, 9 (48)

Cayratia japonica (Not known) Vitaceae
```
A   B   C
01  03  09
```
OC: Aphids H-1, I-1 (1241)   Pieris rapae H-1, I-1 (1241)

Ceanothus americanus (White snowball) Rhamnaceae
```
A   B   C   D   J   M   M   N   O   O
01  02  09  03  08  01  05  07  02  04
```
OC: Popillia japonica H-4, I-7, 8 (105, 1243)
AP: Alk=I-2 (coenothine)(1392); Alk=I-6, 7, 8 (1392)

68

Cecropia mexicana   (Not known)   Moraceae
   A   B   C   D   J
   01 01 09 01 03
OC: Spodoptera litura H-2a, I-1 (101)

Cedrela ciliata   (Not known)   Meliaceae
   A   B   C   D
   01 01 09 01
OC: Epilachna varivestis H-1, I-? (143)   Hypsipyla grandella H-1, I-? (143)

Cedrus deodora   (Himalayan cedar)   Pinaceae
   A   B   C   D   M   M   O
   01 01 09 06 04 05 12
OC: Termites H-8, I-5 (1352)   Tobacco mosaic virus H-21, I-7 (184)
OR: H-20, I-12 (1352)
AP: Alk=I-10 (1373)

Celastrus angulatus   (Bitter tree)   Celastraceae
   A   B   B   C   D   J   J   J   K   K
   01 02 03 09 10 03 04 11 05 06
OC: Caterpillars H-2b, I-2, 5 (101)   Malacosoma neustria H-1, I-? (1012)
   Colaphellus bowringi H-2, I-2 (105)   Phaedon brassicae H-2, I-2, 7 (105)
   Hymenia recurvalis H-2, I-2 (105, 1252) Plagiodera versicolora H-1, I-? (1012)
   Locusta migratoria H-2, I-2 (105, 1012) Spodoptera eridania H-1, I-? (1252)
   Locusts H-2, I-2 (101)   Udea rubigalis H-1, I-? (1252)

Celastrus orbiculatus   (Oriental bittersweet)   Celastraceae
   A   B   C   D   M
   01 03 09 10 13
OC: Anomala cupripes H-4, I-? (162)   Tryporyza incertulas H-1, I-4 (162)
   Hymenia recurvalis H-2, I-2 (101)   Udea rubigalis H-2, I-2 (101)
AP: Alk=I-10 (1370)

Celastrus scandens   (American bittersweet)   Celastraceae
   A   B   C   D   J   J   M   N
   01 03 09 03 02 03 01 04
OC: Aedes aegypti H-1, I-7 (165)   Erwinia carotovora H-18, I-9, 14 (585)

Celosia argentea   (Woolflower genus)   Amaranthaceae
   A   B   C   D   J   L   M   M   M   N   N   O
   03 04 09 01 03 05 01 05 13 07 15 10
OR: H-24, I-2, 10 (1184, 1210)
AP: Alk=I-7 (1390, 1392); Alk=I-10 (1372); **Tri**=I-7 (1390)

Celosia plumosa   (Woolflower genus)   Amaranthaceae
   B   C   D   J
   04 09 01 03
OC: Tobacco necrotic virus H-21, I-7 (1316)

Celtis laevigata   (Mississippi hackberry)   Ulmaceae
   A   B   C   D   J
   01 01 09 03 03

OC: Andropogon gerardi H-24, I-7 (576)   Schizachyrium scoparium H-24, I-7 (576)
    Panicum virgatum H-24, I-7 (576)   Sorghastrum nutans H-24, I-7 (576)
OR: H-24, I-7 (1207, 1214, 1219)

Celtis occidentalis   (Hackberry)   Ulmaceae
    A   B   C   D   M   M   M   N
    01 01 09 03 01 04 16 09
OC: Phyllobius oblongus H-4, I-7 (531)
OR: H-24, I-7 (1207)

Cenchrus ciliaris   (Not known)   Podaceae
    B   C   D   D   J   M
    06 09 01 04 03 02
OR: H-24, I-2 (1203)

Centaurea diffusa   (Diffuse knapweed)   Asteraceae
    A   A   B   C   D   D   L
    01 03 06 09 03 06 05
OR: H-24, I-? (919)
AP: Alk=I-? (1392)

Centaurea maculosa   (Spotted knapweed)   Asteraceae
    A   A   B   C   D   J
    01 02 06 09 03 14
OR: H-17, I-7 (622); H-24, I-? (919)
AP: Alk=I-2, 6, 7, 8 (1392)

Centaurea nigra   (Black knapweed)   Asteraceae
    A   B   C   D   M   O
    01 06 09 03 05 07
OC: Locusta migratoria H-4, 5, I-7 (596)

Centaurea pannonica   (Knapweed genus)   Asteraceae
    B   C   D
    06 09 03
OC: Leptinotarsa decemlineata H-4, I-7 (531)   Pieris brassicae H-4, I-7 (531)
    Phytodecta fornicata H-4, I-7 (531)

Centaurea repens   (Russian knapweed)   Asteraceae
    B   C   D
    06 09 03
OR: H-24, I-? (919)

Centaurium erythraea   (Centaury gentian)   Gentianaceae
    A   B   C   D   D   J   M   O
    01 06 09 03 06 03 05 08
OC: Lice H-2, I-1 (105)

Centella asiatica   (Indian hydrocotyle)   Apiaceae
    A   B   B   C   D   J   L   M   M   M   M   N   O   O
    01 04 05 09 01 04 03 01 05 09 21 07 07 14
OC: Aphis fabae H-2, I-? (101)   Epilachna varivestis H-2, I-? (101)
OR: H-33, I-? (220)
AP: Alk, Sap, Tan=I-6, 7 (1318)

Centipeda orbicularis　　(Not known)　　Asteraceae
　　B　C　D
　　06　09　01
OR: H-2, I-? (1276)

Centratherum anthelminticum　　(Kinka oil ironweed)　　Asteraceae
　　A　B　D　D　J　M　M　O
　　03　06　01　02　09　05　13　07
OR: H-2, I-? (220); H-22, I-10 (1395)

Centrosema pubescens　　(Centrosema)　　Fabaceae
　　B　C　D　J　M　M
　　04　09　01　03　10　12
OC: Meloidogyne incognita H-16, I-7 (115)
AP: Alk=I-6 (1392)

Centrosema virginianum　　(Butterfly pea)　　Fabaceae
　　A　B　C　D　J　M　M
　　01　06　09　01　03　12　17
OC: Oncopeltus fasciatus H-2, I-1 (48)

Cephalaria syriaca　　(Not known)　　Dipsacaceae
　　B　C　D　J　J　J
　　06　09　06　02　03　04
OC: Tylenchulus semipenetrans H-16, I-10 (540)

Cephalotaxus fortunii　　(Chinese plum yew)　　Cephalotaxaceae
　　A　B　C　D
　　01　02　09　03
OC: Oncopeltus fasciatus H-2, I-10 (958)

Cephalotaxus harringtonia　　(Harrington plum yew)　　Cephalotaxaceae
　　B　B　C　J
　　01　02　09　03
OC: Drosophila hydei H-2, I-6, 7 (101)

Ceratonia siliqua　　(Carob)　　Caesalpiniaceae
　　A　B　C　D　D　M　M　M　M　M　N　N
　　01　01　09　02　06　01　02　04　12　19　10　11
OR: H-2, I-? (105, 1276)

Ceratophyllum demersum　　(Hornwort genus)　　Ceratophyllaceae
　　B　B　C　D　J
　　06　09　09　03　16
OC: Fusarium roseum H-14, I-1 (837)
OR: H-19, I-1 (837)

Ceratotheca integribracteata　　(Not known)　　Pedaliaceae
　　B　C　D
　　06　09　01
OR: H-2, I-? (104)

71

Ceratotheca sesamoides   (Not known)   Pedaliaceae
    B   C   D   J   J   J   M   M   N
    06 09 01 03 08 11 01 28 07
OC: Attagenus piceus H-4, I-1, 10 (48)

Ceropegia dichotoma   (Not known)   Asclepiadaceae
    A   B   C   D   J   M
    01 02 09 02 03 13
OC: Blattella germanica H-2, I-6 (48)

Cestrum cuneatum   (Tinto)   Solanaceae
    A   B   C   D   J   K
    01 02 09 01 01 01
OR: H-2, I-1 (118)

Cestrum diurnum   (Day jasmine)   Solanaceae
    A   B   B   C   D   J   J   J   M
    01 01 02 09 01 03 06 09 13
OC: Alternaria brassicae H-14, I-7, 12 (858)   Fusarium oxysporum H-14, I-7, 12 (858)
    Alternaria tenuis H-14, I-7, 12 (858)   Fusarium solani H-14, I-7, 12 (858)
    Amaranthus spinosus H-24, I-7 (159)   Phoma sp. H-14, I-7, 12 (858)
    Colletotrichum capsici H-14, I-7, 12 (858)   Pyricularia oryzae H-14, I-7, 12 (858)
    Colletotrichum gloeosporides H-14, I-7, 12   Rhizoctonia bataticola H-14, I-7, 12 (858)
      (858)   Rhizoctonia solani H-14, I-7, 12 (858)
    Curvularia lunata H-14, I-7, 12 (858)   Ustilago hordei H-14, I-7, 12 (181)
    Drechslera graminea H-14, I-7, 12 (858)   Ustilago tritici H-14, I-7, 12 (181)
    Fusarium moniliforme H-14, I-7, 12 (858)

Cestrum occidentalis   (Not known)   Solanaceae
    A   B   B   C   D   J
    01 01 02 09 01 09
OC: Amaranthus spinosus H-24, I-7 (159)

Chaenactis douglasii   (Not known)   Asteraceae
    A   A   B   C   D   D   J   J
    01 02 06 09 02 03 03 11
OC: Attagenus piceus H-4, I-1 (48)

Chamaecyparis formosensis   (Japanese false cypress)   Cupressaceae
    A   B   C   D   M   M
    01 01 09 03 04 17
OC: Termites H-2, I-? (143)

Chamaecyparis funebris   (Mourning cypress)   Cupressaceae
    A   B   C   D   J
    01 01 09 03 09
OC: Rats H-26, I-16 (1393)

Chamaecyparis lawsoniana   (Lawson cypress)   Cupressaceae
    A   B   C   D   J   J   M   M   M   O
    01 01 09 03 3a 14 04 05 13 13
OC: Tenebrio molitor H-3, I-10 (958)   Tobacco mosaic virus H-21, I-7 (184)

Chamaemelum nobile   (Common chamomile)   Asteraceae
   A   B   C   D   D   M
   01 05 09 03 06 17
OR: H-2, I-8 (104, 105)

Chara foetida   (Not known)   Characeae
   A   B   C
   04 09 02
OC: Mosquitoes H-2, I-1 (105)

Chara fragilis   (Algae)   Characeae
   A   B   C   J   J
   04 09 02 01 02
OC: Mosquitoes H-2, I-1 (105)

Chara vulgaris   (Algae)   Characeae
   A   C   J   J   J
   04 02 01 02 16
OC: Fusarium roseum H-14, I-1 (837)

Cheiranthus cheiri   (Wallflower)   Brassicaceae
   A   B   C   D   J   J   L   M   M   O
   01 06 09 06 03 04 03 05 13 17 08
OC: Phaedon cochleariae H-4, I-7 (1134)   Phyllotreta undulata H-4, I-7 (1134)
   Phyllotreta tetrastigma H-4, I-7 (1134)
OR: H-19, I-1 (581)
AP: **Alk**=I-7, 8, 9, 10 (cheiroline)(1392); **Alk**=I-10 (cheirinine)(1392); **Glu**=I-12 (cheiranthin)
   (504)

Cheiridopsis aspera   (Lobster-claws)   Aizoaceae
   C   D
   09 02
OC: Tobacco mosaic virus H-21, I-7 (184)

Chelidonium majus   (Celandine)   Papaveraceae
   A   A   B   C   J   M
   01 02 06 09 09 04
OC: Athalia rosae H-4, I-7 (531)   Fusarium oxysporum
   Botrytis cinerea H-14, I-2 (946)   f. sp. lycopersici H-14, I-2 (946)
   Ceratocystis ulmi H-14, I-2, 8, 16   Hyphantria cunea H-4, I-7 (531)
    (945, 946)   Pieris brassicae H-4, I-7 (531)
AP: **Alk**=I-2, 6, 7, 16 (allocryptopine, berberine, chelerythrine, chelidamine, chelidonine,
   chelilutine, chelirubine, coptisine, homochelidonine, methoxychelidonine, oxychelidonine,
   protopine, sanguinarine, sparteine, stylopine, tetrahydrocoptisine)(1392); **Alk**=I-6, 7
   (1373); **Alk**=I-6, 7, 8 (1371)

Chelone glabra   (Turtle head)   Scrophulariaceae
   A   B   C   D   E   J   J   J   M   M   O
   01 06 09 03 11 02 03 08 02 05 07
OC: Popillia japonica H-5, I-7 (105, 1243)
OR: H-19, I-? (592)

Chenopodium album   (Lamb's-quarters)   Chenopodiaceae
```
A   B   C   D   D   F   J   L   M   N   N
03  06  09  01  03  07  09  05  01  07  10
```
OC: Alternaria solani H-14, I-1 (944)   Phyllobius oblongus H-4, I-7 (531)
    Athalia rosae H-4, I-7 (531)   Phytodecta fornicata H-4, I-7 (531)
    Colletotrichum lindemuthianum H-14, I-14   Pieris brassicae H-4, I-7 (531)
      (1105)   Potato virus X H-21, I-1, 14 (1136)
    Leptinotarsa decemlineata H-5, I-7 (548)   Tobacco mosaic virus H-21, I-7 (1061, 1313)
    Monilinia fructicola H-14, I-14 (1105)   Tobacco necrotic virus H-21, I-7 (1316)
AP: **Alk**=I-? (chenopodine)(1392); **Alk**=I-6, 7 (1392)

Chenopodium amaranticolor   (Pigweed genus)   Chenopodiaceae
```
A   B   C   D   J   L   M   N
03  06  09  03  3a  05  01  07
```
OC: Potato virus X H-21, I-14 (1136)   Tobacco necrotic virus H-21, I-7, 14
    Tobacco mosiac virus H-21, I-14 (1312, 1313)   (1311, 1316)

Chenopodium ambrosioides   (Mexican tea)   Chenopodiaceae
```
A   A   B   C   D   J   J   J   M   M   M   O
01  03  06  09  10  03  06  11  05  09  21  12
```
OC: Attagenus piceus H-4, I-1 (48)   Popillia japonica H-1, I-1, 12 (105, 1090)
    Cochliomyia hominivorax H-5, I-12 (105)   Sunnhemp rosette virus H-21, I-7 (1314)
    Meloidogyne incognita H-16, I-7 (542)   Tobacco mosaic virus H-21, I-7 (1314)
    Mosquitoes H-2, I-10, 12 (105)
OR: H-17, I-7, 9, 12 (220, 1321, 1348); H-19, I-7, 12 (906, 982); H-24, I-9 (569)
AP: **Alk**=I-2 (1379); **Alk**=I-6 (1378); **Sap**=I-6, 7 (1321); **Tan**=I-9, 10 (1391)

Chenopodium botrys   (Feather geranium)   Chenopodiaceae
```
A   B   C   D   L
03  06  09  03  05
```
OC: Moths H-5, I-? (87)
OR: H-17, 33, I-? (220)

Chenopodium ficifolium   (Pigweed genus)   Chenopodiaceae
```
B   C   D   J
06  09  03  03
```
OC: Tobacco necrotic virus H-21, I-7 (1316)

Chenopodium foetidum   (Pigweed genus)   Chenopodiaceae
```
A   B   C   D
01  06  09  03
```
OC: Ants H-2, I-10 (504)

Chenopodium hybridum   (Pigweed genus)   Chenopodiaceae
```
B   C   D   L
06  09  03  05
```
OC: Leptinotarsa decemlineata H-4, I-7 (531)   Pieris brassicae H-4, I-7 (531)
    Phytodecta fornicata H-4, I-7 (531)

Chenopodium opulifolium   (Pigweed genus)   Chenopodiaceae
```
B   C   D   J   L
06  09  03  03  05
```
OC: Tobacco necrotic virus H-21, I-7 (1316)

Chenopodium quinoa (Quinoa) Chenopodiaceae
    B   C   D   D   L   M   N
    06 09 01 08 05 01 10
OC: Apple chlorotic leaf spot virus H-21, I-7 (1309)

Chenopodium rubrum (Pigweed genus) Chenopodiaceae
    B   C   D   J   L   M   N
    06 09 03 03 05 01 07
OC: Tobacco necrotic virus H-21, I-7 (1316)

Chenopodium urbicum (Pigweed genus) Chenopodiaceae
    B   C   D   J   L
    06 09 03 03 05
OC: Tobacco necrotic virus H-21, I-7 (1316)

Chenopodium vulvaria (Pigweed genus) Chenopodiaceae
    B   C   D   L
    06 09 03 05
OC: Hyphantria cunea H-4, I-7 (531)   Pieris brassicae H-4, I-7 (531)

Chloranthus japonicus (Not known) Chloranthaceae
    B   C   D
    02 09 03
OC: Aphids H-1, I-7 (1247)

Chloris gayana (Rhodes grass) Poaceae
    A   B   C   D   M
    01 06 09 02 02
OR: H-24, I-7 (1247)

Chlorogalum pomeridianum (Soap plant) Liliaceae
    A   B   C   D   J   J   M   M
    01 06 09 03 02 14 03 15
OC: Ceratitis capitata H-6, I-3 (956)   Pseudaletia unipuncta H-1, I-3 (101, 1252)
    Diaphania hyalinata H-2, I-3 (101, 185, 1252)

Chlorophora tinctoria (Fustic) Moraceae
    A   B   C   D   M   M
    01 01 09 01 16 19
OC: Popillia japonica H-5, I-1 (105, 1243)

Chromolaena odorata (Hagonoy) Asteraceae
    C   J
    09 03
OC: Meloidogyne incognita H-16, I-7 (115)

Chrysanthemum balsamita (Mint geranium) Asteraceae
    A   B   D   M   M
    01 06 03 14 17
OC: Mosquitoes H-2, I-6, 7 (99)
OR: H-2, I-6, 7 (87)

Chrysanthemum caucasium   (Not known)   Asteraceae
     B   C
     06 09
OR: H-2, I-? (105)

Chrysanthemum cinerariifolium   (Pyrethrum)   Asteraceae
     A   B   C   D   E   F   F   F   J   J   J   J   J   J   J   K   K   L   M
     01 06 09 10 03 01 03 05 02 03 04 07 08 10 15 05 06 03 13
OC: Acyrthosiphum pisum H-2a, I-8 (1)   Laspeyresia molesta H-2a, I-8 (209)
    Aedes aegypti H-2a, I-8 (290)   Laspeyresia pomonella H-2a, I-8 (250, 265)
    Anopheles quadrimaculatus H-2a, I-8 (290)   Leptinotarsa decemlineata H-2a, I-8 (104)
    Antestiopsis lineaticollis H-2a, I-8   Leucinodes orbonalis H-2a, I-8 (126)
       (571, 573, 636, 638)   Lipaphis erysimi H-2a, I-8 (206)
    Aphids H-2a, I-8 (104)   Lygus elisus H-2a, I-8 (268)
    Aphis fabae H-2a, I-8 (41, 195, 464, 492)   Lygus hesperus H-2a, I-8 (268)
    Aphis pomi H-2a, I-8 (213)   Macrosiphoniella sanborni H-2a, I-8 (195)
    Aphis sorbi H-2a, I-8 (213)   Macrosiphum rosae H-2a, I-8 (195)
    Aphis spiraecola H-2a, I-8 (221)   Macrosteles divisus H-2a, I-8 (241)
    Asphondylia sp. H-2a, I-8 (104, 126)   Malacosoma americana H-2a, I-8 (195)
    Autographa brassicae H-2a, I-8 (224, 232)   Mamestra picta H-2a, I-8 (237)
    Bombyx mori H-2a, I-8 (104)   Mineola scitulella H-2a, I-8 (214)
    Brevicoryne brassicae H-2a, I-8 (221)   Mosquitoes H-2a, I-8 (117)
    Cladius pectinicornis H-2a, I-8 (195)   Murgantia histrionica H-2a, I-8 (224)
    Cockroaches H-2a, I-8 (104)   Musca domestica H-2a, I-8 (248, 266, 812)
    Diabrotica duodecimpunctata H-2a, I-8 (206)   Ophiomyia reticulipennis H-2a, I-8 (126)
    Diabrotica punctata H-2a, I-8 (260)   Periphyllus lyropictus H-2a, I-8 (221)
    Doryphora 10-lineata H-2a, I-8 (671)   Phaedon cochleariae H-2a, I-8 (1)
    Earias fabia H-2a, I-8 (126)   Phyllotreta vittata H-2a, I-8 (221)
    Empoasca devastans H-2a, I-8 (93, 126, 162)   Pieris brassicae H-2a, I-8 (1)
    Empoasca fabae H-2a, I-8 (367)   Pieris rapae H-2a, I-8 (75, 195, 671)
    Endelomyia rosae H-2a, I-8 (195)   Plodia interpunctella H-2a, I-8 (32)
    Ephestia elutella H-2a, I-8 (17, 32)   Plutella xylostella H-2a, I-8 (147, 223)
    Epicauta pennsylvanica H-2a, I-8 (250)   Pteronus ribesii H-2a, I-8 (195)
    Epilachna varivestis H-2a, I-8 (206)   Pyrausta nubilalis H-2a, I-8 (210)
    Eriosoma tesselatum H-2a, I-8 (671)   Sitophilus oryzae H-2a, I-8 (598)
    Erythroneura comes H-2a, I-8 (221)   Sitotroga cerealella H-2a, I-8 (456)
    Eutettix tenellus H-2a, I-8 (249, 486)   Tetranychus cinnabarinus H-11, I-8
    Flies H-2a, I-8 (104, 117)   (91, 205, 215)
    Galerucella luteola H-2a, I-8 (221)   Thermobia domestica H-2a, I-8 (290)
    Gargaphia solani H-2a, I-8 (195)   Thrips H-2a, I-8 (147)
    Gnorimoschema lycopersicella H-1, I-? (228)   Toxoptera aurantii H-2a, I-8 (681)
    Grasshoppers H-2a, I-8 (104)   Wireworms H-2a, I-8 (207)
OR: H-2, I-8 (3, 7, 8, 11, 19, 20, 28, 34, 35, 42, 71, 80, 85, 93, 180, 220, 656, 764, 893,
    1088, 1091, 1099, 1122); H-2a, I-19 (144); H-30, I-9 (101, 110, 465, 818)
AP: Alk=I-? (stachydrine)(1392)

Chrysanthemum coccineum   (Painted daisy)   Asteraceae
     A   B   C   D   D   D   J   K   M
     01 06 09 02 03 06 02 05 13
OC: Manduca sexta H-4, 5, I-8 (669)   Panonychus citri H-4, 5, 11, I-8 (669)
OR: H-2a, I-8 (8, 28, 43, 87, 104, 504)

Chrysanthemum coronarium   (Crown daisy)   Asteraceae
    A   B   C   D   M   N   N
    03 06 09 03 01 07 08
OR: H-5, I-8 (1345)

Chrysanthemum corymbosum   (Not known)   Asteraceae
    A   B   C   D   J   J   K
    01 06 09 03 02 04 05
OR: H-2, I-8 (104)

Chrysanthemum frutescens   (Paris daisy)   Asteraceae
    A   B   C   D   J   J   J   K
    01 06 09 02 02 04 09 05
OR: H-2, I-8 (104); H-19, I-2, 15 (832, 1262)

Chrysanthemum indicum   (Winter aster)   Asteraceae
    A   B   C   D   J   K   M   M   M   N   O   O
    03 06 09 01 03 05 01 05 13 08 01 08
OC: Leptinotarsa decemlineata H-4, I-7 (531)
OR: H-15, I-8 (1320)

Chrysanthemum marshallianum   (Not known)   Asteraceae
    B   C
    06 09
OR: H-2a, I-8 (105)

Chrysanthemum morifolium   (Mum)   Asteraceae
    A   B   C   D   K   M
    01 02 09 03 05 13
OC: Meloidogyne indica H-16, I-? (999)   Pratylenchus alleni H-16, I-? (999)
OR: H-24, I-7 (1207)

Chrysanthemum parthenium   (Feverfew)   Asteraceae
    A   B   C   D   D   J   K   M
    01 06 09 03 06 02 05 14
OC: Cockroaches H-2a, I-8 (105)
OR: H-2, I-8 (504); H-19, I-6, 7, 8 (940)

Chrysanthemum segetum   (Corn marigold)   Asteraceae
    A   B   C   D   D   K
    03 06 09 03 06 05
OC: Venturia inaequalis H-14, I-? (16)
OR: H-2a, I-8 (104); H-19, I-? (580)

Chrysanthemum seticuspe   (Chrysanthemum genus)   Asteraceae
    B   C   D
    06 09 03
OR: H-5, I-8 (1345)

Chrysophyllum oliviforme   (Satin leaf)   Sapotaceae
    A   B   C   D   J   M   M   N
    01 01 09 01 03 01 19 09
OC: Drechslera oryzae H-14, I-7 (113)

Chrysopogon aucheri   (Not known)   Poaceae
   B   C   J
   06 09 03
OR: H-24, I-2 (1203)

Chrysopsis villosa   (Golden aster genus)   Asteraceae
   B   C   D   D
   07 09 03 04
OC: Melanoplus femurrubrum H-1, I-? (87)

Chrysosplenium flagelliferum   (Not known)   Saxifragaceae
   B   B   C   D   J   M
   05 06 09 03 03 09
OC: Drosophila hydei H-2, I-2, 6, 7 (101)

Chrysosplenium yesoense   (Not known)   Saxifragaceae
   B   B   C   D   D
   05 06 09 03 10
OC: Drosophila hydei H-1, I-2, 6, 7 (101)

Chrysothamnus nauseosus   (Rabbitbrush)   Asteraceae
   A   B   C   D   M
   01 02 09 03 26
OC: Leptinotarsa decemlineata H-1, I-7 (997, 1118)

Chrysothamnus viscidiflorus   (Rabbitbrush)   Asteraceae
   A   B   C   D   D   J
   01 02 09 03 04 03
OC: Agropyron spicatum H-24, I-7 (1174)   Stipa thurberiana H-24, I-7 (1174)
   Sitanion hystrix H-24, I-7 (1174)

Cibotium barometz   (Scythian-lamb)   Dicksoniaceae
   A   B   C   D   J   J
   01 01 06 01 03 3a
OC: Aphids H-1, I-1 (1241)   Spider mites H-11, I-1 (1241)

Cibotium chamissoi   (Hawaiian tree fern)   Dicksoniaceae
   A   B   C   D
   01 01 06 01
OC: Erwinia carotovora H-18, I-7 (854)   Xanthomonas campestris
   Pseudomonas solanacearum H-18, I-7 (854)   pv. phaseoli-sojensis H-18, I-7 (854)
OR: H-19, I-? (172)
AP: Alk=I-6, 7 (1362)

Cibotium glaucum   (Tree fern genus)   Dicksoniaceae
   A   B   C   D
   01 01 06 01
OC: Pseudomonas solanacearum H-18, I-7 (854)
   Xanthomonas campestris pv. phaseoli-sojensis H-18, I-7 (854)

Cibotium schieldei   (Mexican tree fern)   Dicksoniaceae
   A   B   C   D   M
   01 01 06 01 13
OC: Erwinia carotovora H-18, I-7 (854)   Xanthomonas campestris
   Pseudomonas solanacearum H-18, I-7 (854)   pv. phaseoli-sojensis H-18, I-7 (854)

Cichorium intybus   (Chicory)   Asteraceae
```
A   B   C   D   D   J   K   M   M   M   N   N   O
01  06  09  03  06  04  05  01  02  05  02  07  02
```
OC: Cassida nebulosa H-4, I-7 (531)   Monilinia fructicola H-14, I-14 (1105)
    Dermacentor marginatus H-13, I-8 (87, 1133) Phytodecta fornicata H-4, I-7 (531)
    Haemaphysalis punctata H-13, I-8 (87, 1133) Pieris brassicae H-4, I-7 (531)
    Ixodes redikorzevi H-13, I-8 (87, 1133)   Rhipicephalus rossicus H-13, I-8, 9
    Leptinotarsa decemlineata H-4, I-7 (531)   (87, 1133)
OR: H-19, I-1, 14 (704)

Cicuta maculata   (Beaver poison)   Apiaceae
```
A   B   C   D
01  06  09  03
```
OR: H-5, I-2 (1345)

Cimicifuga foetida   (Fetid bugbane)   Ranunculaceae
```
A   B   C   D   D   K   L   M   O
01  06  09  03  07  03  03  05  02
```
OC: "Bugs" H-5, I-2 (105, 1276)   Fleas H-5, I-2 (105)
OR: H-5, 32, 33, I-2 (220)

Cinchona calisaya   (Peruvian bark)   Rubiaceae
```
A   B   D   D   J   K   M   M   O
01  01  01  08  03  03  04  05  04
```
OC: Diaphania hyalinata H-2, I-2, 4, 5 (101)   Plutella xylostella H-2, I-2, 4, 5 (101)
OR: H-19, I-? (411); H-22, I-7 (882); H-32, 33, I-4 (73, 220)
AP: Alk=I-4 (1371, 1392); Alk=I-4 (aricine, chairamidine)(1371); Alk=I-4 (chairamine,
    cinchamidine, cinchonamine, cinchonicine, cinchonidine, cinchonine, cinchotine,
    conchairamidine, conchairamine, concusconine, conquinamine, cupreine, cusconine,
    dicinchonine, diconquinine, epiquinidine, epiquinine, hydrocinchonidine, hydroquinidine,
    hydroquinine, javanine, paricine, quinamine, quinicine, quinidine, quinine, sclerotium)
    (1392)

Cinchona officinalis   (Cascarilla)   Rubiaceae
```
A   B   C   D   D   J
01  01  09  01  08  03
```
OC: Clothes moths H-5, I-? (163)
OR: H-32, I-? (73, 220)
AP: Alk=I-2 (cinchonidine, cinchonine, javanine, quinidine, quinine)(1392)

Cinchona pubescens   (Not known)   Rubiaceae
```
A   B   C   D   D
01  01  09  01  08
```
OC: Moths H-5, I-4 (1297)
OR: H-32, 33, I-7 (73, 220)
AP: Alk=I-? (quinomine)(1392); Alk=I-4 (cinchonidine, cinchonine, cinchotine, conquinamine,
    dicinchonine, paricine, quinidine, quinine)(1392)

Cinnamomum camphora   (Camphor tree)   Lauraceae
```
A   B   C   D   D   J   M   M   M   O   O
01  01  09  02  03  06  05  17  19  04  07
```
OC: Aphids H-1, I-12 (119)   Cochliomyia hominivorax H-1, I-? (105)
    Chrysomya macellaria H-1, I-12 (199)   Pericallia ricini H-1, I-12 (943)
    Clothes moths H-5, I-? (105, 163, 220)   Pseudaletia unipuncta H-1, I-12 (119)
OR: H-1, I-? (73, 220); H-20, I-12 (399); H-30, I-? (101)

Cinnamomum cassia   (Chinese cinnamon)   Lauraceae
   A   B   C   D   M
   01 01 09 01 14
OC: Cochliomyia hominivorax H-5, I-? (105)   Cockroaches H-5, I-? (105)

Cinnamomum cecicodaphne   (Malagiri)   Lauraceae
   A   B   C   D   M   M
   01 01 09 08 17 19
OC: Termites H-1, I-? (1352)
OR: H-5, I-5 (504)

Cinnamomum merkadoi   (Kalingag)   Lauraceae
   A   B   C   D   J   K   M   M   O   O   O
   01 01 09 01 04 02 05 14 04 07 12
OC: Dacus dorsalis H-6, I-4 (1302)
OR: H-2, I-? (1302)

Cinnamomum zeylanicum   (Cinnamon)   Lauraceae
   A   B   C   D   M
   01 01 09 01 14
OC: Alternaria solani H-14, I-12 (522)   Drechslera oryzae H-14, I-12 (522)
   Bombyx mori H-4, I-4 (1063)   Fusarium solani H-14, I-12 (522)
   Callosobruchus maculatus H-1, I-12 (645)   Lentinus lepideus H-14, I-12 (428)
   Cochliomyia hominivorax H-1, I-? (105)   Lenzites trabea H-14, I-12 (428)
   Curvularia lunata H-14, I-12 (522)   Polyporus versicolor H-14, I-12 (428)
OR: H-19, I-12 (1161); H-20, I-12 (399)
AP: Str=I-4 (B-sitosterol)(1384)

Cirsium arvense   (Canadian thistle)   Asteraceae
   A   B   C   D   J   J
   01 06 09 03 03 08
OC: Cassida nebulosa H-4, I-7 (531)   Phytodecta fornicata H-4, I-7 (531)
   Hordeum distichon H-24, I-2, 7 (531)   Pieris brassicae H-4, I-7 (531)
   Lolium perenne H-24, I-2, 7 (534)   Trifolium subterraneum H-24, I-2, 7 (534)
   Nematodes H-16, I-? (100)
OR: H-24, I-6, 7, 8 (1392)
AP: Alk=I-6, 7 (1392)

Cirsium discolor   (Thistle genus)   Asteraceae
   B   C   D   J
   06 09 03 03
OR: H-24, I-16 (1254)

Cirsium ehrenbergii   (Thistle genus)   Asteraceae
   B   C   D   J
   06 09 03 03
OC: Spodoptera frugiperda H-2, I-1 (176)

Cirsium japonicum   (Japanese thistle)   Asteraceae
   A   B   C   D   M   M   N   O   O   O
   01 06 09 10 01 05 07 02 04 06 07
OC: Bursaphelenchus xylophilus H-16, I-2 (1249)

Cirsium lipskyi   (Not known)   Asteraceae
   B   C   D
   06 09 03
OC: Nematodes H-16, I-? (100)

Cissus producta   (Grape ivy genus)   Vitaceae
   B   C   D
   06 09 01
OC: Glossina sp. H-5, I-2 (1345)

Cissus rheifolia   (Ivy genus)   Vitaceae
   B   B   C   D   D
   02 03 09 01 02
OC: Ants H-5, I-? (80)

Cissus rhombifolia   (Venezuela treebine vine)   Vitaceae
   A   B   C   D   M   M
   01 04 09 01 03 09
OC: Plutella xylostella H-4, I-7 (87)
OR: H-2a, I-6 (118)

Citrofortunella mitis   (Panama orange)   Rutaceae
   A   B   C   D   M   M   N
   01 02 09 01 01 13 09
OC: Snails H-28, I-9 (662)
OR: H-32, I-9 (662)

Citrullus colocynthis   (Bitter gourd)   Cucurbitaceae
   A   B   B   D   D   E   J   J   J   M   M   M   N   O
   01 04 05 05 09 07 02 03 04 01 02 05 06 09 09
OC: Callosobruchus chinensis H-2, I-? (605)   Moths H-5, I-9 (1353)
   Caterpillars H-1, I-? (105)   Plutella xylostella H-2, I-2 (105)
   Coccus viridis H-1, I-9 (1029)   Rodents H-25, I-9 (1345)
   Epilachna sp. H-1, I-9 (1029)   Tylenchulus semipenetrans H-16, I-9 (540)
   Euproctis fraterna H-2, I-7 (105)   Vermin H-27, I-9 (1353)
OR: H-2, I-2 (605, 1276); H-2, I-9 (1345); H-33, I-? (220)
AP: Alk=I-9 (1392); Glu=I-9 (colocynthin)(1353)

Citrullus lanatus   (Watermelon)   Cucurbitaceae
   A   B   C   D   D   M   N
   03 04 09 01 02 01 09
OC: Pratylenchus zea H-16, I-? (1274)
OR: H-19, I-7 (982); H-22, I-10 (969)
AP: Alk=I-? (1391)

Citrus aurantiifolia   (Lime)   Rutaceae
   A   B   B   C   D   M   M   M   N   O
   01 01 02 09 02 01 05 15 09 07
OC: Callosobruchus maculatus H-1, I-12 (1124)   Pediculus humanus humanus H-1, I-7 (1353)
   Gloeosporium limetticola H-14, I-7 (680)
OR: H-17, I-7 (1353); H-19, I-9 (172)

Citrus aurantium   (Sour orange)   Rutaceae
   A   B   C   D   D   M   M   M   N
   01 01 09 01 02 01 13 17 09
OC: Callosobruchus phaseoli H-2, I-12 (1071)   Mosquitoes H-5, I-12 (105)
   Cochliomyia hominivorax H-5, I-12 (105)   Plutella xylostella H-4, I-7 (87)
   Dysdercus cingulatus H-3, I-7 (441)   Vermin H-5, 27, I-12 (105)
   Locusta migratoria H-4, 5, I-7 (596)
AP: Alk=I-7 (stachydrine)(1392); Alk=I-9 (narcotine)(1392)

- -

Citrus limon   (Lemon)   Rutaceae
   A   B   C   D   D   J   M   N
   01 01 09 01 02 06 01 09
OC: Ants H-5, I-12 (105, 193)   Ixodes sp. H-13, I-? (101)
   Anuraphis maidiradicis H-1, I-12 (193)   Lasius americanus H-2, I-9, 12 (193)
   Aphids H-1, I-12 (193)   Lice H-5, I-12 (105)
   Aphis fabae H-2, I-9, 12 (105)   Sitophilus oryzae H-2, I-9, 12 (330)
   Blattella germanica H-1, I-? (101)
   Callosobruchus maculatus H-2, I-9, 12 (330, 331, 1124)
AP: Alk=I-6, 7, 8 (1374)

- -

Citrus maxima   (Pomelo)   Rutaceae
   A   B   C   D   J   J   M   M   N
   01 01 09 02 03 06 01 13 09
OC: Acromyrmex cephalotes H-6, I-9, 12 (470)   Callosobruchus maculatus H-2, I-12
   Acromyrmex octospinosus H-6, I-9, 12 (470)   (330, 1124)
OR: H-19, I-4 (982)
AP: Alk=I-6, 7 (1374)

- -

Citrus reticulata   (Mandarin orange)   Rutaceae
   A   B   C   D   D   J   J   M   N
   01 01 09 01 02 03 06 01 09
OC: Callosobruchus maculatus H-2, I-9, 12 (331) Meloidogyne javanica H-16, I-9 (100, 616)
   Meloidogyne incognita H-16, I-9 (100, 616)
AP: Alk=I-6, 7 (1374)

- -

Citrus sinensis   (Sweet orange)   Rutaceae
   A   B   C   D   D   J   M   N
   01 01 09 01 02 03 01 09
OC: Leaf-cutting ants H-6, I-9 (101)
AP: Alk=I-? (narcotine)(1392)

- -

Clausena anisata   (Samanobere)   Rutaceae
   A   B   C   D   D   J   M   M
   01 01 09 01 05 11 05 13
OC: Mosquitoes H-5, I-7 (101, 105, 869, 1353)
   Spodoptera exempta H-4, I-4 (171); H-5, I-? (58)
OR: H-4, I-? (132); H-5, I-7 (504); H-17, I-7 (171, 1353)
AP: Alk=I-6 (1374)

- -

Clausena excavata   (Not known)   Rutaceae
   A   B   C   D   J
   01 01 09 01 03
OC: Rice leafhoppers H-1, I-2 (1241)

Claytonia virginica (Virginia spring beauty) Portulaceae
   A   B   C   D   J   M   M   N
   01 06 09 03 08 01 13 03
OC: Popillia japonica H-5, I-1 (105, 1243)

Cleistanthus collinus (Karada) Euphorbiaceae
   A   B   C   D   J   J
   01 01 09 01 03 04
OC: Callosobruchus chinensis H-2, I-7 (95, 475) Termites H-5, I-4 (104)
   Maggots H-2, I-4 (104)
OR: H-2, I-2, 4, 7, 9 (73); H-32, I-4, 10 (104, 220, 504); H-33, I-? (220); H-33, I-9 (504)

Clematis aethusifolis (Vase vine genus) Ranunculaceae
   A   B   C   D   J
   01 03 09 03 03
OC: Aphids H-1, I-2, 7 (1241)   Puccinia rubigavera H-14, I-2, 7 (1241)
   Pieris rapae H-1, I-2, 7 (1241)   Spider mites H-11, I-2, 7 (1241)
   Puccinia graminis tritici H-14, I-2, 7   Verticillium albo-atrum H-14, I-2, 7 (1241)
    (1241)

Clematis armandii (Vase vine genus) Ranunculaceae
   A   B   C   D   J
   01 03 09 03 03
OC: Agrotis sp. H-1, I-2, 6, 7 (1241)   Pieris rapae H-1, I-2, 6, 7 (1241)
   Caterpillars H-1, I-2, 6, 7 (1241)
OR: H-19, I-? (186)

Clematis chinensis (Vase vine genus) Ranunculaceae
   A   B   C   D
   01 03 09 03
OC: Agrotis sp. H-1, I-2, 6, 7 (1241)   Pieris rapae H-1, I-2, 6, 7 (1241)
   Ascotis selenaria H-1, I-2, 6, 7 (1241)

Clematis dioica (Honduras fish poison) Ranunculaceae
   B   C   D   J   M
   03 09 01 03 05
OC: Bombyx mori H-2, I-? (105)
OR: H-32, I-? (1276)

Clematis gouriana (Leather flower genus) Ranunculaceae
   A   B   C   D   J   J   L   M   M   M
   01 03 09 03 03 09 03 05 06 07
OC: Alternaria tenuis H-14, I-6, 7   Fusarium nivale H-14, I-6, 7
   (182, 926, 1295)   (182, 926, 1295)
   Curvularia lunata H-14, I-6, 7   Ustilago hordei H-14, I-7 (181)
   (182, 926, 1295)   Ustilago tritici H-14, I-7 (181)
   Drechslera graminea H-14, I-6, 7 (182, 507, 774, 926, 1295)
OR: H-22, I-16 (881); H-33, I-? (220)

Clematis terniflora (Leather flower genus) Ranunculaceae
   A   B   C   D   J
   01 03 09 03 03
OC: Drosophila hydei H-2, I-4, 7, 10 (101)

Clematis virginiana   (Devil's darning needle)   Ranunculaceae
    A   B   C   D   J   J   M
    01   03   09   03   3a   14   13
OC: Ceratitis capitata H-6, I-6, 7, 8, 9 (956) Monilinia fructicola H-14, I-14 (1105)

Clematis vitalba   (Leather flower genus)   Ranunculaceae
    A   B   C   D   D   M   N
    '01 03   09   03   06   01   15
OC: Cassida nebulosa H-4, I-7 (531)   Phytodecta fornicata H-4, I-7 (531)
    Hyphantria cunea H-4, I-7 (531)   Pieris brassicae H-4, I-7 (531)
    Leptinotarsa decemlineata H-4, I-7 (531) Stored grain pests H-5, I-6, 7, 8 (105)
    Phyllobius oblongus H-4, I-7 (531)   Tanymecus dilaticollis H-4, I-7 (531)
AP: Alk=I-2 (clematine)(1392)

Cleome brachycarpa   (Spider plant genus)   Capparaceae
    B   C   D   D   E
    07   09   01   05   07
OC: Clothes moths H-5, I-7 (1345)

Cleome gynandra   (Spider wisp)   Capparaceae
    A   B   C   D   D   K   L   M   M   N   O   O
    03   06   09   01   02   06   05   01   05   07   10   12
OC: Pediculus humanus capitis H-2, I-10, 12 (87, 105)
OR: H-2, 32, I-? (220); H-17, I-10, 12 (504); H-32, I-1 (983)
AP: Alk=I-7 (1375, 1392)

Cleome monophylla   (Spider plant genus)   Capparaceae
    B   C   D
    06   09   01
OC: Locusta migratoria H-4, 5, I-7 (596)

Cleome pentaphylla   (Cat's whiskers)   Capparaceae
    B   C   D   L   M   N
    06   09   01   05   01   07
OC: Vermin H-27, I-12 (1353)

Cleome viscosa   (Hurhuria)   Capparaceae
    A   B   C   D   F   K   L   M   N   N
    03   06   09   01   06   06   05   01   07   11
OC: Exserohilum turcicum H-14, I-15 (827)   Spodoptera litura H-5, I-7 (126)
OR: H-24, I-10 (1184)

Clerodendrum bungei   (Kashmir bouquet genus)   Verbenaceae
    A   B   C   D
    01   02   09   03
OC: Aphids H-1, I-7 (1241)   Spider mites H-11, I-7 (1241)
    Phytophthora infestans H-14, I-7 (1241)

Clerodendrum calamitosum   (Kashmir bouquet genus)   Verbenaceae
    B   C   D   D   J
    02   09   01   02   07
OC: Spodoptera litura H-4, I-6, 7 (143, 378, 867)
OR: H-19, I-? (982)

Clerodendrum cryptophyllum   (Kashmir bouquet genus)   Verbenaceae
    A   B   C   D   J
    01 02 09 02 07
OC: Spodoptera litura H-4, I-6, 7 (378, 867)

Clerodendrum glabrum   (Kashmir bouquet genus)   Verbenaceae
    A   B   B   C   D   D   J   M   O
    01 01 02 09 01 02 03 05 04
OC: Beetles H-5, I-7 (87)   Maggots H-5, I-7 (87)

Clerodendrum imame   (Banjui)   Verbenaceae
    A   B   C   D   J   J
    01 02 09 01 03 04
OC: Idiocerus sp. H-2, I-6 (105)

Clerodendrum indicum   (Tube flower)   Verbenaceae
    A   B   B   C   D   J
    01 02 06 09 01 03
OC: Pyricularia oryzae H-14, I-7 (114)

Clerodendrum inerme   (Kashmir bouquet genus)   Verbenaceae
    B   C   D   J   J   J   J   K   M   O
    02 09 02 02 03 07 10 07 05 02
OC: Aedes aegypti H-1, I-7 (1328)
OR: H-4, I-6 (378); H-5, I-? (605)

Clerodendrum infortunatum   (Bhant)   Verbenaceae
    A   C   D   J   K
    01 09 01 02 09
OC: Oryctes rhinoceros H-5, I-7 (119)   Stored rice pests H-5, I-7 (119)
    Sitotroga cerealella H-1, I-7 (121, 1124)
OR: H-1, I-1, 7, 8 (127)

Clerodendrum myricoides   (Kashmir bouquet genus)   Verbenaceae
    A   B   C   D   D
    01 02 09 01 02
OC: Spodoptera exempta H-4, I-? (94, 1037)

Clerodendrum phillipinum   (Kashmir bouquet genus)   Verbenaceae
    A   B   C   D   D   M
    01 02 09 02 03 13
OC: Spodoptera litura H-4, I-6, 7 (378, 867)

Clerodendrum serratum   (Kashmir bouquet genus)   Verbenaceae
    A   B   C   D   J   M   N   N
    01 02 09 02 09 01 07 08
OC: Rats H-26, I-16 (867)

Clerodendrum trichotomum   (Kashmir bouquet genus)   Verbenaceae
    A   B   D   D   J
    01 02 02 03 03
OC: Cicadella viridis H-4, I-7 (379)   Spodoptera litura H-4, I-? (379, 867)
    Euproctis pseudoconspersa H-4, I-7 (379)

Clethra alnifolia (Sweet pepperbush) Clethraceae
  A   B   C   D   D   J
  01 02 09 02 03 14
OC: Oncopeltus fasciatus H-3, I-6, 7, 8, 9 (958)

Clibodium arboreum (Not known) Asteraceae
  B   C   D   J
  06 09 01 03
OC: Attagenus piceus H-4, I-7 (48)

Clibodium erosum (Not known) Asteraceae
  B   C   D
  06 09 01
OC: Mosquitoes H-2, I-7, 10 (101); H-32, I-? (1276)

Clibodium surinamense (Not known) Asteraceae
  B   C   J   L
  06 09 03 03
OC: Bombyx mori H-2, I-16 (101, 284)
OR: H-32, 33, I-6, 7 (504)

Clibodium sylvestre (Barbasco) Asteraceae
  A   B   B   C   D   J   K
  01 02 07 09 01 02 03
OR: H-5, I-6, 7 (118); H-32, I-? (105, 1276)

Clinopodium vulgare (Basil) Lamiaceae
  A   B   C   D   F   M   M   O
  01 06 09 03 10 05 16 07
OC: Venturia inaequalis H-14, I-? (16)

Clitoria arborescens (Butterfly pea genus) Fabaceae
  A   B   C   D   J   M   M
  01 02 09 02 11 12 13
OC: Attagenus piceus H-4, I-10 (48)
AP: Alk=I-7 (1392)

Clitoria ternatea (Butterfly pea genus) Fabaceae
  A   B   C   D   D   M   M   M   M   M   M   N   O
  01 06 09 01 02 01 02 05 09 13 16 11 10
OR: H-1, I-1 (138); H-15, I-2 (220)
AP: Alk=I-? (1391); Alk=I-7, 10 (1392)

Cnicus benedictus (Blessed thistle) Asteraceae
  A   B   C   D   D   J   M
  03 06 09 03 06 03 02
OC: Mosquitoes H-2, I-1 (87)

Cnidium monnieri (Not known) Apiaceae
  C
  09
OC: Aphis gossypii H-1, I-1, 10 (1241)   Puccinia rubigavera H-14, I-1, 10 (1241)
    Phytophthora infestans H-14, I-1, 10 (1241) Pyricularia oryzae H-14, I-1, 10 (1241)
    Puccinia graminis tritici H-14, I-1, 10 (1241)

Cnidoscolus texanus   (Bull nettle)   Euphorbiaceae

   A   B   C   D   J
   01 01 09 01 03

OC: Heliothis virescens H-1, I-7 (1240)

Cnidoscolus urens   (Spurge nettle)   Euphorbiaceae

   A   B   C   D   J
   01 01 09 01 04

OC: Hymenia recurvalis H-2, I-9 (101, 190, 304)
    Pachyzancla bipunctalis H-2, I-9 (101, 190, 304)

Cocculus indicus   (Not known)   Menispermaceae

   B   C   D   D   J   J
   03 09 01 02 03 08

OC: Attagenus piceus H-4, I-10 (48)

Cocculus trilobus   (Not known)   Menispermaceae

   A   B   C   D   J   J
   01 02 09 10 03 12

OC: Abraxas miranda H-4, I-7 (867)   Psylla pyricola H-4, I-7 (391)
    Callosobruchus chinensis H-2, I-7 (87, 381) Spodoptera litura H-2, I-7
    Drosophila hydei H-2, I-2, 7 (101)   (87, 382, 391, 867)
    Nephotettix bipunctatus H-2, I-7 (87, 381) Trimeresia miranda H-2, I-7 (87, 382, 391)
    Parabenzoin trilobum H-23, I-? (867)

OR: H-4, I-? (132)

AP: **Alk**=I-? (isotrilobine, magnoflorine, menisioline, mensine, normenisarine, trilobamin)
    (1392); **Alk**=I-2 (trilobine)(1392); **Alk**=I-6 (tetrandrine)(1392); **Alk**=I-6, 7 (1368);
    Alk=I-7 (cocculolidine)(87)

Cochlospermum religiosum   (Silk cotton tree)   Cochlospermaceae

   A   B   C   D   D   M   M
   01 01 09 05 09 03 13

OC: Frankliniella fusca H-2, I-13 (105)   Macrosiphum ambrosiae H-5, I-13 (105)

OR: H-30, I-13 (105)

AP: **Sap, Tan, Ter**=I-7 (1359)

Cocos nucifera   (Coconut palm)   Arecaceae

   A   B   C   D   F   J   J   K   M   M   M   M   M   M   N
   01 01 09 01 03 06 07 06 01 03 05 06 13 16 21 10

OC: Mites H-11, I-1 (504)   Zabrotes subfasciatus H-2, I-12 (334)
    Sitophilus oryzae H-1, I-12 (105)

OR: H-3, I-? (132); H-17, I-9 (1321); H-20, I-4 (504)

AP: **Sfr, Tan**=I-7 (1321); **Tan**=I-9 (1353)

Coffea arabica   (Arabian coffee)   Rubiaceae

   A   B   C   D   J   J   J   M   N
   01 02 09 01 09 12 14 01 10

OC: Amaranthus spinosus H-23, I-10 (159, 1197) Echinochloa colonum H-23, I-10 (159)
    Avena fatua H-23, I-10 (159, 1197)   Echinochloa crus-galli H-23, I-10 (159)
    Bipolaris maydis H-14, I-10 (857, 1197)   Lathyrus aphaca H-23, I-10 (159)
    Callosobruchus chinensis H-1, I-10 (1197) Vicia sativa H-23, I-10 (159)
    Ceratitis capitata H-6, I-4 (956)

OR: H-24, I-10 (569, 1234)

AP: **Alk**=I-7, 8 (caffeine)(1392); **Alk**=I-10 (caffeine, trigonelline)(1392)

Coffea robusta   (Robusta coffee)   Rubiaceae
```
    A   B   B   C   D   J
    01  01  02  09  01  14
```
OC: Ceratitis capitata H-6, I-? (956)   Dacus dorsalis H-6, I-? (956)
    Dacus cucurbitae H-6, I-? (956)   Oncopeltus fasciatus H-2, I-4 (958)
AP: Alk=I-4, 7 (caffeine, theobromine)(1392); Alk=I-10 (caffeine)(1392)

Coffea sp.   (Coffee genus)   Rubiaceae
```
    A   B   C   D   M   N
    01  01  09  01  01  10
```
OC: Fusarium solani f. sp. phaseoli H-14, I-10 (360)

Colchicum autumnale   (Meadow saffron)   Liliaceae
```
    A   B   C   D   D   G   J   K   M   M   O   O
    01  06  09  03  06  05  03  03  05  13  03  10
```
OC: Aphids H-1, I-2, 4 (105)   Flies H-1, I-7 (1025)
    Beetles H-1, I-7 (1025)   Leptinotarsa decemlineata H-2, I-? (87)
    Cochliomyia hominivorax H-2, I-? (87)   Musca domestica H-3, I-? (87)
    Conotrachelus nenuphar H-2, I-? (87)   Spodoptera litura H-4, I-2, 10 (105)
    Culex quinquefasciatus H-2, I-10 (463)
AP: Alk=I-3, 7, 8 (colchiceine, colchicene)(1392); Alk=I-3, 10 (colchicum)(1392)

Collinsonia anisata   (Horseweed)   Lamiaceae
```
    A   B   C   D   J   M
    01  06  09  03  11  17
```
OC: Attagenus piceus H-4, I-7 (48)

Collinsonia canadensis   (Stoneroot)   Lamiaceae
```
    A   B   C   D   J   M   O
    01  06  09  03  09  05  02
```
OC: Rodents H-25, I-1 (944)   Sclerotinia fructicola H-14, I-1 (944)

Colocasia esculenta var. antiquorum   (Egyptian taro)   Araceae
```
    A   B   C   D
    01  06  09  01
```
OC: Dacus fergugineus H-6, I-1 (473)   Dacus zonatus H-6, I-1 (473)
AP: Alk=I-7 (1375)

Colubrina asiatica   (Kabatiti)   Rhamnaceae
```
    A   B   C   D   J   J   J   M   M   M   M   N   O
    01  01  09  01  03  08  11  01  05  10  15  07  09
```
OC: Attagenus piceus H-4, I-6, 7 (48)
OR: H-20, I-7 (1349); H-32, I-9 (983, 1349)
AP: Alk=I-2, 4, 7, 8 (1392)

Comandra umbellata   (Bastard toadflax)   Santalaceae
```
    B   C   J   M   N
    06  09  08  01  09
```
OC: Popillia japonica H-5, I-1 (105)

Combretum caoucia   (Not known)   Combretaceae
```
    A   B   C   D   J   M
    01  01  09  01  11  13
```
OC: Attagenus piceus H-4, I-9 (48)

Combretum nigricans   (Bush willow genus)   Combretaceae
   A   B   C   D   M
   01   01   09   01   13
OC: Locusta migratoria H-4, 5, I-7 (596)

Commelina communis   (Day flower genus)   Commelinaceae
   A   B   C   D   L
   03   06   09   03   05
OC: Aphis gossypii H-1, I-? (1241)

Commiphora abyssinica   (Abyssinian myrrh)   Burseraceae
   C   D   M
   09   05   17
OC: Mosquitoes H-5, I-13 (87)

Commiphora africana   (African myrrh)   Burseraceae
   C   D   M
   09   05   17
OC: Termites H-2, 5, I-13 (87)

Comptonia aspleniifolia   (Sweet fern)   Myricaceae
   A   B   C   D   J   M   O
   01   06   09   07   09   05   07
OC: Rodents H-25, I-1 (944)

Coniogramme japonica   (Bamboo fern)   Polypodiaceae
   A   C   D   J
   01   06   03   03
OC: Drosophila hydei H-4, I-7 (101)

Conium maculatum   (Poison hemlock)   Apiaceae
   A   B   C   D   J   J   L   M   O
   02   03   09   03   03   04   03   05   09
OC: Blow flies H-2, I-7, 8 (105)   Manduca sexta H-4, I-7 (669)
    Dermacentor marginatus H-13, I-7 (87, 1133) Panonychus citri H-4, 11, I-7 (669)
    Haemaphysalis punctata H-13, I-7 (87, 1133) Rhipicephalus rossicus H-13, I-7
    Ixodes redikorzevi H-13, I-7 (87, 1133)   (87, 1133)
OR: H-32, I-2 (401); H-33, I-1, 2, 9, 10 (504, 507, 620); H-34, I-9 (504)
AP: Alk=I-1 (620); Alk=I-6, 7, 8 (1374); Alk=I-6, 7, 8, 9 (conhydrine, coniceine, coniine,
    N-methylconiine, piperidine, 2-methylpiredine)(1392)

Conringia orientalis   (Not known)   Brassicaceae
   B   C   D   J   J   M
   06   09   06   02   03   28
OC: Aedes aegypti H-2, I-7, 8, 9 (77, 165, 843)
    Drosophila melanogaster H-2, I-9 (165)

Consolida ambigua   (Rocket larkspur)   Ranunculaceae
   A   B   C   D   D
   03   06   09   03   06
OC: Cimex lectularius H-2, I-10, 12 (105, 800) Pediculus humanus humanus H-2, I-10 (87)
    Maggots H-2, I-2, 6, 7 (1241)   Tylenchulus semipenetrans H-16, I-7 (540)
OR: H-2, I-10 (504)
AP: Alk=I-10 (ajacine, ajacinine, ajacinoidine)(1392)

Consolida regalis (Field larkspur) Ranunculaceae
    A   B   C   D   J
    03 06 09 03 14
OC: Aleyrodes vaporariorum H-2, I-10, 12 (203)   Myzus cerasi H-2, I-10, 12 (203)
    Aphids H-2, I-10, 12 (105, 800)   Myzus persicae H-2, I-10, 12 (203)
    Aphis fabae H-2, I-10, 12 (203)   Paratetranychus pilosus H-11, I-10, 12
    Aphis pomi H-2, I-10, 12 (203)   (203)   )
    Aspidiotus perniciosa H-2, I-10, 12 (203)   Pediculus humanus capitis H-2, I-10 (105
    Chrysomphalus sp. H-2, I-10, 12 (203)   Termites H-2, I-10 (105)
    Hyphantria cunea H-2, I-10 (105)   Tetranychus cinnabarinus H-11, I-10, 12
    Lice H-2, I-10, 12 (87)   (105, 203, 800)
    Macrosiphum rosae H-2, I-10, 12 (203)   Thrips H-2, I-10 (105)
OR: H-2, I-7, 10 (504); H-33, I-10, 12 (800)
AP: Alk=I-10 (anthranoyllycoctonine, consolidine, delcosine, delsoline, delsonine)(1392)

Convallaria majalis (Lily-of-the-valley) Liliaceae
    A   B   C   D   J   M
    01 06 09 10 03 13
OC: Attagenus piceus H-4, I-1 (48)   Reticulitermes flavipes H-4, I-2 (87)

Convolvulus arvensis (Field bindweed) Convolvulaceae
    A   B   C   D   J   L
    01 06 09 03 08 05
OC: Aedes aegypti H-2, I-2, 6, 7, 8, 9   Erwinia carotovora H-18, I-6, 7, 14 (585)
    (77, 843)   Leptinotarsa decemlineata H-4, I-7 (531)
    Agrobacterium tumefaciens H-18, I-2,   Phyllobius oblongus H-4, I-7 (531)
    6, 7, 14 (585)   Phytodecta fornicata H-4, I-7 (531)
    Athalia rosae H-4, I-7 (531)   Pieris brassicae H-4, I-7 (531)
    Cassida nebulosa H-4, I-7 (531)   Tanymecus dilaticollis H-4, I-7 (531)

Copaifera lansdorfii (Copaiba) Caesalpiniaceae
    A   B   C   D   F   M   M   M   M   M   M   M   O   O
    01 01 09 01 07 04 05 12 17 19 20 21 13 17
OC: Cochliomyia hominivorax H-5, I-12 (105)

Copaifera officinalis (African copaifera) Caesalpiniaceae
    A   B   C   D   M   M
    01 01 09 01 12 20
OC: Ceratitis rosa H-6, I-12 (105)

Corchorus capsularis (Jute) Tiliaceae
    A   B   C   D   F   J   K   M   M   M   N   N   O   O
    03 02 09 01 04 04 01 01 03 05 07 15 07 10
OC: Rice field insects H-2, I-10 (155)

Corchorus olitorius (Jew's mallow) Tiliaceae
    A   B   C   D   J   L   M   M   M   N   N   O   O
    03 02 09 01 03 05 01 03 05 07 15 07 10
OC: Drechslera oryzae H-14, I-7 (113)   Locusta migratoria H-4, 5, I-7 (596)
AP: Alk, Sap=I-6, 7, 9 (1360); Sfr=I-2, 6 (1321)

Cordia dichotoma (Anonang) Boraginaceae
    A   B   C   D   M   M   M   M   O
    01 01 09 01 02 03 05 13 04
OC: Termites H-6, I-6 (137)

Cordyline roxburghiana   (Tigue)   Agavaceae
   A   C   D   D   M   M   M   O
   01 09 01 02 03 05 13 02
OC: Rice field insects H-5, I-7 (137)

Coreopsis grandiflora   (Tickseed genus)   Asteraceae
   A   B   C   D   J   J
   01 06 09 03 03 08
OC: Popillia japonica H-5, I-1 (105, 1243)

Coreopsis lanceolata   (Lanced-leaved tickseed)   Asteraceae
   A   B   C   D   D   M
   01 03 09 02 03 17
OC: Drosophila melanogaster H-3, I-2, 6, 7, 8 (372)

Coriandrum sativum   (Coriander)   Apiaceae
   A   B   D   K   M   M   M   M   N   N   O
   03 06 03 06 01 05 14 15 17 07 10 10
OC: Aphids H-1, I-? (74, 75)   Lenzites trabea H-14, I-? (428)
   Aphis gossypii H-2, I-12 (105)   Leptinotarsa decemlineata H-1, I-? (75)
   Cladosporium fulvum H-14, I-12 (185) Polyporus versicolor H-14, I-? (428)
   Cochliomyia hominivorax H-5, I-? (105) Tetranychus cinnabarinus H-11, I-12 (105)
   Lentinus lepideus H-14, I-? (428)   Tribolium castaneum H-5, I-? (695)
OR: H-24, I-9 (569)

Coriaria sinica   (Not known)   Coriariaceae
   A   C   D   J   J   K
   01 09 01 03 04 06
OC: Aphids H-1, I-6, 7, 8 (1241)   Pieris rapae H-1, I-6, 7, 8 (1241)
   Maggots H-1, I-6, 7, 8 (1241)   Rice borers H-1, I-6, 7, 8 (1241)
   Phytophthora infestans H-14, I-6, 7, 8 Spider mites H-11, I-6, 7, 8 (1241)
     (1241)

Corispermum hyssopifolium   (Not known)   Chenopodiaceae
   C   J
   09 03
OC: Aedes aegypti H-2a, I-1 (48)   Attagenus piceus H-4, I-1 (48)

Cornus sanguinea   (Bloodtwig dogwood)   Cornaceae
   A   B   C   D   M
   01 02 09 03 04
OC: Venturia inaequalis H-14, I-? (16)

Coronilla varia   (Crown vetch)   Fabaceae
   B   C   D   J   M   M
   06 09 03 03 09 12
OR: H-24, I-7 (1204)
AP: Alk=I-6, 7 (1392); Alk=I-10 (cytisine)(1392)

Corydalis aurea   (Not known)   Fumariaceae
   A   A   B   C   D   J   J
   01 03 06 09 03 02 03
OC: Aedes aegypti H-2, I-7, 8 (165)
AP: Alk=I-2, 6, 7 (protopine)(1392); Alk=I-6, 7 (allocryptopine, aurotensine, bicucine, bicuculline, capauridine, capaurine, cordrastine)(1392); Alk=I-10 (corypalline)(1392)

Corypha elata   (Gebang palm)   Arecaceae
   A   B   C   D   M   M   N   N
   01 01 09 01 01 03 10 14
OC: Rice field insects H-5, I-7 (137)

Couroupita guianensis   (Cannonball tree)   Lecythidaceae
   A   B   C   D
   01 01 09 01
OC: Dacus dorsalis H-6, I-7 (473)

Crassocephalum crepedioides   (Not known)   Asteraceae
   A   B   C   D   F   G   J   K
   03 06 09 01 06 04 12 06
OC: Dysdercus cingulatus H-2, I-7 (125)

Crassula arborescens   (Silver jade plant)   Crassulaceae
   A   B   C   D   D   J   M
   01 02 09 01 02 03 13
OC: Tobacco mosaic virus H-21, I-7 (184)

Crassula argentea   (Jade tree)   Crassulaceae
   A   B   C   D   J   M
   01 02 09 02 03 13
OC: Tobacco mosaic virus H-21, I-7 (184)

Crassula falcata   (Scarlet paint brush)   Crassulaceae
   A   B   C   D   J
   01 02 09 01 03
OC: Tobacco mosaic virus H-21, I-7 (184)

Crassula multicava   (Jade plant genus)   Crassulaceae
   A   C   D   J
   01 09 01 03
OC: Tobacco mosaic virus H-21, I-7 (184)

Crassula prealtum   (Jade plant genus)   Crassulaceae
   A   C   D   J
   01 09 01 03
OC: Tobacco mosaic virus H-21, I-7 (184)

Crassula rupestris   (Rosary vine)   Crassulaceae
   A   C   D   J
   01 09 01 03
OC: Tobacco mosaic virus H-21, I-7 (184)

Crepis bursifolia   (Not known)   Asteraceae
   B   C   D   J
   06 09 10 11
OC: Attagenus piceus H-4, I-6, 7 (48)

Crinum bulbispermum   (Spider lily genus)   Amaryllidaceae
   A   B   C   D   M
   01 06 09 03 13
OC: Athalia proxima H-5, I-7 (1011)   Schistocerca gregaria H-4, I-7 (615)
OR: H-5, I-? (100)

Crinum defixum (Spider lily genus) Amaryllidaceae
    A   B   C   D
    01   06   09   02
OC: Athalia proxima H-1, I-? (438); H-4, I-7, 14 (607)
AP: **Alk**=I-3, 10 (caranine, crinamine, crinidine, galanthamine, galanthine, haemanthamine, hippeastrine, lycorine)(1392)

Crossosoma bigelovii (Not known) Crossosomataceae
    A   B   C   D   J   M
    01   02   09   04   03   13
OC: Attagenus piceus H-4, I-2, 6, 7 (48)   Blattella germanica H-2a, I-2, 6, 7 (48)

Crotalaria anagyroides (Rattlebox genus) Fabaceae
    A   B   C   D
    01   02   09   01
OC: Fusarium oxysporum f. sp. cubense H-14, I-? (952)
AP: **Alk**=I-6, 7 (1392); **Alk, Tri**=I-7 (1390)

Crotalaria breviflora (Rattlebox genus) Fabaceae
    B   D   J   K   M
    06   01   01   01   12
OC: Meloidogyne sp. H-16, I-2 (479)

Crotalaria juncea (Junjunia) Fabaceae
    A   B   C   D   D   J   J   J   K   M   M   M
    03   02   09   01   02   03   11   15   01   03   10   12
OC: Meloidogyne incognita H-16, I-2, 7 (147, 922)
    Sitophilus oryzae H-2a, I-8 (87, 416, 1124)
AP: **Alk**=I-? (1391); **Alk**=I-10 (junceine, riddelliine, senecionine, seneciphylline, trichodesmine)(1392)

Crotalaria medicagenea (Rattlebox genus) Fabaceae
    B   D   J   M
    06   01   03   12
OR: H-25, I-1 (1212)

Crotalaria mucronata (Rattlebox genus) Fabaceae
    B   C   D   D   J   K   M   M
    06   09   01   02   01   01   10   12
OC: Meloidogyne sp. H-16, I-2 (479)   Radopholus similis H-16, I-? (1275)
    Pratylenchus zea H-16, I-? (1274)
AP: **Alk**=I-? (1372); **Alk**=I-2, 6, 7, 9 (1369); **Alk**=I-6, 7 (1368); **Alk**=I-7 (1390); **Alk**=I-7, 9, 10 (1392); **Alk**=I-8 (1373); **Sap**=I-7 (1380)

Crotalaria paniculata (Rattlebox genus) Fabaceae
    B   C   D   M
    02   09   01   12
OR: H-2, 32, I-? (105, 1276)

Crotalaria pumila (Rattlebox genus) Fabaceae
    B   C   D   J   K   M
    06   09   01   01   01   12
OC: Meloidogyne sp. H-16, I-2 (479)

Crotalaria retusa (Bangla) Fabaceae

    A   B   C   D   M
    01 02 09 01 12

OR: H-2, I-1 (127)
AP: Alk=I-10 (monocrotaline, retusamine, retusine)(1392); **Alk, Sap**=I-6, 7, 9 (1360)

Crotalaria sagittalis (Not known) Fabaceae

    B   B   C   D   M
    02 06 09 01 12

OC: Attagenus piceus H-4, I-1, 9 (48, 78)
AP: Alk=I-10 (1391, 1392)

Crotalaria spectabilis (Rattlebox) Fabaceae

    B   B   C   D   D   J   L   M   M   M
    02 06 09 01 02 03 02 09 10 12

OC: Belonolaimus longicaudatus H-16, I-? (955) Meloidogyne javanica H-16, I-? (955)
    Criconemoides ornatum H-16, I-? (955)   Radopholus similis H-16, I-? (1275)
    Meloidogyne arenaria H-16, I-? (955)   Trichodorus christiei H-16, I-? (955)
    Meloidogyne incognita H-16, I-7 (115, 955) Xiphinema americanum H-16, I-? (955)
OR: H-33, I-7, 10 (398)
AP: Alk=I-2, 8, 7, 10, 16 (monocrotaline, spectabiline)(1392)

Crotalaria trifaliastruam (Not known) Fabaceae

    A   B   C   D   M
    01 06 09 01 12

OR: H-2, I-2 (127)
AP: Alk=I-2 (1392)

Crotalaria verrucosa (Rattlebox genus) Fabaceae

    A   C   D   M
    01 09 01 12

OR: H-4, I-? (132); H-24, I-15 (1184)
AP: Alk=I-6, 7, 9 (1392)

Croton bonplandianum (Kala bhangra) Euphorbiaceae

    A   B   C   D   J   K   M
    03 02 09 02 03 03 05

OC: Cyperus rotundus H-23, I-7, 9 (149)

Croton californicus (Not known) Euphorbiaceae

    C   D   D
    09 01 05

OC: Aphids H-1, I-? (106)
OR: H-32, I-? (106)

Croton caudatus (Not known) Euphorbiaceae

    C   D   J   M   O
    09 01 09 05 02

OC: Musca domestica H-1, I-16 (785)
OR: H-2, I-? (100)

Croton eluteria (Cascarilla) Euphorbiaceae

    A   C   D   D   K
    01 09 01 02 07

OC: Mosquitoes H-5, I-4 (104)

Croton flavens   (Not known)   Euphorbiaceae
    C   D   D
    09   01   02
OR: H-2, I-? (104); H-5, I-? (101)

Croton klotzschianus   (Not known)   Euphorbiaceae
    B   C   D   D   J
    02   09   01   02   04
OC: Aphids H-2, I-10 (119)   Mites H-11, I-10 (119)

Croton macrostachys   (Not known)   Euphorbiaceae
    A   B   C   D
    01   01   09   01
OR: H-4, I-? (132); H-17, I-7, 10, 14 (504, 1143)
AP: Alk=I-4, 6 (1372)

Croton oblongifolius   (Not known)   Euphorbiaceae
    A   B   C   D   J   J
    01   01   09   01   03   04
OR: H-2, I-? (73, 220); H-2, I-10 (105, 504); H-32, 33, I-10 (220, 504)

Croton sparsiflorum   (Croton)   Euphorbiaceae
    A   B   C   D   D   E   F   G   J   K   M
    03   06   09   01   02   01   01   05   05   01   05
OC: Meloidogyne sp. H-16, I-1 (161)   Rotylenchulus sp. H-16, I-1 (161)
AP: Alk=I-7 (1379); Alk=I-10 (1392)

Croton texensis   (Skunkweed)   Euphorbiaceae
    C   D   J
    09   03   11
OC: Attagenus piceus H-4, I-1 (48)
OR: H-2, I-16 (504)

Croton tiglium   (Purging croton)   Euphorbiaceae
    A   B   B   C   D   D   F   J   J   J   K   K   M   O   O
    01   01   02   09   01   02   07   03   04   11   05   06   05   10   12
OC: Aphids H-2, I-10 (101, 105, 1012)   Mosquitoes H-2, I-10 (91)
    Bombyx mori H-2, I-10 (101)   Musca domestica H-2, I-10 (87)
    Dactynotus carthani H-2, I-10 (91)   Rondotia menciana H-2, I-10, 12 (105, 1012)
    Dysdercus koenigii H-2, I-10 (91)   Spodoptera exigua H-1, I-? (84)
    Lipaphis erysimi H-2, I-10 (91)   Spodoptera litura H-1, I-? (84)
    Lymnaea auricularia rubiginosa H-28, 1-10 (662)
OR: H-1, I-10, 12 (73, 80, 97); H-2, I-7, 8, 10, 12 (96, 116, 871); H-32, I-4, 7, 9, 10
    (192, 220, 408, 504, 801, 864, 873, 874, 968, 983, 1349); H-33, I-2, 9 (220, 457, 983)
AP: Alk=I-10 (ricinine)(1392); Sap=I-7 (968)

Cryptostegia grandiflora   (Rubber vine)   Asclepiadaceae
    A   B   C   D   J   M   M
    01   03   09   01   03   13   26
OC: Attagenus piceus H-4, I-6 (48)
OR: H-33, I-7 (220)
AP: Alk=I-7 (1392)

Crysophilia argentea   (Not known)   Arecaceae
   A B C D J
   01 01 09 01 03
OC: Oncopeltus fasciatus H-2, I-5 (48)

Cucumis callosus   (Melon genus)   Cucurbitaceae
   A B B C J
   03 05 06 09 03
OR: H-24, I-1 (1212)

Cucumis melo   (Musk melon)   Cucurbitaceae
   A B B C D M M M N
   03 05 06 09 01 01 02 10 09
OC: Colletotrichum lindemuthianum H-14, I-14 (1105)
   Monilinia fructicola H-14, I-14 (1105)

Cucumis prophetarum   (Melon genus)   Cucurbitaceae
   A B B C D J M O
   03 05 06 09 01 03 05 08
OC: Alternaria tenuis H-14, I-8 (480)   Helminthosporium sp. H-14, I-8 (480)
   Curvularia penniseti H-14, I-8 (480)
OR: H-19, I-9 (1353)
AP: Alk=I-6, 7, 9 (1369); Alk=I-9 (1370)

Cucumis sativus   (Cucumber)   Cucurbitaceae
   A B B C D J M N
   03 04 06 09 10 03 01 09
OC: Cockroaches H-2b, I-9, 14 (87, 105)   Panicum miliaceum H-24, I-2 (1294)
   Cucumber mosaic virus H-21, I-?   Tetranychus urticae H-11, I-? (1153)
    (1072, 1132)   Tomato aucuba mosaic virus H-21, I-? (1132)
   Leptinotarsa decemlineata H-4, I-7 (531) Tomato bushy stunt virus H-21, I-? (1132)
OR: H-19, I-9 (883, 982); H-24, I-2, 7 (569, 1397)

Cucurbita foetidissima   (Wild gourd)   Cucurbitaceae
   A B C D J J M M M N O
   01 07 09 03 08 11 01 05 15 10 02
OC: Attagenus piceus H-4, I-2 (48)

Cucurbita moschata   (Winter crookneck squash)   Cucurbitaceae
   B C D M N
   04 09 01 01 09
OC: Aphids H-1, I-7, 10 (1241)
OR: H-17, I-7, 10 (1241)
AP: Alk=I-10 (1391)

Cucurbita pepo   (Summer squash and Autumn pumpkin)   Cucurbitaceae
   A B C D D J M M N O O
   03 06 09 01 02 10 01 05 09 08 09
OC: Flies H-5, I-7 (105)   Mosquitoes H-2, I-10 (105)
AP: Alk=I-10 (1372, 1392)

Cudrania tricuspidata   (Not known)   Moraceae
   A B C D M M N
   01 01 09 03 01 13 09
OC: Fusarium roseum H-14, I-15 (917)

Cuminum cyminum   (Cumin)   Apiaceae
```
A    B    C    D    M    M    O
03   06   09   06   05   14   10
```
OC: Alternaria tenuissima H-14, I-? (383)   Fusarium poae H-14, I-? (383)
    Cochliomyia americana H-1, I-10, 12 (244)   Phytophthora parasitica H-14, I-? (911)
    Curvularia lunata H-14, I-? (383)   Rhizoctonia solani H-14, I-? (383)
    Fusarium chlamydosporum H-14, I-? (383)

Cupressus lusitanica   (Cypress)   Cupressaceae
```
A    B    C    D    J    M    O
01   01   08   01   03   05   04
```
OC: Rumex crispulus H-23, I-9 (154)

Curcuma amada   (Hidden lily genus)   Zingiberaceae
```
A    B    C    D    J    J    M    M    M    N    O
01   06   09   01   03   10   01   02   05   02   02
```
OC: Alternaria tenuis H-14, I-2 (1295)   Mucor sp. H-14, I-? (797)
    Aspergillus niger H-14, I-2 (410, 797)   Phytophthora parasitica H-14, I-? (797)
    Curvularia lunata H-14, I-2 (1295)   Pythium sp. H-14, I-? (797)
    Curvularia sp. H-14, I-? (797)   Sclerotium sp. H-14, I-? (797)
    Drechslera oryzae H-14, I-? (797)

Curcuma angustifolia   (East Indian arrowroot)   Zingiberaceae
```
A    B    C    D    M    N
01   06   09   01   01   02
```
OC: Aspergillus niger H-14, I-2 (903)   Drechslera oryzae H-14, I-2 (903)
    Curvularia oryzae H-14, I-2 (903)   Trichoderma viride H-14, I-2 (903)

Curcuma aromatica   (Hidden lily genus)   Zingiberaceae
```
A    B    C    D    J    M    N
01   06   09   01   03   01   02
```
OC: Aspergillus niger H-14, I-2 (899)   Drechslera oryzae H-14, I-2 (899)
    Curvularia oryzae H-14, I-2 (899)   Trichoderma viride H-14, I-2 (899)
OR: H-7, I-2 (411); H-15, 19, I-? (411); H-22, I-16 (977)

Curcuma domestica   (Turmeric)   Zingiberaceae
```
A    B    C    D    J    J    J    K    M    M    M    N    O    O
01   06   09   10   02   04   11   05   01   05   14   16   02   02   14
```
OC: Aedes aegypti H-5, I-2, 12 (825)   Rhizoctonia solani H-14, I-2 (1297)
    Callosobruchus maculatus H-4, I-? (644)   Rhizopertha dominica H-5, I-2 (179, 695)
    Helminthosporium sp. H-14, I-2 (1297)   Sclerotium oryzae H-14, I-2 (1297)
    Meloidogyne incognita H-16, I-2 (616)   Sclerotium rolfsii H-14, I-2 (1297)
    Meloidogyne javanica H-16, I-2 (616)   Sitophilus oryzae H-5, I-2 (112)
    Mites H-11, I-? (105)   Tribolium castaneum H-5, I-2 (179, 695)
    Pyricularia oryzae H-14, I-2 (1297)   Tribolium confusum H-5, I-2 (179, 695)
    Rats H-25, I-2 (799)
OR: H-17, I-2 (1318); H-20, I-2 (1353)
AP: Gly, Sap=I-2 (1318)

Curcuma zedoaria   (Zedoary)   Zingiberaceae
```
A    B    C    D    J    J    J    M    M    N    O
01   06   09   01   09   10   11   01   05   02   02
```
OC: Alternaria tenuis H-14, I-2 (1295)   Fusarium oxysporum H-14, I-2 (1149)
    Aspergillus niger H-14, I-2 (429, 786)   Sclerotium rolfsii H-14, I-2 (1149)
    Drechslera oryzae H-14, I-2 (1149)

OR: H-19, I-? (429)
AP: Alk=I-2 (1366)

Cuscuta americana   (Dodder genus)   Cuscutaceae
    B   B   C   D   J   L
    04 10 09 01 03 05
OC: Attagenus piceus H-4, I-6, 8 (48)   Blattella germanica H-2, I-6, 8 (48)

Cuscuta racemosa   (Dodder genus)   Cuscutaceae
    B   B   C   D   J   L
    04 10 09 03 11 05
OC: Attagenus piceus H-4, I-6 (48)

Cuscuta reflexa   (Dodder genus)   Cuscutaceae
    B   C   D   D   J   L
    10 09 01 04 03 05
OC: Meloidogyne incognita H-16, I-8 (542)   Tobacco ring spot virus H-21, I-6 (557)
    Tobacco mosiac virus H-21, I-6 (557)
OR: H-22, I-16 (1395)
AP: Alk=I-6, 7, 8, 9 (1380)

Cyamopsis tetragonolobus   (Cluster bean)   Fabaceae
    B   C   D   D   M   M   M   M   N
    06 09 01 04 01 02 12 22 11
OC: Schistocerca gregaria H-4, 5, I-16 (611)

Cyathula capitata   (Not known)   Amaranthaceae
    B   C
    06 09
OC: Chilo suppressalis H-1, I-? (1003)   Pieris rapae H-1, I-? (1003)
    Culex pipiens molestus H-1, I-? (1003)   Pluttella xylostella H-1, I-? (1003)
    Musca domestica H-1, I-? (1003)
OR: H-3, I-? (853)

Cyathula prostata   (Not known)   Amaranthaceae
    B   C   D   J   M
    06 09 01 17 05
OR: H-3, I-2 (853); H-12, I-1 (1353)

Cycas circinalis   (Sago palm)   Cycadaceae
    A   B   C   D   J   K   M   N   N
    01 01 07 01 01 01 01 05 10
OR: H-2, I-7 (105); H-19, I-10 (776); H-35, I-10 (105)

Cycas revoluta   (Japanese fern palm)   Cycadaceae
    A   C   D   D   M   M   N
    01 07 01 02 01 13 10
OC: Manduca sexta H-4, I-? (669)   Panonychus citri H-11, I-? (669)

Cyclamen elegans   (Not known)   Primulaceae
    B   C   D   J
    06 09 06 03
OC: Mites H-11, I-3 (101)
OR: H-2, I-3 (105)
AP: Alk=I-2 (1392)

Cymbopogon citratus (Lemongrass) Poaceae

A	B	C	D	D	J	M	M	M	M	M	M	M	N	O
01	06	09	01	02	03	01	05	08	09	13	14	17	07	07

OC: Aedes aegypti H-2, I-12 (87)
    Alternaria tenuis H-14, I-12 (426)
    Aphis gossypii H-2, I-12 (105)
    Botrytis allii H-14, I-12 (426)
    Ceratocystis ulmi H-14, I-12 (426)
    Cladosporium fulvum H-14, I-12 (426)
    Claviceps purpurea H-14, I-12 (426)
    Culex fatigans H-2, I-12 (87, 781)
    Diplodia maydis H-14, I-12 (426)
    Fusarium oxysporum
      f. sp. conglutinans H-14, I-12 (426)

Fusarium oxysporum
    f. sp. lycopersici H-14, I-12 (426)
Gibberella fujikuroi H-14, I-12 (426)
Glossina sp. H-5, I-7 (1353)
Meloidogyne incognita H-16, I-7 (115)
Musca domestica H-2, I-12 (87)
Musca nebulo H-1, I-? (781)
Tetranychus cinnabarinus H-11, I-10 (105)
Ustilago avenae H-14, I-12 (426)
Verticillium albo-atrum H-14, I-12 (426)

OR: H-5, I-? (80, 101); H-34, I-12 (1353)

Cymbopogon flexuosus (East Indian lemongrass) Poaceae

A	B	C	D	M	O
01	06	09	01	05	12

OC: Glossina sp. H-5, I-? (101)
    Helicotylenchus indicus H-16, I-? (1329)
    Hoplolaimus indicus H-16, I-? (1329)

Meloidogyne incognita H-16, I-? (1329)
Rotylenchulus reniformis H-16, I-? (1329)
Tylenchorhynchus brassicae H-16, I-? (1329)

Cymbopogon marginatus (Oil grass genus) Poaceae

A	B	C	D	M
01	06	09	01	17

OC: Dacus dorsalis H-6, I-? (87)   Moths H-5, I-2 (87)

Cymbopogon martinii (Gingergrass) Poaceae

A	B	C	D	D	M
01	06	09	01	04	15

OC: Alternaria brassicae H-14, I-? (417)
    Alternaria solani H-14, I-? (417)
    Colletotrichum falcatum H-14, I-? (417)
    Colletotrichum lindemuthianum H-14, I-?
    (417)

Drechslera oryzae H-14, I-? (417)
Fusarium moniliforme H-14, I-? (417)
Fusarium solani H-14, I-? (417)
Rhizoctonia solani H-14, I-? (417)

AP: Alk=I-6 (1360)

Cymbopogon nardus (Citronella grass) Poaceae

A	B	C	D	D	F	G	J	K	M	M	M	O	O
01	06	09	01	02	03	04	06	06	05	06	17	07	12

OC: Aspergillus niger H-14, I-12 (146, 1300)
    Beetles H-6, I-? (1098)
    Chrysomya macellaria H-5, I-12 (199)
    Cochliomyia hominivorax H-5, I-? (105)
    Cockroaches H-5, I-? (105)
    Dacus diversus H-6, I-12 (105, 661)
    Dacus zonatus H-6, I-12 (105, 661)
    Erwinia carotovora H-18, I-? (146, 1300)

Flies H-2, I-? (101)
Lentinus lepideus H-14, I-12 (428)
Lenzites trabea H-14, I-12 (428)
Macrophomina phaseolina H-14, I-? (146)
Mosquitoes H-2, I-12 (101); H-5, I-?
    (73, 101, 105)
Pericallia ricini H-1, I-12 (943)
Polyporus versicolor H-14, I-12 (428)

OR: H-2, I-12 (80, 504); H-4, I-? (132); H-19, I-12 (904)

Cymbopogon oliverii (Lemongrass genus) Poaceae

A	B	C	D
01	06	09	01

OC: Alternaria brassicae H-14, I-? (417)   Drechslera oryzae H-14, I-? (417)
    Alternaria solani H-14, I-? (417)   Fusarium moniliforme H-14, I-? (417)
    Colletotrichum falcatum H-14, I-? (417)   Fusarium solani H-14, I-? (417)
    Colletotrichum lindemuthianum H-14, I-?   Rhizoctonia solani H-14, I-? (417)
      (417)

Cymbopogon sp.   (Lemongrass genus)   Poaceae
    A   B   C   D
    01 06 09 01
OC: Drechslera oryzae H-14, I-? (1297)   Sclerotium oryzae H-14, I-? (1297)
    Pyricularia oryzae H-14, I-? (1297)   Sclerotium rolfsii H-14, I-? (1297)
    Rhizoctonia solani H-14, I-? (1297)

Cymbopogon winterianus   (Lemongrass genus)   Poaceae
    A   C   D
    01 09 01
OC: Mosquitoes H-5, I-7 (1345)

Cynanchum arnottianum   (Bhankalink)   Asclepiadaceae
    A   B   C   D   J   J   L
    01 06 09 02 02 04 05
OC: Maggots H-2, I-7 (73, 105, 220); H-24, I-7, 14 (569)

Cynanchum auriculatum   (Not known)   Asclepiadaceae
    A   B   C   J   K
    01 06 09 03 02
OC: Aphids H-1, I-2 (1241)   Rats H-25, I-2 (1241)

Cynanchum macrorrhizum   (Not known)   Asclepiadaceae
    A   B   C   D   J   K   L
    01 06 09 01 01 01 05
OR: H-10, I-8 (105)

Cynodon dactylon   (Bermuda grass)   Poaceae
    A   B   C   D   D   J   L   M   M   M   L
    01 06 09 01 02 03 05 02 08 13 05
OC: Meloidogyne incognita H-16, I-7 (115)   Mites H-11, I-14 (105)
OR: H-22, I-1 (881, 1040); H-24, I-2 (1194, 1247)

Cynoglossum officinale   (Hound's tongue genus)   Boraginaceae
    A   B   C   D   J   K   L   M   M   N
    02 06 09 03 03 03 05 01 05 07
OC: Attagenus piceus H-4, I-2 (48)
    Drosophila melanogaster H-1, I-2, 6, 7 (165)
OR: H-2, I-2, 7 (504)
AP: Alk=I-2, 16 (cynoglossine, cynoglossophine)(1392)

Cyperus articulatus   (Umbrella sedge)   Cyperaceae
    B   C   D   M   M
    06 09 01 17 24
OR: H-5, I-2 (1345)

Cyperus esculentus   (Yellow nutsedge)   Cyperaceae
    A   B   C   D   L   M   N   N   N
    01 06 09 01 05 01 02 10 12
OR: H-24, I-2 (1256)

Cyperus rotundus   (Purple nutsedge)   Cyperaceae
    A   B   C   D   J   L   M   M   M   M   N   O
    01 06 09 01 03 05 01 02 05 17 02 02
OC: Drechslera graminea H-14, I-2 (1295)   Meloidogyne incognita H-16, I-7 (115)
    Fusarium nivale H-14, I-2 (1295)   Stored grain pests H-5, I-? (80)
    Grasshoppers H-2a, I-? (101)
OR: H-2, 5, I-2 (166, 1353); H-5, I-3 (87); H-24, I-2 (1140, 1191, 1195)
AP: Alk=I-2 (1392)

Cyperus scaviosus   (Not known)   Cyperaceae
    B   C   D
    06 09 01
OR: H-3, I-? (100)
AP: Alk=I-2 (1392)

Cyphostemma kilimandscharia   (Not known)   Vitaceae
    B   C   G
    03 09 04
OC: Oncopeltus fasciatus H-1, I-6 (958)

Cyrilla racemiflora   (Ironwood)   Cyrillaceae
    A   B   C   D   J   J   L   M
    01 02 09 03 03 08 01 18
OC: Attagenus piceus H-4, I-6, 7 (48)

Cyrtomium falcatum   (Holly fern)   Polypodiaceae
    A   C   D   J   M
    01 06 01 17 13
OC: Pseudomonas solanacearum H-18, I-7 (854)
    Xanthomonas campestris pv. phaseoli-sojensis H-18, I-7 (854)

Cyrtomium flacerum   (Holly fern genus)   Polypodiaceae
    A   C
    01 .06
OC: Tobacco mosaic virus H-21, I-7 (1195)

Cyrtomium fortunei   (Holly fern genus)   Polypodiaceae
    A   C   D
    01 06 03
OC: Aphids H-1, I-1 (1241)   Puccinia graminis tritici H-14, I-1 (1241)
OR: H-21, I-16 (977)

Cytisus laburnum   (Golden chain)   Fabaceae
    A   B   C   D   D   M   M
    01 01 09 03 06 12 13
OC: Aphids H-2, I-? (38)
AP: Alk=I-2, 6, 7, 15 (cytisine, genisteine, laburnine, lupanine, N-methylcytisine, saro-
    thamnine, sparteine)(1392)

Cytisus scoparius   (Scotch broom)   Fabaceae
    A   B   C   D   D   L   M
    01 02 09 03 06 05 12
OC: Aphids H-2, I-? (38)
                                          Pieris brassicae H-1, I-15 (104)
OR: H-33, I-? (220)

101

AP: Alk=I-6 (hydroxytyramine, tyramine)(1392); Alk=I-6, 7 (1373); Alk=I-6, 7, 8, 9 (1368);
    Alk=I-6, 7, 9 (1370); Alk=I-6, 7, 9 (genisteine, sarothamnine, sparteine)(1392); Alk=I-10
    (cytisine, hydroxylupanine, lupanine)(1392); Tan=I-10 (1391)

Dacrydium franklinii   (Huon pine)   Podocarpaceae
    A   B   C   D   M   M
    01 01 08 01 12 14
OC: Blow flies H-5, I-12 (101)
OR: H-19, I-12 (775, 821); H-30, I-12 (101)

Dacryodes hexandra   (Not known)   Burseraceae
    A   B   C   D   J
    01 01 09 01 11
OC: Attagenus piceus H-4, I-1 (48)

Daemia tomentosa   (Not known)   Asclepiadaceae
    C   J
    09 03
OC: Blattella germanica H-2, I-1 (48)

Dahlia pinnata   (Dahlia genus)   Asteraceae
    A   B   C   D   D   M
    01 06 09 01 08 13
OC: Leptinotarsa decemlineata H-4, I-7 (531)   Tomato aucuba mosaic virus H-21, I-? (1132)
    Pieris brassicae H-4, I-7 (531)   Tomato bushy stunt virus H-21, I-? (1132)
OR: H-22, I-9 (969)
AP: Alk=I-2 (trigonelline)(1392)

Dalbergia latifolia   (Indian rosewood)   Fabaceae
    A   B   C   D   J   M   M   M
    01 01 09 01 09 04 12 19
OC: Termites H-8, I-5 (1352)
OR: H-17, I-4, 6 (881)
AP: Alk, Tan=I-9 (1360)

Dalbergia retusa   (Nicaragua rosewood)   Fabaceae
    A   B   D   M   M
    01 09 01 12 19
OC: Anopheles quadrimaculatus H-3, I-? (392)   Tribolium confusum H-3, I-? (392)
OR: H-32, I-? (392)

Daniellia oliveri   (Not known)   Caesalpiniaceae
    A   B   C   D   D   M
    01 01 09 01 02 12
OC: Termites H-8, I-13 (1345)

Daphne cneorum   (Garland flower)   Thymelaeaceae
    A   B   C   D   D   J   J   J
    01 02 09 03 06 03 08 11
OC: Attagenus piceus H-4, I-1 (48)
OR: H-32, I-4 (504)

Daphne genkwa   (Daphne genus)   Thymelaeaceae
    A   B   C   D   M   M   O
    01 02 09 03 05 13 07

OC: Agrotis sp. H-1, I-6 (1241)   Gryllotalpa sp. H-1, I-6 (1241)
    Aphids H-1, I-6 (1241)   Puccinia graminis tritici H-14, I-6 (1241)
    Glomerella gossypii H-14, I-6 (1241)
OR: H-17, I-7 (504)

Daphne mezereum   (Spurge laurel)   Thymelaeaceae
    A   B   C   D   D   J   K   L   M   M   O
    01 02 09 06 10 01 01 03 05 14 04
OC: Beetles H-2, I-16 (105, 1025)   Flies H-2, I-16 (105)
    Fleas H-2, I-16 (1025)
OR: H-29, I-1 (105); H-33, I-1, 10 (507, 1345)

Daphne odora   (Winter Daphne)   Thymelaeaceae
    A   B   C   D   M   J
    01 02 09 03 17 13
OC: Aphelencoides besseyi H-16, I-2 (52, 187, 376)
    Nematodes H-16, I-? (100)

Datura candida   (Safford)   Solanaceae
    C   D   J   L
    09 01 03 03
OC: Mosquitoes H-2, I-7 (101)   Musca domestica H-2, I-7 (101)
OR: H-33, 35, I-7 (504)

Datura metel   (Angel trumpet)   Solanaceae
    A   B   C   D   D   J   J   J   K   K   L   M   M   M   O   O   O
    03 06 09 01 02 04 08 11 03 09 03 05 16 17 07 08 10
OC: Aphis gossypii H-1, I-9 (120)   Pericallia ricini H-2, I-7 (105)
    Chiggers H-11, I-7 (105, 1353)   Potato virus X H-21, I-14 (1136)
    Crocidolomia binotalis H-2, I-7, 9 (105) Rats H-25, I-7, 10 (1345)
    Cucumis virus H-21, I-7 (368)   Spodoptera litura H-2, I-7 (105)
    Epilachna sp. H-1, I-7 (1029)   Sunnhemp rosette virus H-21, I-7 (368)
    Euproctis fraterna H-2, I-7, 9 (105)   Tobacco mosaic virus H-21, I-7
    Fleas H-5, I-7 (87)   (368, 557, 1308)
    Gomphrena mosaic virus H-21, I-7 (368) Tobacco ring spot virus H-21, I-7
    Lice H-5, I-7 (105)   (368, 557, 1308)
    Lipaphis erysimi H-1, I-10 (84, 91)   Vermin H-5, 27, I-7 (105)
OR: H-2, 15, 24, 33, I-7 (1350); H-2, I-10 (871); H-17, I-1 (881); H-20, I-19 (1148);
    H-32, I-? (1276); H-33, I-1 (127); H-33, 36, I-7, 10 (1353); H-35, I-7, 10 (1353)
AP: Alk=I-2, 7, 8, 10 (atropine, hyoscine, hyoscyamine, norhyoscyamine, scopolamine)(1392);
    Alk=I-6, 7 (1318, 1366); Alk=I-6, 7, 9 (1371); Alk=I-6, 10 (atropine, hyoscine, hyo-
    scyamine, scopolamine)(1392); Alk=I-7 (1390); Alk=I-10 (1391)

Datura stramonium   (Jimsonweed)   Solanaceae
    A   B   C   D   J   J   J   J   J   K   K   L   M   O   O
    03 06 09 03 02 03 04 09 12 05 06 03 05 07 10
OC: Alternaria tenuis H-14, I-8 (480, 523)   Dysdercus cingulatus H-3, I-1 (435, 441)
    Aphids H-1, I-7, 9 (87); H-2, I-6, 7, 10 Helminthosporium sp. H-14, I-8 (480, 523)
      (105, 1028, 1241)   Linum usitatissimum H-24, I-7, 10 (1196)
    Aulacophora abdominalis H-2, I-7 (105)   Maggots H-1, I-7, 9 (87)
    Caterpillars H-2, I-6, 7, 10 (105, 1028) Meloidogyne incognita H-16, I-7 (100, 616)
    Cnaphalocrocis medinalis H-1, I-1 (1241) Meloidogyne javanica H-16, I-7 (100, 616)
    Colaphellus bowringi H-1, I-1 (1241)   Pericallia ricini H-2, I-? (88, 435)
    Cucumber mosaic virus H-21, I-? (1132)   Phytodecta fornicata H-4, I-7 (435, 531)
    Curvularia penniseti H-14, I-8 (480)   Pieris rapae H-1, I-1 (1241)

Potato virus X H-21, I-? (1136) Stored grain pests H-5, I-7 (533)
Rice borers H-1, I-1 (1241) Tanymecus dilaticollis H-4, I-7
Rice leafhoppers H-1, I-1 (1241) (435, 441, 531)
Rondotia menciana H-1, I-1 (1241) Tobacco mosaic virus H-21, I-?
Spodoptera litura H-1, I-1 (435, 1241); (1131, 1132, 1306)
  H-2a, I-8 (127) Tomato bushy stunt virus H-21, I-? (1132)
OR: H-2, I-7 (1026); H-2, I-10 (1353); H-4, I-? (132); H-20, I-19 (1148, 1209); H-15, I-7
(1143); H-17, I-1 (881); H-19, I-? (594); H-22, I-16 (977, 1043); H-24, I-10 (1192);
H-35, I-7, 10 (507)
AP: **Alk**=I-1, 2, 10 (hyoscine, hyoscyamine)(1392); **Alk**=I-2, 7 (hyoscyamine)(1353); **Alk**=I-2, 9
(scopolamine)(1353, 1392); **Alk**=I-7 (atropine, hyoscyamine)(1392); **Alk**=I-9 (1392)

Daucus carota (Carrot genus) Apiaceae
  A A B C D D J M M N
  02 03 06 09 02 03 03 01 14 02
OC: Rats H-26, I-10 (799)
OR: H-22, I-6, 7 (969); H-24, I-7, 14 (569, 1204)
AP: **Alk**=I-7 (daucine, pyrrolidine)(1392); **Alk**=I-8, 9 (1374); **Alk**=I-9 (1369)

Davallia pentaphylla (Squirrel's foot fern genus) Polypodiaceae
  A B B C D
  01 04 10 06 01
OC: Erwinia carotovora H-18, I-7 (764) Pseudomonas solanacearum H-18, I-7 (764)

Davidia involucrata (Handkerchief tree) Nyssaceae
  A B C D
  01 01 09 03
OC: Agrobacterium tumefaciens H-18, I-? (957) Monilinia fructicola H-14, I-? (957)
  Colletotrichum lindemuthianum H-14, I-? Pseudomonas syringae H-18, I-? (957)
   (957) Xanthomonas campestris
  Endothia parasitica H-14, I-? (957) pv. phaseoli H-18, I-? (957)
  Erwinia amylovora H-18, I-? (957) pv. pruni H-18, I-? (957)
  Erwinia carotovora H-18, I-? (957) pv. vesicatoria H-18, I-? (957)

Deeringia celosioides (Not known) Amaranthaceae
  B C D J
  06 09 01 11
OC: Attagenus piceus H-4, I-1 (48)

Delonix regia (Royal poinciana) Caesalpiniaceae
  A B D J M M
  01 01 01 15 12 13
OC: Beetles H-2, I-8 (87) Sitophilus oryzae H-2, I-8 (400, 598)
  Caterpillars H-2, I-8 (87) Weevils H-2, I-8 (87)

Delphinium brunonianum (Musk larkspur) Ranunculaceae
  B C D L
  06 09 03 03
OC: Ticks H-13, I-7 (73, 104)
OR: H-2, I-10, 16 (504); H-2, 32, 33, I-? (220)

Delphinium cheilantum (Garland larkspur) Ranunculaceae
  B C D
  06 09 03

OC: Cockroaches H-2, I-7, 9 (101)   Lice H-2, I-7, 9 (101)
    Flies H-2, I-7, 9 (101)   Mosquitoes H-2, I-7, 9 (101)
OR: H-2, I-7, 10 (504)

Delphinium coeruleum   (Dhakangu)   Ranunculaceae
    B   C   D   D
    06   09   03   08
OC: Maggots H-2, I-2 (73, 104, 504)
OR: H-1, 32, I-? (220); H-2, I-7, 10 (504)

Delphinium delavayi   (Larkspur genus)   Ranunculaceae
    B   C   D   J   K
    06   09   03   04   05
OC: Epilachna varivestis H-2, I-1 (29, 101, 1012)

Delphinium dictyocarpum   (Not known)   Ranunculaceae
    B   C   D   J
    06   09   03   03
OC: Cockroaches H-2, I-8, 9 (101)   Lice H-2, I-8, 9 (101)
    Flies H-2, I-8, 9 (101)   Mosquitoes H-2, I-8, 9 (101)
OR: H-2, I-7, 10 (504)

Delphinium elatum   (Candle larkspur)   Ranunculaceae
    A   B   C   D
    01   06   09   03
OC: Maggots H-1, I-? (105)
OR: H-1, 32, I-? (220); H-2, I-7, 10 (73, 504); H-33, I-2, 7 (620)
AP: Alk=I-10 (delatine, delphelatine, delpheline, elatidine, elatine, eldeline, methyl-
    lycaconitine)(1392)

Delphinium formosum   (Larkspur genus)   Ranunculaceae
    A   B   C   D
    01   06   09   03
OC: Cimex lectularius H-2, I-10 (101)   Lice H-2, I-10 (101)
OR: H-2, I-7, 10 (504)

Delphinium glaucum   (Not known)   Ranunculaceae
    B   C   D   J   M
    06   09   03   06   13
OC: Mosquitoes H-2b, 4, I-? (800)
    Leptinotarsa decemlineata H-2b, 5, I-10, 12 (101)
OR: H-2, I-7, 10 (504)
AP: Alk=I-? (methyllycaconitine)(1392)

Delphinium grandiflorum   (Bouquet larkspur)   Ranunculaceae
    A   B   C   D   J
    01   06   09   03   03
OC: Cockroaches H-2, I-8, 9 (101)   Mosquitoes H-2, I-8, 9 (101)
    Flies H-2, I-8, 9 (101)   Plutella xylostella H-4, I-7 (87)
    Lice H-2, I-8, 9 (101)
OR: H-2, I-7, 10 (504)

Delphinium hybridum   (Larkspur genus)   Ranunculaceae
    A   B   C   D
    01   06   09   03
OC: Pieris brassicae H-4, I-? (531)

Delphinium laxiflorum   (Larkspur genus)   Ranunculaceae
   B C D J
  06 09 03 03
OC: Ants H-2, I-8, 9 (101)   Lice H-2, I-8, 9 (101)
    Cockroaches H-2, I-8, 9 (101)   Mosquitoes H-2, I-8, 9 (101)
    Flies H-2, I-8, 9 (101)
OR: H-2, I-7, 10 (504)

Delphinium orientale   (Larkspur genus)   Ranunculaceae
   A B C D
  03 06 09 03
OC: Musca domestica H-2, I-10 (87)
OR: H-2, I-7, 10 (504)

Delphinium retropilosum   (Larkspur genus)   Ranunculaceae
   B C D J
  06 09 03 03
OC: Cockroaches H-2, I-8, 9 (101)   Lice H-2, I-8, 9 (101)
    Flies H-2, I-8, 9 (101)   Mosquitoes H-2, I-8, 9 (101)
OR: H-2, I-7, 10 (504)

Delphinium semibarbatum   (Larkspur genus)   Ranunculaceae
   A B C D M
  01 06 09 03 16
OR: H-2, I-10 (504)

Delphinium sp.   (Larkspur)   Ranunculaceae
   B C D D J J K
  06 09 02 03 03 04 05
OC: Diaphania hyalinata H-2, I-2 (101)   Locusts H-2a, I-7 (105)
    Evergestis rimosalis H-2, I-2 (101)   Ostrinia nubilalis H-2, I-2 (101)
    Flies H-1, I-? (105)   Pediculus humanus capitis H-2, I-2 (101)
    Grasshoppers H-2, I-1 (105)   Trichoplusia ni H-2, I-2 (101)
    Haematopinus eurysternus H-1, I-7 (105)   Urbanus proteus H-2, I-2 (101)
    Lice H-1, I-7 (105)
OR: H-2, I-7, 10 (504)

Delphinium staphisagria   (Lousewort)   Ranunculaceae
   A B C D J J
  02 06 09 03 03 11
OC: Aleyrodes vaporariorum H-2, I-10, 12 (203) Mites H-11, I-? (104)
    Aphids H-2, I-10, 12 (800)   Myzus cerasi H-2, I-10, 12 (203)
    Aphis fabae H-2, I-10, 12 (203)   Myzus persicae H-2, I-10, 12 (203)
    Aphis pomi H-2, I-10, 12 (203)   Paratetranychus pilosus H-11, I-10, 12
    Aspidiotus perniciosa H-2, I-10, 12 (203)   (203)
    Attagenus piceus H-4, I-10 (48)   Pediculus humanus capitis H-2, I-? (104)
    Chrysomphalus sp. H-2, I-10, 12 (203)   Tetranychus cinnabarinus H-11, I-10, 12
    Lice H-1, I-? (87)   (203, 800)
    Macrosiphum rosae H-2, I-10, 12 (203)   Vermin H-27, I-? (104)
AP: Alk=I-10 (delphinine, delphinoidine, delphisine, staphisagroine, staphisine)(1392)

Delphinium vestitum   (Not known)   Ranunculaceae
   A B C D D
  01 06 09 03 08

OC: Maggots H-2, I-? (105)
OR: H-2, I-7, 10 (504); H-2, 32, I-? (220)

Delphinium virescens   (Prairie larkspur)   Ranunculaceae
    A   B   C   D   J   J   M
    01   06   09   03   02   03   12
OC: Aedes aegypti H-2, I-2, 8 (165)   Oncopeltus fasciatus H-2, I-10 (48)
    Blattella germanica H-2, I-10 (48)
OR: H-2, I-10 (504)

Delphinium yunnanense   (Larkspur genus)   Ranunculaceae
    B   C
    06   09
OC: Maggots H-1, I-2, 6, 7 (1241)

Dendrobium aggregatum   (Not known)   Orchidaceae
    B   B   C   D
    06   10   09   01
OC: Flies H-1, I-6 (1241)

Dendrocalamus strictus   (Calcutta bamboo)   Poaceae
    A   C   D   J   M   M   M   M   M
    01   09   01   10   02   04   06   19   22
OC: Dysdercus koenigii H-3, I-6 (795)

Dennettia tripetala   (Not known)   Annonaceae
    A   C   D   M
    01   09   01   14
OC: Periplaneta americana H-2, I-10, 12 (337)   Zonocerus variegatus H-2, I-10, 12 (337)

Dennstaedtia punctilobula   (Hay-scented fern)   Polypodiaceae
    A   C   D   D
    01   06   01   02
OC: Pseudomonas solanacearum H-18, I-7 (854)
    Xanthomonas campestris pv. phaseoli-sojensis H-18, I-7 (854)

Derris elliptica   (Derris)   Fabaceae
    A   B   C   D   F   J   J   J   K   K   M
    01   02   09   01   13   03   04   07   05   06   05
OC: Aphis fabae H-2, I-2 (21, 33, 38, 673)   Menopon biseriatum H-2, I-2 (201)
    Aphis medicaginis H-2, I-2 (526)   Mosquitoes H-2, I-2 (526)
    Autographa brassicae H-2, I-2 (276)   Myzus persicae H-2, I-2 (485)
    Bombyx mori H-2, I-2 (40)   Oryzaephilus surinamensis H-2, I-2
    Busseola fusca H-2, I-2 (256)   (21, 674, 1124)
    Callosobruchus chinensis H-2, I-2 (605)   Phalera bucephala H-2, I-2 (40)
    Ceratitis capitata H-2, I-2 (111)   Phymatocera aterrima H-2, I-2 (40)
    Coccus viridis H-2, I-2 (1029)   Pieris brassicae H-2, I-2 (40)
    Crocidolomia binotalis H-2, I-2 (605)   Pieris rapae H-2, I-2 (276)
    Epilachna varivestis H-2, I-2 (276)   Plutella xylostella H-2, I-2 (276, 605)
    Euphydryas chalcedona H-2, I-2 (201)   Pteronus ribesii H-2, I-2 (40)
    Hadena oleracea H-2, I-2 (40)   Pyrausta nubilalis H-2, I-2 (276)
    Macrosiphum liriodendri H-2, I-2 (485)   Pyricularia oryzae H-14, I-7 (114)
    Malacosoma neustria H-2, I-2 (40)   Rhopalosiphum persicae H-2, I-2 (201)
    Meloidogyne incognita H-16, I-7 (115)   Spodoptera litura H-2, I-2 (126, 605)
OR: H-2, I-2 (10, 11, 33, 38, 73, 80, 82, 83, 87, 101, 256, 511, 1352); H-32, 33, I-?
    (220, 459, 983)

Derris fordii (Indian tuba root) Fabaceae

A B C D J J J M
01 03 09 01 02 04 10 12

OC: Aphis fabae H-2a, I-2 (101)   Plutella xylostella H-2a, I-2 (101, 276)
    Oryzaephilus surinamensis H-2a, I-2 (101)

Derris malaccensis (Malaccan derris) Fabaceae

A B C D G J J K M M
01 03 09 01 04 02 03 06 07 12

OC: Aphis citri H-2, 5, I-2 (157)   Henosepilachna sparsa H-2, 5, I-2 (157)
    Autographa brassicae H-2, I-2 (276)   Macrosiphum solanifolii H-2, I-2 (276)
    Busseola fusca H-2, I-2 (256)   Pieris rapae H-2, I-2 (276)
    Crocidolomia binotalis H-2, 5, I-2 (157)   Plutella xylostella H-2, 5, I-2 (157)
    Epilachna varivestis H-2, I-2 (276)   Pyrausta nubilalis H-2, I-2 (276)

OR: H-2, I-2 (82, 83, 105, 509, 511); H-32, I-2 (459)

Derris malaccensis var. sarawakensis (Sarawak derris) Fabaceae

A B B C D J J M
01 01 03 09 01 02 04 12

OC: Aphis fabae H-2, I-2 (33)   Ceratitis capitata H-2a, I-2 (111)

Derris metaloides (Flame tree genus) Fabaceae

A B C D M
01 03 09 01 12

OC: Leptinotarsa decemlineata H-4, I-? (548)

Derris philippinensis (Philippine derris) Fabaceae

A B C J M
01 03 09 04 12

OC: Aphids H-2, I-2 (105)   Mosquitoes H-2, I-2 (526)
    Aphis medicaginis H-2, I-2 (526)

OR: H-32, I-2 (983)

Derris polyantha (Flame tree genus) Fabaceae

A B C D J J K M
01 01 09 01 03 04 05 12

OC: Aphids H-2, I-2 (105)   Aphis fabae H-2, I-2 (33)

OR: H-32, I-2 (459, 528, 864, 983)

Derris sp. (Not known) Fabaceae

A B C D J J M
01 01 09 01 02 04 12

OC: Acalymma vittata H-2, I-2 (255)   Epilachna varivestis H-2, I-2
    Acyrthosiphum pisum H-2, I-1, 2   (227, 258, 262, 488)
       (229, 230, 247)   Erythroneura comes H-2, I-2 (221)
    Anasa tristis H-2, I-2 (9, 239, 262)   Evergestis rimosalis H-2, I-2 (262)
    Aphids H-2, I-2 (25, 85)   Galerucella luteola H-2, I-2 (221)
    Aphis fabae H-2, I-2 (204)   Gnorimoschema lycopersicella H-2, I-2 (228)
    Autographa brassicae H-2, I-2 (222, 262)   Laspeyresia pomonella H-2, I-2 (255)
    Brevicoryne brassicae H-2, I-2 (204, 262)   Leptinotarsa decemlineata H-2, I-2 (262)
    Bruchus pisorum H-2, I-? (251)   Lipaphis erysimi H-2, I-2 (262)
    Ceratomia catalpae H-2, I-2 (262)   Macrosteles divisus H-2, I-2 (241)
    Chrysochus auritus H-2, I-2 (262)   Murgantia histrionica H-2, I-2 (224, 262)
    Epicauta pennsylvanica H-2, I-2 (252)   Musca domestica H-2, I-2 (248, 255)

Nezara viridula H-2, I-2 (262)
Oncopeltus fasciatus H-2, I-2 (9, 239)
Parasa herbifera H-2, I-2 (105)
Periphyllus lyropictus H-2, I-2 (221)
Phyllotreta vittata H-2, I-2 (221)
Pieris rapae H-2, I-2 (222, 226, 232, 259)
Plutella xylostella H-2, I-2
  (222, 224, 226)

Popillia japonica H-2, 5, I-2 (934)
Spodoptera eridania H-2, I-2 (255)
Taeniothrips gladioli H-2, I-2 (490)
Tetranychus cinnabarinus H-11, I-2
  (204, 490)
Thrips nigropilosus H-2, I-2 (490)
Thrips tabaci H-2, I-2 (204, 490)
Trialeurodes vaporariorum H-2, I-2 (204)

Derris thyrsiflora   (Flame tree genus)   Fabaceae
   A   B   B   C   D   J   M
   01 01 03 09 01 02 12
OR: H-2, I-2 (105)

Derris trifoliata   (Flame tree genus)   Fabaceae
   A   B   C   D   J
   01 01 09 01 03
OC: Rice leafhoppers H-1, I-2 (1241)

Derris tubli   (Tuble)   Fabaceae
   A   B   C   D   J   M
   01 03 09 01 02 12
OR: H-2, I-2 (504)

Derris uliginosa   (Not known)   Fabaceae
   A   B   C   D   J   J   K   L   M   M
   01 03 09 01 02 03 05 04 12 21
OC: Aphids H-2, I-2 (104)   Macrosiphum liriodendri H-1, I-? (485)
    Bombyx mori H-2, I-2 (104)   Mosquitoes H-2, I-2 (135)
    Cockroaches H-2, I-2 (104)   Musca domestica H-2, I-2 (104)
    Flies H-2, I-2 (104)   Myzus persicae H-1, I-? (485)
OR: H-2, I-4 (163); H-32, I-2, 4, 6 (73, 87, 220, 459, 504, 515); H-33, I-6 (983)
AP: Alk=I-4 (1392); Tan=I-6, 7 (1377)

Descurainia pinnata   (Not known)   Brassicaceae
   B   C   D   M   N   N
   06 09 03 01 07 10
OC: Mosquitoes H-10, I-10 (916)

Descurainia sophia   (Fixweed)   Brassicaceae
   B   C   D   D   M
   06 09 03 06 14
OC: Mosquitoes H-10, I-10 (916)
AP: Alk=I-2, 6, 7, 8 (1370)

Desmodium caudatum   (Not known)   Fabaceae
   B   C
   06 09
OC: Aphids H-1, I-7, 10 (1241)   Rice borers H-1, I-7, 10 (1241)
    Maggots H-1, I-7, 10 (1241)

Desmodium laburnifolium   (Tick clover)   Fabaceae
   B   B   C   D   D   M
   02 06 09 01 02 12
OR: H-2, I-7 (105)

Desmodium paniculatum   (Beggar-ticks genus)   Fabaceae
   B   C   D   J   M
   06 09 01 09 12
OC: Botrytis cinerea H-14, I-1 (946)

Desmodium tortuosum   (Beggarweed)   Fabaceae
   A   B   C   D   M   M   M
   01 06 09 01 09 10 12
OC: Criconemoides ornatum H-16, I-? (955)   Meloidogyne javanica H-16, I-? (955)
    Meloidogyne arenaria H-16, I-? (955)   Xiphinema americanum H-16, I-? (955)
    Meloidogyne incognita H-16, I-? (955)

Detarium senegalense   (Tallow tree)   Caesalpiniaceae
   A   B   C   D   J   K   L   M   M   M   N   N
   01 01 09 01 03 07 03 01 12 17 02 11
OC: Mosquitoes H-5, I-10 (1345)   Termites H-8, I-5 (504)
OR: H-33, I-4 (504)

Dianella ensifolia   (Umbrella dracaena)   Liliaceae
   A   B   C   D
   01 06 09 01
OC: Rats H-25, I-? (504)

Dianella nemorosa   (Not known)   Liliaceae
   A   B   C   D   M   M
   01 06 09 01 16 17
OR: H-1, I-2, 20 (504)

Dianthus barbatus   (Sweet William)   Caryophyllaceae
   A   A   B   C   D
   01 03 06 09 03
OC: Cucumber mosaic virus H-21, I-? (1060)   Tobacco mosaic virus H-21, I-? (1060)
    Tobacco etch virus H-21, I-? (1060)   Tobacco ring spot virus H-21, I-? (1060)

Dianthus caryophyllus   (Carnation)   Caryophyllaceae
   A   B   C   D   D   J   M   M
   01 06 09 03 06 03 13 17
OC: Tobacco mosaic virus H-21, I-7, 14 (210, 312, 339, 1058, 1130)

Dichapetalum cymosum   (Not known)   Dichapetalaceae
   B   B   C   D   J   J   J
   02 05 09 01 03 08 11
OC: Attagenus piceus H-4, I-1 (48)
AP: Alk=I-? (trigonelline)(1392)

Dichapetalum ruhlandii   (Not known)   Dichapetalaceae
   B   C   D   L
   03 09 01 03
OC: Aphids H-2, I-? (681)
OR: H-33, I-? (105)

Dichapetalum toxicaria   (West African ratsbane)   Dichapetalaceae
   B   C   D   J   J   J   L
   03 09 01 03 08 11 03

OC: Attagenus piceus H-4, I-1, 2, 6, 7, 9 (48) Rats H-25, I-10 (504)
    Pediculus humanus capitis H-2, I-? (105)   Tineola bisselliella H-4, I-2 (48)
OR: H-32, 33, I-10 (504)

Dichroa febrifuga   (Not known)   Saxifragaceae
    B   C   D   M   O   O
    06 09 03 05 02 07
OC: Aphids H-1, I-1 (1241)   Spider mites H-11, I-1 (1241)
    Rice borers H-1, I-1 (1241)
AP: Alk=I-2, 7 (dichroidine, dichroine, febrifugine, isofebrifugine, 4-ketodihydro-
    quinazoline)(1392)

Dichrostachys cinera   (Not known)   Mimosaceae
    A   B   C   D   D   M   M   M   M   M   O
    01 02 09 01 05 04 05 12 19 21 02
OC: Gasterophilus intestinalis H-1, I-7 (105)
AP: Tan=I-4 (504)

Dictamnus dasycarpus   (Burning bush genus)   Ranunculaceae
    A   B   C
    01 06 09
OC: Aphids H-1, I-9 (1241)

Dictyosphaerium divaricata   (Green algae)   Scenedesmaceae
    A   B   C
    04 09 02
OC: Mucor racemosus H-14, I-1 (1024)   Rhizopus oryzae H-14, I-1 (1024)

Dictyosphaerium favulosa   (Green algae)   Scenedesmaceae
    A   B   C
    04 09 02
OC: Mucor racemosus H-14, I-1 (1024)

Dictyosphaerium indica   (Green algae)   Scenedesmaceae
    A   B   C
    04 09 02
OC: Mucor racemosus H-14, I-1 (1024)

Didymocarpus pedicellata   (Not known)   Gesneriaceae
    A   B   C   D   J   J   J
    01 06 09 01 11 13 16
OC: Alternaria sp. H-14, I-? (1300)   Xanthomonas campestris
    Fusarium sp. H-14, I-? (1300)   pv. campestris H-18, I-7 (439)
    Helminthosporium sp. H-14, I-? (1300)

Didymopanax morototoni   (Matchwood)   Araliaceae
    A   B   C   D   J   M   M
    01 01 09 10 08 04 22
OC: Attagenus piceus H-4, I-6, 7 (48)

Didymopanax tremulum   (Not known)   Araliaceae
    B   C   D   J   J   J
    01 09 01 03 08 11
OC: Attagenus piceus H-4, I-6, 7 (48)

Dieffenbachia sequine   (Mother-in-law plant)   Araceae
```
     B   C   D   J   J   M
     06  09  01  08  11  13
```
OC: Attagenus piceus H-4, I-6, 7 (48)
OR: H-33, I-2, 6 (620)

Digera alternifolia   (Amaranth genus)   Amaranthaceae
```
     A   B   C   D   D   J
     03  06  09  05  09  03
```
OR: H-24, I-2, 6, 7 (1206)

Digera arvensis   (Amaranth genus)   Amaranthaceae
```
     A   B   C   D   J   M   M   N
     03  06  09  01  03  01  02  07
```
OR: H-24, I-2, 15 (1184, 1208)

Digitalis grandiflora   (Yellow foxglove)   Scrophulariceae
```
     B   C   D   D   J   J   K   K   M   O
     07  09  03  06  03  06  05  06  05  07
```
OC: Aphids H-2, I-6, 7 (105)   Musca domestica H-2a, I-8 (87)
    Flea beetles H-2, I-6, 7 (105)

Digitalis lanata   (Grecian foxglove)   Scrophulariaceae
```
     A   B   C   D   D   K   M   O
     01  06  09  03  06  06  05  07
```
OC: Reticulitermes flavipes H-4, I-10 (87)

Digitalis purpurea   (Common foxglove)   Scrophulariaceae
```
     A   A   B   C   J   J   L   M   O
     01  02  06  09  03  09  03  05  07
```
OC: Aphids H-2, I-7 (105)   Reticulitermes flavipes H-4, I-? (87)
OR: H-19, I-6, 7, 8, 9 (413, 940); H-33, I-? (220)
AP: Alk=I-10 (1374)

Digitaria decumbens   (Pangola grass)   Poaceae
```
     A   B   B   C
     01  06  09  02
```
OR: H-24, I-7 (1247)

Digitaria sanguinalis   (Hairy crabgrass)   Poaceae
```
     A   B   C   D   L
     03  06  09  03  05
```
OC: Amaranthus retroflexus H-24, I-2 (582)   Aristida oligantha H-24, I-2 (582)
    Ambrosia elatior H-24, I-2 (582)   Bromus japonicus H-24, I-2 (582)

Dillenia indica   (Elephant apple)   Dilleniaceae
```
     A   B   C   D   J   M   M   M   N
     01  01  09  10  08  01  04  19  09
```
OC: Coccus viridis H-1, I-? (1029)   Euproctis fraterna H-2, I-7 (105)
    Crocidolomia binotalis H-2, I-7 (105)   Spodoptera litura H-2, I-7 (105)
    Epilachna sp. H-1, I-? (1029)

Dioclea megacarpa   (Not known)   Fabaceae
```
     B   C   D   M
     04  09  01  12
```

OC: Manduca sexta H-2, I-? (191)
OR: H-2, I-10 (132, 724)

Dioscorea balcana   (Yam genus)   Dioscoreaceae
   B   C   D   D
  04 09 01 04
OC: Locusta migratoria H-4, 5, I-7 (596)

Dioscorea batatas   (Chinese yam)   Dioscoreaceae
   B   C   D   D   M   N
  04 09 03 04 01 02
OC: Trimeresia miranda H-1, I-7 (382)   Spodoptera littoralis H-1, I-7 (382)

Dioscorea bulbifera   (Air potato)   Dioscoreaceae
   B   C   D   D   J   M   N
  04 09 01 02 11 01 02
OC: Attagenus piceus H-4, I-2 (48)
AP: Alk=I-2, 6, 7 (1363); Alk=I-6, 9 (1360)

Dioscorea cylindrica   (Not known)   Dioscoreaceae
   B   C   D   J
  04 09 01 04
OC: Maggots H-2, I-2 (105)

Dioscorea deltoidea   (Kilari)   Dioscoreaceae
   B   C   D   D   D   J
  03 09 01 02 08 03
OC: Lice H-2, I-4 (504)
OR: H-32, I-2 (504)

Dioscorea hispida   (Nami)   Dioscoreaceae
   B   C   D   L
  03 09 01 03
OC: Aphids H-1, I-2 (87)   Maggots H-2, I-2 (82, 967)
OR: H-32, I-2 (983); H-33, I-2 (504, 967, 983)
AP: Alk=I-2 (dioscorine)(1366, 1392)

Dioscorea latifolia   (Yam genus)   Dioscoreaceae
   B   C   D
  04 09 02
OC: Chiggers H-5, 11, I-3 (1345)

Dioscorea nipponica   (Yam genus)   Dioscoreaceae
   B   C   D   J
  04 09 03 03
OC: Aphids H-1, I-2, 7 (1241)

Dioscorea piscatorum   (Sakut)   Dioscoreaceae
   B   C   D
  04 09 01
OC: Parasa herbifera H-2, I-? (105)
OR: H-2, I-14 (105); H-32, I-? (105)

Dioscorea prazeri   (Yam genus)   Dioscoreaceae
    B   C   D   L
    04   09   03   03
OC: Lice H-2, I-2 (73, 504)

Diospyros argentea   (Persimmon genus)   Ebenaceae
    A   B   B   C   D   K
    01   01   02   09   01   07
OC: Mosquitoes H-5, I-7 (83)

Diospyros discolor   (Mabala)   Ebenaceae
    A   C   C   J   J   M   N
    01   01   09   08   14   01   09
OC: Ceratitis capitata H-6, I-4, 5, 6 (956)   Dacus dorsalis H-6, I-4, 5, 6 (956)
    Dacus cucurbitae H-6, I-4, 5, 6 (956)
OR: H-19, I-7 (982)

Diospyros ebenum   (Ebony)   Ebenaceae
    A   B   C   D   J   M   M   N
    01   01   09   02   03   01   19   09
OC: Termites H-1, I-? (1352)
OR: H-2, I-? (100); H-32, I-? (73, 220)

Diospyros insignis   (Bombay ebony)   Ebenaceae
    A   B   C   D   J
    01   01   09   01   09
OC: Rats H-26, I-16 (1393)

Diospyros kaki   (Date plum)   Ebenaceae
    A   B   C   D
    01   01   09   03
OC: Fusarium roseum H-14, I-15 (917)   Fusarium solani f. sp. mori H-14, I-15 (917)
AP: Tan=I-9 (1350)

Diospyros malacapai   (White ebony)   Ebenaceae
    A   B   C   D
    01   01   09   01
OR: H-5, I-5 (1345)

Diospyros montana   (Date plum)   Ebenaceae
    A   B   C   D   J   J   M
    01   01   09   01   03   08   06
OC: Achaea janata H-2, I-7 (105)   Epilachna sp. H-2, I-7 (105)
    Diacrisia obliqua H-2, I-7 (105)
OR: H-32, I-7, 9 (73, 605); H-33, I-9 (507)
AP: Tri=I-6 (lupeol)(1385)

Dipcadi cowanii   (Rongolo voalova)   Liliaceae
    B   C   D
    06   09   01
OC: Rats H-25, I-3 (504)

Dipsacus fullonum   (Common teasel)   Dipsacaceae
    A   C   D   J
    02 06 03 09
OC: Sclerotinia fructicola H-14, I-1, 9 (944)

Dipterocarpus turbinatus   (Not known)   Dipterocarpaceae
    C   D
    09 01
OR: H-5, I-12 (105)

Disanthus cercidifolius   (Not known)   Hamamelidaceae
    A   B   C
    01 02 03
OC: Spodoptera littoralis H-1, I-7 (382)   Trimeresia miranda H-1, I-7 (382)

Dodonaea viscosa   (Switchsorrel)   Sapindaceae
    A   B   C   D   M   M   M   M   M   N   O   O
    01 02 09 01 03 05 06 08 21 28 04 13
OC: Termites H-8, I-5 (1353)
OR: H-2, 32, I-? (220); H-24, I-10, 15 (1184); H-32, I-7 (73, 983, 1353)
AP: Alk=I-7 (1392); Sap=I-5 (1365); Tan=I-4 (1352)

Dolichos buchani   (Not known)   Fabaceae
    B   C   D   J   J   J   M   M   M   M   N   N
    04 09 01 03 08 11 01 10 12 13 10 11
OC: Aedes aegypti H-4, I-2 (48)   Oncopeltus fasciatus H-4, I-2 (48)
    Attagenus piceus H-4, I-2 (48)   Tribolium castaneum H-4, I-2 (48)

Dolichos kilimandshauricus   (Indian butter bean)   Fabaceae
    B   B   D   M
    04 06 01 12
OC: Tribolium castaneum H-2, I-7 (1124)

Dolichos lablab   (Lab-lab bean)   Fabaceae
    A   B   C   D   J   J   J   L   M   M   M   M   M   M   N   N   N   O   O
    01 03 09 01 03 08 11 03 1a 02 05 10 12 27 07 10 11 07 10
OC: Attagenus piceus H-4, I-2, 10 (48)   Fusarium oxysporum
    Fusarium nivale H-14, I-1, 14 (1048)   f. sp. cubense H-14, I-? (952)
OR: H-22, I-16 (977); H-33, I-2 (1321)
AP: Alk, Sfr=I-6 (1321)

Dolichos pseudopachyrrhyzus   (Not known)   Fabaceae
    B   C   D   J   M
    04 09 01 03 12
OC: Ticks H-13, I-2 (105)
OR: H-2, I-7, 15 (87)

Doodia media   (Hacksaw fern genus)   Polypodiaceae
    A   C   D
    01 06 01
OC: Erwinia carotovora H-18, I-7 (854)
    Xanthomonas campestris pv. phaseoli-sojensis H-18, I-7 (854)
AP: Sap=I-7 (1364)

Dorycnium rectum   (Not known)   Fabaceae
    B   C   D   J   M   M
    06 09 06 14 12 13
OC: Ceratitis capitata H-6, I-6, 7, 8 (956)

Doryphora sassafras   (Sassafras)   Monimiaceae
    A   B   C   J   J   J   M
    01 01 09 06 09 11 17
OC: Aedes sp. H-1, I-? (101)   Mosquitoes H-30, I-? (101)
OR: H-19, I-12 (775, 776, 822)
AP: Alk=I-4 (doryphorine)(1392); Alk=I-4, 6 (1371)

Dracocephalum moldavica   (Dragonhead genus)   Lamiaceae
    A   B   C   D   M
    03 06 09 03 17
OC: Aphis gossypii H-2, I-12 (105)
    Tetranychus cinnabarinus H-11, I-12 (1030)

Dracunculus vulgaris   (Not known)   Araceae
    A   B   C   D   D   K
    01 06 09 03 06 07
OC: Blow flies H-6, I-? (105)   Livestock pests H-1, I-20 (504)

Drimia cowanni   (Not known)   Liliaceae
    B   C
    06 09
OC: Rats H-25, I-3 (504)

Drimys lanceolata   (Pepper tree)   Winteraceae
    A   B   C   D   M   M   N   N
    01 01 09 02 01 14 09 10
OR: H-4, I-? (146); H-19, I-12 (775)

Drymaria pachyphylla   (Not known)   Caryophyllaceae
    B   C   D   J   J
    06 09 03 03 11
OC: Attagenus piceus H-4, I-1 (48)   Oncopeltus fasciatus H-4, I-1 (48)
    Blattella germanica H-4, I-1 (48)

Drynaria quercifolia   (Oak-leaved fern)   Polypodiaceae
    A   B   C   D   M   O
    01 10 06 01 05 02
OC: Erwinia carotovora H-18, I-7 (854)
    Xanthomonas campestris pv. phaseoli-sojensis H-18, I-7 (854)
OR: H-19, I-7 (982)

Dryopteris austriaca   (Spinulosa wood)   Polypodiaceae
    A   C   D   J
    01 06 03 09
OC: Ceratocystis ulmi H-14, I-9, 16 (946)   Rodents H-25, I-1 (944)
    Fusarium oxysporum f. sp. lycopersici H-14, I-1 (944)

Dryopteris bissetiana   (Wood fern genus)   Polypodiaceae
   A   C   D   D
   01 06 01 03
OC: Drosophila hydei H-1, I-? (507)

Dryopteris cochealata   (Wood fern genus)   Polypodiaceae
   A   C   D
   01 06 10
OC: Drechslera oryzae H-14, I-? (419)

Dryopteris crassirhizoma   (Wood fern genus)   Polypodiaceae
   A   C   D   M   O
   01 06 03 05 01
OC: Phytophthora infestans H-14, I-1 (1241)   Puccinia graminis tritici H-14, I-1 (1241)
   Pseudoperonospora cubensis H-14, I-1 (1241)

Dryopteris erythrosora   (Japanese shield fern)   Polypodiaceae
   A   C   J
   01 06 03
OC: Drosophila hydei H-2, I-7 (101)

Dryopteris filix-mas   (Male fern)   Polypodiaceae
   A   C   D   J   J   M   O
   01 06 03 02 10 05 02
OC: Aphis fabae H-2, I-2 (105, 467)   Pseudaletia unipuncta H-2, I-2 (101)
   Culex quinquefasciatus H-2, I-2 (105)   Pseudomonas solanacearum H-18, 1-2 (854)
   Mosquitoes H-2, I-2 (98)   Xanthomonas campestris
   Musca domestica H-2, I-2 (101)   pv. phaseoli H-18, I-2 (854)
OR: H-17, I-2 (504)

Dryopteris marginalis   (Leather wood fern)   Polypodiaceae
   A   C   D
   01 06 03
OC: Leptinotarsa decemlineata H-4, I-7 (548)

Duboisia hopwoodii   (Not known)   Solanaceae
   A   B   C   D
   01 02 09 01
OC: Aphids H-1, I-? (81)   Culex quinquefasciatus H-2, I-? (105)
   Aphis fabae H-2, I-? (105, 279)   Mealy bugs H-2, I-? (81)
AP: Alk=I-7 (nicotine, nornicotine)(1392)

Duboisia myoporoides   (Corkwood)   Solanaceae
   A   B   D   J
   01 02 01 03
OC: Tobacco mosaic virus H-21, I-? (1126)
OR: H-32, I-? (504)
AP: Alk=I-? (anabasine, hyoscine, hyoscyamine, isopelletierine, isoporoidine, nicotine,
   norhyoscyamine, nornicotine, poroidine, scopolamine, valeroidine)(1392); Alk=I-2 (1374)

Duchesnea indica   (Indian strawberry)   Rosaceae
   A   B   C   D   L   M
   01 06 09 01 05 09
OC: Aphids H-1, I-1 (1241)   Maggots H-1, I-1 (1241)

Duranta repens   (Pigeon berry)   Verbenaceae
```
A    B    C    D    D    J    J    J    J    J    K
01   01   09   01   02   02   03   04   08   11   06
```
OC: Aedes aegypti H-2, I-7 (165)   Mosquitoes H-2, I-9 (101, 105, 1345)
    Attagenus piceus H-4, I-9 (48)

Durio zibethinus   (Durian)   Bombacaceae
```
A    B    C    D    M    M    N    O
01   01   09   01   01   05   09   10   02
```
OR: H-2, I-? (100)

Dysoxylum procerum   (Not known)   Meliaceae
```
A    B    C    D
01   01   09   01
```
OR: H-2, I-? (504)

Echinacea angustifolia   (Rattlesnake)   Asteraceae
```
A    B    C    D    J    M    O
01   06   09   03   14   05   02
```
OC: Mosquitoes H-1, I-2 (818)   Tenebrio molitor H-1, I-1 (1150);
    Musca domestica H-2, I-2 (47, 87)   H-3, I-2 (958)
OR: H-19, I-? (818)

Echinacea pallida   (Purple coneflower)   Asteraceae
```
A    B    C    D    M    M    O
01   06   09   03   05   13   02
```
OC: Pseudomonas solanacearum H-18, I-1, 8 (945)

Echinochloa crus-galli   (Barnyard grass)   Poaceae
```
A    B    C    D    D    L
03   06   09   01   02   05
```
OC: Leptinotarsa decemlineata H-4, I-7 (531)   Pieris brassicae H-4, I-7 (531)
    Nilaparvata lugens H-1, I-7 (560, 561)

Echinops echinatus   (Not known)   Asteraceae
```
B    C    D    D    J    J
06   09   02   06   03   04
```
OC: Argemone mexicana H-24, I-15 (1185)   Pediculus humanus capitis H-1, I-2 (104)
    Maggots H-1, I-2 (104)

Echinops setifer   (Globe thistle genus)   Asteraceae
```
B    C    D
06   09   06
```
OC: Tobacco mosaic virus H-21, I-? (1126)

Eclipta alba   (Eclipta)   Asteraceae
```
A    B    C    D    F    G    J    J    J    K    M    M    M    M    O
03   06   09   01   01   05   03   05   11   01   05   16   21   27   07
```
OC: Attagenus piceus H-4, I-1, 8 (48)   Nilaparvata lugens H-1, I-2 (1289)
    Meloidogyne sp. H-16, I-? (161)   Rotylenchulus sp. H-16, I-? (161)
OR: H-22, I-1 (881)
AP: Alk=I-1 (nicotine)(1392); Tan=I-2, 7 (1321); Sfr=I-6 (1321)

Ehretia dicksonii   (Not known)   Boraginaceae
```
A   B   C   D
01  01  09  03
```
OC: Aphis gossypii H-1, I-7, 9 (1241)   Spider mites H-11, I-7, 9 (1241)

Elaeis guineensis   (African oil palm)   Arecaceae
```
A   B   C   D   M   M   M   M   N   N   N
01  01  09  01  01  02  15  29  09  10  12
```
OC: Macrophomina phaseolina H-14, I-12 (1300)
     Xanthomonas campestris pv. campestris H-18, I-12 (1300)
     Xanthomonas campestris pv. citri H-18, I-12 (1300)

Eleocharis colorodoensis   (Dwarf spike rush)   Cyperaceae
```
B   C
06  09
```
OR: H-24, I-2 (1257)

Eleocharis obtusa   (Not known)   Cyperaceae
```
B   C   J
06  09  09
```
OC: Botrytis cinerea H-14, I-1, 9 (946)

Elephantopus scaber   (Elephant's foot)   Asteraceae
```
A   B   C   D   F   G   J   K   L   M   O   O
03  06  09  01  06  04  06  06  05  05  02  07
```
OC: Dysdercus koenigii H-3, I-6, 7, 12 (795)   Tribolium castaneum H-2a, I-7 (125)
AP: Alk, Sap=I-2, 6 (1360); Tri=I-1 (lupeol)(1384, 1386, 1391)

Elodea canadensis   (Water weed)   Hydrocharitaceae
```
B   B   C   D   J   J   L
06  09  09  03  03  16  05
```
OC: Fusarium roseum H-14, I-1 (837)
OR: H-19, I-1 (837)

Elsholtzia blanda   (Not known)   Lamiaceae
```
A   B   C   D
03  06  09  03
```
OC: Mosquitoes H-5, I-7 (1297)

Elymus canadensis   (Canada wild rye)   Poaceae
```
A   B   C   D   E   M   M
01  06  09  03  07  02  08
```
OC: Melanoplus femurrubrum H-1, I-7 (87)
AP: Alk=I-2, 6, 7, 8 (1370)

Embelia ribes   (Baberang)   Myrsinaceae
```
C   D   J
09  03  11
```
OC: Dysdercus koenigii H-2, I-? (91)   Rats H-26, I-2 (799)
     Lipaphis erysimi H-2, I-? (91)
OR: H-17, I-10 (504, 1135)

Embelia viridiflora   (Not known)   Myrsinaceae
```
A   D   J
01  02  09
```
OC: Rats H-26, I-16 (1393)

Emblica officinalis   (Emblic myrobalan)   Myrsinaceae
```
A   B   D   J   M   M   M   M   M   M   N
01  01  01  09  01  04  06  15  16  21  09
```
OC: Colletotrichum falcatum H-14, I-9 (1047)   Pyricularia oryzae H-14, I-9 (1047)
AP: Sap, Str, Tan=I-6 (1359); Tan=I-4, 7, 9 (1354); Tan=I-7 (1360); Tan=I-9 (1047)

Emilia tuberosa   (Not known)   Asteraceae
```
B   C   D
06  09  01
```
OC: Caterpillars H-2, I-? (101)   Musca domestica H-2, I-? (101)
    Mosquitoes H-2, I-? (101)
OR: H-2a, I-? (101)

Eminium intortum   (Not known)   Araceae
```
A   B   C   D   D   J   J
01  06  09  01  06  03  04
```
OC: Tylenchulus semipenetrans H-16, I-7 (540)

Encelia farinosa   (Brittle bush)   Asteraceae
```
A   B   C   D   J   M   M   N
01  02  09  05  03  01  17  13
```
OC: Diaphania hyalinata H-2, I-? (1252)
OR: H-5, I-? (173); H-24, I-7 (553, 555, 556, 559)
AP: Ter=I-? (173)

Encelia flagellaris var. radians   (Not known)   Asteraceae
```
A   B   B   C   D
01  01  02  09  04
```
OC: Diaphania hyalinata H-1, I-? (1252)

Encelia frutescens   (Brittle bush genus)   Asteraceae
```
A   B   C   D   J
01  02  09  05  03
```
OR: H-24, I-7 (554, 559)

Endymion non-scriptus   (Blue bell)   Liliaceae
```
A   B   C   D   L
01  06  09  01  03
```
OC: Venturia inaequalis H-14, I-? (16)
OR: H-33, I-3 (504)

Engelhardtia spicata   (Not known)   Juglandaceae
```
A   B   C   D   M
01  01  09  10  21
```
OR: H-2, I-? (100)
AP: Tan=I-4 (504)

Entada gigas (St. Thomas bean) Mimosaceae
  A   B   C   D   J   M   M   M   M   N   N
  01 03 09 01 03 1a 03 12 15 07 10
OC: Snails H-28, I-4 (662)
OR: H-32, I-4, 7, 10 (73, 751); H-32, 33, I-? (220, 864)
AP: Alk=I-10 (1392)

Entada polystachia (Paria rosa) Mimosaceae
  A   B   C   D   J   M   M   M
  01 02 09 01 03 03 12 15
OC: Spodoptera litura H-1, I-? (101)

Enterolobium contortisiliquum (Not known) Mimosaceae
  A   B   C   D   J   J   J   M   M   M   M
  01 01 09 01 03 08 11 02 04 12 13
OC: Attagenus piceus H-4, I-6 (48)

Enterolobium cyclocarpum (Elephant's ear) Mimoscaceae
  A   B   C   D   J   J   M   M   M   M   M   O
  01 01 09 01 08 11 04 05 12 15 19 04
OC: Anthonomus grandis H-2a, I-9 (101)   Pachyzancla bipunctalis H-1, I-4 (101)
    Attagenus piceus H-4, I-2 (48)   Periplaneta americana H-1, I-? (101)
    Oncopeltus fasciatus H-2a, I-2 (48)   Spodoptera litura H-2a, I-9 (101)

Enterolobium saman (Acacia) Mimosaceae
  A   B   D   M   M   M
  01 01 01 02 12 13
OC: Rice field insects H-5, I-6, 7 (137)

Epilobium angustifolium (Fireweed) Onagraceae
  A   B   C   D   M   N   N
  01 06 09 03 01 07 15
OC: Locusta migratoria H-4, 5, I-7 (596)

Epilobium glandulosum (Fireweed genus) Onagraceae
  B   C   D   J
  06 09 03 14
OC: Oncopeltus fasciatus H-1, I-2, 6, 7, 8 (958)

Epimedium sagittatum (Not known) Berberidaceae
  A   B   B   C   D   M
  01 06 10 09 03 13
OC: Aphids H-1, I-1 (1241)   Puccinia rubigavera H-14, I-1 (1241)
    Phytophthora infestans H-14, I-1 (1241)   Verticillium albo-atrum H-14, I-1 (1241)
    Puccinia graminis tritici H-14, I-1 (1241)

Equisetum arvense (Horsetail genus) Equisetaceae
  A   B   C   D   E   M   M   N   N   O
  01 06 05 03 11 01 05 02 16 16
OC: Pieris rapae H-1, I-1 (1241)
AP: Alk=I-1 (methoxypyridine, nicotine, palustrine)(1392)

Equisetum fluviatile   (Not known)   Equisetaceae
    A   B   C   J   J
    01 06 09 02 03
OC: Aedes aegypti H-2, I-7, 10 (165)

Eragrostis amabilis   (Feather love grass)   Poaceae
    A   B   C   D   J
    03 06 09 10 03
OC: Meloidogyne incognita H-16, I-7 (115)

Eragrostis cilianenis   (Stink grass)   Poaceae
    A   B   C   D   J   J   M
    01 06 09 01 03 11 10
OC: Attagenus piceus H-4, I-1 (48)

Eragrostis curvula   (Weeping love grass)   Poaceae
    B   C   D   D   M   M   M   M
    06 09 02 04 02 08 09 13
OR: H-24, I-7 (1247)

Eranthis hyemalis   (Winter aconite)   Ranunculaceae
    A   B   C   D   J
    01 06 09 03 04
OC: Plutella xylostella H-2, I-2 (101)
AP: Alk=I-? (1392)

Eremocarpus setigerus   (Dove weed)   Euphorbiaceae
    A   B   C   D   J   L   M
    03 06 09 04 03 05 17
OC: Aphids H-1, I-? (106)   Evergestis rimosalis H-2, I-? (101)
OR: H-32, I-16 (106, 192)
AP: Alk=I-7 (1392)

Eremophila mitchelli   (Rosewood)   Myoporaceae
    A   B   C   D   M   M   M
    01 01 09 01 04 07 19
OR: H-19, I-12 (775); H-30, I-5, 12 (101)
AP: Alk=I-7 (1392)

Erica arborea   (Tree heath)   Ericaceae
    C   J   M   M
    09 03 18 19
OR: H-24, I-16 (1189)

Erica australis   (Heath genus)   Ericaceae
    A   B   C   D   J   M
    01 02 09 06 03 18
OR: H-24, I-16 (1189)

Erica scoparia   (Besom heath)   Ericaceae
    A   B   B   C   D   J   M
    01 01 02 09 06 03 18
OR: H-24, I-7 (1179)

Erigeron affinis (Fleabane genus) Asteraceae
   B   C   D   J   M   O
   06 09 03 11 05 07
OC: Laspeyresia pomonella H-2, I-2 (810, 811)   Musca domestica H-2, I-2 (810, 811)
   Mosquitoes H-2, I-2 (810, 811)
OR: H-34, I-7 (504)

Erigeron annuus (Daisy fleabane) Asteraceae
   A   B   C   D   J
   03 06 09 03 11
OC: Attagenus piceus H-4, I-16 (48)

Erigeron bellidiastrum (Western fleabane) Asteraceae
   A   B   C   D   J
   03 06 09 03 04
OC: Diaphania hyalinata H-2, I-1 (101)   Pachyzancla bipunctalis H-2, I-1 (101)

Erigeron canadensis (Canadian fleabane) Asteraceae
   B   C   D   J   J   J   M   O
   06 09 03 03 04 08 05 07
OC: Anasa tristis H-1, I-? (1252)   Exserohilum turcicum H-14, I-? (827)
   Athalia rosae H-4, I-7 (517)   Pieris brassicae H-4, I-? (531)
   Autographa oo H-1, I-? (1252)   Pieris rapae H-1, I-? (1252)
   Diaphania hyalinata H-2, I-7 (101, 1252)   Popillia japonica H-5, I-7, 16 (105, 1243)
AP: Sap=I-2 (1378)

Erigeron flagellaris (Running fleabane) Asteraceae
   A   A   B   B   D
   01 02 06 09 03
OC: Anasa tristis H-1, I-? (1252)   Pieris rapae H-1, I-? (1252)
   Autographa oo H-1, I-? (1252)   Udea rubigalis H-1, I-? (1252)
   Diaphania hyalinata H-1, I-? (1252)

Erigeron glabellus (Not known) Asteraceae
   A   B   C   D   J
   02 06 09 03 03
OC: Attagenus piceus H-4, I-1 (48)

Erigeron linifolius (Fleabane genus) Asteraceae
   B   C   D   G   J   L
   06 09 03 04 03 05
OC: Alternaria solani H-14, I-7 (859)   Glomerella cingulata H-14, I-7 (859)
   Bipolaris maydis H-14, I-7 (859)   Pythium aphanidermatum H-14, I-7 (859)
   Drechslera oryzae H-14, I-7 (859)   Sclerotium rolfsii H-14, I-7 (859)
   Exserohilum turcicum H-14, I-7, 15 (827, 859)

Erigeron philadelphicus (Fleabane genus) Asteraceae
   B   C   D   J
   06 09 03 09
OC: Ceratocystis ulmi H-14, I-1 (836)

Erigeron repens (Not known) Asteraceae
   B   C   D   J
   06 09 03 04
OC: Ostrinia nubilalis H-2, I-? (101)

Erigeron sphaerocephalcum   (Fleabane genus)   Asteraceae
    B   C   D
    06 09 03
OC: Leptinotarsa decemlineata H-4, I-? (1118)

Eriobotrya japonica   (Loquat)   Rosaceae
    A   B   C   D   J   M   M   N
    01 01 09 10 03 01 13 09
OC: Drosophila sp. H-2, I-8 (101)   Ustilago tritici H-14, I-7 (181)
OR: H-24, I-7 (569)

Eriolaena quinquelocularis   (Not known)   Sterculiaceae
    A   C   D   J
    01 09 01 09
OC: Rats H-26, I-16 (1393)

Eriosema psoraleoides   (Not known)   Fabaceae
    B   C   D   D   M
    02 09 01 02 12
OC: Lice H-1, I-7 (87, 105)

Erodium cicutarium   (Stork's bill)   Geraniaceae
    A   B   C   D   D   J   J   J   M   N
    03 06 09 01 06 03 08 11 01 15
OC: Attagenus piceus H-4, I-1 (48)   Spodoptera frugiperda H-1, I-1 (176)
AP: Alk=I-1 (caffeine, tyramine)(1392)

Eruca vesicaria   (Rocket salad)   Brassicaceae
    A   B   D   J   M   M   M   M   N   N   O
    01 09 01 09 01 05 06 29 07 15 07
OC: Alternaria tenuis H-14, I-8 (480)   Helminthosporium sp. H-14, I-8 (480)
    Curvularia penniseti H-14, I-8 (480)   Hoplolaimus indicus H-16, I-? (612)

Erysimum cheiranthoides   (Wormseed mustard)   Brassicaceae
    A   B   C   D
    03 06 09 03
OC: Phaedon cochleariae H-4, I-7 (1134)   Phyllotreta undulata H-4, I-7 (1134)
    Phyllotreta tetrastigma H-4, I-7 (1134)

Erysimum hieraciifolium   (Wallflower genus)   Brassicaceae
    A   A   B   B   C
    01 02 03 06 09
OC: Phaedon cochleariae H-4, I-7 (1134)   Phyllotreta undulata H-4, I-7 (1134)
    Phyllotreta tetrastigma H-4, I-7 (1134)

Erysimum perofskianum   (Afgan bittercress)   Brassicaceae
    A   B   C   D   M
    03 06 09 03 13
OC: Popillia japonica H-5, I-1 (105)

Erythrina americana   (Colorin)   Fabaceae
    A   B   C   D   F   J   L   M
    01 01 09 01 08 03 05 12

OR: H-5, I-? (118)
AP: **Alk**=I-10 (erysodine, erysopine, erysothiopine, erysothiovine, erysovine, erythroidine, hipaphorine)(1392)

Erythrina flabelliformis   (Coralina)   Fabaceae
   A   C   D   D   M   M
   01 09 01 04 12 19
OC: Callosobruchus maculatus H-4, I-? (644)
AP: **Alk**=I-2 (1376); **Alk**=I-10 (erysodine, erysopine, erysothiopine, erysothiovine, erysovine, hypophorine)(1392)

Erythrina fusca   (Swamp immortelle)   Fabaceae
   A   B   C   D   M   M
   01 01 09 01 12 13
OC: Periplaneta americana H-2, I-10 (78)
OR: H-2, I-? (101)
AP: **Alk**=I-10 (erysodine, erysopine, erysovine, erythraline, hypaphorine)(1392)

Erythrina senegalensis   (Coral tree genus)   Fabaceae
   A   B   B   C   D   M   M   M   M   M   O   O
   01 01 02 09 01 03 05 12 13 23 04 07
OC: Beetles H-6, I-? (1353)
AP: **Alk**=I-10 (erysodine, erysopine, hypophorine)(1392)

Erythrina variegata   (Coral tree)   Fabaceae
   A   B   C   D   M   M   M   M   M   M   M   M   M   N
   01 01 09 01 01 02 03 04 12 13 16 19 21 22 07
OC: Dysdercus cingulatus H-3, I-? (712)   Maggots H-2, I-7, 14 (105)
AP: **Alk**=I-4, 7 (1392); **Alk**=I-6, 7 (1392); **Alk**=I-10 (erythraline, hypaphorine)(1373, 1392); **Tan**=I-4 (1352)

Erythronium americanum   (Trout lily)   Liliaceae
   A   B   C   D   J
   01 06 09 03 04
OC: Cochliomyia hominivorax H-5, I-7 (105)   Erwinia carotovora H-18, I-7, 8 (585)
AP: **Alk**=I-6, 7 (1392)

Erythronium dens-canis   (Dog-tooth violet)   Liliaceae
   A   B   C   D   J   M   N   N
   01 06 09 03 03 01 02 03 07
OC: Drosophila hydei H-2, I-2, 7 (101)

Erythrophleum couminga   (Not known)   Caesalpiniaceae
   A   B   C   D   J   J   L   M
   01 01 09 01 08 11 03 12
OC: Attagenus piceus H-4, I-4 (48)
OR: H-33, I-4 (504)

Erythrophleum ivorense   (Not known)   Caesalpiniaceae
   A   B   C   D   J   J   J   L   M
   01 01 09 01 03 08 11 03 12
OC: Attagenus piceus H-4, I-2, 4 (48)
OR: H-33, I-4 (504)

Erythrophleum suaveolans   (Red water tree)   Caesalpiniaceae
   A   B   C   D   J   J   J   K   L   L   M   M   M   M   M   M   O
   01 01 09 01 03 07 11 09 02 03 04 05 06 12 19 21 28 04
OC: Attagenus piceus H-4, I-5 (48)   Stored grain pests H-1, I-7 (87, 1353);
   Botrytis cinerea H-14, I-4 (945)   H-5, I-? (533)
   Rats H-25, I-4 (1353)   Termites H-8, I-5 (1353)
OR: H-5, I-7 (1353); H-20, I-4 (1353); H-33, I-4, 10 (504, 507, 835, 1353); H-34, I-4 (1353)
AP: **Alk**=I-4 (cassaidine, cassaine, cassamine, erythrophlamine, erythrophleine, homophleine)
   (1392); **Alk**=I-4 (erythrophleine)(1353); **Tan**=I-4 (1353)

Erythroxylum coca   (Cocaine plant)   Erythroxylaceae
   A   B   C   D   D   J   L   M   M   M   O
   01 02 09 01 08 03 03 05 21 28 07
OC: Lice H-1, I-7 (105)   Pyricularia oryzae H-14, I-7 (114)
OR: H-34, 35, I-7 (114)
AP: **Alk**=I-7 (benzoylegonine, benzoyltropine, cinnamylcocaine, cocaine, cuscohygrine, dihy-
   droxytropane, hygrine, methyloocaine, methylecgonedine, tropacocaine, truxilline)(1392)

Ethulia conyzoides   (Not known)   Asteraceae
   B   C   D   J   M   O
   06 09 01 11 05 07
OC: Attagenus piceus H-4, I-1 (48)

Eucalyptus botryoides   (Blue gum)   Myrtaceae
   A   B   C   D   J   M
   01 01 09 02 04 04
OC: Aedes punctor H-2, I-7 (101)
   Xanthomonas campestris pv. campestris H-18, I-7 (1357)

Eucalyptus camaldulensis   (Murray red gum)   Myrtaceae
   A   B   C   D   J   J   M   M
   01 01 09 01 03 14 02 22
OC: Avena fatua H-24, I-7 (668)   Hypochoeris glabra H-24, I-7 (668)
   Bromus mollis H-24, I-7 (668)   Lolium sp. H-24, I-7 (668)
   Bromus rigidus H-24, I-7 (668)   Oncopeltus fasciatus H-3, I-2, 5 (958)
   Erodium medicago H-24, I-7 (668)   Spergula arvensis H-24, I-7 (668)
   Hordeum leporinum H-24, I-7 (668)   Trifolium hirtum H-24, I-7 (668)
   Hordeum stebbinsii H-24, I-7 (668)
OR: H-19, I-12 (775)

Eucalyptus cinerea   (Spiral eucalyptus)   Myrtaceae
   A   B   C   D
   01 01 09 01
OC: Manduca sexta H-4, I-7 (669)   Panonychus citri H-4, 11, I-7 (669)
OR: H-19, I-7, 12 (1041)

Eucalyptus cloeziana   (Queensland-messmate)   Myrtaceae
   A   B   C   D   J
   01 01 09 01 09
OC: Rats H-26, I-16 (1393)

Eucalyptus dalrympeana   (Mountain gum)   Myrtaceae
   A   B   C   D   M   M
   01 01 09 01 04 19
OC: Locusta migratoria H-4, 5, I-7 (596)

Eucalyptus globulus (Tasmanian blue gum) Myrtaceae
   A   B   C   D   D   D   F   J   K   M
   01   01   09   01   02   06   14   06   05   04
OC: Aphids H-5, I-? (75)   Mosquitoes H-5, I-? (104)
OR: H-2, I-7 (116); H-2, I-12 (220); H-5, I-7 (80, 1345); H-24, I-? (1207)

Eucalyptus paniculata (Grey iron bark) Myrtaceae
   A   B   C   D   J
   01   01   09   01   04
OC: Aedes punctor H-2, I-7 (101)

Eucalyptus robusta (Swamp mahogany) Myrtaceae
   A   B   C   D
   01   01   09   01
OR: H-24, I-? (1247)

Eucalyptus sp. (Eucalyptus) Myrtaceae
   A   B   C
   01   01   09
OC: Pericallia ricini H-1, I-12 (943)
OR: H-3, 17, 19, I-? (172, 441, 865, 908, 1041, 1081, 1161); H-24, I-7 (1117)

Eugenia formosa (Not known) Myrtaceae
   A   B   C   D   M   M   N
   01   01   09   01   01   13   09
OR: H-2, I-? (100)

Eugenia haitiensis (Haitian Eugenia) Myrtaceae
   A   B   C   D   J
   01   01   09   01   06
OC: Cimex lectularius H-1, I-7, 12 (772)   Mosquitoes H-1, I-7, 12 (772)
    Flies H-1, I-7, 12 (722)

Eugenia heyneana (Not known) Myrtaceae
   A   B   C   D
   01   01   09   01
OC: Fusarium culmorum H-14, I-12 (901)   Pythium aphanidermatum H-14, I-12 (901)
    Fusarium oxysporum H-14, I-12 (901)   Trichoderma viride H-14, I-12 (901)
    Helminthosporium sp. H-14, I-12 (901)

Euonymus atropurpurea (Wahoo) Celastraceae
   A   B   C   D   J   M   O
   01   01   09   03   03   05   04
OC: Attagenus piceus H-4, I-2 (48)   Pediculus humanus capitis H-1, I-10 (104)

Euonymus europaea (European spindle tree) Celastraceae
   A   B   C   D   D   J   J   J   J   K   M   M
   01   01   09   02   03   03   04   08   11   05   15   16
OC: Ahasverus advena H-1, I-9 (101)   Macrosiphoniella sanborni H-1, I-9 (101)
    Attagenus piceus H-4, I-10 (48)   Oryzaephilus surinamensis H-1, I-9 (101)
    Lice H-2, I-9 (105)   Phyllobius oblongus H-4, I-7 (531)
AP: Alk=I-10 (1392)

Euonymus japonica (Spindle tree) Celastraceae
   A B B C D M
  01 01 02 09 06 26
OC: Drosophila sp. H-3, 4, I-7 (779)   Plutella xylostella H-4, I-7 (87)

Eupatorium aromaticum (White snake root) Asteraceae
   A B C D D J M M O
  01 06 09 02 03 11 05 17 05
OC: Attagenus piceus H-4, I-1 (48)

Eupatorium ayapana (Ayapana) Asteraceae
   A B C D M M N O
  01 06 09 01 01 05 07 07
OC: Fusarium culmorum H-14, I-12 (901)   Pythium aphanidermatum H-14, I-12 (901)
    Fusarium oxysporum H-14, I-12 (901)   Trichoderma viride H-14, I-12 (901)
    Helminthosporium sp. H-14, I-12 (901)
OR: H-7, I-? (170)

Eupatorium capillifolium (Dog fennel) Asteraceae
   A B C D K
  01 01 09 03 03
OR: H-5, I-16 (104, 504)

Eupatorium compositifolium (Not known) Asteraceae
   B C M K
  06 09 03 03
OR: H-5, I-7 (504)

Eupatorium glandulosum (Not known) Asteraceae
   B C D E J
  06 09 01 05 09
OC: Musca domestica H-1, I-1 (785)
OR: H-2, I-? (100)

Eupatorium hyssopifolium (Thoroughwort) Asteraceae
   A B C D J
  01 06 09 03 08
OC: Popillia japonica H-5, I-7, 8 (105, 1243)

Eupatorium japonicum (Not known) Asteraceae
   B C J
  06 09 12
OC: Drosophila melanogaster H-2, 3, I-7 (143, 370)
    Drosophila sp. H-3, 4, I-? (779)

Eupatorium maculatum (Snake root) Asteraceae
   A B C D D E J J M O O O
  01 06 09 03 06 05 03 08 05 02 07 08
OC: Aedes aegypti H-2, I-6, 7, 8, 9   Botrytis cinerea H-14, I-16 (946)
   (77, 156, 843)   Culex tarsalis H-3, I-? (156)

Eupatorium odoratum (Not known) Asteraceae
   A B C D J
  01 02 09 01 03
OC: Locusta migratoria H-4, 5, I-7 (596)

OR: H-32, I-1 (73, 220, 459)
AP: Alk=I-2 (1392)

Eupatorium perfoliatum   (Common boneset)   Asteraceae
   A   C   D   J   M   O   O
   01 09 03 03 05 07 08
OC: Drosophila melanogaster H-1, I-7 (165)
OR: H-19, I-? (592, 1171)

Eupatorium purpureum   (Joe-pye weed)   Asteraceae
   A   B   C   D   M   N
   01 06 09 01 01 07
OC: Pseudomonas solanacearum H-18, I-1, 8 (945)
OR: H-19, I-2 (1171)

Eupatorium staechadosmum   (Ayapana du Tonkin)   Asteraceae
   A   B   C   D   J
   01 06 09 01 13
OC: Spodoptera litura H-4, I-7 (1064)

Euphorbia adenochlora   (Spurge genus)   Euphorbiaceae
   A   B   C   D   J
   01 06 09 01 03
OC: Drosophila hydei H-2, I-6, 7 (101)

Euphorbia antiquorum   (Not known)   Euphorbiaceae
   A   B   D   L   M   N
   01 02 01 03 01 06
OC: Maggots H-2, I-12 (105)
OR: H-2, I-? (73, 220); H-32, I-? (220); H-33, I-14 (504)

Euphorbia balsamifera   (Balsam spurge)   Euphorbiaceae
   C   D   M   N
   09 01 01 15
OC: Glossina sp. H-5, I-14 (1345)

Euphorbia biglandulosa   (Spurge genus)   Euphorbiaceae
   A   C   D   J   M
   01 09 06 03 13
OR: H-1, I-? (105); H-1, I-7 (1026)

Euphorbia cyparissias   (Cypress spurge)   Euphorbiaceae
   A   B   C   D   J   L   M
   01 06 09 03 07 05 09
OC: Mole crickets H-1, I-12 (105)   Musca domestica H-2a, I-1 (87)

Euphorbia dendroides   (Not known)   Euphorbiaceae
   C   J   J   J
   09 03 04 10
OC: Ants H-2, I-7, 14 (101)   Mosquitoes H-2, I-7, 14 (101)
  "Bugs" H-2, I-7, 14 (101)   Musca domestica H-2, I-7, 14 (101)
OR: H-1, I-7 (1026); H-1, 32, I-? (754); H-32, I-1 (504)

Euphorbia dentata   (Spurge genus)   Euphorbiaceae
   C   D
   09 10
OC: Ceratocystis ulmi H-14, I-1, 9 (946)

Euphorbia esula   (Wolf's milk)   Euphorbiaceae
   A   B   C   D   J
   01 06 09 03 03
OC: Maggots H-1, I-1 (1241)   Rats H-25, I-1 (1241)
    Puccinia graminis tritici H-14, I-1 (1241) Rodents H-25, I-16 (945)
    Puccinia rubigavera H-14, I-1 (1241)
OR: H-24, I-6, 7 (1258)
AP: Alk=I-6, 7 (1372)

Euphorbia fischeriana   (Not known)   Euphorbiaceae
   C   J
   09 03
OC: Aphis gossypii H-1, I-2 (1241)   Pieris rapae H-1, I-2 (1241)
    Cnaphalocrocis medinalis H-1, I-2 (1241) Rats H-25, I-2 (1241)
    Colaphellus bowringi H-1, I-2 (1241)   Spider mites H-11, I-2 (1241)

Euphorbia helioscopia   (Dhodak)   Euphorbiaceae
   C   J
   09 03
OC: Aphids H-1, I-1 (1241)   Phytophthora infestans H-14, I-1 (1241)
    Ascotis selenaria H-1, I-1 (1241)   Puccinia graminis tritici H-14, I-1 (1241)
    Athalia rosae H-4, I-7 (531)   Spider mites H-11, I-1 (1241)
    Leptinotarsa decemlineata H-4, I-7 (531)
    Phytodecta fornicata H-4, I-7 (531)

Euphorbia hirta   (Garden spurge)   Euphorbiaceae
   A   B   C   D   J   L   M   M   O   O
   03 06 09 01 03 05 05 21 01 07
OC: Drechslera oryzae H-14, I-7 (113)   Rhabdoscelus obscurus H-1, I-? (137)
OR: H-22, I-? (1287); H-34, I-1 (504)
AP: Alk=I-16 (xanthothamnine)(1392)

Euphorbia lateriflora   (Poinsettia genus)   Euphorbiaceae
   B
   09
OC: Pediculis humanus capitis H-1, I-14 (1353)
OR: H-15, I-14 (1353); H-17, I-1 (1353); H-32, I-? (1276)

Euphorbia lathyris   (Caper spurge)   Euphorbiaceae
   A   A   C   D   M   M   N   O
   02 03 09 03 01 05 10 10
OC: Plutella xylostella H-1, I-? (87)
AP: Alk=I-10 (1391)

Euphorbia maculata   (Spurge genus)   Euphorbiaceae
   C   J   J
   09 03 09
OC: Oncopeltus fasciatus H-2, I-6, 7 (101)   Rodents H-25, I-1 (944)

130

Euphorbia marginata   (Snow-on-the-mountain)   Euphorbiaceae

   A   B   C   D   L   M
   03 06 09 03 04 13

OC: Cochliomyia hominivorax H-5, I-14 (105)
OR: H-33, I-14 (507)

Euphorbia neriifolia   (Hedge euphorbia)   Euphorbiaceae

   A   B   B   C   D   J   M   M   M   N   O   O
   01 01 02 09 01 03 1a 05 13 15 07 14

OR: H-2, I-2, 14 (87, 105, 1276); H-19, I-? (982); H-32, I-4, 7, 14 (73, 220, 504, 968, 983);
    H-33, I-7, 14 (968)
AP: Tri=I-6 (friedelan)(1381)

Euphorbia pekinensis   (Not known)   Euphorbiaceae

   C   J
   09 03

OC: Aphids H-1, I-1, 10 (1241)   Puccinia rubigavera H-14, I-10 (1241)
    Pieris rapae H-1, I-1, 10 (1241)   Verticillium albo-atrum H-14, I-10 (1241)
    Puccinia graminis tritici H-14, I-10 (1241)

Euphorbia peplis   (Spurge)   Euphorbiaceae

   C   J
   09 09

OC: Locusta migratoria H-4, 5, I-7 (596)

Euphorbia primulaefolia   (Spurge genus)   Euphorbiaceae

   C   D   K
   09 01 02

OC: Rats H-25, I-7 (504)

Euphorbia pulcherrima   (Poinsettia)   Euphorbiaceae

   A   B   C   D   F   G   J   J   J   L   L   M   M   M   M   N   O   O   O   O
   01 02 09 01 13 04 03 11 15 03 04 01 05 13 21 10 02 04 07 09

OC: Cercospora cruenta H-14, I-7 (1122)   Sitophilus oryzae H-4, I-8
    Drechslera oryzae H-14, I-7 (113)   (87, 400, 598, 1124)
    Manduca sexta H-4, I-? (669)   Ustilago hordei H-14, I-7 (181)
    Panonychus citri H-4, 11, I-? (669)   Ustilago tritici H-14, I-7 (181)
    Pyricularia oryzae H-14, I-7 (114)
OR: H-15, I-? (651); H-19, I-? (172, 413, 651); H-20, I-2, 4, 7 (1319)
AP: Alk=I-4, 6, 7, 9 (1362); **Alk, Tan**=I-7 (1319)

Euphorbia royleana   (Spurge genus)   Euphorbiaceae

   A   B   C   D   F   J   J   J   K   L
   01 01 09 01 07 02 03 09 06 02

OC: Athalia proxima H-1, I-7 (89, 438); H-2a, I-7 (116); H-4, 5, I-7 (607, 1011)
OR: H-5, I-? (100); H-22, I-6 (881); H-32, I-? (73, 220)
AP: Ter=I-6 (1382)

Euphorbia thymifolia   (Golandrina)   Euphorbiaceae

   A   C   D   L   M   O
   03 09 01 05 05 07

OC: Mosquitoes H-5, I-12 (73)
OR: H-1, I-? (220); H-2, I-? (105)

Euphorbia tirucalli   (Milk bush)   Euphorbiaceae
   A   B   B   C   D   J   K   K   L   M   M   O
   01 01 02 09 01 15 04 06 03 05 13 14
OC: Aphids H-1, I-6 (105)   Mosquitoes H-1, 5, I-? (87, 105)
    Glossina sp. H-1, 5, I-15 (754)   Snails H-28, I-15 (754)
    Grasshoppers H-1, 5, I-15 (754)   Toxoptera aurantii H-1, I-6 (105, 681)
OR: H-2, I-14 (73); H-2, 32, I-? (105, 1276); H-5, I-15 (101); H-19, I-? (172); H-32, I-?
    (220, 983); H-32, I-1, 4 (105, 220, 1276); H-33, I-2, 6, 14 (968)

Eurotia lantana   (Winter fat)   Chenopodiaceae
   B   C   D
   02 09 03
OC: Pediculus humanus capitis H-1, I-10 (504)

Euscaphis japonica   (Not known)   Staphyleaceae
   A   B   C   D
   01 01 09 03
OC: Rice borers H-1, I-4, 6, 7 (1241)

Evodia daniellii   (Not known)   Rutaceae
   A   B   C   D   J   J
   01 01 09 03 09 10
OC: Agrobacterium tumefaciens H-18, I-? (957)   Monilinia fructicola H-14, I-? (957)
    Colletotrichum lindemuthianum H-14, I-?   Pseudomonas syringae H-18, I-? (957)
      (957)   Xanthomonas campestris
    Endothia parasitica H-14, I-? (957)   pv. phaseoli H-18, I-? (957)
    Erwinia amylovora H-18, I-? (957)   pv. pruni H-18, I-? (957)
    Erwinia carotovora H-18, I-? (957)   pv. vesicatoria H-18, I-? (957)
AP: Alk=I-6, 7, 9 (1368)

Evodia hupehensis   (Not known)   Rutaceae
   A   B   C   D   J   J
   01 01 09 03 09 11
OC: Attagenus piceus H-4, I-6, 7 (48)
OR: H-19, I-9 (1262)
AP: Alk=I-4, 5, 6 (1374)

Evodia rutaecarpa   (Not known)   Rutaceae
   A   C   D   J   M   O
   01 09 10 03 05 07
OC: Aphids H-1, I-7 (1241)   Maggots H-1, I-7 (1241)
OR: H-22, I-16 (977)
AP: Alk=I-9 (evodiamine, rutaecarpine)(1392)

Evolvulus linifolius   (Not known)   Convolvulaceae
   A   B   C   D   M   O
   03 06 09 01 05 07
OR: H-17, I-? (1353)
AP: Alk=I-2, 6, 7 (1360); Alk=I-6, 7 (1392)

Excoecaria agallocha   (Blind-your-eyes tree)   Euphorbiaceae
   A   B   C   D   L   M
   01 01 09 01 03 17

OC: Maggots H-2, I-14 (105)
OR: H-32, I-4, 6, 7, 14 (73, 968, 983, 1146); H-33, I-14 (220, 968, 983)
AP: Fla, Tan=I-6, 7 (1377)

Fagopyrum esculentum   (Buckwheat)   Polygonaceae

A	B	C	D	E	M	M	M	M	N	N
03	07	09	03	12	01	09	10	18	09	10

OC: Meloidogyne sp. H-16, I-20 (424)   Rhizoctonia solani H-14, I-? (353)
OR: H-24, I-9 (569)

Fagus grandifolia   (American beech)   Fagaceae

A	B	C	D	M	M	M	M	M	M	M	N
01	01	09	03	01	06	13	19	21	22	25	10

OC: Manduca sexta H-4, I-7 (669)   Panonychus citri H-4, 11, I-7 (669)
OR: H-20, I-5 (504)
AP: Alk=I-6, 7 (1392): Tan=I-5 (504)

Ferula assa-foetida   (Not known)   Apiaceae

A	B	C	D	M	M	M	M	N	O
01	06	09	03	01	05	13	14	01	12

OC: Ants H-5, I-? (105)
OR: H-2, 5, I-? (1276)

Ferula leucopyrus   (Not known)   Apiaceae

A	B	C	D
01	06	09	06

OR: H-2, 32, I-4, 6 (1276)

Ferula microcarpa   (Not known)   Apiaceae

A	B	C	D
01	06	09	06

OR: H-2, 32, I-4 (1276)

Festuca elatior   (English bluegrass)   Poaceae

A	B	C	D	J	M
01	06	09	03	03	02

OC: Pratylenchus zea H-16, I-? (1274)
OR: H-24, I-7 (1228)
AP: Alk=I-7 (perloline)(1392)

Fevillea cordifolia   (Not known)   Cucurbitaceae

B	C	D	J	M	M	M	O	O
04	09	01	11	05	06	15	10	12

OC: Attagenus piceus H-4, I-10 (48)

Ficus carica   (Common fig)   Moraceae

A	B	C	D	J	M	N
01	01	09	06	13	01	09

OC: Helicotylenchus indicus H-16, I-15 (1330)   Spodoptera litura H-1, I-7 (378)
    Hoplolaimus indicus H-16, I-15 (1330)   Tylenchorhynchus brassicae H-16, I-15
    Lice H-5, I-9 (1345)   (1330)
    Locusta migratoria H-4, I-7 (694)   Tylenchus filiformis H-16, I-15 (1330)
    Rotylenchulus reniformis H-16, I-15 (1330)
AP: Alk=I-6 (1392)

Ficus elastica   (Indian rubber tree)   Moraceae

A   B   C   D   M
01 01 09 01 26

OC: Helicotylenchus indicus H-16, I-15 (1330)   Tylenchorhynchus brassicae H-16, I-15
    Hoplolaimus indicus H-16, I-15 (1330)   (1330)
    Rotylenchulus reniformis H-16, I-15 (1330)   Tylenchus filiformis H-16, I-15 (1330)

OR: H-19, I-4, 7 (867)

Ficus racemosa   (Cluster fig)   Moraceae

A   B   C   D   M   N
01 01 09 01 01 09

OC: Helicotylenchus indicus H-16, I-15 (1330)   Tylenchorhynchus brassicae H-16, I-15
    Hoplolaimus indicus H-16, I-15 (1330)   (1330)
    Rotylenchulus reniformis H-16, I-15 (1330)   Tylenchus filiformis H-16, I-15 (1330)

Ficus religiosa   (Sacred fig)   Moraceae

A   B   C   D   J   M   M   N
01 01 09 01 03 01 21 15

OC: Alternaria tenuis H-14, I-7 (480)   Helminthosporium sp. H-14, I-7 (480)
    Curvularia penniseti H-14, I-7 (480)

OR: H-17, I-4 (881)

Filipendula vulgaris   (Dropwort)   Rosaceae

A   B   C   D   J   M   M   M   N   N   O   O
01 06 09 03 04 01 02 05 02 07 02 07

OC: Dermacentor marginatus H-13, I-8 (87, 1133) Ixodes redikorzevi H-13, I-8 (87, 1133)
    Haemaphysalis punctata H-13, I-8 (87, 1133) Rhipicephalus rossicus H-13, I-8 (87, 1133)

OR: H-17, I-2, 7, 8 (504)

Firmiana simplex   (Phoenix tree)   Sterculiaceae

A   B   C   D   J   M   M
01 01 09 01 03 03 04

OC: Aphids H-1, I-7 (1241)   Glomerella gossypii H-14, I-7 (1241)

AP: Alk=I-10 (caffeine)(1392)

Fleurya interrupta   (Lipang-aso)   Urticaceae

B   B   C   L   M   O
02 06 09 03 05 02

OC: Nematodes H-16, I-? (100)

OR: H-33, I-17 (968)

Flourensia cernua   (Not known)   Asteraceae

B   C   J   M   O   O
06 09 03 05 07 08

OC: Blattella germanica H-2, I-16 (48)

Fluggea leucopyrus   (Not known)   Euphorbiaceae

A   B   C   D   M   N
01 02 09 01 01 09

OC: Maggots H-1, I-7, 12 (220)

AP: Alk=I-7 (1392)

134

Fluggea melanthesioides   (Not known)   Euphorbiaceae
    A   B   C   D   J
    01 02 09 03 03
OC: Maggots H-1, I-7 (105)
OR: H-2, I-? (73); H-32, I-4 (105, 504)

Fluggea virosa   (Not known)   Euphorbiaceae
    A   B   C   J   J   J   L   M   M   M   N   O
    01 02 09 03 04 08 03 01 05 13 09 07
OC: Maggots H-1, I-7 (220)
OR: H-19, I-7 (982, 1076); H-32, I-4, 7 (73, 105, 220, 504, 983, 1349, 1353); H-33, I-4 (1353)
AP: Alk=I-2, 4 (flueggeine)(1392)

Foeniculum vulgare   (Common fennel)   Apiaceae
    A   B   C   D   D   F   M   M   M   N   N
    01 06 09 03 06 07 01 14 17 06 07
OC: Chrysomya macellaria H-5, I-12 (199)   Lenzites trabea H-14, I-? (428)
    Cochliomyia americana H-1, I-12 (244)   Polyporus versicolor H-14, I-? (428)
    Cochliomyia hominivorax H-5, I-? (105)   Pratylenchus penetrans H-16, I-? (428)
    Lentinus lepideus H-14, I-? (428)
OR: H-15, 19, I-12 (908); H-24, I-9 (569)
AP: Alk=I-6, 7, 8 (1374); Alk=I-6, 7, 10 (1392)

Fontanesiana editorum   (Doghobble)   Ericaceae
    A   B   C   D
    01 02 09 03
OC: Agrobacterium tumefaciens H-18, I-? (935)   Pseudomonas syringae H-18, I-? (935)
    Colletotrichum lindemuthianum H-14, I-?   Xanthomonas campestris
     (935)   pv. phaseoli H-18, I-? (935)
    Endothia parasitica H-14, I-? (935)   pv. pruni H-18, I-? (935)
    Erwinia amylovora H-18, I-? (935)   pv. vesicatoria H-18, I-? (935)
    Erwinia carotovora H-18, I-? (935)

Forsythia viridissima   (Golden bells genus)   Oleaceae
    A   B   C   D
    01 02 09 03
OC: Phyllobius oblongus H-4, I-7 (531)

Fragaria ananassa   (Garden strawberry)   Rosaceae
    A   B   C   D   J   K   M   N
    01 06 09 03 01 01 01 09
OC: Athalia rosae H-4, I-7 (531)   Phytodecta fornicata H-4, I-7 (531)
    Cassida nebulosa H-4, I-7 (531)   Pieris brassicae H-4, I-7 (531)
    Leptinotarsa decemlineata H-4, I-7 (531)

Fragaria chiloensis   (Beach strawberry)   Rosaceae
    A   B   C   D   J   K   M   N
    01 06 09 03 01 01 01 09
OC: Pratylenchus brachyurus H-16, I-? (1274)

Fragaria vesca   (Woodland strawberry)   Rosaceae
    A   B   C   D   M   N
    01 06 09 03 01 09
OC: Tobacco mosaic virus H-21, I-? (1059, 1061)

135

Franseria dumosa   (White bursage)   Asteraceae
    B   J
    09 03
OR: H-24, I-2, 7 (555, 556)
AP: Alk=I-6, 7, 8 (1372)

Fraxinus americana   (White ash)   Oleaceae
    A   B   C   D   M   M
    01   01   09   03   04   19
OC: Lymantria dispar H-5, I-? (340)
AP: Alk=I-4, 6, 7 (1392)

Fraxinus excelsior   (European ash)   Oleaceae
    A   B   C   D   M   M
    01   01   09   03   04   19
OC: Phyllobius oblongus H-4, I-7 (531)

Fumaria schleicheri   (Not known)   Papaveraceae
    A   B   C   D   J
    01   06   09   03   03
OC: Musca domestica H-2a, I-8 (87)
AP: Alk=I-2 (fumaramine, fumaridine, fumarinine, fumaritine, protoprine)(1392)

Funastrum clausum   (Petaquilla)   Asclepiadaceae
    B   C   D
    02   09   01
OC: Flies H-2, I-14 (504, 1345)

Funastrum gracile   (Not known)   Asclepiadaceae
    A   B   C   D   J
    01   02   09   01   11
OC: Attagenus piceus H-4, I-6 (87)

Galinsoga parviflora   (Not known)   Asteraceae
    B   C   D   M   N
    06   09   10   01   01
OC: Athalia rosae H-4, I-7 (531)   Phytodecta fornicata H-4, I-7 (531)
    Cassida nebulosa H-4, I-7 (531)
AP: Sap=I-2 (1379)

Ganoderma lucidum   (Not known)   Polypodiaceae
    A   C   J   K
    04   01   15   06
OC: Sitophilus oryzae H-2, I-9 (87, 1124)

Garcinia indica   (Mangosteen oil tree)   Clusiaceae
    A   B   J   M   M   M   M   N   N   O
    01   09   09   01   05   09   10   09   10   10
OC: Rats H-26, I-16 (1393)

Gardenia campanulata   (Gardenia genus)   Rubiaceae
    A   B   C   D   D   D   G   J   M   O
    01   01   09   01   02   08   04   03   05   09
OR: H-2, I-9 (73, 105); H-11, I-? (100); H-17, 32, I-9 (504); H-32, I-14 (105)

136

Gardenia gummifera   (Gardenia genus)   Rubiaceae
  A   B   C   D   M   M   N   O
  01   01   09   01   01   05   09   13
OR: H-2, I-13 (1276); H-5, I-13 (105, 504, 1320); H-17, 20, I-4 (504)

Gardenia jasminoides   (Cape jasmine)   Rubiaceae
  A   B   C   D   M   M   M   N
  01   02   09   01   01   16   17   09
OC: Drechslera oryzae H-14, I-6, 7, 8 (113, 432)
OR: H-11, I-? (100); H-15, 17, 19, I-9, 13 (1320); H-22, I-16 (977, 1395)
AP: Alk=I-6, 7 (1392); Sap=I-7 (1320)

Gardenia lucida   (Gardenia genus)   Rubiaceae
  A   B   C   D   M   O
  01   01   09   01   05   13
OC: Flies H-5, I-13 (105, 504)
OR: H-20, I-13 (504)

Gaultheria fragrantissima   (Indian wintergreen)   Ericaceae
  A   B   B   C   D   D   J   L   M   O   O
  01   01   02   09   01   08   06   03   05   07   12
OR: H-2, 5, I-7, 12 (73); H-33, I-? (220)

Gaultheria procumbens   (Wintergreen)   Ericaceae
  A   B   B   C   D   J   M   M   M   N   N   O   O
  01   02   05   09   03   06   01   05   14   07   09   07   12
OC: Alternaria tenuis H-14, I-12 (185)   Pericallia ricini H-1, I-12 (943)
    Blatta orientalis H-6, I-12 (105)
OR: H-20, I-12 (504)

Gaura coccinea   (Not known)   Onagraceae
  A   B   C   D   D   J
  01   06   09   02   03   03
OC: Drosophila melanogaster H-2, I-6 (165)

Gelsemium elegans   (Yellow jessamine genus)   Loganiaceae
  A   B   B   C   D   D   L   M
  01   02   03   09   01   02   03   13
OR: H-1, I-? (105); H-33, I-7 (504)
AP: Alk=1-2, 6 (koumicine, koumine, kouminicine, kouminidine, kuminine)(1392); Alk=I-6, 7
    (1366)

Gelsemium sempervirens   (Evening trumpet flower)   Loganiaceae
  A   B   C   D   J   L   M   M
  01   03   09   03   03   01   13   17
OC: Attagenus piceus H-4, I-2 (48)
OR: H-2, I-8 (105)
AP: Alk=I-2 (gelsedine, gelsemicine, gelsemidine, gelsemine, gelseminine, gelsevirine,
    sempervirine)(1392)

Gendarussa vulgaris   (Malabulak)   Acanthaceae
  A   B   C   D   J
  01   02   09   01   02

OC: Callosobruchus chinensis H-2, I-2, 6 (605)   Coccus viridis H-1, I-4 (1209)
OR: H-5, I-2, 6 (605)
AP: Alk=I-2, 10 (1388); Alk=I-6, 7 (1392)

Genipa americana   (Marmalade box)   Rubiaceae
  A   B   C   D   M   M   M   N
  01 01 09 01 01 04 16 09
OR: H-5, I-9, 14 (1345)

Genista germinica   (European broom)   Fabaceae
  A   B   B   C   J   M   M   M
  01 02 03 09 03 12 13 16
OR: H-2, I-6 (105)

Geranium eriostemon var. onoei   (Not known)   Geraniaceae
  A   B   C   D
  01 06 09 03
OC: Drosophila hydei H-2, I-6, 7 (101)

Geranium macrorrhizum   (Geranium genus)   Geraniaceae
  A   B   C   D
  01 06 09 03
OC: Dermestres maculatus H-3, 4, I-? (544)   Graphosoma italicum H-3, 4, I-? (544)
    Dysdercus cingulatus H-3, 4, I-? (544)   Graphosoma mellonella H-3, 4, I-? (544)

Geranium maculatum   (Wild geranium)   Geraniaceae
  A   B   C   D   K   M   O   O
  01 06 09 03 09 05 02 07
OC: Pseudomonas solanacearum H-18, I-2 (946)

Geranium nepalense   (Nepal geranium)   Geraniaceae
  A   B   C   D   M   M   M
  01 06 09 08 12 16 21
OC: Colletotrichum falcatum H-14, I-2 (1047)   Pyricularia oryzae H-14, I-2 (1047)
AP: Alk=I-2 (1047, 1352)

Gerbera piloselloides   (Not known)   Asteraceae
  A   B   C   D   J
  01 06 09 01 03
OC: Aphids H-1, I-1 (1241)

Gerea viscida   (Not known)   Asteraceae
  A   B   C   D   D   J   K
  01 06 09 02 04 01 01
OC: Moths H-5, I-17 (173)

Ginkgo biloba   (Maidenhair tree)   Ginkgoaceae
  A   B   J   L   L   M   M   M   M   N   N   O
  01 01 08 03 04 01 05 06 13 09 10 10
OC: Lepisma saccharina H-5, I-7 (87, 720)   Pyrausta nubilalis H-2, I-2 (720)
    Monilinia fructicola H-14, I-7, 12 (720)   Southern mosaic bean virus H-21, I-2 (720)
    Ostrinia nubilalis H-4, I-2, 7 (87)
OR: H-34, I-10 (504)

Glechoma hederacea   (Ground ivy)   Lamiaceae
   A   B   C   D   L   M   M   N   O
   01 06 09 03 05 01 05 07 07
OC: Athalia rosae H-4, I-7 (531)   Hyphantria cunea H-4, I-7 (531)

Glechoma longituba   (Ground ivy genus)   Lamiaceae
   B   B   C   D   J
   05 06 09 03 03
OC: Aphis gossypii H-1, I-10 (1241)

Gleditsia amorphoides   (Honey locust genus)   Caesalpiniaceae
   A   B   C   D   J   M   M
   01 01 09 01 03 12 19
OC: Attagenus piceus H-4, I-9 (48)   Blattella germanica H-2, I-9 (48)

Gleditsia aquatica   (Swamp locust)   Caesalpiniaceae
   A   B   C   D   D   J   M
   01 01 09 01 02 11 12
OC: Attagenus piceus H-4, I-9 (48)

Gleditsia sinensis   (Honey locust genus)   Caesalpiniaceae
   A   B   C   D   D   J   J   M   M   M
   01 01 09 01 02 03 08 12 15 19
OC: Attagenus piceus H-4, I-6, 7 (48)

Gliricidia sepium   (Madre)   Fabaceae
   A   B   C   D   F   J   J   J   J   K   K   K   L   M   M
   01 01 09 01 06 01 04 11 15 04 05 06 03 12 13
OC: Aphids H-1, I-4 (101)   Nymphula depunctalis H-5, I-16 (137)
   Coccids H-1, I-4 (101)   Rats H-25, I-1, 4, 10 (118, 504)
   Culex sp. H-2, I-9 (190)   Rodents H-25, I-1 (117)
   Diacrisia virginica H-2, I-2, 7 (101, 190)   Spodoptera eridania H-2, I-6, 7, 9
   Fleas H-2, I-7 (163, 1319)   (101, 190, 304)
   Heliothis armigera H-2, I-? (90)   Ticks H-13, I-7 (163, 1319)
   Hydrellia philippia H-5, I-16 (137)   Trichoplusia ni H-2, I-2, 9 (101, 304)
OR: H-2, I-? (80, 105); H-4, I-1 (118); H-33, I-4, 10 (507)

Glochidion fortunei   (Not known)   Euphorbiaceae
   A   C   D   D
   01 09 01 02
OR: H-24, I-7 (1247)

Glochidion puberum   (Not known)   Euphorbiaceae
   C
   09
OC: Aphids H-1, I-7 (1241)   Maggots H-1, I-7 (1241)
   Leafhoppers H-1, I-7 (1241)

Gloriosa simplex   (Glory lily genus)   Liliaceae
   A   B   B   C   D   M
   01 04 06 09 01 13
OC: Lice H-2, I-3 (1345)

Gloriosa superba   (Glory lily)   Liliaceae
```
A    B    C    D    J    L    M    M    O
01   04   09   01   03   03   05   13   02
```
OC: Pediculus humanus capitis H-1, I-7 (105, 1353)
OR: H-2, I-7 (73); H-15, I-2 (504); H-24, I-2 (1211); H-33, I-2 (1353); H-34, I-7 (1353)
AP: **Alk**=I-3, 8 (colchicine, demethylcolchicine, gloriosine, lumicolchicine)(1392); **Alk**=I-7
    (colchine)(1353); **Alk**=I-7 (1360)

Glycine max   (Soybean)   Fabaceae
```
A    B    C    D    J    J    K    M    M    N
03   06   09   03   01   06   03   01   12   10
```
OC: Athalia rosae H-4, I-7 (531)   Pieris brassicae H-4, I-7 (531)
    Laspeyresia pomonella H-2, I-6 (78)   Pratylenchus penetrans H-16, I-18 (1054)
    Leaf-cutting ants H-6, I-? (87)   Rhizoctonia solani H-14, I-20 (353, 422)
    Leptinotarsa decemlineata H-4, I-7 (548) Sphaerothica humuli H-14, I-? (496)
    Nematodes H-16, I-20 (424)   Streptomyces scabies H-18, I-? (359, 366)
    Oregma lanigera H-2, I-6 (78)   Zabrotes subfasciatus H-2, I-10 (334)
    Peronospora tabacina H-14, I-? (482)
OR: H-5, I-12 (334); H-24, I-6 (1239)
AP: **Alk**=I-? (trigonelline)(1392)

Glycosmis pentaphylla   (Jamaican mandarin orange)   Rutaceae
```
A    B    C    D    M    M    N    O
01   02   09   01   01   05   07   07
```
OR: H-2, I-2, 5, 7 (127); H-17, 20, I-7 (504)
AP: **Alk**=I-? (glycosine, glycosinimine, skimmianine)(1392); **Alk**=I-2, 6, 7, 9 (1361); **Alk**=I-6
    (arborinine)(1384); **Alk**=I-7 (arborine)(1383); **Alk**=I-7 (kokusaginine)(1392); **Sap**=I-7 (1380)

Glycyrrhiza glabra   (Licorice)   Fabaceae
```
B    C    D    D    G    J    M
06   09   01   06   04   09   12
```
OC: Reticulitermes flavipes H-2, I-? (87)
OR: H-15, I-1, 2 (1151, 1262); H-15, 19, I-? (411)

Glycyrrhiza lepidota   (Wild licorice)   Fabaceae
```
B    C    D    D    J    M    M    N
06   09   02   03   03   01   12   02
```
OC: Drosophila melanogaster H-2, I-9 (165)

Gmelina arborea   (Malay bushbeech)   Verbenaceae
```
A    B    C    D    J    M    M    M    M    N    O    O
01   01   09   01   09   01   04   05   19   09   02   07
```
OC: Maggots H-1, I-7 (105)
OR: H-17, I-7 (504); H-22, I-4, 5 (881, 1040); H-34, I-2 (504)
AP: **Alk**=I-7 (1359); **Alk**=I-9 (1374)

Gomphrena globosa   (Bachelor's button)   Amaranthaceae
```
A    B    C    D
03   06   09   01
```
OC: Leptinotarsa decemlineata H-4, I-7 (531)
AP: **Tri**=I-7 (1390)

Goniothalamus tapis   (Not known)   Annonaceae
    A   B   C   D
    01 02 09 01
OC: Mosquitoes H-5, I-4 (1345)

Gossypium herbaceum   (Levant cotton)   Malvaceae
    A   B   C   D
    01 06 09 03
OC: Aphids H-1, I-10, 12 (1241)   Spider mites H-11, I-10, 12 (1241)
    Ascotis selenaria H-1, I-10, 12 (1241)   Thrips H-1, I-10, 12 (1241)

Gossypium hirsutum   (Upland cotton)   Malvaceae
    A   A   B   D   J   J   L   M   M   M   O
    01 03 06 01 06 12 03 05 16 21 04
OC: Callosobruchus chinensis H-3, I-12 (516)   Peronospora tabacina H-14, I-? (482)
    Chionaspis salicis-nigrae H-2, I-10, 12   Phenacoccus gossypii H-2, I-10, 12 (101)
      (101)   Pratylenchus penetrans H-16, I-12 (1054)
    Heliothis virescens H-3, I-8, 15 (1007)   Pratylenchus zea H-16, I-? (1274)
    Lepidosaphes ulmi H-2, I-10, 12 (101)   Sphaerothica humuli H-14, I-?
    Locusta migratoria H-4, 5, I-7 (596)   (493, 496, 760)
    Meloidogyne sp. H-16, I-20 (424)   Zabrotes subfasciatus H-5, I-10, 12 (334)
OR: H-33, I-10 (504)
AP: Alk=I-9 (5-hydroxytryptamine)(1392); Tan=I-10 (1391)

Gouania polygama   (Not known)   Rhamnaceae
    B   C   D   D
    04 09 01 02
OR: H-2, I-? (104)

Grewia carpinifolia   (Not known)   Tiliaceae
    A   B   C   D   D   J   M   M   M   N
    01 01 09 01 02 03 01 02 24 09
OC: Pediculus humanus capitis H-1, I-7 (504); H-2, I-14 (105)

Grewia tiliifolia   (Not known)   Tiliaceae
    A   B   C   D   K
    01 01 09 01 09
OC: Rats H-26, I-16 (1393)
AP: Sap, Str, Tan=I-7 (1359); Tri=I-2, 4 (betulin, friedelin, lupeol)(1381)

Grindelia humilis   (Sticky heads)   Asteraceae
    A   B   C   D   E   E
    01 06 09 10 04 11
OC: Aphids H-1, I-? (143)   Schizaphis graminum H-4, I-8 (646)

Grindelia perennis   (Gumweed genus)   Asteraceae
    B   C   J
    06 09 11
OC: Attagenus piceus H-4, I-1 (48)

Grindelia squarrosa   (Curly-cup gumweed)   Asteraceae
    A   B   C   D   J   M   O
    02 06 09 03 11 05 08
OC: Attagenus piceus H-4, I-7, 8, 15 (48)

Guaiacum officinale　　(Lignum-vitae genus)　　Zygophyllaceae

A	B	C	D	E	J	M	M
01	01	09	01	04	09	04	13

OC: Cochliomyia hominivorax H-5, I-? (105)　　　Reticulitermes flavipes H-2, I-5 (87)
　　Cryptotermes brevis H-8, I-5 (87)

Guarea rusbyi　　(Cocillana)　　Meliaceae

A	B	C	D	D	J	M	O
01	01	09	01	08	02	05	04

OC: Popillia japonica H-5, I-4 (105)

Gustavia augusta　　(Not known)　　Lecythidaceae

A	B	C	D	J
01	01	09	01	08

OC: Attagenus piceus H-4, I-9 (48)

Gymnema sylvestre　　(Hyena's vine)　　Asclepiadaceae

B	C	D	M	O
04	01	01	05	07

OR: H-4, I-? (132)

Gymnocladus dioica　　(Kentucky coffee tree)　　Caesalpiniaceae

A	B	C	D	J	J	M	M	M	M	N
01	01	09	10	08	11	01	04	05	10	12

OC: Attagenus piceus H-4, I-6 (48)　　　　Popillia japonica H-5, I-7 (105, 1243)
　　Flies H-2, I-7, 9 (104)　　　　　　　Spodoptera eridania H-2, I-? (143)

Gynocardia odorata　　(Kushthapa)　　Flacourtiaceae

B	C	D	J	J	J	M	O	O
01	09	01	03	08	11	05	10	12

OC: Attagenus piceus H-4, I-10 (48)　　　　Tineola bisselliella H-4, I-10 (48)
OR: H-2, 32, I-9 (105, 220, 1276); H-15, 19, I-? (186, 411); H-32, I-9, 10 (73, 459, 504)

Gynura segetum　　(Not known)　　Asteraceae

B	C	D	J
06	09	01	03

OC: Aphids H-1, I-1 (1241)　　　　　　　Rice borers H-1, I-1 (1241)
　　Leafhoppers H-1, I-1 (1241)

Gypsophila paniculata　　(Baby's breath)　　Caryophyllaceae

A	B	C	D	J	J	J	M	O
01	02	09	03	02	03	08	05	02

OC: Aedes aegypti H-2, I-2, 7, 8, 9 (77, 165, 843)
　　Drosophila melanogaster H-3, I-2 (165)

Haematoxylum campechianum　　(Logwood)　　Caesalpiniaceae

A	B	B	C	D	J	M	M	M	O
01	01	02	09	01	17	05	16	19	05

OC: Popillia japonica H-5, I-? (105)
OR: H-19, I-7 (443, 444)

Halesia carolina　　(Wild olive)　　Styracaceae

A	B	C	D	J	J	K
01	01	09	01	01	11	01

OC: Attagenus piceus H-4, I-4 (48)　　　　Popillia japonica H-2, I-16 (105)

Halimione portulacoides (Not known) Chenopodiaceae
    A B C D E E
    01 06 09 03 04 11
OC: Tobacco necrotic virus H-21, I-7 (1316)

Haplopappus heterophylles (Rayless goldenrod) Asteraceae
    B C D J J
    06 09 03 09 17
OC: Spodoptera frugiperda H-4, I-1, 12 (937)

Haplophyton cimicidum (Cockroach plant) Apocynaceae
    A B C D D D J J J K
    01 02 09 01 02 04 02 03 08 05
OC: Anasa tristis H-2, I-2 (101)   Gargaphia solani H-2, I-6 (101)
    Anastrepha ludens H-2, I-6 (96, 1092)   Grasshoppers H-2, I-6 (101)
    Anopheles quadrimaculatus H-4, I-6, 7 (48) Laspeyresia pomonella H-2, I-6 (101)
    Attagenus piceus H-4, I-6, 7 (48)   Leptinotarsa decemlineata H-2, I-6 (101)
    Autographa oo H-1, I-? (1252)   Musca domestica H-2, I-6 (96)
    Blattella germanica H-2, I-6, 7 (504)   Oncopeltus fasciatus H-2, I-6, 7 (48)
    Cockroaches H-2, I-7 (504)   Ostrinia nubilalis H-2, I-6 (101)
    Diaphania hyalinata H-1, I-? (1252)   Parasites (human) H-2, I-7 (504)
    Epicauta vittata H-2, I-2 (101)   Pieris rapae H-2, I-2 (101, 1252)
    Epilachna varivestis H-2, I-2 (101)   Spodoptera eridania H-2, I-2 (101, 1252)
    Fruit flies H-1, I-7, 14 (105)   Udea rubigalis H-1, I-2 (101, 1252)
AP: Alk=I-16 (cimicidine, haplophytine)(1392)

Hardwickia mannii (Not known) Caesalpiniaceae
    A B C D M M
    01 01 09 01 04 19
OC: Termites H-2, I-6 (78)

Harpullia arborea (Not known) Sapindaceae
    A B C D J J M M O O
    01 01 09 01 08 11 05 15 10 12
OC: Attagenus piceus H-4, I-4 (48)
OR: H-32, I-4 (968, 983, 1349)
AP: Sap=I-4 (968)

Harrisonia abyssinica (Mosabubini) Simaroubaceae
    A B C D
    01 02 09 01
OR: H-2, I-2 (1345); H-4, I-? (132); H-4, 20, I-2 (169)

Hedeoma pulegioides (Pennyroyal) Lamiaceae
    A B D D D J M O O
    03 09 02 03 04 04 05 07 08
OC: Blatta orientalis H-5, I-? (105)   Fleas H-5, I-7 (105)
    Cochliomyia hominivorax H-5, I-? (105)   Flies H-5, I-7 (1297)

Hedera helix (English ivy) Araliaceae
    A B C D D G J J M M M
    01 03 09 01 03 04 03 08 13 15 16

OC: Attagenus piceus H-4, I-7 (48)   Plutella xylostella H-1, I-7 (87)
    Lice H-2, I-7 (220)   Venturia inaequalis H-14, I-7, 8,
    Pediculus humanus capitis H-1, I-7 (87)   9, 10 (16)
AP: Sap=I-7 (220)

Hedychium coronarium   (Butterfly ginger)   Zingiberaceae
    A   B   C   D
    01 06 09 01
OR: H-5, I-2, 6 (1241)

Hedychium spicatum   (Spiked ginger lily)   Zingiberaceae
    A   B   C   D   M   M   M
    01 06 09 01 13 17 19
OC: Clothes moths H-5, I-? (105)
OR: H-15, 19, I-? (411)

Helenium elegans   (Sneezeweed)   Asteraceae
    B   C   D   J
    06 09 03 03
OC: Diaphania hyalinata H-2, I-? (101)

Helenium mexicana   (Yerba de la Pulga)   Asteraceae
    B   C   D   J   J   K   K
    06 09 10 01 04 01 05
OC: Laspeyresia pomonella H-1, I-8 (101)   Pachyzancla bipunctalis H-1, I-8 (101)
OR: H-5, I-2 (105, 895)

Helenium quadridentalum   (Sneezeweed)   Asteraceae
    A   B   C   D   F   J
    03 06 09 01 08 11
OC: Attagenus piceus H-4, I-1 (48)   Lice H-1, I-1 (118)

Helianthus annuus   (Sunflower)   Asteraceae
    A   B   D   F   G   J   J   K   M   M   M   M   M   M   O   O
    03 06 10 03 05 03 06 01 02 05 10 13 16 21 02 08
OC: Acromyrmex cephalotes H-6, I-? (592)   Leaf-cutting ants H-6, I-12 (87, 470)
    Acromyrmex octospinosus H-6, I-? (592)   Leptinotarsa decemlineata H-4, I-7 (531)
    Amaranthus retroflexus H-24, I-2, 7 (574) Lipaphis erysimi H-2, I-? (84)
    Aristida oligantha H-24, I-2, 7 (574)   Meloidogyne incognita H-16, I-2 (147, 158)
    Bromus japonicus H-24, I-2, 7 (574)   Pectinophora gossypiella H-1, I-? (1152)
    Croton glandulosus H-24, I-2, 7 (574)   Phyllobius oblongus H-4, I-7 (531)
    Erigeron canadensis H-24, I-2, 7 (574)   Phytodecta fornicata H-4, I-7 (531)
    Flies H-1, I-? (87)   Pieris brassicae H-4, I-7 (531)
    Haplopappus ciliatus H-24, I-2, 7 (574)   Pseudomonas solanacearum H-18, I-1, 8 (836)
    Heliothis virescens H-5, I-7 (1240)   Tanymecus dilaticollis H-4, I-7 (531)
OR: H-19, I-? (732, 1171); H-24, I-7 (569, 1207, 1228, 1264)
AP: Alk=I-10 (1391); Tan=I-6 (1321)

Helianthus decapetalus   (Thin leaf)   Asteraceae
    A   B   C   D
    01 06 09 03
OC: Sclerotinia fructicola H-14, I-1 (944)
OR: H-19, I-? (1171)

Helianthus giganteus   (Giant sunflower)   Asteraceae
    A   B   C   D
    01   06   09   03
OC: Botrytis cinerea H-14, I-1, 8 (945)
OR: H-19, I-? (1171)

Helianthus mollis   (Ashy sunflower)   Asteraceae
    A   B   C   D   J
    01   06   09   03   03
OR: H-19, I-1 (704); H-24, I-1 (1222)
AP: Alk=I-9, 10 (1391)

Helianthus petiolaris   (Sunflower genus)   Asteraceae
    B   C
    06   09
OC: Melanoplus femurrubrum H-1, I-1 (87)

Helianthus rigidus   (Prairie sunflower)   Asteraceae
    A   B   C   D
    01   06   09   03
OC: Monarda fistulosa H-24, I-2 (584)   Poa pratensis H-24, I-2 (584)

Helichrysum hookeri   (Everlasting genus)   Asteraceae
    A   B   C   D   K   L
    03   06   09   03   06   05
OC: Dysdercus cingulatus H-2, I-? (88)

Heliopsis gracilus   (Oxeye genus)   Asteraceae
    B
    06
OC: Musca domestica H-2, I-7 (101, 766)

Heliopsis helianthoides   (Oxeye genus)   Asteraceae
    A   C   D   L   J
    01   09   03   05   11
OC: Fusarium oxysporum H-14, I-2 (896)   Pseudaletia unipuncta H-2, I-2 (101)
    Musca domestica H-2, I-2, 6 (87, 101, 766, 896)
OR: H-1, I-2 (30); H-19, I-? (1171)

Heliopsis longipes   (Chilcuan)   Asteraceae
    B   C   D   J   J   J   J
    06   09   03   03   07   08   11
OC: Acanthoscelides obtectus H-2, I-2 (87)   Laspeyresia pomonella H-2, I-? (96, 101)
    Aedes aegypti H-4, I-2 (48)   Mosquitoes H-1, I-? (101)
    Anasa tristis H-1, I-? (101, 1252)   Musca domestica H-2, I-2 (96, 101, 896)
    Attagenus piceus H-4, I-2 (48)   Pachyzancla bipunctalis H-2, I-2 (101)
    Diaphania hyalinata H-2, I-2 (101, 1252)   Pediculus humanus capitis H-2, I-2 (101)
    Fusarium oxysporum H-14, I-2 (896)   Tineola bisselliella H-4, I-2 (48)
OR: H-5, I-2 (1345)

Heliopsis parvifolia   (Oxeye genus)   Asteraceae
    B   C   D   J
    06   07   03   11
OC: Flies H-2, I-2, 6 (101)   Musca domestica H-1, I-2 (766)

Heliotropium arborescens   (Heliotrope)   Boraginaceae
   A   C   D   D   F   M
   01 09 01 06 06 17
OC: Pediculus humanus humanus H-2, I-12 (96, 105)

Heliotropium indicum   (Hatisura)   Boraginaceae
   A   B   C   D   D   L   L   M   O
   03 06 09 01 04 03 05 05 07
OR: H-2a, I-1 (118); H-33, I-? (220)
AP: Alk=I-16 (1392)

Heliotropium parviflorum   (Scorpion tail)   Boraginaceae
   A   B   C   D   J   M   O
   03 06 09 01 03 05 07
OR: H-14, I-7 (118)

Heliotropium subulatum   (Heliotrope genus)   Boraginaceae
   B   C
   06 09
OC: Locusta migratoria H-4, 5, I-7 (596)

Helleborus niger   (Christmas rose)   Ranunculaceae
   A   B   C   D   E   G   J   J   K   L   M   O
   01 06 09 10 03 04 04 11 05 03 05 02
OC: Popillia japonica H-4, I-2 (1243)   Tribolium castaneum H-5, I-10 (179)
AP: Alk=I-2 (1374)

Helonias bullate   (Swamp pink)   Liliaceae
   A   B   C   D   E   M
   01 06 09 03 11 13
OC: Popillia japonica H-5, I-? (105, 1243)

Hemarthria altissima   (Not known)   Poaceae
   B   C
   06 09
OR: H-24, I-2 (1241)

Hemerocallis dumortieri   (Day lily)   Liliaceae
   A   B   C   D
   01 06 09 03
OC: Plutella xylostella H-4, I-? (87)

Hepatica nobilis var. nipponica   (Liver leaf genus)   Ranunculaceae
   A   B   C   D   J
   01 06 09 03 03
OC: Drosophila hydei H-2, I-2, 6, 7, 8 (101)

Heracleum laciniatum   (Cow parsnip genus)   Apiaceae
   B   C   D
   06 09 03
OR: H-24, I-10 (1245)

Heuchera americana   (Rock geranium)   Saxifragaceae
    A   B   C   D
    01   06   09   03
OC: Pseudomonas solanacearum H-18, I-1, 8, 9 (945)

Heuchera sanguinea   (Coral bells)   Saxifragaceae
    A   B   C   D   D
    01   06   09   03   08
OC: Plutella xylostella H-4, I-? (87)

Hibiscus moscheutos   (Common mallow)   Malvaceae
    A   B   C   D
    01   06   09   03
OC: Pseudomonas solanacearum H-18, I-8, 16 (945)

Hibiscus rosa-sinensis   (Rose-of-China)   Malvaceae
    A   B   C   D   D   J   J   M   M   O   O
    01   02   09   01   02   11   15   05   13   07   15
OC: Sitophilus oryzae H-2, I-8 (87, 400, 1124)

Hibiscus syriacus   (Rose-of-Sharon)   Malvaceae
    A   B   C   D   J   M   N
    01   02   09   03   03   01   07
OC: Anthonomus grandis H-4, I-8 (87, 321)   Pieris brassicae H-4, I-7 (531)
    Hyphantria cunea H-4, I-7 (531)
AP: Fla=I-8 (1378)

Hibiscus vitifolius   (Mallow genus)   Malvaceae
    C   D   M
    09   01   03
OC: Pediculus humanus capitis H-2, I-2 (105)

Hieracium auranthiacum   (Devil's paint brush)   Asteraceae
    A   B   C   D   E   L
    01   06   09   03   12   05
OC: Pseudomonas solanacearum H-18, I-1, 8 (945)

Hieracium boreale   (Hawkweed genus)   Asteraceae
    A   B   C   D   D   E   F
    01   06   09   03   08   12   14
OC: Venturia inaequalis H-14, I-? (16)

Hieracium japonicum   (Hawkweed genus)   Asteraceae
    A   B   C   D   E   J   L
    01   06   09   03   12   03   05
OC: Drosophila hydei H-2, I-2, 6, 7 (78, 101)

Hieracium pilosella   (Mouse-ear hawkweed)   Asteraceae
    A   B   C   D   E
    01   06   09   03   12
OC: Musca domestica H-2, I-? (87)

Hieracium pratense   (Hawkweed genus)   Asteraceae
    A   B   C   J   L
    01   06   09   08   05
OC: Popillia japonica H-5, I-1 (105, 1243)

Hierochloe odorata   (Sweet grass)   Poaceae
    A   B   C   D   M   M
    01   06   09   03   17   24
OR: H-5, I-7 (1345)

Hiptage benghalensis   (Not known)   Malpighiaceae
    A   B   C   D   J   M   M   M   M   O
    01   03   09   01   11   05   13   17   21   07
OC: Attagenus piceus H-4, I-9 (48)
OR: H-20, I-7 (504)
AP: Sap=I-6, 7, 8 (1360); Tan=I-7 (1360)

Holarrhena antidysenterica   (Tellicherry bark)   Apocynaceae
    A   B   C   D   J   M   O
    01   01   09   01   03   05   10
OC: Meloidogyne incognita H-16, I-10 (616)   Meloidogyne javanica H-16, I-10 (616)
OR: H-15, 19, I-? (411); H-17, I-? (220)
AP: Alk=I-4 (conamine, conarrhimine, connessidine, connessimine, connessine, conimine, con-
    kurchine, conkurchinine, holarrhessimine, holarrhidine, holarrhimine, isoconessimine,
    kurchamine, kurchine, lettocine, monomethylholarrhimine I & II, norconessine, tetra-
    methylholarrhimine, trimethylconkurchine)(1392); Alk=I-6, 7 (1380); Alk=I-10 (1372);
    Alk, Tan=I-7 (1360)

Holigarna grahamii   (Not known)   Anacardiaceae
    A   B   C   D
    01   09   01   01
OC: Rats H-26, I-16 (1393)
OR: H-33, I-14 (220)

Hordeum vulgare   (Barley)   Poaceae
    A   B   C   D   J   M   N
    03   06   09   03   03   01   10
OC: Athalia rosae H-4, I-7 (531)   Schizaphis graminum H-1, I-? (1175)
    Fusarium solani   Thielaviopis basicola H-14, I-? (960)
      f. sp. phaseoli H-14, I-? (422, 960)   Verticillium albo-atrum H-14, I-? (361)
    Rhizoctonia solani H-14, I-? (422, 960)
OR: H-22, I-10 (881); H-24, I-7 (1228, 1229)
AP: Alk=I-2 (hordenine, N-methyltyramine, leaf-gramine)(1392); Alk=I-7 (hordenine)(1392)

Hormothamnion enteromorphoides   (Algae)   Nostocaceae
    A   B   C
    04   09   09
OC: Mucor racemosus H-14, I-1 (1024)   Rhizopus oryzae H-14, I-1 (1024)

Hosta minor   (Plantain lily genus)   Liliaceae
    A   B   C   D
    01   06   09   03
OC: Colletotrichum lindemuthianum H-14, I-?   Endothia parasitica H-14, I-? (957)
      (957)   Monilinia fructicola H-14, I-? (957)

148

Houttuynia cordata   (Tsi)   Saururaceae
```
A   B   C   D   J   M   M   N   O
01  06  09  03  3a  01  05  07  07
```
OC: Aphids H-1, I-1 (1241)   Rondotia menciana H-1, I-1 (1241)
    Caterpillars H-1, I-1 (1241)   Spider mites H-11, I-1 (1241)
    Puccinia graminis tritici H-14, I-1 (1241)
OR: H-20, I-1 (504)

Humboldtia brunonis   (Not known)   Fabaceae
```
C   K   M
09  09  12
```
OC: Rats H-26, I-16 (1393)

Humulus lupulus   (Common hop)   Cannabaceae
```
A   B   B   C   D   F   J   J   M   M   M   N
01  04  06  09  03  07  02  04  01  05  13  07
```
OC: Alternaria citri H-14, I-13 (566)   Pythium sp. H-14, I-13 (566)
    Aspergillus oryzae H-14, I-13 (566)   Rhizoctonia solani H-14, I-13 (566)
    Cochliomyia hominivorax H-2, I-2 (105)   Rhizopus nigricans H-14, I-13 (566)
    Diaphania hyalinata H-2, I-7 (101, 1252) Sclerotinia fructicola H-14, I-13 (566)
    Fusarium oxysporum   Sclerotium bataticola H-14, I-13 (566)
      f. sp. lycopersici H-14, I-13 (566)   Spodoptera eridania H-2, I-7 (101)
    Phytophthora citrophthora H-14, I-13 (566) Spodoptera litura H-4, I-7 (1064)
OR: H-19, I-8 (594, 653, 732, 1022)
AP: Alk=I-? (chopeine, codeine, coniine, morphine) (1392)

Hura crepitans   (Monkey's dinner bell)   Euphorbiaceae
```
A   B   C   D   J   J   M
01  01  09  01  10  11  13
```
OC: Aphis spiraecola H-2, I-14 (105)   Spodoptera eridania H-2, I-6 (101)
    Culex sp. H-2, I-9 (190, 304)
OR: H-19, I-8 (172); H-32, I-14 (1147, 1276); H-33, I-14 (620)

Hura polyandra   (Avilla tree)   Euphorbiaceae
```
A   B   C   J   J   L   M
01  01  09  03  10  03  04
```
OC: Mosquitoes H-2, I-10 (101)
OR: H-20, I-14 (105); H-32, I-14 (101, 504, 1276); H-33, I-10 (504)

Hybanthus yucatensis   (Not known)   Violaceae
```
B   C   D   D   J   J   J
06  09  01  02  03  08  11
```
OC: Attagenus piceus H-4, I-2, 6, 7 (48)

Hydnocarpus anthelminthicus   (Chaulmogra tree)   Flacourtiaceae
```
A   B   C   D   M
01  01  09  01  05
```
OR: H-2, I-10 (104)

Hydnocarpus kurzii   (Not known)   Flacourtiaceae
```
A   B   C   M   O
01  01  09  05  12
```
OC: Aspergillus niger H-14, I-12 (1300)
    Xanthomonas campestris pv. campestris H-18, I-12 (1300)
    Xanthomonas campestris pv. citri H-18, I-12 (1300)

OR: H-1, 32, I-9, 12 (220); H-19, I-? (186); H-32, I-5, 9, 12 (73, 220)
AP: Glu=I-12 (220)

Hydnocarpus laurifolia   (Marotti)   Flacourtiaceae
A   B   D   J   J   K   M   M   O   O
01 01 01 03 10 06 05 06 10 12
OC: Meloidogyne incognita H-16, I-18 (100, 521, 616)
    Meloidogyne javanica H-16, I-18 (100, 521, 616)
OR: H-32, I-9, 10 (73)
AP: Glu=I-12 (220)

Hydnocarpus venenata   (Not known)   Flacourtiaceae
A   B   D   M   O
01 01 01 05 12
OR: H-2, 32, I-9 (105, 1276)

Hydnocarpus wightiana   (Maravitti tree)   Flacourtiaceae
A   B   D   J   K   K   M   O   O
01 01 01 04 04 06 05 10 12
OC: Ants H-5, I-18 (87, 105)   Xylorycetes jamaicensis H-5, I-18 (87, 105)
    Toxoptera aurantii H-1, I-? (681)
OR: H-2, I-10 (1345); H-32, I-9 (105, 459)
AP: Glu=I-12 (220)

Hydrangea arborescens   (Hydrangea root)   Saxifragaceae
B   D   D   J   J   J   M   M   O
02 01 08 07 10 17 05 13 02
OC: Mosquitoes H-2, I-2 (98)
OR: H-19, I-8 (413)
AP: Alk=I-6, 7, 8 (1369)

Hydrastis canadensis   (Golden seal)   Ranunculaceae
A   B   D   M   M   O
01 06 03 05 16 02
OC: Mosquitoes H-2, I-2 (98, 105)
AP: Alk=I-2 (berberine, canadine, hydrostine)(1392)

Hydrocotyle americana   (Water pennywort genus)   Apiaceae
A   B   B   C
01 05 06 09
OC: Manduca sexta H-4, I-7 (669)   Panonychus citri H-4, 11, I-7 (669)

Hydrocotyle asiatica   (Tankuni)   Apiaceae
A   B   D   D   L   M   M   N
03 06 04 10 05 01 09 07
OR: H-2, I-1 (127); H-15, 19, I-? (411); H-33, I-? (220)
AP: Alk=I-6, 7 (hydrocotyline)(1392)

Hydrocotyle javanica   (Water pennywort genus)   Apiaceae
A   B   B   C   D   J
01 05 06 06 10 03
OC: Drosophila hydei H-2, I-6, 7 (101)
OR: H-32, I-? (73, 220)

Hydrocotyle podantha   (Water pennywort genus)   Apiaceae
    A   B   B   C   J
   01 05 06 09 09
OC: Rats H-26, I-1 (1393)

Hymenaea courbaril   (Locust tree)   Caesalpiniaceae
   A   B   C   D   M   N   N
   01 01 09 01 01 11 20
OC: Atta cephalotes H-5, I-7 (138, 175)
OR: H-15, I-7 (138, 175)
AP: Ter=I-7 (caryophyllene)(138)

Hymenocallis littoralis   (Spider lily)   Amaryllidaceae
   A   B   C   D   J
   01 06 09 01 09
OC: Schistocerca gregaria H-4, I-2, 7, 8, 9 (1009, 1051)
OR: H-22, I-7 (847)
AP: Alk=I-2 (lycorine)(1392); Alk=I-3 (1051); Alk=I-3 (tazettine)(1392)

Hypoestes verticillaris   (Flamingo plant genus)   Acanthaceae
   A   C   D   J   M
   01 09 01 14 16
OC: Oncopeltus fasciatus H-2, I-2 (958)

Hypoxis latifolia   (Stargrass genus)   Hypoxidaceae
   B   C   D   K   M
   06 09 01 09 02
OC: Rodents H-25, I-3 (1345)

Hyoscyamus albus   (Not known)   Solanaceae
   A   A   B   C   D   D   J   M   O
   01 03 06 09 03 06 03 05 07
OR: H-1, I-7 (105, 1026); H-34, 35, I-7 (504)
AP: Alk=I-2, 7, 10 (hyoscyamine, hyposcine)(1026, 1392)

Hyoscyamus major   (Not known)   Solanaceae
   B   C   D   D   M   M   O
   06 09 04 06 05 28 07
OR: H-2, I-7 (1026)
AP: Alk=I-7 (hyoscine, hyoscyamine)(504)

Hyoscyamus niger   (Black henbane)   Solanaceae
   A   A   B   C   D   D   D   J   L   M   O
   02 03 06 09 02 03 06 03 03 05 07
OC: Aphids H-2, I-2, 6, 7, 8, 9 (87, 105)   Pteronus ribesii H-1, I-1 (1035)
   Malacosoma neustria H-1, I-1 (1035)   Rats H-25, I-7 (1345)
OR: H-1, I-7 (1026); H-33, I-1 (220, 507); H-34, 35, I-7 (504, 507)
AP: Alk=I-2, 6, 7, 10 (atropine, cuscohygrine, hyoscine, hyoscyamine)(1392)

Hypericum perforatum   (St. John's wort)   Hypericaceae
   A   B   C   D   D   D   J   J   L   M   M   N
   01 06 09 06 08 10 04 09 03 01 16 07
OC: Dermacentor marginatus H-13, I-7 (87, 1133) Ixodes redikorzevi H-13, I-7 (87, 1133)
   Haemaphysalis punctata H-13, I-7 (87, 1133) Rhipicephalus rossicus H-13, I-7 (87, 1133)

OR: H-17, I-7 (504); H-19, I-? (732, 1101, 1171); H-20, I-1 (504); H-22, I-2, 6, 7 (969)
AP: Alk=I-? (1392); Sap=I-? (220)

Hyptis spicigera   (Not known)   Lamiaceae
```
A    B    C    D    K    K    M    M    N
03   06   09   01   07   09   01   20   10
```
OC: Caryedon serratus H-1, I-7 (1124)   Stored grain pests H-5, I-16 (105, 533)
    Mosquitoes H-5, I-16 (87, 1353)   Termites H-5, I-16 (87, 105, 1353)

Hyptis suaveolens   (Desert lavender)   Lamiaceae
```
B    C    D    J    J    K    M    M    N    O
02   06   01   01   03   03   01   05   07   07
```
OC: Cimex lectularius H-5, I-7 (163, 166, 504) Culex pipiens H-5, I-7 (163, 166)

Iberis amara   (Rocket candytuft)   Brassicaceae
```
A    B    C    D    M    M    N    O
03   02   09   03   01   05   10   07
```
OC: Phyllotreta nemorum H-1, I-? (143); H-4, I-7 (1134, 1156)

Iberis umbellata   (Globe candytuft)   Brassicaceae
```
A    B    C    D
03   07   09   06
```
OC: Phyllotreta nemorum H-4, I-7 (1134)   Phyllotreta undulata H-4, I-7 (1134)
    Phyllotreta tetrastigma H-4, I-7 (1134)
OR: H-19, I-? (581)

Ilex decidua   (Possum haw)   Aquifoliaceae
```
A    B    B    C    D    J
01   01   02   09   03   3a
```
OC: Erwinia carotovora H-18, I-9 (584)

Ilex opaca   (American holly)   Aquifoliaceae
```
A    B    C    D    J    M
01   01   09   03   09   19
```
OC: Popillia japonica H-5, I-7 (1243)

Ilex verticillata   (Winterberry)   Aquifoliaceae
```
A    B    C    D    J    J    M    N
01   02   09   10   08   11   01   07
```
OC: Attagenus piceus H-4, I-9 (48)

Illicium lanceolatum   (Anise tree genus)   Illiciaceae
```
A    C    J
01   09   03
```
OC: Agrotis sp. H-1, I-2, 4, 7, 9 (1241)   Caterpillars H-1, I-2, 4, 7, 9 (1241)

Impatiens balsamina   (Rose balsam)   Balsaminaceae
```
A    B    C    D    E    F    G    J    M
03   06   09   01   07   13   05   03   13
```
OC: Drechslera oryzae H-14, I-7 (113, 432)   Pyricularia oryzae H-14, I-7 (114, 432)
OR: H-19, I-14 (670, 982)

Impatiens capensis   (Spotted touch-me-not)   Balsaminaceae
```
A    C    D    J
03   09   03   09
```

OC: Bipolaris sorokiniana H-14, I-1 (944)   Glomerella cingulata H-14, I-1 (944)
   Colletotrichum lindemuthianum H-14, I-14   Sclerotinia fructicola H-14, I-14 (1105)
      (1105)
   Fusarium oxysporum f. sp. lycopersici H-14, I-1 (944)
OR: H-19, I-? (1171)

Impatiens longipes   (Not known)   Balsaminaceae
   B   C   D   D   E   J   M
   06   09   01   02   07   09   13
OC: Musca domestica H-1, I-1 (785)   Tribolium castaneum H-1, I-1 (785)
OR: H-2, I-? (100)

Impatiens pallida   (Pale touch-me-not)   Balsaminaceae
   A   B   C   D   K
   03   06   09   03   09
OC: Botrytis cinerea H-14, I-1 (944)
OR: H-7, I-1 (944)

Impatiens parviflora   (Touch-me-not genus)   Balsaminaceae
   B   C
   07   09
OC: Athalia rosae H-4, I-7 (531)   Pieris brassicae H-4, I-7 (531)

Impatiens wallerana   (Busy Lizzy)   Balsaminaceae
   A   B   C   E   M
   01   06   09   07   13
OC: Pluttela xylostella H-4, I-? (87)

Imperata cylindrica   (Kogon grass)   Poaceae
   A   B   C   D   D   L   M
   01   06   09   01   02   03   09
OC: Meloidogyne incognita H-16, I-7 (115)
OR: H-19, I-2 (982); H-22, I-16 (790)
AP: Alk=I-7 (1392); Alk, Tri=I-7 (1390)

Indigofera cordifolia   (Indigo genus)   Fabaceae
   B   B   J   M
   06   09   03   12
OR: H-24, I-1 (1212)

Indigofera hirsuta   (Indigo genus)   Fabaceae
   B   C   D   M
   06   09   01   12
OC: Belonolaimus longicaudatus H-16, I-? (955) Trichodorus christiei H-16, I-? (955)
   Criconemoides ornatum H-16, I-? (955)   Xiphinema americanum H-16, I-? (955)
AP: Alk=I-7 (1392)

Indigofera lespedezioides   (Anil)   Fabaceae
   A   B   C   D   D   F   J   J   L   M   M
   03   06   09   01   02   08   01   02   03   05   13
OC: H-2a, I-1 (118)

Indigofera suffruticosa   (Indigo genus)   Fabaceae
   A   C   D   J   M
   01   09   01   04   12
OC: Fleas H-2, I-2, 10 (1345)

Indigofera tinctoria   (Indigo plant)   Fabaceae
    A   B   C   D   D   M   M
    03 06 09 01 02 12 16
OC: Lice H-2, I-10 (87)   Vermin H-27, I-2 (87, 104)

Inula conyza   (Cinnamon root)   Asteraceae
    B   C
    07 09
OR: H-2, I-? (104)

Inula helenium   (Horseheal)   Asteraceae
    A   B   C   D   J   M   M   M   O
    01 01 09 03 10 05 15 16 02
OC: Cimex lectularius H-1, I-? (101)   Lice H-1, I-? (101)
    Clothes moths H-1, I-? (105)   Mosquitoes H-2, I-2 (98, 101)
    Flies H-1, I-? (101)
OR: H-19, I-2, 12 (504, 1171)

Inula viscosa   (Not known)   Asteraceae
    A   B   C   D   K   M   M
    01 02 09 06 07 16 17
OC: Mosquitoes H-5, I-? (104)

Ipomoea aquatica   (Potato vine)   Convolvulaceae
    A   B   B   C   D   K   M   O
    03 04 05 09 01 03 05 15
OC: Drechslera oryzae H-14, I-7 (113)

Ipomoea batatas   (Sweet potato)   Convolvulaceae
    A   B   B   C   D   J   L   M   M   M   M   N   N   O   O   O
    01 04 05 09 01 03 02 01 02 05 16 21 02 07 06 07 15
OC: Drechslera oryzae H-14, I-2, 6, 7 (432)   Pratylenchus brachyurus H-16, I-? (1274)
    Pediculus humanus capitis H-1, I-7 (1353)   Pyricularia oryzae H-14, I-7 (114, 432)
OR: H-19, I-9, 19 (172, 851); H-24, I-14 (569)
AP: Tan=I-6, 7 (1321)

Ipomoea cornea   (Morning glory genus)   Convolvulaceae
    A   B   B   C   D   F   J   J   K   K
    01 04 05 09 01 07 02 04 05 06
OC: Callosobruchus chinensis H-2, I-7 (89, 116); H-5, I-? (599)

Ipomoea fistulosa   (Not known)   Convolvulaceae
    A   B   C   D
    01 04 09 01
OC: Helicotylenchus indicus H-16, I-15 (1330)   Tylenchorhynchus brassicae H-16, I-15
    Hoplolaimus indicus H-16, I-15 (1330)   (1330)
    Rotylenchulus reniformis H-16, I-15 (1330)   Tylenchus filiformis H-16, I-15 (1330)

Ipomoea hederacea   (Morning glory genus)   Convolvulaceae
    A   B   C   D   J   J   K   L   M   O
    03 04 09 01 04 11 05 05 05 10
OC: Tribolium castaneum H-5, I-10 (179)
OR: H-17, I-10 (504)

Ipomoea muricata   (Morning glory genus)   Convolvulaceae
    B   C   D   M   O
    04 09 01 05 10
OC: "Bugs" H-2, I-14 (105)
OR: H-2, I-14 (73, 90)

Ipomoea nil   (Morning glory genus)   Convolvulaceae
    A   A   B   B   C   D   J   M   O
    01 03 04 05 09 02 04 05 10
OC: Aphids H-2, I-10 (101)   Tobacco mosaic virus H-21, I-? (1126)

Ipomoea pandurata   (Wild potato vine)   Convolvulaceae
    A   B   B   C   D
    01 04 06 09 03
OR: H-5, I-14 (1297)

Ipomoea pes-tigridis   (Seaside morning glory)   Convolvulaceae
    B   B   C   D   J   M   M   N   O
    04 06 09 01 03 01 05 02 10
OR: H-24, I-1 (1212)

Ipomoea purpurea   (Common morning glory)   Convolvulaceae
    A   B   C   D   D   J   J   J
    03 04 09 01 02 03 09 15
OC: Aphids H-2, I-7 (105)   Locusta migratoria H-4, I-7 (596)
    Caterpillars H-2, I-7 (105)   Rodents H-25, I-1 (944)
    Flea beetles H-2, I-7 (105)   Scale insects H-2, I-7 (105)
AP: Alk=I-10 (1391)

Ipomoea quamoclit   (Cypress vine)   Convolvulaceae
    A   B   C   D   J   J
    03 06 09 01 03 04
OC: Aphids H-2, I-10 (101)   Heliothis virescens H-5, I-7 (1240)
AP: Alk=I-4, 6, 7, 9 (1392)

Ipomoea sindica   (Morning glory genus)   Convolvulaceae
    B   B   C   J
    04 06 09 03
OR: H-24, I-16 (1212)

Ipomopsis aggregata   (Scarlet gilia)   Polemoniaceae
    A   B   C   D   D   J   M   M   N   O
    02 03 09 03 08 03 01 05 02 07
OC: Attagenus piceus H-4, I-1 (48)

Iris dichotoma   (Iris genus)   Iridaceae
    A   B   C   D
    01 06 09 03
OC: Pieris rapae H-1, I-1 (1241)

Iris douglasiona   (Iris genus)   Iridaceae
    A   B   C   D
    01 06 09 03
OC: Oncopeltus fasciatus H-3, I-2, 6, 9 (958)

Iris ensata   (Sword-leaved iris)   Iridaceae
   A   B   C   D   M   M   N   O
   03 06 09 03 01 05 02 02
OR: H-3, I-? (698)

Iris japonica   (Japanese iris)   Iridaceae
   A   B   C   D   M   M   M   N
   01 06 09 03 01 13 17 02
OC: Aphelencoides besseyi H-16, I-? (52, 187)

Isopyrum stoloniferum   (Not known)   Ranunculaceae
   A   B   C   D   J
   01 06 09 03 03
OC: Drosophila hydei H-2, I-2, 6, 7 (101)

Iva axillaris   (Poverty weed)   Asteraceae
   A   B   C   D   J   L
   01 06 09 03 14 05
OC: Ceratitis capitata H-6, I-6, 7, 8 (956)   Dacus dorsalis H-6, I-6, 7, 8 (956)
   Dacus cucurbitae H-6, I-6, 7, 8 (956)

Ixora coccinea   (Flame-of-the-woods)   Rubiaceae
   A   B   D   J   J   M   M   M   M   O
   01 02 01 03 09 05 13 21 27 07
OC: Drechslera oryzae H-14, I-7 (113)   Rats H-26, I-16 (1393)
AP: Sap=I-2, 6, 7 (1319); Sfr=I-6, 7 (1319); Tan=I-6, 7 (1319)

Jacaranda obtusifolia var. rhombifolia   (Not known)   Bignoniaceae
   A   B   D   J   K   M
   01 01 01 04 05 13
OC: Flies H-2, I-? (101)   Mosquitoes H-2, I-? (101)
OR: H-2, I-? (104)

Jacquinia aristata   (Not known)   Theophrastaceae
   A   B   D   D   D   J   J
   01 02 01 02 04 03 04
OC: Diaphania hyalinata H-2, I-2 (101, 1108)   Plutella xylostella H-2, I-2 (101)
OR: H-32, I-9 (504)

Jacquinia auriantaca   (Palo de animas)   Theophrastaceae
   A   B   C   D   F   L
   01 02 09 01 08 02
OR: H-2a, I-1 (118); H-32, I-2, 6, 7, 9 (504, 1276)

Jacquinia barbasco   (Barbasco)   Theophrastaceae
   B   B   D   D   J   J
   01 02 01 04 03 08
OC: Attagenus piceus H-4, I-6, 7 (48)
OR: H-32, I-? (507)

Jaquemontia tamnifolia   (Not known)   Convolvulaceae
   A   B   B   C   D
   01 04 07 09 03
OR: H-2, I-10 (1345)

Jasminum arborescence   (Jasmine genus)   Oleaceae
    A   B   C   D
    01 02 09 01
OC: Meloidogyne incognita H-16, I-7 (542)

Jatropha angustidens   (Not known)   Euphorbiaceae
    C   D   D
    09 01 02
OC: Attagenus piceus H-4, I-4 (48)

Jatropha curcas   (Barbados nut)   Euphorbiaceae
    A   B   C   D   J   J   L   M   M   M   M   O   O   O   O   O
    01 01 09 01 03 06 03 05 06 15 16 02 07 10 12 14
OC: Aulacophora foveicollis H-1, I-? (600)   Mosquitoes H-1, I-? (101)
    Lipaphis erysimi H-1, I-? (600)   Musca domestica H-1, I-? (101)
    Mites H-11, I-6, 14 (1321)   Snails H-28, I-10 (662)
OR: H-2, I-5, 7, 9, 10 (101, 127, 1345); H-15, I-6, 7 (1321); H-17, I-7 (1353); H-20, I-10, 12
    (1321, 1353); H-32, I-4, 6, 7, 14 (73, 220, 983, 1353); H-33, I-10 (983, 1349, 1352)
AP: Alk=I-6, 7 (1392); Alk=I-10 (1369); Sap=I-7 (1321)

Jatropha glanduli   (Lalbherada)   Euphorbiaceae
    A   B   C   D   D   M   O
    01 01 09 01 02 05 01
OR: H-2, I-4, 7, 10 (127)

Jatropha gossypifolia   (Wild cassada)   Euphorbiaceae
    A   B   C   D   F   G   M   M   M   O   O
    01 02 09 01 03 04 05 13 21 02 07
OC: Drechslera oryzae H-14, I-7 (113)   Pyricularia oryzae H-14, I-7 (114)
OR: H-2, I-4, 7, 10 (127)

Jatropha macrorrhiza   (Not known)   Euphorbiaceae
    A   C   D   D   J   J   J
    01 09 01 02 03 08 11
OC: Attagenus piceus H-4, I-10 (48)

Jatropha multifida   (Coral plant)   Euphorbiaceae
    A   B   B   C   D   L   M   M   O   O
    01 02 07 09 01 02 05 13 10 14
OC: Mites H-11, I-7 (1318)
OR: H-32, I-4 (983); H-33, I-10 (504)
AP: Sap, Tan=I-6, 7 (1318)

Jatropha podagrica   (Australian bottle plant)   Euphorbiaceae
    B   C   J
    02 09 03
OC: Drechslera oryzae H-14, I-7 (113)

Joannesia princeps   (Not known)   Euphorbiaceae
    A   B   C   D   J   M   O   O
    01 01 09 01 03 05 10 12
OC: Attagenus piceus H-4, I-10 (48)
OR: H-20, I-10, 12 (504)

157

Juglans cinerea   (White walnut)   Juglandaceae
    A   B   C   D   M   M   M   M   N   N   O
    01 01 09 03 01 05 16 19 10 14 02
OR: H-24, I-? (1207)

Juglans mandschurica   (Chinese walnut)   Juglandaceae
    A   B   C   D   J   M
    01 01 09 03 03 19
OC: Aphids H-1, I-7 (1241)

Juglans nigra   (Black walnut)   Juglandaceae
    A   B   C   D   J   J   K   M   M   N
    01 01 09 03 03 14 01 01 19 10
OC: Acalymma vittata H-5, I-7 (105)   Dacus cucurbitae H-6, I-4, 7 (956)
    Caterpillars H-5, I-7 (105)   Flies H-5, I-7 (105)
    Ceratitis capitata H-6, I-4, 7 (956)   Scolytus multistriatus H-2, I-? (143)
OR: H-2, I-7, 10 (1345); H-5, I-7 (101); H-15, I-7 (651); H-24, I-2, 4, 7, 9 (551, 950, 1106,
    1200, 1207, 1236)

Juglans regia   (English walnut)   Juglandaceae
    A   B   C   D   J   J   J   M   M   M   M   M   N   N   O
    01 01 09 01 03 04 09 01 05 16 19 20 21 10 12 09
OC: Athalia rosae H-4, I-7 (531)   Ixodes redikorzevi H-13, I-7 (87, 1133)
    Dermacentor marginatus H-13, I-7 (87, 1133) Pieris brassicae H-4, I-7 (531)
    Haemaphysalis punctata H-13, I-7 (87, 1133) Rhipicephalus rossicus H-13, I-7 (87, 1133)
OR: H-5, I-7 (504); H-17, I-9 (504); H-22, I-16 (882); H-24, I-2 (1207); H-32, I-9 (73, 220)
AP: Tan=I-9 (1352)

Juniperus communis   (Juniper)   Cupressaceae
    A   B   B   C   D   J   M   M   M   M   N   N   O   O
    01 01 02 08 08 03 01 03 13 14 09 10 09 10
OC: Tobacco mosaic virus H-21, I-7 (184)

Juniperus horizontalis   (Creeping juniper)   Cupressaceae
    A   B   C   D   K
    01 02 08 03 07
OR: H-5, I-15 (1345)

Juniperus oxycedrus   (Prickly juniper)   Cupressaceae
    A   B   B   C   D   D   J
    01 01 02 09 03 06 06
OC: Cochliomyia hominivorax H-6, I-? (105)   Phylloxera sp. H-2, I-? (105)
OR: H-20, I-5, 12 (504)

Juniperus recurva   (Drooping juniper)   Cupressaceae
    A   B   C   J   M
    01 01 08 07 17
OC: Culex pipiens pallens H-1, I-5 (1066)
AP: Fla=I-6 (1380)

Juniperus sabina   (Needle juniper)   Cupressaceae
    A   B   C   D   J   J   M
    01 02 09 03 03 06 17

OC: Clothes moths H-5, I-7 (504)   Pediculus humanus humanus H-1, I-7 (504)
   Cochliomyia americana H-1, I-12 (244)
OR: H-2, I-16 (104); H-17, I-6, 12 (504)

Juniperus virginiana   (Red cedar)   Cupressaceae
   A   B   C   D   J   K   M   M   M
   01 01 09 03 06 07 04 05 17
OC: Clothes moths H-2, I-12 (101, 1098)   Moths H-5, I-5, 12 (885)
   Cochliomyia hominivorax H-2, I-? (105)   Termites H-1, I-? (87)
   Fleas H-1, I-2 (105, 885)   Tineola bisselliella H-2, I-? (308);
   Mites H-11, I-5, 12 (885)   H-5, I-5 (504, 1089)

Justicia adhatoda   (Malabar nut tree)   Acanthaceae
   A   B   C   D   D   F   G   J   J   J   J   J   K   K   M   M   M   M   M   M   O
   01 02 09 01 02 01 02 02 03 04 06 08 05 09 05 06 16 18 19 30 07
OC: Chirida bipunctata H-1, I-7 (966)   Rhizopertha dominica H-4, I-7
   Fleas H-1, I-? (104)   (88, 124, 1124)
   Flies H-1, I-? (104)   Sitotroga cerealella H-4, I-7
   Meloidogyne incognita H-16, I-7 (616)   (88, 121, 124, 1124)
   Meloidogyne javanica H-16, I-7 (616)   Stored grain pests H-1, I-7 (120, 826)
   Mosquitoes H-1, I-? (104)   Tribolium castaneum H-1, I-7 (87, 179, 1124)
OR: H-1, I-2, 7 (127); H-2, 14, 19, 24, I-7 (751, 1352); H-5, I-? (80, 117); H-22, I-2 (881)
AP: Alk=I-6, 7, 8 (1360); Alk=I-7 (vasicine)(1379, 1392); Sap=I-2 (1379)

Justicia gendarussa   (Water willow genus)   Acanthaceae
   B   C   D   D   L   M   O   O
   02 09 01 02 05 05 02 07
OC: Callosobruchus chinensis H-2, I-2,   Euproctis fraterna H-2, I-7 (105)
   6, 7 (105)   Spodoptera litura H-2, I-7 (105)
   Clothes moths H-5, I-7 (105)
AP: Alk=I-7 (1366, 1392); Str=I-2 (B-sitosterol)(1384)

Justicia schrimperiana   (Water willow genus)   Acanthaceae
   A   B   C   D
   01 02 09 01
OC: Pediculus humanus humanus H-1, I-6, 7 (1143)
AP: Alk=I-6 (1375)

Kaempferia galanga   (Not known)   Zingiberaceae
   A   B   C   M   M   M   M   M   O   O
   01 06 09 02 05 07 14 15 02 12
OC: Sitophilus oryzae H-5, I- ? (517)   Tribolium castaneum H-5, I-2 (517, 876)
OR: H-4, I-? (132)
AP: Alk=I-2, 7 (1320)

Kalanchoe beharensis   (Velvet leaf)   Crassulaceae
   A   B   C   D   J
   01 02 09 01 03
OC: Tobacco mosaic virus H-21, I-7 (184)

Kalanchoe daigremontiana   (Devil's-backbone)   Crassulaceae
   A   B   C   D
   01 06 09 01
OR: H-24, I-2, 6 (1190)

Kalanchoe marmorata   (Pen wiper)   Crassulaceae

    A   B   C   D
    01   02   09   10
OC: Tobacco mosiac virus H-21, I-7 (184)
OR: H-5, I-? (87); H-5, I-14 (1345)

Kalanchoe pinnata   (Air plant)   Crassulaceae

    A   B   C   D   D   J   M   M   M   M   O
    01   06   09   01   02   03   05   13   15   21   07
OC: Pyricularia oryzae H-14, I-7 (114, 432)   Ustilago tritici H-14, I-7 (181)
    Ustilago hordei H-14, I-7 (181)
OR: H-19, I-? (172); H-20, I-7 (1319)
AP: Tan=I-6, 7 (1319)

Kalanchoe spathulata   (Palm beach bells)   Crassulaceae

    A   B   C   D   L
    01   02   09   01   03
OR: H-1, I-7 (72, 105); H-1, 33, I-7 (220)

Kallstroemia maxima   (Not known)   Zygophyllaceae

    B   C   D   M   N
    06   09   01   01   16
OC: Ceratitis capitata H-6, I-2, 6, 7, 8, 9 (956)

Kalmia latifolia   (Mountain laurel)   Ericaceae

    A   B   B   C   J   L   M   O
    01   01   02   09   09   03   05   07
OC: Lymantria dispar H-1, I-7 (830); H-5, I-? (340)
OR: H-19, I-7 (1262); H-33, 34, 35, I-7 (504)

Khaya nyasica   (Red mahogany)   Meliaceae

    A   B   C   D   J   J   M
    01   01   09   01   03   06   04
OC: Pediculus humanus capitis H-2, I-10, 12   Termites H-8, I-5 (504)
    (504)   Vermin H-27, I-10, 12 (87)

Kirchneriella irregularis   (Green algae)   Selenastraceae

    A   B   C
    04   09   02
OC: Aedes sp. H-2, I-? (772)   Culex quinquefasciatus H-2, I-? (772)
    Anopheles quadrimaculatus H-2, I-? (772)

Kleinhovia hospita   (Not known)   Sterculiaceae

    A   B   C   D   J   M   M   M   M   N   O
    01   01   09   01   03   01   03   05   13   07   07
OC: Attagenus piceus H-4, I-4 (48)   Rice field insects H-5, I-6, 7 (137)
OR: H-2, I-4 (163); H-32, I-7 (983)
AP: Alk=I-7 (1392); Tri=I-7 (1390)

Koelreuteria paniculata   (Varnish tree)   Sapindaceae

    A   B   C   D   J   M   M   M   N   N
    01   01   09   03   10   01   16   17   07   09
OC: Mosquitoes H-2, I-7, 10 (99)

Kosteletzky virginica   (Seashore mallow)   Malvaceae

A   B   C   D   D
01 06 09 01 03
OC: Pseudomonas solanacearum H-18, I-1, 8 (945)

Lachnanthes caroliana   (Redroot)   Haemodoraceae

A   B   C   D   D   E   M   M   O   O
01 06 09 01 02 11 05 16 02 07
OC: Popillia japonica H-5, I-1 (105)

Lactuca sativa   (Garden lettuce)   Asteraceae

A   A   B   C   D   J   K   M   N
02 03 06 09 03 02 06 01 07
OC: Aphids H-1, I-7 (105)   Manduca sexta H-2, I-7 (87)
Delia brassicae H-1, I-? (498)   Pratylenchus brachyurus H-16, I-? (1274)
Fruit flies H-1, I-? (84)
AP: Alk=I-9 (1369)

Laetia calophylla   (Not known)   Flacourtiaceae

A   B   C   D   J
01 06 09 01 11
OC: Attagenus piceus H-4, I-2, 6 (48)

Lagenandra ovata   (Not known)   Araceae

B   C   D   E
06 09 01 11
OR: H-2, I-? (105, 220)

Lagenandra toxicaria   (Not known)   Araceae

B   C   D   E
06 09 01 11
OR: H-2, I-? (73)

Lamium maculatum   (Spotted dead nettle)   Lamiaceae

A   B   C   D
01 06 09 03
OC: Hyphantria cunea H-4, I-7 (531)   Pieris brassicae H-4, I-7 (531)

Lansium domesticum   (Lanzones)   Meliaceae

A   B   C   D   F   J   K   K   M   M   N
01 01 09 01 01 02 06 07 01 17 09
OC: Mosquitoes H-5, I-9 (163)
OR: H-33, I-4, 10 (983)

Lantana camara   (Common lantana)   Verbenaceae

A   B   B   C   D   D   F   J   J   J   J   K   L   M   M   M   N   O   O   O
01 02 03 09 01 02 04 03 04 10 12 05 05 01 05 13 09 02 07 08
OC: Aphis fabae H-1, I-? (33)   Manduca sexta H-4, I-7 (669)
Athalia proxima H-2a, 4, 5, I-7, 14   Musca domestica H-1, I-? (90)
  (116, 438, 607, 1101)   Ostrinia furnacalis H-1, I-? (90)
Dysdercus cingulatus H-3, I-6 (690)   Panonychus citri H-4, 11, I-7 (669)
Dysdercus koenigii H-3, I-6, 7, 8, 12 (795) Plutella xylostella H-4, I-7 (87, 669)
Lipaphis erysimi H-1, I-? (84)   Sitophilus oryzae H-1, I-? (90)
OR: H-1, I-7, 8 (127); H-5, I-? (100); H-19, I-7 (1076)
AP: Alk=I-? (lantanine)(1353); **Alk**=I-6, 7 (1369, 1392); **Fla**=I-2 (1379); **Tri**=I-7 (1390)

Lantana horrida   (Shrub verbena)   Verbenaceae
   B   C   D   D   J   J   J
   02 09 01 02 03 08 11
OC: Attagenus piceus H-4, I-2 (48)

Lantana rugulosa   (Verbena genus)   Verbenaceae
   A   B   C   D
   01 02 09 01
OC: Stored grain pests H-5, I-7 (533)

Laportea canadensis   (Wood nettle)   Urticaceae
   C   D   J   J   M
   09 03 02 03 03
OC: Aedes aegypti H-2, I-7 (165)

Larix laricina   (American larch)   Pinaceae
   A   B   C   D   M   M   O
   01 01 08 03 04 05 04
OC: Pyrrhocoris apterus H-3, I-5 (684)
OR: H-20, I-4 (504)

Lasiosiphon eriocephalus   (Not known)   Thymelaeaceae
   B   C   D   D   J   K
   02 09 01 02 09 02
OC: Aphids H-1, I-7 (105)
OR: H-2, 32, I-4, 9 (105); H-22, I-6 (881); H-32, I-4, 7 (73, 105, 220, 605, 1276)

Lasiosiphon krausii   (Not known)   Thymelaeaceae
   B   C   D   D   J   K   L
   02 09 03 08 03 02 03
OC: Rodents H-25, I-2 (1345)
OR: H-32, 33, I-2, 7 (504)

Lathyrus cicera   (Vetch)   Fabaceae
   B   C   D   J   M   M   N
   06 09 03 03 01 12 07
OR: H-24, I-? (951); H-33, I-11 (504)

Lathyrus sylvestris   (Flat pea)   Fabaceae
   A   B   C   D   D   M   M
   01 03 09 03 08 02 12
OC: Blattella germanica H-2, I-6, 7 (78)

Lathyrus tuberosus   (Earthnut pea)   Fabaceae
   A   B   C   D   D   M
   01 04 06 09 03 12
OC: Phytodecta fornicata H-4, I-7 (531)

Launaea nudicaulis   (Not known)   Asteraceae
   B   C   D   J
   06 09 10 09
OC: Musca domestica H-1, I-? (785)   Tribolium castaneum H-1, I-? (785)
OR: H-2, I-? (100)

Laurus nobilis   (Noble laurel)   Lauraceae
```
A   B   C   D   M   M   M   N
01  01  09  06  01  14  15  07
```
OC: Alternaria tenuis H-14, I-12 (426)   Fusarium oxysporum
   Botrytis allii H-14, I-12 (426)   f. sp. lycopersici H-14, I-12 (426)
   Ceratocystis ulmi H-14, I-12 (426)   Gibberella fujikuroi H-14, I-12 (426)
   Cladosporium fulvum H-14, I-12 (426)   Lentinus lepideus H-14, I-12 (428)
   Claviceps purpurea H-14, I-12 (426)   Lenzites trabea H-14, I-12 (428)
   Diplodia maydis H-14, I-12 (426)   Livestock pests H-1, I-9 (87)
   Flies H-5, I-7, 14 (105)   Polyporus versicolor H-14, I-12 (428)
   Fusarium oxysporum   Ustilago avenae H-14, I-? (426)
     f. sp. conglutinans H-14, I-12 (426) Verticillium albo-atrum H-14, I-12 (426)
OR: H-2, I-9 (87); H-19, I-12 (904, 908)

Lavandula angustifolia   (English lavender)   Lamiaceae
```
B   C   D   D   J   K
02  09  03  06  06  06
```
OC: Aphis gossypii H-2, I-12 (105)   Locusta migratoria H-4, 5, I-7 (596)
   Cockroaches H-5, I-12 (105)   Polyporus versicolor H-14, I-12 (428)
   Lentinus lepideus H-14, I-12 (428)   Tetranychus cinnabarinus H-11, I-? (1030)
   Lenzites trabea H-14, I-12 (428)
OR: H-2, I-7, 8 (504)

Lawsonia inermis   (Henna)   Lythraceae
```
A   B   B   C   D   D   J   M   M   M   O   O
01  01  02  09  01  06  03  05  16  17  04  07
```
OC: Alternaria solani H-14, I-7 (709)   Helminthosporium sp. H-14, I-7 (480)
   Alternaria tenuis H-14, I-7 (480, 523)   Ustilago hordei H-14, I-7 (181)
   Curvularia penniseti H-14, I-7 (480)   Ustilago tritici H-14, I-7 (181)
   Exserohilum turcicum H-14, I-15 (827)
OR: H-17, I-2, 7 (1353); H-20, I-7 (1353)
AP: Tan=I-7 (1353)

Lechea maritima   (Not known)   Cistaceae
```
C   C   D   J
06  09  03  14
```
OC: Ceratitis capitata H-6, I-2, 6, 7, 9 (956)

Ledum groenlandicum   (True labrador tea)   Ericaceae
```
B   C   E   E   M   N
02  09  07  11  01  07
```
OC: Lice H-2, I-? (104)

Ledum palustre   (Crystal tea)   Ericaceae
```
B   C   D   D   M   M   N   O
02  09  03  07  01  05  07  07
```
OC: Cimex lectularius H-2, I-2 (104)   Lice H-2, I-2 (104)
   Fleas H-2, I-2 (104)   Moths H-2, I-2 (104)

Leea aequata   (Giang)   Leeaceae
```
A   B   B   C   D   J
01  01  02  09  01  09
```
OC: Rats H-26, I-16 (1393)
OR: H-20, I-5, 7 (504)

Leersia hexandra   (Rice grass)   Poaceae
    A   B   C   D   E   J   M
   01   06   09   03   11   03   02
OR: H-24, I-2 (527, 559)

Lemaireocereus gummosus   (Dagger cactus)   Cactaceae
    A   B   C   D   J   J   M   N
   01   08   09   05   03   11   01   07
OC: Attagenus piceus   H-4, I-6 (48)
OR: H-32, I-6, 14 (504)

Lemna minor   (Duckweed)   Lemnaceae
    A   B   B   C   D   D   J   J   M
   01   06   09   09   01   02   09   16   02
OC: Alternaria sp.   H-14, I-1 (837)   Fusarium roseum   H-14, I-1 (837)
OR: H-19, I-1 (837)

Leonotis leonurus   (Lion's ear)   Lamiaceae
    A   B   B   C   D   J   M   O   O
   01   02   06   09   01   03   05   07   08
OC: Attagenus piceus   H-4, I-2 (48)   Oncopeltus fasciatus   H-2a, I-2 (48)
OR: H-17, 20, I-8 (504)

Leonotis nepetifolia   (Lion's ear)   Lamiaceae
    A   B   C   D   D   J   J   M   O
   03   06   09   03   06   04   14   05   07
OC: Cerotoma trifurcata   H-2a, I-6, 10 (78, 101)   Oncopeltus fasciatus   H-1, I-6, 7 (958)
     Diaphania hyalinata   H-2a, I-6, 10 (78, 101)   Vermin   H-27, I-1 (1353)
     Dysdercus flavidus   H-1, I-6 (78)

Leonurus sibiricus   (Siberian motherwort)   Lamiaceae
    A   B   C   D   J   M   M   O
   02   06   09   03   09   05   13   07
OC: Aphis gossypii   H-1, I-1 (1241)   Puccinia rubigavera   H-14, I-1 (1241)
     Phytophthora infestans   H-14, I-1 (1241)   Pyricularia oryzae   H-14, I-1 (1241)
     Pseudoperonospora cubensis   H-14, I-1 (1241)
OR: H-22, I-1 (1393)
AP: Alk=I-1 (leoninine, leonurinine)(1392)

Lepechinia calycina   (Pitcher sage)   Lamiaceae
    A   B   C   D
   01   02   09   03
OR: H-24, I-7 (988)

Lepidium draba   (Peppergrass genus)   Brassicaceae
    B   C   D   M   M   N
   06   09   03   01   14   14
OC: Agrobacterium tumefaciens   H-18, I-14, 16   Phytodecta fornicata   H-4, I-7 (531)
     (585)   Tanymecus dilaticollis   H-4, I-7 (531)
     Cassida nebulosa   H-4, I-7 (531)   Tylenchulus semipenetrans   H-16, I-10 (540)
     Hyphantria cunea   H-4, I-7 (531)
OR: H-1, 32, I-? (220)

Lepidium flavum   (Peppergrass genus)   Brassicaceae
   B   C   D
   06 09 10
OC: Mosquitoes H-10, I-10 (916)

Lepidium ruderale   (Wild peppergrass)   Brassicaceae
   B   C   D   J   K   L
   06 09 03 04 07 05
OC: Aphids H-1, I-1 (105)   Mites H-11, I-1 (105)
   Flea beetles H-1, I-1 (105)

Lepidium sativum   (Garden cress)   Brassicaceae
   B   C   D   D   M   M   N   N
   06 09 03 06 01 27 07 12
OC: Meloidogyne incognita H-16, I-? (1138)   Musca domestica H-6, I-2 (932)
AP: Alk=I-10 (1372); Sfr=I-7 (1116)

Lepidium virginicum   (Virginia pepperweed)   Brassicaceae
   B   C   J
   06 09 03
OC: Coronilla varia H-24, I-? (364)   Pythium ultimum H-14, I-? (364)
   Fusarium oxysporum H-14, I-? (364)   Rhizoctonia sp. H-14, I-? (364)
   Helminthosporium sp. H-14, I-? (364)   Sclerotium rolfsii H-14, I-? (364)
AP: Alk=I-6, 7, 8 (1392)

Lepiota procera   (Parasol mushroom)   Agaricaceae
   B   C   D   M   N
   02 01 03 01 09
OC: Culex pipiens H-2, I-1 (101)

Leptospermum scoparium   (Not known)   Myrtaceae
   A   B   B   C
   01 01 02 09
OC: Musca domestica H-1, 30, I-12 (101)
OR: H-19, I-12 (775)

Lespedeza cuneata   (Chinese lespedeza)   Fabaceae
   A   B   D   J   M   M   M   M
   01 02 03 03 02 09 12 21
OC: Pratylenchus zea H-16, I-? (1274)

Lespedeza intermedia   (Bush clover genus)   Fabaceae
   A   B   C   D   M
   01 07 09 03 12
OC: Rodents H-25, I-1, 9 (945)

Leucaena glauca   (White popinac)   Mimosaceae
   A   B   C   D   J   L   M   M   M   M   M   M   M   N   N   N   N
   01 02 09 01 03 05 01 06 07 08 10 12 23 07 08 09 15
OR: H-5, I-6 (118)
AP: Alk=I-? (1391); Alk=I-6, 7, 8, 9 (1363); Alk=I-10 (leucenol, mimosine)(1392)

Leucaena leucocephala   (Ipil ipil)   Mimosaceae
   A   B   B   C   D   J   J   M   M   M   N
   01 01 02 09 01 03 3a 01 06 09 12 10

OC: Cercospora cruenta H-14, I-7 (1122)   Meloidogyne incognita H-16, I-7 (115)
    Imperata cylindrica H-23, I-1 (1350)

Leucas aspera   (Dandhakalas)   Lamiaceae
    A   B   C   D   J   K   M   M   M   N   O
    01 06 09 01 04 06 01 05 14 07 07
OC: Dysdercus cingulatus H-2a, I-7 (124); H-3, I-1 (441)
OR: H-1, I-1 (780); H-2, I-? (87, 88, 127); H-20, I-7 (504)
AP: Alk=I-7 (1392)

Leucas cephalotes   (Not known)   Lamiaceae
    B   C   D
    06 09 01
OC: Mites H-11, I-14 (105)
OR: H-2, 17, I-? (87)

Leucas martinicensis   (Wild tea bush)   Lamiaceae
    B   C   J   K   M   O
    02 09 03 07 05 07
OC: Mosquitoes H-5, I-? (87)
OR: H-5, I-7 (504)

Leucas procumbens   (Not known)   Lamiaceae
    B   C   D   J
    06 09 01 03
OC: Ustilago hordei H-14, I-7 (181)

Leucas zeylanica   (Guma-guma)   Lamiaceae
    B   C   D   J   M   O   O
    06 09 01 03 05 01 07
OR: H-2, I-1 (78, 90, 100); H-20, I-1 (504)
AP: Tri=I-7 (1390)

Leucosyke capitellata   (Alagasi)   Urticaceae
    A   B   C   D   J   M
    01 01 09 01 03 03
OC: Drechslera oryzae H-14, I-7 (113)

Leucothoe axillaris   (Fetterbush genus)   Ericaceae
    A   B   C   D   J   K   M
    01 02 09 03 03 03 05
OC: Attagenus piceus H-4, I-6, 7 (48)   Tineola bisselliella H-4, I-6, 7 (48)

Leucothoe grayana   (Fetterbush genus)   Ericaceae
    A   B   C   D   J   J
    01 02 09 03 02 03
OC: Phaedon brassicae H-2, I-7 (105)   Sitophilus oryzae H-2, I-7 (101)

Leucothoe keiskei   (Fetterbush genus)   Ericaceae
    A   B   C   D
    01 02 09 03
OC: Drosophila hydei H-2, I-6, 7, 8 (101)

Levisticum officinale   (Lovage)   Apiaceae
    A   B   C   D   J   K   M   M   M   M   N
    01 06 09 03 04 05 01 13 14 17 07

OC: Dermacentor marginatus H-13, I-7 (87, 1133) Ixodes redikorzevi H-13, I-7 (87, 1133)
    Haemaphysalis punctata H-13, I-7 (87, 1133) Rhipicephalus rossicus H-13, I-7 (87, 1133)
AP: Alk=I-6, 7, 8 (1392)

Lewisia rediviva (Bitter root) Portulacaceae
   A   B   C   D   D   J   M   N
   01 06 09 03 08 11 01 02
OC: Attagenus piceus H-4, I-1 (48)

Liatris punctata (Blazing-star genus) Asteraceae
   A   B   C   D   M   N
   01 06 09 03 01 02
OC: Melanoplus femurrubrum H-1, I-? (87)
OR: H-22, I-? (975)
AP: Alk=I-2, 3, 6, 7, 8 (1369)

Libanotis ugoensis (Not known) Apiaceae
   C   J
   09 03
OC: Drosophila hydei H-2, I-2, 6, 7 (101)

Licania salicifolia (Not known) Chrysobalanaceae
   A   B   C   J
   01 01 01 14
OC: Ceratitis capitata H-6, I-6, 7 (956)   Dacus dorsalis H-6, I-6, 7 (956)

Ligustrum obtusifolium (Privet genus) Oleaceae
   A   B   C   D   J   M
   01 02 09 01 03 13
OC: Drosophila hydei H-2, I-7 (101)

Ligustrum vulgare (Privet) Oleaceae
   A   B   D   D   J   M   M   M   M
   01 02 03 06 07 04 06 16 24
OC: Botrytis cinerea H-14, I-7 (739)   Phyllobius oblongus H-4, I-? (531)
    Locusta migratoria H-4, 6, I-7 (596)

Lilium longiflorum (White trumpet) Liliaceae
   A   B   C   D   M
   01 06 09 03 17
OC: Plutella xylostella H-4, I-? (87)

Limonium carolinianum (Marsh Rosemary) Plumbaginaceae
   A   B   D
   01 06 03
OC: Pseudomonas solanacearum H-18, I-1, 8 (945)

Linaria vulgaris (Toadflax) Scrophulariaceae
   A   C   D   J   M   O
   01 09 01 07 05 07
OC: Flies H-2, I-8, 14 (87, 104)
OR: H-20, I-7 (504)

Lindera strychnifolia (Spicebush genus) Lauraceae
   A   C
   01 09

OC: Aphids H-1, I-2, 7 (1241)   Puccinia rubigavera H-14, I-2, 7 (1241)
    Phytophthora infestans H-14, I-2, 7 (1241)
    Puccinia graminis tritici H-14, I-2, 7 (1241)

Linum usitatissimum   (Flax)   Linaceae
    A   B   C   D   L   M   M   M   N
    03 02 09 10 03 01 02 03 20 10
OC: Acromyrmex cephalotes H-6, I-10, 12 (470)   Meloidogyne javanica H-16, I-18 (925, 992)
    Acromyrmex octospinosus H-6, I-10, 12 (470) Peronospora tabacina H-14, I-? (482)
    Leaf-cutting ants H-6, I-? (87)   Sphaerothica humuli H-14, I-? (496)
OR: H-33, I-? (220)

Lipia geminata   (Wild sage)   Verbenaceae
    B   C   D   M   N
    02 09 01 10 01
OC: Drechslera oryzae H-14, I-7 (1332)   Stored grain pests H-1, I-7 (1332)
    Fusarium moniliforme H-14, I-7 (1332)

Liquidambar orientali   (Oriental sweet gum)   Hamamelidaceae
    A   B   C   D   D   M   M   M   O
    01 01 09 03 06 05 13 17 13
OC: Mites H-11, I-13 (504)
OR: H-20, I-13 (504)

Liquidambar taiwaniana   (Taiwanese gum)   Hamamelidaceae
    A   B   C   D   D   J   M   M   M
    01 01 09 02 03 03 13 17 19
OC: Caterpillars H-1, I-7 (1241)   Puccinia graminis tritici H-14, I-7 (1241)
OR: H-22, I-16 (977)

Liriodendron tulipifera   (Tulip tree)   Magnoliaceae
    A   B   C   D   D   J   J   M   O
    01 01 09 01 02 03 09 05 04
OC: Lymantria dispar H-4, I-7 (143, 816)   Mosquitoes H-2, I-7 (98, 105)
OR: H-4, I-? (146); H-19, I-7 (651)
AP: Alk=I-2, 4 (tulipiferine)(1392)

Liriope spicata   (Creeping lilyturf)   Liliaceae
    A   B   C   D   D   J   J   M   O
    01 06 09 01 03 09 10 05 02
OC: Colletotrichum lindemuthianum H-14, I-?   Endothia parasitica H-14, I-? (957)
    (957)   Monilinia fructicola H-14, I-? (957)

Litsea cubeba   (Not known)   Lauraceae
    A   C   J   M   M   O
    01 09 03 05 14 04
OC: Aphis gossypii H-1, I-4, 7 (1241)   Rice borers H-1, I-4, 7 (1241)
    Mosquitoes H-5, I-4, 7 (1241)
AP: Alk=I-4 (laurotetanine, N-methyllaurotetanine)(1392); Alk=I-4, 7 (1366)

Litsea quatemalensis   (Not known)   Lauraceae
    A   B   B   D
    01 01 09 01
OC: Ants H-2, I-7 (105)

Lobelia chinensis   (Lobelia genus)   Lobeliaceae
   C   J
   09 03
OC: Aphids H-1, I-1 (1241)   Spider mites H-11, I-1 (1241)
    Maggots H-1, I-1 (1241)

Lobelia decurrens   (Not known)   Lobeliaceae
   B   C
   06 09
OR: H-2, I-? (87)
AP: Alk=I-? (lobeline)(87)

Lobelia inflata   (Indian tobacco)   Lobeliaceae
   A   B   C   D   L   M   O   O
   03 06 09 03 03 05 06 07
OC: Aphis fabae H-2, I-? (105)
OR: H-33, I-? (504)
AP: Alk=I-? (lobeline)(105); Alk=I-1 (isolobinanidine, isolobinine, lelobanidinis I-II,
    lelobanine, lobelanidine, lobelanine, lobeline, lobinanidine, lobinine, norleloban-
    idine, norlobelanine, norlobelonidine)(1392)

Lobelia siphilitica   (Great lobelia)   Lobeliaceae
   A   B   C   D   J   J
   01 06 09 10 02 03
OC: Aedes aegypti H-2, I-7, 8, 10 (165)
AP: Alk=I-? (lobeline, lophilacrine, lophiline)(1392)

Lolium multiflorum   (Italian ryegrass)   Poaceae
   A   B   C   D   J   M
   01 06 09 03 03 02
OR: H-24, I-4, 7 (1244)
AP: Alk=I-2 (annuldine)(1392); Alk=I-7 (perlolidine, perloline)(1392)

Lolium perenne   (English ryegrass)   Poaceae
   A   B   C   D
   01 06 09 03
OC: Heteronychus arator H-4, I-2 (642)
AP: Alk=I-7 (perlolidine, perloline, o-picoline)(1392)

Lomatia silaifolia   (Parsley fern)   Proteaceae
   A   B   C   D   K   M
   01 02 09 01 07 13
OC: Flies H-2, I-8 (105)

Lomatium dissectum   (Lace-leaved leptotaenia)   Apiaceae
   A   B   C   D
   01 06 09 03
OC: Fusarium sp. H-14, I-2 (624)   Rhizoctonia solani H-14, I-2 (624)
    Rhizoctonia oryzae H-14, I-2 (624)
OR: H-2, I-2 (1345); H-19, I-2 (445)

Lonchocarpus chrysophyllus   (Black haiari)   Fabaceae
   A   B   C   D   M
   01 01 09 01 12

OC: Aphis fabae H-2, I-2 (38, 676)   Orgyia antiqua H-2, I-6 (34)
    Cheimatobia brumata H-2, I-2 (44)   Selenia tetralunaria H-2, I-6 (34)
OR: H-2, I-2 (105, 514); H-32, I-2 (105, 507)

Lonchocarpus densiflorus   (White haiari)   Fabaceae
    A   B   C   D   J   M
    01 04 09 01 04 12
OC: Aphis fabae H-2, I-2, 6 (38)   Selenia tetralunaria H-2, I-2, 6 (44)
    Cheimatobia brumata H-2, I-2, 6 (44)

Lonchocarpus nicou   (Lancepod genus)   Fabaceae
    A   B   C   D   J   J   M
    01 04 09 01 02 04 12
OR: H-2, I-? (26); H-32, I-2 (504)

Lonchocarpus urucu   (Not known)   Fabaceae
    A   B   C   D   J   M
    01 01 09 01 03 12
OC: Epilachna varivestis H-1, I-? (284)   Oncopeltus fasciatus H-2, I-2 (48)
OR: H-2, I-2 (105); H-32, I-? (549)

Lonchocarpus utilis   (Cube)   Fabaceae
    A   B   C   D   M
    01 01 09 01 12
OC: Anasa tristis H-2, I-2 (262)   Evergestis rimosalis H-2, I-2 (262)
    Autographa brassicae H-2, I-2 (262)   Leptinotarsa decemlineata H-2, I-2 (262)
    Brevicoryne brassicae H-2, I-2 (262)   Lipaphis erysimi H-2, I-2 (262)
    Ceratomia catalpae H-2, I-2 (262)   Murgantia histrionica H-2, I-2 (262)
    Chrysochus auritus H-2, I-2 (262)   Nezara viridula H-2, I-2 (262)
    Epilachna varivestis H-2, I-2 (262, 284)

Lonicera caprifolium   (Honeysuckle)   Caprifoliaceae
    A   B   C   D   J   M   O
    01 03 09 03 03 05 08
OC: Tobacco mosaic virus H-21, I-? (184)

Lonicera japonica   (Japanese honeysuckle)   Caprifoliaceae
    A   B   C   L
    01 02 03 05
OC: Botrytis cinerea H-14, I-? (945)
AP: Alk=I-6, 7 (1369); Alk=I-6, 7, 9 (1363)

Lonicera periclymenum   (Woodbine)   Caprifoliaceae
    A   B   C   D   D   F
    01 03 09 03 06 13
OC: Venturia inaequalis H-14, I-7 (16)

Lonicera tatarica   (Tatarian honeysuckle)   Caprifoliaceae
    A   B   D
    01 02 03
OC: Phyllobius oblongus H-4, I-9 (531)
OR: H-19, I-9 (729); H-24, I-7 (1201)

Lophira alata (African oak)   Ochnaceae
   A   B   C   D   M   M   M   N
   01 01 09 01 01 05 15 12
OC: Rodents H-25, I-6 (945)
OR: H-19, I-4 (836)

Lophopetalum toxicum   (Not known)   Celastraceae
   A   B   C   D   J   J   L
   01 01 09 01 03 11 03
OC: Attagenus piceus H-4, I-4 (48)
OR: H-33, I-4 (863, 983)
AP: Alk=I-? (1392)

Lotus corniculatus   (Bird's-foot treefoil)   Fabaceae
   A   B   C   D   M   M
   01 06 09 03 12 21
OC: Heteronychus arator H-4, I-2 (642)   Schistocerca gregaria H-1, I-? (1370)
    Locusta migratoria H-4, 6, I-7 (596)

Lotus pedunculatus   (Lotus major)   Fabaceae
   A   B   C   E   J   J   M
   01 06 03 12 3a 12 12
OC: Castelytra zealandica H-4, I-2 (631, 640, 641)
    Heteronychus arator H-1, I-? (642)
OR: H-4, I-? (100)

Luffa acutangula   (Angled loofah)   Cucurbitaceae
   A   B   C   D   J   J   K   M   M   N
   03 04 09 10 13 15 06 01 13 09
OC: Attagenus piceus H-4, I-10 (48)   Tineola bisselliella H-4, I-10 (48)
    Sitophilus oryzae H-2, I-8 (87, 598, 1124)
AP: Alk=I-10 (1391)

Luffa aegyptiaca   (Sponge gourd)   Cucurbitaceae
   A   B   C   D   J   M   M   N
   03 04 09 01 04 01 13 09
OC: Pieris rapae H-1, I-1, 10 (1241)   Stored grain pests H-6, I-7 (533)
    Spider mites H-11, I-1, 10 (1241)
OR: H-17, I-10 (1321); H-19, I-7 (982); H-20, I-7 (1321); H-32, I-? (983)
AP: Sap=I-7 (1321)

Luina hypoleuca   (Not known)   Asteraceae
   A   B   C   D   D   J   J
   01 06 09 03 08 08 11
OC: Attagenus piceus H-4, I-1 (48)

Lupinus angustifolius   (Lupine genus)   Fabaceae
   A   B   C   D   M
   03 06 09 06 12
OC: Heteronychus arator H-4, I-2 (642)
AP: Alk=I-6, 7 (lupanine)(1392); Alk=I-6, 7, 8 (1368); Alk=I-10 (angustifoline, hydroxy-
   lupanine, isolupanine, lupanine, matrine)(1392)

Lupinus argenteus   (Not known)   Fabaceae
A   C   D   E   J   J   L   M   M   M   N
01 09 02 12 02 03 03 1a 02 12 10
OC: Aedes aegypti H-2, I-1, 7, 8, 10 (165)   Drosophila melanogaster H-2, I-1, 8 (165)

Lupinus gentryanus   (Lupine genus)   Fabaceae
B   C   D   D   M
06 09 01 03 12
OR: H-2, I-10 (1345)

Lupinus mutabilis   (Not known)   Fabaceae
A   C   D   E   J   M   M
03 09 01 12 03 12 13
OC: Attagenus piceus H-4, I-10 (48)
OR: H-2, I-10 (1345)

Luvungia scandens   (Not known)   Rutaceae
A   B   C   D
01 03 09 01
OC: Phytophthora parasitica H-14, I-? (911)

Lychnis coronaria   (Dusty miller)   Caryophyllaceae
A   A   B   C   D   J
02 03 06 09 06 10
OC: Culex quinquefasciatus H-2, I-1 (463)
AP: Alk=I-10 (1372)

Lycium chinense   (Chinese matrimony vine)   Solanaceae
A   B   D   M   M   N
01 02 03 01 17 07
OC: Tobacco mosaic virus H-21, I-? (1126)
AP: Alk=I-2, 6, 7 (1392)

Lycium halimifolium   (Matrimony vine)   Solanaceae
A   B   C   D   D   J   M   O
01 02 09 03 06 04 05 07
OC: Grasshoppers H-2b, I-? (105)   Leptinotarsa decemlineata H-5, I-? (548)
AP: Alk=I-7 (1392)

Lycopersicon hirsutum   (Not known)   Solanaceae
A   B   C   D   J
01 06 09 10 03
OC: Aphis gossypii H-2, I-7 (716)   Tetranychus cinnabarinus H-11, I-7 (939)
    Heliothis zea H-2, I-7 (716, 1144)   Tetranychus urticae H-11, I-7 (939)
    Manduca sexta H-2, I-7 (716)
OR: H-2, 14, 18, I-7 (87)
AP: Alk=I-16 (tomatidine)(1392)

Lycopersicon lycopersicum   (Tomato)   Solanaceae
A   B   C   D   F   F   J   J   J   J   K   M   N
03 06 09 10 06 14 03 04 08 12 06 01 09
OC: Alternaria tenuis H-14, I-7 (182)   Caterpillars H-2, I-7 (105, 117)
    Aphids H-5, I-7 (105)   Cockroaches H-2, I-6 (104)

Curvularia lunata H-14, I-7 (182)
Drechslera graminea H-14, I-7 (182)
Flies H-2, I-6 (104)
Fusarium nivale H-14, I-7, 14 (182, 1048)
Fusarium oxysporum H-14, I-2, 6, 7, 9
  (563, 564, 734, 1097)
Fusarium oxysporum
  f. sp. conglutinans H-14, I-? (1097)
  f. sp. lycopersici H-14, I-2, 6, 9
  (563, 564, 734, 1097)
  f. sp. pisi H-14, I-? (1097)
Grasshoppers H-2, I-6 (104)
Hyphantria cunea H-4, I-7 (531)

Leptinotarsa decemlineata H-5, 6, I-7 (87)
Manduca sexta H-5, I-7 (87)
Meloidogyne sp. H-16, I-20 (424)
Pieris brassicae H-4, 5, I-7 (531, 1002)
Pieris napi H-5, I-? (1002)
Pieris rapae H-5, I-? (1002)
Plutella xylostella H-4, 5, I-7 (87, 1004)
Potato virus X H-1, I-14 (1136)
Pratylenchus zea H-16, I-? (1274)
Pseudomonas solanacearum H-18, I-? (158)
Ustilago hordei H-14, I-7 (181)
Ustilago triciti H-14, I-7 (181)
Venturia inaequalis H-14, I-7 (16)

OR: H-4, I-? (132); H-15, 19, I-7 (87, 883); H-15, 20, I-7, 9 (626, 1321)
AP: Alk=I-7 (solanidine, tomatidine)(1392); Alk=I-9 (narcotine, tomatidine)(1392)

Lycopersicon peruvianum   (Tomato genus)   Solanaceae
    A   B   C   D   D
    01 06 09 01 08
OR: H-2, 15, 19, I-? (87)
AP: Alk=I-16 (tomatidine)(1392)

Lycopersicon pimpinellifolium   (Currant tomato)   Solanaceae
    A   B   C   D   D   M   N
    01 06 09 08 10 01 09
OC: Leptinotarsa decemlineata H-1, I-7 (87)
OR: H-15, 19, I-7 (87)
AP: Alk=I-16 (tomatidine)(1392)

Lycopodium clavatum   (Ground pine)   Lycopodiaceae
    A   B   B   D   D   J   J   M
    01 05 06 03 08 03 11 05
OC: Attagenus piceus H-4, I-1 (48)
AP: Alk=I-7 (annotine, clavatine, clavatoxine, lycopodine, nicotine)(1392)

Lycopodium complanatum   (Ground cedar)   Lycopodiaceae
    A   B   C   D   J
    01 02 04 08 03
OC: Lice H-2, I-? (104)
AP: Alk=I-7 (complanatine, lycopodine, nicotine, obscurine)(1392)

Lycopodium obscurum var. dendroideum   (Princess pine)   Lycopodiaceae
    A   B   C   C   J
    01 06 05 09 09
OC: Ceratocystis ulmi H-14, I-1 (946)

Lycopodium selago   (Fire club moss)   Lycopodiaceae
    A   B   C   D   D
    01 06 04 03 08
OR: H-2, I-? (104)
AP: Alk=I-? (acrifolin, lycopodine, selagine)(1392)

Lycoris africana   (Golden spider lily)   Amaryllidaceae
    B   C   D   D   J
    06 09 01 02 03

OC: Agrotis sp. H-1, I-3 (1241)   Pieris rapae H-1, I-3 (1241)
    Aphids H-1, I-3 (1241)   Rondotia menciana H-1, I-3 (1241)
    Grasshoppers H-1, I-3 (1241)

Lycoris radiata   (Spider lily)   Amaryllidaceae
    A   B   C   D   J   J   M
    01   06   09   03   03   11   05
OC: Attagenus piceus H-4, I-3 (48)   Drosophila hydei H-2, I-2, 6 (101)
AP: **Alk**=I-2, 7 (1366); **Alk**=I-3 (dimethylhomolycorine, galanthamine, homolycorine,
    lycoramine, lycoremine, lycorenine, lycorine, norpluviine, pluviine, sekisanine,
    suisenine, tazettine)(1392)

Lygodium japonicum   (Japanese climbing fern)   Schizaeaceae
    A   B   C   D
    01   03   06   03
OC: Pseudomonas solanacearum H-18, I-7 (854)
    Xanthomonas campestris pv. phaseoli-sojensis H-18, I-7 (854)

Lyonia ovalifola   (Not known)   Ericaceae
    A   B   B   C   D   D   E   E   J   L
    01   01   02   09   03   08   07   11   03   02
OR: H-2, I-7 (105, 220, 1297)

Lysimachia ciliata   (Loosestrife genus)   Primulaceae
    B   D   J
    06   03   09
OC: Rodents H-25, I-1 (944)

Lysimachia hybridae   (Not known)   Primulaceae
    B   C
    06   09
OC: Aedes aegypti H-2, I-10 (165)

Lysimachia mauritiana   (Not known)   Primulaceae
    A   B   C   D
    01   06   09   03
OC: Drosophila hydei H-2, I-2, 7 (101)

Lysimachia nummularia   (Moneywort)   Primulaceae
    A   B   B   C   D   M   M   N   O
    01   05   06   09   03   01   05   07   07
OC: Stored grain pests H-2, I-6, 8 (104)

Macaranga hypoleuca   (Not known)   Euphorbiaceae
    A   C   D
    01   09   01
OR: H-2, I-6 (1345)

Macaranga peltata   (Not known)   Euphorbiaceae
    A   B   C   D
    01   01   09   01
OC: Dysdercus cingulatus H-3, I-? (712)

Machaeranthera varians   (Not known)   Asteraceae
```
    B   C   D   J   J
    06  09  03  03  11
```
OC: Attagenus piceus H-4, I-8, 15 (48)

Macleaya cordata   (Pink plume poppy)   Papaveraceae
```
    A   B   C   D   J   M   M   M
    01  01  09  03  03  05  13  16
```
OC: Culex pipiens H-2, I-7 (101)   Maggots H-1, I-1 (1241)
    Drosophila hydei H-2, I-7 (101)   Puccinia graminis tritici H-14, I-1 (1241)
    Euproctis pseudoconspersa H-1, I-1 (1241)
OR: H-2, I-? (105); H-34, I-4 (101)
AP: Alk=I-? (allocryptopine, B-homochelidonine, chelerythrine, protopine, sanguinarine)(1392)

Maclura aurantiaca   (Osage orange)   Moraceae
```
    A   B   B   C   M   M
    01  01  09  01  04  16
```
OC: Leptinotarsa decemlineata H-1, I-7 (531)   Phyllobius oblongus H-1, I-7 (531)
OR: H-4, I-7 (531)

Maclura pomifera   (Osage orange)   Moraceae
```
    A   B   C   D   E   J
    01  01  09  03  12  05
```
OC: Ceratocystis ulmi H-14, I-4, 7 (946)   Cryptotermes brevis H-1, I-? (101)
OR: H-5, I-2, 4, 5 (101); H-22, I-? (969)
AP: Alk=I-10 (1391)

Macrosiphonia hypoleuca   (Rosa de San Juan)   Apocynaceae
```
    B   C   D   J   J   K
    06  09  10  03  04  02
```
OC: Cockroaches H-2, I-2 (504)   Oncopeltus fasciatus H-2, I-6 (48)

Madhuca butyracea   (Illipe butter tree)   Sapotaceae
```
    A   B   C   D   M
    01  01  09  01  28
```
OR: H-2, 32, I-4 (105)

Madhuca indica   (Mowra)   Sapotaceae
```
    A   B   C   D   D   F   G   J   J   J   K   K   M   M   M   M   M   M   M   N   N
    01  01  09  01  02  06  04  03  06  09  02  06  01  04  06  10  15  19  28  10  12
```
OC: Alternaria tenuis H-14, I-? (539)   Macrophomina phaseolina H-14, I-? (146)
    Aphelenchus avenae H-16, I-? (100)   Meloidogyne incognita H-16, I-18
    Colletotrichum atramentarium H-14, I-18   (100, 539, 541, 923)
      (923)   Meloidogyne javanica H-16, I-18 (992)
    Ditylenchus cypei H-16, I-? (100)   Musca domestica H-1, I-4, 6 (785)
    Erwinia carotovora H-18, I-7, 9, 12   Myzus persicae H-2, I-? (126)
      (146, 1300)   Rhizoctonia solani H-14, I-18
    Fusarium oxysporum H-14, I-18 (541)   (539, 541, 923)
    Fusarium oxysporum   Rotylenchulus reniformis H-16, I-18
      f. sp. lycopersici H-14, I-? (539)   (539, 923)
    Fusarium sp. H-14, I-18 (923)   Sclerotium rolfsii H-14, I-12 (146, 1300)
    Helicotylenchus erythrinae H-16, I-18   Tylenchorhynchus brassicae H-16, I-18
      (541, 923)   (100, 539, 541, 617, 923)
    Hoplolaimus indicus H-16, I-18 (100, 539, 541, 923)

OR: H-2, I-? (100); H-2, 32, I-18 (1352)
AP: Tan=I-4 (1352)

Madhuca latifolia   (Indian butter tree)   Sapotaceae

A	B	C	D	J	J	J	K	K	M	M	M	N	N	N	N
01	01	09	01	03	04	08	05	07	01	18	28	08	09	10	12

OC: Callosobruchus chinensis H-2, I-2, 4, 6, 7   Plutella xylostella H-2, I-6, 7 (105, 120)
      (105, 120, 516, 605, 1124)   Rats H-25, I-18 (105)
      Crocidolomia binotalis H-2, I-6, 7 (105)   Spodoptera litura H-2, I-6 (105, 120)
      Euproctis fraterna H-2, I-6, 7 (105, 120)   Weevils H-2, I-12 (105)
OR: H-2, I-7, 10 (73); H-2, I-18 (220); H-32, I-6 (459); H-32, I-18 (220)
AP: Alk=I-7 (1392)

Madhuca longifolia   (Butter tree)   Sapotaceae

A	B	C	D	J	M	M	M	M	M	M	M	N	O
01	01	09	01	06	01	05	06	10	15	19	21	12	12

OR: H-2, I-10 (73); H-2, 32, I-18 (220); H-20, I-10 (504)

Madia glomerata   (Tarweed genus)   Asteraceae

B	C	D	J	M	M	N
06	09	03	03	01	17	09

OC: Blattella germanica H-2, I-1, 6, 8 (48)
OR: H-24, I-1 (1260)

Maesa indica   (Not known)   Myrsinaceae

A	B	C	D	J	J	J	M	M	N
01	02	09	01	03	08	11	01	14	09

OC: Attagenus piceus H-4, I-4, 6, 7 (48)   Tineola bisselliella H-4, I-4, 6, 7 (48)
OR: H-32, I-7, 12 (73, 220, 459)

Maesa lanceolata   (Not known)   Myrsinaceae

A	B	D
01	02	01

OC: Mollusks H-28, I-7, 10 (1143)
OR: H-17, I-7, 10 (1143); H-24, I-? (1143)

Maesa rufescens   (Not known)   Myrsinaceae

B	C	D	J	J
02	09	01	02	11

OC: Attagenus piceus H-4, I-2, 4, 6 (48)   Tineola bisselliella H-4, I-2, 4, 6 (48)

Magnolia grandiflora   (Southern magnolia)   Magnoliaceae

A	B	C	D	M	M	O
01	01	09	02	05	13	04

OC: Agrobacterium tumefaciens H-18, I-? (957)   Monilinia fructicola H-14, I-? (957)
      Colletotrichum lindemuthianum H-14, I-?   Pseudomonas syringae H-18, I-? (957)
        (957)   Xanthomonas campestris
      Endothia parasitica H-14, I-? (957)   pv. phaseoli H-18, I-? (957)
      Erwinia amylovora H-18, I-? (957)   pv. pruni H-18, I-? (957)
      Erwinia carotovora H-18, I-? (957)   pv. vesicatoria H-18, I-? (957)
AP: Alk=I-2 (condicine, solicifoline)(1392); Alk=I-4 (magnoflorine)(1392)

Magnolia virginiana   (Sweet bay)   Magnoliaceae

A	B	B	C	D	M
01	01	02	09	03	13

OC: Popillia japonica H-5, I-7 (105, 1243)

176

Mahonia bealei   (Holly grape genus)   Berberidaceae
   A   B   C   D   J
   01 02 09 03 03
OC: Cnaphalocrocis medinalis H-1, I-2, 6, 7 (1241)

Mahonia swaseyi   (Holly grape genus)   Berberidaceae
   A   B   C   D
   01 02 09 01
OC: Phymatotrichum omnivorum H-14, I-? (552)
AP: Alk=I-? (berbamine)(1392); Alk=I-2, 6 (berberine)(1392)

Mahonia trifoliata   (Holly grape genus)   Berberidaceae
   A   B   C   D   J   M   M   M   N
   01 02 09 10 09 01 16 21 09
OC: Phymatotrichum omnivorum H-14, I-2 (552)
AP: Alk=I-2, 6 (berberine)(1392)

Maianthemum canadense   (Wild lily-of-the-valley)   Liliaceae
   A   B   C   D
   01 06 09 03
OC: Popillia japonica H-5, I-? (105, 1243)

Maillardia bordonica   (Not known)   Moraceae
   B   C   J
   02 09 11
OC: Attagenus piceus H-4, I-2 (48)

Malcomia maritima   (Virginia stock)   Brassicaceae
   A   B   C   D   J
   03 06 09 06 03
OR: H-19, I-? (581); H-24, I-2 (1207)

Mallotus apelta   (Not known)   Euphorbiaceae
   C
   09
OR: H-2, 14, I-7 (1276)

Mallotus philippinensis   (Kamala tree)   Euphorbiaceae
   A   B   C   D   J   J   M
   01 01 09 01 03 08 16
OR: H-5, I-10 (1241); H-17, 20, I-9 (504); H-19, I-7, 9 (411, 881, 982); H-32, I-? (983)
AP: Alk=I-7 (1392); Fla=I-4 (1361); Sap=I-? (1361)

Malouetia obtusiloba   (Not known)   Apocynaceae
   A   B   C   D   J   J
   01 01 09 01 03 11
OC: Attagenus piceus H-4, I-4, 6, 7 (48)   Tineola bisselliella H-4, I-6, 7 (48)

Malouetia tamaquarina   (Spoon tree)   Apocynaceae
   A   B   C   J
   01 01 09 03
OC: Attagenus piceus H-4, I-4, 6 (48)

Malus pumila　　(Common apple)　　Rosaceae
　　A　B　C　D　M　M　N
　　01　01　09　03　01　04　09
OC: Locusta migratoria　H-4, 5, I-7 (596)

Malus sylvestris　　(Crab apple)　　Rosaceae
　　A　B　B　C　D　J　M　N
　　01　01　02　09　03　07　01　09
OC: Manduca sexta　H-4, I-7 (669)　　　　Panonychus citri　H-4, 11, I-7 (669)

Malva neglecta　　(Musk mallow genus)　　Malvaceae
　　B　C　D　D　M　N
　　06　09　03　06　01　09
OC: Cassida nebulosa　H-4, I-7 (531)

Malvastrum coromandelvianum　　(Kinay-lumpang)　　Malvaceae
　　A　B　C　D　L　M　M　O
　　01　02　09　01　05　05　21　07
OR: H-24, I-10 (1184)
AP: Tan=I-6 (1321)

Malvastrum tricumpisdatum　　(Not known)　　Malvaceae
　　C
　　09
OC: Schistocerca gregaria　H-4, I-7 (615)
AP: Alk=I-6, 7, 10 (1392)

Mammea americana　　(Mammee apple tree)　　Clusiaceae
　　A　B　C　D　D　F　J　J　J　J　K　L　M　M　N　O
　　01　01　09　01　02　08　03　08　10　11　06　03　01　05　09　10
OC: Andrector ruficornis　H-2, I-10 (101, 1108)　　Oncopeltus fasciatus　H-2a, I-4, 7, 9 (48)
　　Ascia monuste　H-2, I-10 (101, 278)　　　　　　Pachyzancla bipunctalis　H-2, I-10
　　Attagenus piceus　H-4, I-4, 5, 6, 7, 9 (48)　　　　(101, 1252)
　　Blattella germanica　H-2, I-? (278)　　　　　　Peridroma saucia　H-2, I-10 (101, 1252)
　　Cerotoma ruficornis　H-2, I-10 (101, 278)　　　Periplaneta americana　H-2, I-10 (278, 1252)
　　Ctenocephalides canis　H-2, I-10 (101)　　　　Pieris rapae　H-2, I-10 (101, 304)
　　Culex sp.　H-2, I-7, 9 (190, 304)　　　　　　　Plutella xylostella　H-2, I-10 (278, 1108)
　　Diabrotica bivittata　H-2, I-10 (101, 768)　　Prenolepis longicornis　H-2, I-10 (278)
　　Diaphania hyalinata　H-1, I-10　　　　　　　　Pseudaletia unipuncta　H-2, I-10 (101, 1252)
　　　(101, 278, 768, 1108, 1252)　　　　　　　　Rhipicephalus sanguineus　H-13, I-9 (532)
　　Fleas　H-2, I-9 (532)　　　　　　　　　　　　Sitophilus oryzae　H-2, I-10 (101)
　　Laphygma frugiperda　H-2, I-10 (278, 1108)　　Spodoptera eridania　H-2, I-10 (101)
　　Lice　H-2, I-9 (532)　　　　　　　　　　　　　Spodoptera frugiperda　H-2, I-? (101, 1108)
　　Macrosiphum sonchi　H-2, I-10 (101, 278)　　　Tineola bisselliella　H-4, I-2, 4, 9 (48)
　　Myzus persicae　H-2, I-10 (101, 278)
OR: H-2, I-10 (307, 478); H-5, I-10 (118); H-17, I-10 (504)

Mandevilla foliosa　　(Not known)　　Apocynaceae
　　B　C　D　J　M
　　02　09　01　01　13
OC: Blattella germanica　H-2a, I-6, 7 (48)　　　　Tineola bisselliella　H-4, I-6, 7 (48)

Mandevilla mollissima   (Not known)   Apocynaceae
    B   C   D   D   J   M
    02   09   01   02   11   13
OC: Attagenus piceus H-4, I-1 (48)

Mangifera indica   (Mango)   Anacardiaceae
    A   B   C   D   D   J   J   M   M   M   M   N   O   O
    01   01   09   01   02   01   04   01   05   21   27   09   09   13
OC: Dacus diversus H-6, I-8 (473)   Musca domestica H-1, I-? (785)
    Mites H-11, I-12, 13 (105)   Stored rice pests H-5, I-7 (155)
    Mosquitoes H-1, I-8 (87, 105)
OR: H-3, I-? (712); H-17, I-7, 10 (504, 1319); H-19, I-9 (172, 836); H-20, I-13 (1319);
    H-24, I-7, 14 (569)
AP: Glu=I-7 (1319); Sap, Tan=I-6, 7 (1319); Sfr=I-6 (1319)

Manihot esculenta   (Cassava)   Euphorbiaceae
    A   B   C   D   M   M   M   N   O
    01   02   09   01   01   05   21   02   04
OC: Dysdercus cingulatus H-3, I-? (712)   Ustilago tritici H-14, I-7 (181)
OR: H-17, I-4 (1319)
AP: Glu, Tan=I-6, 7 (1319); Sap=I-7 (1319); Tri=I-7 (1390)

Marah fabaceus   (Wild cucumber)   Cucurbitaceae
    A   B   C   D   J
    01   04   09   02   04
OC: Hymenia recurvalis H-2, I-? (1252)   Ostrinia nubilalis H-2, I-2 (101)
AP: Alk=I-9, 10 (1371, 1391)

Maranta arundinaceae   (Arrowroot)   Marantaceae
    A   B   C   D   M   N
    01   06   09   01   01   02
OC: Callosobruchus chinensis H-5, I-? (518)   Stegobium paniceum H-5, I-? (518)
    Rhizopertha dominica H-5, I-? (518)   Tribolium castaneum H-5, I-? (517, 518)
    Sitophilus oryzae H-5, I-? (517, 518)   Trogoderma granarium H-5, I-? (518)

Markhamia stipulata   (Not known)   Bignoniaceae
    A   B   B   C   D   J
    01   01   02   09   01   09
OC: Reticulitermes flavipes H-2, I-? (87)

Marsdenia clausa   (Not known)   Asclepiadaceae
    A   B   C   D   D   J   J   J
    01   03   09   01   02   03   08   11
OC: Attagenus piceus H-4, I-7 (48)

Matricaria matricarioides   (Rayless chamomile)   Asteraceae
    A   B   C   D   J   L   M
    03   06   09   03   04   05   17
OC: Plutella xylostella H-2, I-8 (101)
OR: H-5, I-8 (1345)

Matricaria recutita   (German false chamomile)   Asteraceae
    A   B   C   C   K   M   M   M   O   O
    03   03   03   09   02   05   15   17   07   08

179

OC: Blatta orientalis H-1, I-8 (104, 1133)   Meloidogyne incognita H-16, I-8 (1133, 1302)
    Dermacentor marginatus H-13, I-8 (87, 1133) Rhipicephalus rossicus H-13, I-8 (87, 1133)
    Haemaphysalis punctata H-13, I-8 (87, 1133) Spodoptera litura H-4, I-7 (1064)
    Ixodes redikorzevi H-13, I-8 (87, 1133)
OR: H-17, I-7 (504)

Maughania chappar   (Flame tree genus)   Fabaceae
    B   C   D   M
    06   09   01   12
OC: Alternaria sp. H-14, I-? (1300)   Helminthosporium sp. H-14, I-? (1300)
    Fusarium sp. H-14, I-? (1300)

Maytenus senegalensis   (Not known)   Celastraceae
    A   B   C   D   D   J   J   J   M   O
    01   01   09   01   06   03   04   14   05   02
OC: Oncopeltus fasciatus H-1, I-4, 5 (958)   Pediculus humanus capitis H-2, I-4 (105)
OR: H-20, I-2 (504)

Medeola virginiana   (Indian cucumber)   Liliaceae
    A   B   C   D   M   N
    01   06   09   03   01   02
OC: Erwinia carotovora H-18, I-2, 6, 7 (584)

Medicago lupulina   (Black medic)   Fabaceae
    A   B   D   M   M   M   N
    03   03   03   01   02   12   10
OC: Venturia inaequalis H-14, I-? (16)

Medicago sativa   (Alfalfa)   Fabaceae
    A   B   C   D   D   J   J   M   M   M
    01   06   09   03   04   03   09   02   12   18
OC: Bipolaris sorokiniana H-14, I-6, 7 (593)   Phyllobius oblongus H-4, I-7 (531)
    Botrytis cinerea H-14, I-1 (944)   Phytophthora cinnamoni H-14, I-? (357)
    Castelytra zealandica H-2b, I-1 (639, 641) Pratylenchus zea H-16, I-? (1274)
    Exserohilum turcicum H-14, I-7 (757)   Rhizoctonia solani H-14, I-20 (422)
    Heteronychus arator H-1, I-? (642)   Tylenchulus semipenetrans H-16, I-20 (423)
    Leptinotarsa decemlineata H-4, I-7 (531)   Verticillium albo-atrum H-14, I-? (361)
    Locusta migratoria H-4, I-7 (596)   Verticillium dahliae H-14, I-10 (948)
    Melanoplus mexicana H-1, I-7 (87)
OR: H-24, I-1, 20 (1232)
AP: Alk=I-6, 7, 10 (homostachydrine, stachydrine)(1392)

Melaleuca bracteata   (Black tea tree)   Myrtaceae
    A   B   C   D
    01   01   09   01
OC: Aedes sp. H-5, I-? (105)   Musca domestica H-2, 30, I-12 (101)
OR: H-5, I-12 (504); H-19, I-12 (775)
AP: Alk=I-7 (1392)

Melaleuca leucadendron   (Cajaput tree)   Myrtaceae
    A   B   C   D   J   J   K   M   M   O
    01   01   09   01   06   09   09   05   22   12
OC: Cimex lectularius H-1, I-7, 12 (87)   Mosquitoes H-5, I-6, 7, 12 (73, 220)
    Lymantria dispar H-4, I-7 (1018)
OR: H-1, I-? (220); H-5, I-12 (73, 80, 105); H-15, 17, I-12 (504, 908); H-19, I-7, 8, 9, 12
    (775, 908, 1262)

Melanthium virginicum   (Bunchflower)   Liliaceae
    A   B   C   D
    01   06   09   03
OC: Flies H-2, I-2, 8 (105, 1345)
AP: **Alk**=I-2, 6, 7, 8 (1392)

Melia azedarach   (Chinaberry)   Meliaceae
    A   B   C   D   D   F   J   J   J   J   J   K   K   K   K   L   M   M   M   M   M   O   O
    01   01   09   01   02   05   03   04   08   09   11   03   05   06   09   03   04   05   15   19   23   07   09
OC: Alternaria tenuis H-14, I-7 (480)
    Anomala cupripes H-4, I-12 (162, 183)
    Aphis citri H-4, I-12 (183)
    Attagenus piceus H-4, I-4, 9 (48)
    Aulacophora foveicollis H-2, I-? (600, 877)
    Bagrada cruciferarum H-1, I-7 (608)
    Bagrada picta H-2, I-? (600)
    Blattella germanica H-2, I-4, 6, 7, 9 (48)
    Bombyx mori H-2, I-7 (104); H-3, I-? (108)
    Brevicoryne brassicae H-2, I-9 (105)
    Chrotogonus trachypterus H-1, I-? (877)
    Cockroaches H-1, I-9 (104)
    Crickets H-5, I-7 (686)
    Curvularia penniseti H-14, I-7 (480)
    Diaphorina citri H-4, I-10, 12 (162)
    Fleas H-2, I-9 (1352)
    Grasshoppers H-5, I-6, 7, 9 (101)
    Helicotylenchus indicus H-16, I-4, 7,
      8, 9 (1123)
    Heliothis virescens H-3, 5, I-7 (108, 1240)
    Heliothis zea H-3, 4, I-7 (87, 108, 324)
    Helminthosporum sp. H-14, I-7 (480)
    Holotrichia ovata H-4, I-12 (183)
    Hoplolaimus indicus H-16, I-4, 7, 8, 9
      (1123)
    Leucania venalba H-4, I-10, 12 (183)
    Lipaphis erysimi H-1, I-9 (91); H-2, I-?
      (84, 600); H-4, I-10 (145)
    Locusta migratoria H-2a, 5, I-7 (116);
      H-4, 5, I-7 (596)
    Locusts H-5, I-6, 7, 9 (96, 101, 105)
    Meloidogyne javanica H-16, I-7 (923)
    Myzus persicae H-4, I-10, 12 (145)
    Nephotettix virescens H-2a, I-10, 12 (141)

    Nilaparvata lugens H-2a, 4, I-10, 12
      (141, 162)
    Oncopeltus fasciatus H-2, I-4, 6, 7, 9 (48)
    Ostrinia furnacalis H-4, I-10, 12 (162)
    Panonychus citri H-4, 11, I-10, 12 (162)
    Pectinophora gossypiella H-3, I-? (108)
    Phyllocnistis citrella H-4, I-10, 12 (162)
    Pieris brassicae H-1, I-7, 10, 12
      (89, 600, 877); H-2, I-9 (87)
    Pieris rapae H-4, I-10, 12 (162)
    Rephidopalpa foveicollis H-1, I-6, 7, 9
      (1278)
    Rhizopertha dominica H-1, I-4 (862)
    Schistocerca gregaria H-4, I-? (615)
    Sitotroga cerealella H-1, I-7, 9
      (457, 613, 1124)
    Sogatella furcifera H-2a, I-10, 12 (141)
    Spodoptera abyssina H-4, I-10, 12 (162, 183)
    Spodoptera frugiperda H-3, 4, I-7, 9
      (87, 108, 324)
    Spodoptera litura H-4, I-10, 12 (162)
    Stored grain pests H-2, I-7, 10, 12
      (89, 861)
    Stored rice pests H-2, I-7, 10, 12
      (89, 861)
    Termites H-1, I-7, 10, 12 (96, 105); H-8,
      I-5 (1352)
    Tineola bisselliella H-4, I-4, 6, 7, 9 (48)
    Tribolium castaneum H-1, I-4 (862)
    Trogoderma granarium H-1, I-? (458)
    Tryporyza incertulas H-4, I-10, 12
      (133, 184)
    Tylenchus filiformis H-16, I-4, 7, 8, 9
      (1123)
OR: H-17, I-7 (504, 1135); H-22, I-10 (1395); H-32, I-? (983); H-33, I-8 (220, 824)
AP: **Alk**=I-4, 7, 9 (azadirine, margosine)(1392); **Alk**=I-6, 7, 9 (1371); **Alk**=I-8 (1367);
    Alk=I-10 (1373, 1391)

Melia toosendan   (Chinaberry)   Meliaceae
    A   B   C   D   J   J   J
    01   01   09   03   3a   09   11
OC: Diaphorina citri H-4, 5, I-10, 12 (162)
    Leucania venalba H-4, I-10, 12 (162)
    Nilaparvata lugens H-4, I-10, 12 (162)
    Orseolia oryzae H-1, I-12 (162)
    Ostrinia furnacalis H-4, I-10, 12 (162)

    Panonychus citri H-11, I-10, 12 (162)
    Phyllocnistis citrella H-4, I-10, 12 (162)
    Pieris rapae H-4, I-10, 12 (162)
    Spodoptera abyssina H-4, I-10, 12 (162)
    Tryporyza incertulas H-4, I-10, 12 (162)

Melicope erythrococca   (Not known)   Rutaceae
    A   B   C   D   J
    01   01   09   01   03
OC: Attagenus piceus H-4, I-6, 7 (48)

Melilotus alba   (White sweet clover)   Fabaceae
    A   B   C   D   M
    02   06   09   03   12
OC: Leptinotarsa decemlineata H-4, I-7 (531)   Phyllobius oblongus H-4, I-7 (531)
OR: H-24, I-? (220)

Melilotus infesta   (Sweet clover genus)   Fabaceae
    B   C   D   J   M
    06   09   02   03   12
OC: Sitona cylindricollis H-4, I-7 (442, 715)

Melilotus officinalis   (Yellow sweet clover)   Fabaceae
    A   A   B   C   D   J   M   M   M   M   M   M   M
    02   03   06   09   10   03   02   10   11   12   14   17   18
OC: Bipolaris sorokiniana H-14, I-? (593)   Phytodecta fornicata H-4, I-7 (531, 593)
    Moths H-5, I-? (105)   Sitona cylindricollis H-4, I-7 (442)
AP: Alk=I-6, 7 (1370)

Melinis minutiflora   (Molasses grass)   Poaceae
    A   B   C   D   D   J   K   M
    01   06   09   01   02   01   01   02
OC: Ants H-10, I-? (101)   Mosquitoes H-5, I-12 (504); H-6a, 10,
    Aphids H-10, I-? (105)   I-7 (105)
    Chiggers H-10, I-? (101)   Pyrilla perpusilla H-6a, 10, I-7 (105)
    Glossina sp. H-10, I-? (101)   Ticks H-5, 6a, 10, 13, I-7 (101, 105)

Melissa officinalis   (Bee balm)   Lamiaceae
    A   B   C   D   D   J   M   M   O
    01   03   09   06   10   08   05   17   07
OC: H-2, I-? (105); H-22, I-7 (979, 1054, 1120)

Menabea venenata   (Ksopo)   Asclepiadaceae
    A   B   C   D   J   J   J   L
    01   01   09   01   03   08   11   03
OC: Attagenus piceus H-4, I-2 (48)
OR: H-33, I-2 (504)

Menispermum canadense   (Yellow parilla)   Menispermaceae
    A   B   C   D   J   J   J   M   O
    01   03   09   03   02   03   08   05   02
OC: Aedes aegypti H-2, I-6, 7 (77, 165, 843)   Drosophila melanogaster H-1, I-6 (165)
AP: Alk=I-2 (dauricine)(1392); Alk=I-6, 7 (1392)

Menispermum cocculus   (Fish berry)   Menispermaceae
    A   B   C   D   J   J   L   M   M   O
    01   03   09   10   03   08   03   03   05   09
OC: Attagenus piceus H-4, I-10 (48)   Pediculus humanus humanus H-1, I-9, 10, 14
    Euproctis fraterna H-2, I-2, 6, 7, 9 (105)   (105, 1348)
    Idiocerus sp. H-2, I-9 (105)   Snails H-28, I-? (662)
    Oncopeltus fasciatus H-2, I-6, 10 (48)

OR: H-1, I-7 (73, 101); H-5, I-? (1276); H-17, I-10 (1348); H-32, I-9, 10 (105, 459, 504, 968, 983, 1348); H-33, I-10 (504, 1348)
AP: Alk=I-9 (cocculine, menispermine, paramenispermine)(1392)

Menispermum dauricum   (Moonseed genus)   Menispermaceae
   A   B   C   D   J
  01 03 09 03 03
OC: Aphids H-1, I-2, 6, 7 (1241)   Rice borers H-1, I-2, 6, 7 (1241)
AP: Alk=I-? (tetronolrine)(1392); Alk=I-2 (daunicine, menisperine)(1392)

Mentha arvensis   (Field mint)   Lamiaceae
   A   B   C   D   J   L   M   M   M   M
  01 02 09 10 02 03 05 14 17 21
OC: Dermacentor marginatus H-13, I-7 (87, 1133) Phytophthora parasitica H-14, I-? (911)
   Drechslera oryzae H-14, I-? (113, 432)   Rats H-25, I-7 (799)
   Haemaphysalis punctata H-13, I-7 (87, 1133) Rhipicephalus rossicus H-13, I-7 (87, 1133)
   Ixodes redikorzevi H-13, I-7 (87, 1133)   Tribolium castaneum H-5, I-4 (179)
OR: H-19, I-? (1081)
AP: Tan=I-6, 7 (1318)

Mentha haplocalyx   (Not known)   Lamiaceae
   A   B   C   D   J
  01 06 09 01 03
OC: Aphis gossypii H-1, I-1 (1241)

Mentha longifolia   (Horse mint)   Lamiaceae
   A   B   C   D   D   J   J   K   M   M   O
  01 06 09 03 06 03 04 05 05 17 07
OC: Aphis gossypii H-5, I-? (105)   Tribolium castaneum H-5, I-1 (179)
OR: H-20, I-7 (504)

Mentha piperita   (Peppermint)   Lamiaceae
   A   B   C   D   J   J   M   M   M   M   N   O
  01 06 09 03 03 06 01 05 14 17 07 07
OC: Alternaria tenuis H-14, I-7 (480)   Curvularia penniseti H-14, I-7 (480)
   Botrytis allii H-14, I-12 (185)   Helminthosporium sp. H-14, I-6 (523)
   Cladosporium fulvum H-14, I-12 (185)   Leptinotarsa decemlineata H-4, I-7 (531)
OR: H-22, I-7 (1121)

Mentha pulegium   (Pennyroyal)   Lamiaceae
   A   B   C   D   J   K   M   M   O
  01 03 09 03 06 07 05 17 12
OC: Alternaria tenuis H-14, I-12 (185)   Mosquitoes H-2, I-7 (104)
   Cochliomyia americana H-1, I-12 (244, 428) Spodoptera frugiperda H-4, I-12 (937)
   Leptinotarsa decemlineata H-1, I-? (185)
OR: H-4, I-? (132)

Mentha spicata   (Spearmint)   Lamiaceae
 · A   B   C   D   J   J   K   M   M   O
  01 06 09 03 04 06 09 05 17 07
OC: Cochliomyia hominivorax H-1, I-? (105)   Sitophilus oryzae H-1, I-7 (455, 1124)

Mercurialis annua   (Not known)   Euphorbiaceae
   B   C   D
  06 09 03

OC: Athalia rosae H-4, I-7 (531)   Phytodecta fornicata H-4, I-7 (531)
AP: Alk=I-? (1392)

Merremia aegyptia   (Not known)   Convolvulaceae
    B   C   J
    04 09 03
OR: H-24, I-1 (1212)

Mertensia lanceolata   (Bluebell genus)   Boraginaceae
    A   B   C   D   J   J   M
    01 03 09 03 03 11 13
OC: Attagenus piceus H-4, I-2 (48)

Mesembryanthemum caprohetum   (Ice plant genus)   Aizoaceae
    B   C   D   D   J
    06 09 03 06 03
OC: Potato virus Y H-21, I-7 (184)   Tobacco mosaic virus H-21, I-7 (184)

Mesembryanthemum crystallinum   (Ice plant genus)   Aizoaceae
    B   C   D   D   E   M   N
    06 09 03 06 04 01 07
OR: H-24, I-? (1182)

Mesua ferrea   (Ironwood)   Clusiaceae
    A   B   C   D   M   M
    01 01 09 01 13 17
OC: Dysdercus similis H-3, I-? (93)   Termites H-1, I-? (1352)

Metaplexis japonica   (Not known)   Asclepiadaceae
    C   D   J
    09 03 03
OC: Aphids H-1, I-1 (1241)

Metasequoia glyptostroboides   (Dawn redwood)   Taxodiaceae
    A   C   D   J
    01 08 01 09
OC: Monochamus alternatus H-4, I-7 (633)
OR: H-19, I-7 (1262)

Microlepia strigosa   (Not known)   Polypodiaceae
    A   C   D   M
    01 06 01 13
OC: Erwinia carotovora H-18, I-7 (854)   Xanthomonas campestris
    Pseudomonas solanacearum H-18, I-7 (854)   pv. phaseoli-sojensis H-18, I-7 (854)

Miliusa velutina   (Not known)   Apiaceae
    A   B   C   D   J
    01 01 09 01 09
OC: Musca domestica H-1, I-6, 7 (785)   Tribolium castaneum H-1, I-6, 7 (785)
OR: H-2, I-? (100)

Millettia auriculata   (Not known)   Fabaceae
    A   B   C   D   D   D   J   M
    01 03 09 01 02 08 03 12
OC: Livestock pests H-2, I-2, 6 (105, 504)   Rodents H-25, I-? (105)
OR: H-2, I-2 (73, 80, 220); H-32, I-? (220)

Millettia lasiopetala   (Not known)   Fabaceae
    A   C   D   M
    01 09 01 12
OR: H-2, 14, I-7 (1276); H-32, I-7 (1276)

Millettia pachycarpa   (Fish poison climber)   Fabaceae
    A   B   B   C   D   D   J   J   K   K   M
    01 01 03 09 01 02 03 04 05 06 12
OC: Acraea issoria H-2, I-? (1012)   Epilachna varivestis H-2, I-6 (29)
    Aphis fabae H-2, I-6 (29)   Macrosiphum granarium H-2, I-? (1012)
    Aulacophora cattigarensis H-2, I-? (1012)   Malacosoma neustria H-2, I-? (1012)
    Autoserica sp. H-2, I-? (1012)   Musca domestica H-2, I-2, 10 (270, 1012)
    Bombyx mori H-2, I-2, 6, 10   Mylabris phalerata H-2, I-? (1012)
      (29, 270, 1012)   Oides decimpunctata H-2, I-? (1012)
    Cicadella viridis H-2, I-? (1012)   Oregma lanigera H-2, I-? (1012)
    Colaphellus bowringi H-2, I-? (1012) Pareva vesta H-2, I-2, 10 (270)
    Coptosoma cribraria var. punctatissima Pieris rapae H-2, I-2, 10 (270, 1012)
      H-2, I-? (1012)   Spodoptera litura H-2, I-? (1012)
    Dysdercus megalopygus H-2, I-? (1012) Tetroda histeroides H-2, I-? (1012)
OR: H-2, I-? (220, 1079); H-32, I-2 (220, 270, 459, 504, 873, 1079)

Millettia piscidia   (Not known)   Fabaceae
    A   B   B   C   D   D   J   J   M
    01 01 03 09 01 02 03 04 12
OR: H-2, I-4, 6, 8 (105, 1276); H-32, I-2, 6 (220, 504, 1276)

Millettia reticulata   (Not known)   Fabaceae
    B   D   M
    03 03 12
OC: Colaphellus bowringi H-2, I-2 (105)

Millettia taiwania   (Not known)   Fabaceae
    A   B   B   C   D   D   J   J   M
    01 01 03 09 01 02 03 04 12
OR: H-2, I-2 (105); H-32, I-2 (504)

Millingtonia hortensis   (Not known)   Bignoniaceae
    A   B   C   D
    01 01 09 01
OC: Ustilago hordei H-14, I-7 (181)   Ustilago tritici H-14, I-7 (181)

Mimosa pudica   (Sensitive plant)   Mimosaceae
    A   B   C   D   D   J   J   M   M   M   M   O   O
    01 02 09 01 02 03 09 02 05 09 12 25 01 07
OC: Meloidogyne incognita H-16, I-7 (115)
OR: H-22, I-1 (1395)
AP: Alk=I-2, 6, 7 (mimosine)(1392); Alk=I-10 (1391)

Mirabilis jalapa   (Four-o'clock plant)   Nyctaginaceae
    A   B   D   F   G   J   M   M   O
    01 06 01 13 04 03 05 13 07
OC: Cercospora cruenta H-14, I-7 (1122)   Mosquitoes H-5, I-8 (105, 1353)
    Drechslera oryzae H-14, I-7 (113)
OR: H-19, I-7 (982)
AP: Alk=I-2 (trigonelline)(1366, 1392); Alk=I-10 (1373); Sap=I-6, 7 (1319)

Mirabilis nyctaginea   (Umbrellawort genus)   Nyctaginaceae
    B   C   D   J
    06 09 03 03
OC: Flies H-2, I-7 (1345)

Miscanthus floridulus   (Not known)   Poaceae
    A   B   C   D
    01 06 09 10
OR: H-24, I-2, 7 (1247)

Mollugo pentaphylla   (Not known)   Aizoaceae
    C   D   J   M   O
    09 10 03 05 07
OC: Exserohilum turcicum H-14, I-15 (827)
OR: H-20, I-7 (504)

Momordica charantia   (Balsam pear)   Cucurbitaceae
    A   B   C   D   D   J   J   L   M   M   N   N   N   O   O   O
    03 04 09 01 02 03 08 03 01 05 07 09 15 07 09 14
OC: Athalia rosae H-2, 4, I-10, 12 (610)   Meloidogyne javanica H-16, I-? (100)
    Attagenus piceus H-4, I-1 (48)   Rats H-25, I-9 (406)
    Meloidogyne incognita H-16, I-9, 10 (100, 542, 616)
OR: H-2, I-2, 7, 12, 14 (87, 105, 1276); H-2, 32, 33, I-9 (406); H-17, I-7, 10, 14
    (504, 1353); H-20, I-1 (1318); H-32, I-? (1276); H-33, I-7, 9 (983)
AP: **Alk**=I-? (momordicine)(1392); **Alk**=I-7 (1318); **Alk**=I-10 (1391); **Glu**=I-6, 7 (1318)

Momordica cochinchinensis   (Not known)   Cucurbitaceae
    B   C   D   M   M   N
    04 09 01 01 15 09
OC: Aphis gossypii H-1, I-7, 8 (1241)   Spider mites H-11, I-7, 8 (1241)
    Pieris rapae H-1, I-7, 8 (1241)

Momordica foetida   (Not known)   Cucurbitaceae
    B   C   D   M   M   N
    04 09 01 01 13 09
OC: Ants H-2, I-? (87)   Weevils H-2, I-? (87)
    Moths H-2, I-? (87)

Momordica schimperiana   (Not known)   Cucurbitaceae
    B   C   D   M   N
    04 09 01 01 09
OR: H-1, I-? (504); H-2, I-9 (87, 105)

Monotropa uniflora   (Corpse plant)   Pyrolaceae
    A   B   D   J   M   N
    01 06 03 11 05 02
OC: Attagenus piceus H-4, I-1 (48)
OR: H-19, I-? (592)

Montanoa grandiflora   (Not known)   Asteraceae
    A   B   C   D   J   J
    01 02 09 01 02 03
OC: Spodoptera frugiperda H-1, I-1 (176)

Moringa pterygosperma (Horseradish tree) Moringaceae

A	B	C	D	D	E	E	J	L	M	M	M	M	M	M	M	N	N	N	N	O
01	01	09	01	04	07	12	03	03	01	02	03	05	06	10	17	02	07	08	09	02
						08		20	21	27						11	12			

OC: Alternaria solani H-14, I-? (452) Meloidogyne incognita H-16, I-7 (115)
   Aspergillus niger H-14, I-? (452)   Rhizopus nigricans H-14, I-? (452)
   Fusarium oxysporum f. sp. lycopersici H-14, I-? (452)
OR: H-1, 30, I-? (100); H-17, I-11 (1319); H-19, I-2, 4, 7, 8, 9 (451, 745, 982, 1076, 1160, 1352); H-22, I-4, 9 (881, 1040); H-33, I-2, 4 (220)
AP: Alk=I-2, 4 (moringine)(220, 1392); Alk=I-10 (1373, 1396); Sfr, Tan=I-6 (1319); Tan=I-4 (1353); Tri=I-4 (baurenol)(1381)

Morus alba (White mulberry) Moraceae

A	B	C	D	M	N
01	01	09	03	01	09

OC: Locusta migratoria H-1, I-7 (596)
OR: H-15, I-4 (918); H-15, I-7 (500)
AP: Alk=I-? (trigonelline)(1392); Alk=I-6, 7 (1392)

Morus bombycis (Mulberry genus) Moraceae

A	B	C
01	01	09

OC: Fusarium lateritium f. sp. mori H-14, I-15 (917)
   Fusarium roseum H-14, I-15 (917)

Mucuna deeringiana (Velvet bean) Fabaceae

A	B	C	D	M	M	M	M
01	03	09	01	02	09	10	12

OC: Fusarium oxysporum f. sp. cubense H-14, I-? (952)

Mundulea suberosa (Sweetcane) Fabaceae

A	B	C	D	J	J	J	J	J	K	M
01	02	09	01	03	04	07	08	09	05	12

OC: Aphis fabae H-2, I-6, 10, 11 (34, 676) Flies H-2, I-4 (105)
   Aphis tavaresi H-2, I-4, 10   Hypsa ficus H-2, I-4 (105)
     (23, 25, 69, 678, 679)   Idiocerus sp. H-2, I-4, 10, 11 (105)
   Beetles H-2, I-4, 10 (105)   Leptinotarsa decemlineata H-2, I-4 (105)
   Brithys pancratii H-2, I-4 (25, 678) Macrosiphoniella sanborni H-2, I-2, 4,
   Bruchid sp. H-2, I-4 (105)   6, 7 (674)
   Callosobruchus chinensis H-2, I-4 (605) Oncopeltus fasciatus H-2, I-4 (958)
   Citrus psylla H-2, I-4, 10 (25, 678) Orthezia insignis H-2, I-4, 10 (25, 678)
   Coccus viridis H-2, I-4 (1029)   Oryzaephilus surinamensis H-2, I-2
   Cockroaches H-2, I-4 (105)   (674, 1124)
   Crocidolomia binotalis H-2, I-4 (605) Pericallia ricini H-2, I-4 (605)
   Epacromia tamulus H-2, I-4 (605)   Plutella xylostella H-2, I-4 (605)
   Epilachna viginti-octopunctata H-2, I-4 Spodoptera litura H-2, I-4 (605)
     (605)   Toxoptera aurantii H-2, I-4, 10 (25, 678)
   Euproctis fraterna H-2, I-4 (605)
OR: H-2, I-4, 10, 11 (22, 69, 73, 96, 105, 958); H-22, I-16 (882); H-32, I-2, 4, 7, 10 (220, 459, 504, 1276, 1353); H-33, I-4 (1353)

Murraya paniculata (Orange jasmine) Rutaceae

A	B	C	D
01	01	09	01

OC: Trichoderma viride H-14, I-12 (901)
OR: H-15, I-12 (901); H-19, I-7, 8 (982, 1262)
AP: Alk=I-4, 6, 7 (1366, 1374, 1392)

Musa paradisiaca   (Edible banana plantain)   Musaceae
   A B C D M M M
   01 07 09 01 04 13 14
OR: H-5, I-9 (1345)

Mussaenda kajewskii   (Not known)   Rubiaceae
   B C D
   02 09 01
OC: Lice H-5, I-7, 9 (1345)

Myrica cerifera   (Waxberry)   Myricaceae
   A B C D E M M M O
   01 02 09 01 11 05 06 13 02
OC: Fleas H-5, I-6 (104)   Moths H-5, I-6 (104)

Myrica gale   (Sweet gale)   Myricaceae
   A B B C D E M M O
   01 01 02 09 03 11 05 14 17 07
OC: "Bugs" H-2, I-? (105)   Lice H-2, I-? (105)
OR: H-19, I-? (592); H-20, I-7 (504)

Myrica mexicana   (Arbol de la cera)   Myricaceae
   A C D L M M M O
   01 09 01 05 02 05 06 02
OR: H-2a, I-? (118)

Myrica rubra   (Loobai)   Myricaceae
   A B C D D M N
   01 01 09 01 02 01 10
OC: Aphids H-1, I-1, 4 (1241)   Mosquitoes H-5, I-1, 4 (1241)
   Euproctis pseudoconspersa H-1, I-1 (1241)   Spider mites H-11, I-1, 4 (1241)
   Maggots H-1, I-1, 4 (1241)
OR: H-32, I-4 (983)

Myriophyllum spicatum   (Water milfoil)   Haloragaceae
   B B C D D J J J
   06 09 09 02 03 03 09 16
OC: Fusarium roseum H-14, I-1 (837)
OR: H-19, I-1 (837)

Myristica acuminata   (Nutmeg)   Myristicaceae
   A B C D M
   01 01 09 01 14
OC: Callosobruchus maculatus H-1, I-? (644)   Lenzites trabea H-14, I-? (428)
   Lentinus lepideus H-14, I-? (428)   Polyporus versicolor H-14, I-? (428)

Myristica fragrans   (Nutmeg)   Myristicaceae
   A B C D L M
   01 01 09 01 03 14
OC: Bombyx mori H-3, I-10 (380)   Cochliomyia hominivorax H-5, I-? (105)
   Callosobruchus maculatus H-4, I-? (644)   Musca domestica H-2, 30, I-12 (101)
OR: H-33, I-? (504)

Myrrhis odorata   (Myrrh)   Apiaceae
    A   B   C   D
    01   06   09   02
OC: Ustilago avenae H-14, I-12 (185)

Nandina domestica   (Sacred bamboo)   Berberidaceae
    A   B   C   D
    01   02   09   01
OR: H-4, I-? (132)
AP: Alk=I-? (isodomesticine, nandazurine, nantenine)(1392); Alk=I-2 (nandinine)(1392);
    Alk=I-2, 4 (berberine, domesticine, jatrorrhizine)(1392); Alk=I-6 (magnoflorine,
    menisperine)(1392); Alk=I-9 (domesticine, domestine)(1392); Alk=I-10 (protopine)
    (1392); Tan=I-10 (1391)

Narcissus tazetta   (Polyanthus narcissus)   Amaryllidaceae
    B   C   D   D   J   M
    06   09   03   06   09   13
OR: H-22, I-3 (921, 1015); H-25, I-7 (847)
AP: Alk=I-3 (clycorine, fiancine, galanthamine, galanthine, haemanthamine, hippeastrine,
    homolycorine, narcissidine, nartazine, narzettine, pluvine, suisenine, tazattine)(1392)

Nasturtium officinale   (Watercress)   Brassicaceae
    A   B   B   C   D   M   N   N
    01   06   09   09   10   01   07   15
OC: Botrytis cinerea H-14, I-7 (945)   Heterodera rostochiensis H-16, I-? (677)
AP: Alk=I-2, 6, 7, 8, 9 (1370); Alk=I-6, 7, 8 (1371)

Nelumbo lutea   (American lotus)   Nymphaeaceae
    B   B   C   M   N   N
    06   09   09   01   02   10
OC: Cockroaches H-2, I-? (104)

Neorautanenia fisifolia   (Not known)   Fabaceae
    B   D   J   J   M   M   N   N
    09   01   03   08   01   12   02   10
OC: Aphis fabae H-2, I-2 (34, 105, 676)
OR: H-32, I-2 (87)

Neorautanenia pseudopachyrriza   (Not known)   Fabaceae
    B   C   D   M
    09   01   01   12
OR: H-2, I-2 (87)

Nepeta cataria   (Catnip)   Lamiaceae
    B   C   D   D   J   J   J   J   M   M   M   M   M   N   O   O
    06   09   03   04   03   08   11   13   01   05   07   09   14   30   07   07   08
OC: Aedes aegypti H-2, I-2, 7, 8 (77, 843)   Leptinotarsa decemlineata H-5, I-? (75)
    Aphids H-5, I-? (75)   Pieris rapae H-6, I-? (75)
    Cicindela trifasciata H-5, I-? (722)   Spodoptera litura H-4, I-7 (1064)
    Leichenum canoliculatum H-5, I-7 (722)
OR: H-5, I-? (87); H-24, I-16 (1254)
AP: Alk=I-9, 10 (1391)

Nepeta subsessilis   (Catnip genus)   Lamiaceae
    B   C   D   D
    06 09 03 04
OC: Drosophila sp. H-2, I-2, 6, 7 (78)

Nephrolepis biserrata   (Sword fern)   Polypodiaceae
    A   C   D   M
    01 06 01 13
OC: Erwinia carotovora H-18, I-7 (854)
    Xanthomonas campestris pv. phaseoli-sojensis H-18, I-7 (854)

Nephrolepis exaltata   (Sword fern genus)   Polypodiaceae
    A   C   D   D   J   K   M
    01 06 01 02 12 06 13
OC: Dysdercus cingulatus H-2, I-7 (88, 125, 441)
    Xanthomonas campestris pv. phaseoli-sojensis H-18, I-7 (854)
OR: H-2a, I-? (124)

Nerium oleander   (Oleander)   Apocynaceae
    A   A   B   C   D   D   D   J   J   J   J   J   J   K   L   M   M   M   O   O
    01 02 02 09 05 06 10 03 3a 04 08 11 15 05 03 05 13 18 02 07
OC: Attagenus piceus H-4, I-2, 6, 8 (48)   Drechslera oryzae H-14, I-7 (113)
    Bagrada picta H-2, I-7 (120)   Drosophila hydei H-2, I-6, 7 (101)
    Caterpillars H-1, I-7 (1029)   Empoasca devastans H-2, I-7 (120)
    Helicotylenchus indicus H-16, I-15 (1330) Sitophilus oryzae H-2, I-8 (87, 400, 1124)
    Hoplolaimus indicus H-16, I-15 (1330)   Tylenchorhynchus brassicae H-16, I-15
    Plutella xylostella H-4, I-7 (87)   (1330)
    Rats H-25, I-2 (504)   Tylenchus filiformis H-16, I-15 (1330)
    Rotylenchulus reniformis H-16, I-15 (1330)
OR: H-1, I-2, 4, 14 (104, 105, 163, 871); H-2, 15, I-4, 7 (1318); H-15, 25, 33, I-2 (504);
    H-22, I-7 (945); H-33, I-1 (220, 507, 1318)
AP: Alk=I-6, 7, 8 (1392); Glu=I-6, 7 (1318)

Nicandra physalodes   (Apple-of-Peru)   Solanaceae
    A   B   C   D   J   J   L
    03 06 09 01 03 07 05
OC: Athalia rosae H-4, I-7 (531)   Manduca sexta H-2, I-7, 8 (87, 389)
    Cassida nebulosa H-4, I-7 (531)   Musca domestica H-2, I-7 (87, 105, 389)
    Epilachna varivestis H-1, I-7 (143, 349) Phytodecta fornicata H-4, I-7 (531)
    Hyphantria cunea H-4, I-7 (531)   Potato virus X H-21, I-14 (1136)
    Leptinotarsa decemlineata H-4, I-7   Trialeurodes vaporariorum H-5, I-16
      (531, 548)   (105, 106)
OR: H-1, I-? (220)
AP: Alk=I-? (1392); Alk=I-6, 7, 9 (1374)

Nicotiana benthamiana   (Tobacco genus)   Solanaceae
    A   B   C   D   J
    03 06 09 10 03
OC: Manduca sexta H-2b, I-7 (327, 328)
AP: Alk=I-7 (anabasine, nicotine, nornicotine)(1392)

Nicotiana debneyi   (Not known)   Solanaceae
    A   B   C   D   D
    01 06 09 01 03

190

OR: H-2, I-? (30)
AP: Alk=I-7 (anabasine, nicotine, nornicotine)(1392)

Nicotiana glauca (Tree tobacco) Solanaceae
```
B    C    L    M
02   09   02   03
```
OC: Aphids H-1, I-? (504)
OR: H-2, I-2, 6, 15 (81, 104, 769)
AP: Alk=I-2, 7 (anabasine, nicotine, nornicotine)(1392); Alk=I-6, 7 (1374); Alk=I-6, 15 (81)
Alk=I-8 (1371)

Nicotiana glutinosa (Tobacco genus) Solanaceae
```
A    B    C    D
03   06   09   10
```
OC: Tobacco mosaic virus H-21, I-7 (1129)   Uramyces sp. H-14, I-7 (170)
OR: H-22, I-? (1043)
AP: Alk=I-2 (anabasine, anatabine)(1392); Alk=I-2, 7 (nicotine, nornicotine)(1392)

Nicotiana repanda (Tobacco genus) Solanaceae
```
B    C    D
06   09   10
```
OC: Manduca sexta H-2b, I-7 (327, 328)
AP: Alk=I-2, 6, 7, 8 (1371); Alk=I-7 (anabasine, nicotine, nornicotine)(1392)

Nicotiana rustica (Wild tobacco) Solanaceae
```
A    B    C    D    D    D    F    J    J    K    L
03   06   09   01   04   08   07   02   03   06   03
```
OC: Aedes aegypti H-2, I-7 (165)   Jassid sp. H-2, I-7 (148)
Aphids H-2, I-7 (148)   Leafhoppers H-2, I-7 (105)
Bemisia tabaci H-2, I-7 (123)   Leeches H-28, I-? (220)
Centrococcus insolitus H-2, I-7 (123)   Leptinotarsa decemlineata H-1, I-? (548)
Cheimatobia brumata H-2, I-7 (105)   Malacosoma neustria H-2, I-7 (105)
Flea beetles H-2, I-7 (105)   Psylla mali H-2, I-7 (105)
OR: H-1, I-? (220); H-2, I-? (73, 504); H-24, I-7, 14 (569); H-33, I-7 (620)
AP: Alk=I-7 (nicotine)(1392)

Nicotiana sp. (Tobacco) Solanaceae
```
A    A    B    C    D    F    J    J    J    K    K    K    L
01   03   06   09   10   06   01   03   04   05   06   07   03
```
OC: Acyrthosiphum pisum H-2, I-7 (235)   Citrus psylla H-2, I-7 (678)
Anasa tristis H-2, I-7 (239)   Culex quinquefasciatus H-2, I-7 (217)
Aphidius phorodontis H-2, I-7 (267)   Diabrotica duodecimpunctata H-2, I-7 (206)
Aphids H-2, I-7 (89)   Empoasca fabae H-2, I-7 (267)
Aphis fabae H-2, I-7   Empoasca maligna H-2, I-7 (235)
  (4, 231, 235, 316, 317, 489)   Epilachna varivestis H-2, I-7 (206)
Aphis hederae H-2, I-7 (200)   Eriosoma lanigerum H-2, I-7 (253)
Aphis pomi H-2, I-7 (253, 489)   Gnorimoschema lycopersicella H-2, I-7 (228)
Aphis populifoliae H-2, I-7 (483)   Hyphantria cunea H-2, I-7 (483)
Aphis spiraecola H-2, I-7 (206)   Laspeyresia pomonella H-2, I-7
Aphis tavaresi H-2, I-7 (678)   (223, 267, 297)
Autographa brassicae H-2, I-7 (232, 235) Leptinotarsa decemlineata H-2, I-7
Bombyx mori H-2, I-7 (267)   (235, 483)
Brevicoryne brassicae H-2, I-7 (245)   Lipaphis erysimi H-2, I-7 (206)
Brithys pancratii H-2, I-7 (678)   Macrosiphum rosae H-2, I-7 (489)
```
```

Musca domestica H-2, I-7 (483)
Myzus persicae H-2, I-7
  (194, 197, 198, 273, 315)
Oncopeltus fasciatus H-2, I-7 (267)
Orthezia insignis H-2, I-7 (483, 678)
Pieris brassicae H-2, I-7 (4)
Pieris rapae H-2, I-7 (232, 235)
Psylla pyricola H-2, I-7 (303)
Pyrausta nubilalis H-2, I-7 (210, 303)
Reticulitermes flavipes H-2, I-7 (267)

Rhopalosiphum persicae H-2, I-7 (200)
Rhopalosiphum rufomaculata H-2, I-7 (489)
Stored grain pests H-5, I-7 (533)
Tetranychus cinnabarinus H-11, I-7
  (206, 215)
Therioaphis maculata H-2, I-7 (60)
Thrips tabaci H-2, I-7 (212)
Toxoptera aurantii H-2, I-7 (678)
Trialeurodes packardii H-2, I-7 (267)

OR: H-22, I-? (1013); H-24, I-? (621); H-33, I-? (1163)

Nicotiana stocktoni   (Tobacco genus)   Solanaceae
  B C D J
  06 09 10 03
OC: Manduca sexta H-2b, I-7 (327, 328)

Nicotiana sylvestris   (Not known)   Solanaceae
  A B C D D L
  01 03 09 02 03 03
OC: Aphis fabae H-2, I-7 (105)
AP: Alk=I-7 (nornicotine)(105); Alk=I-7 (anabasine, nicotine, nornicotine)(1392)

Nicotiana tabacum   (Tobacco)   Solanaceae
  A A B C D D F G J J J K K L L
  01 03 06 09 01 04 06 04 03 04 09 05 06 03 05
OC: Amaranthus spinosus H-24, I-10 (159)
  Aphids H-2, I-? (86); H-2a, I-1 (117)
  Aphis craccivora H-2a, I-1 (123)
  Empoasca biguttula H-2a, I-1 (123)
  Myzus persicae H-1, I-7 (93, 126)
  Phyllocnistis citrella H-1, I-7 (93, 126)
  Pieris brassicae H-4, I-7 (531)
  Sitotroga cerealella H-1, I-1 (95)

  Hoplolaimus sp. H-16, I-6, 7 (1271)
  Leptinotarsa decemlineata H-4, I-7
    (531, 548)
  Mites H-11, I-1 (117)
  Snails H-28, I-7 (662)
  Tobacco mosaic virus H-21, I-7 (1126)
  Tylenchorhynchus dubius H-16, I-6, 7 (1271)

OR: H-1, I-? (80, 82, 92, 504); H-2a, I-6, 7 (111, 117, 126, 162, 166); H-5, I-? (118);
  H-22, I-? (1199); H-32, I-7 (662)
AP: Alk=I-2, 16 (anabasine, anatobine, myosinine, nicoteine, nicotelline, nicotine, nico-
  tyrine, nornicotine, piperidine, pyrrolidine)(1392)

Nigella sativa   (Black cumin)   Ranunculaceae
  A B D D J J K M M O
  03 06 03 06 03 06 06 05 14 10
OC: Aspergillus niger H-14, I-10, 12 (798)
  Bruchid sp. H-1, I-10, 12 (476)
  Callosobruchus chinensis H-1, I-10, 12
    (476)
  Clothes moths H-5, I-? (87, 105)
  Curvularia lunata H-14, I-10, 12
    (87, 105, 798)
  Curvularia oryzae H-14, I-10, 12 (798)
  Drechslera graminea H-4, I-10, 12 (1295)

  Fleas H-2, I-20 (87, 105)
  Flies H-2, I-20 (87, 105)
  Meloidogyne incognita H-16, I-? (1138)
  Mosquitoes H-2, I-20 (87, 105)
  Popillia japonica H-5, I-? (105, 1243)
  Sitophilus oryzae H-1, I-10, 12 (476)
  Stegobium paniceum H-1, I-10, 12 (476)
  Tribolium castaneum H-5, I-10, 12 (179, 476)

OR: H-2, 17, I-10 (73, 504, 1345); H-15, 19, I-10, 12 (411, 798, 897)
AP: Alk=I-10 (cornigelline, nigelline)(1392)

Nuphar advena   (Spatter dock)   Nymphaeaceae

A	B	B	C	D	J	M	N	N	N
01	06	09	09	03	09	01	02	07	10

OC: Pseudomonas solanacearum H-18, I-2 (946)
OR: H-19, I-? (1081)
AP: Alk=I-2 (nupharidine)(1392); Alk=I-2, 6, 7, 9 (1392)

Nuphar luteum   (Yellow water lily)   Nymphaeaceae

A	B	C	D
01	09	09	06

OC: Corynebacterium michiganense H-18, I-? (1081)

Nuphar variegatum   (Cow lily genus)   Nymphaeaceae

A	B	B	C	D	J	J
01	06	09	09	03	09	16

OC: Fusarium roseum H-14, I-1 (837)
OR: H-19, I-1 (837)

Nymphaea odorata   (Pond lily)   Nymphaeaceae

A	B	B	C	D
01	06	09	09	10

OC: Agrobacterium tumefaciens H-18, I-? (957)   Xanthomonas campestris
    Erwinia amylovora H-18, I-? (957)   pv. phaseoli H-18, I-? (957)
    Erwinia carotovora H-18, I-? (957)   pv. pruni H-18, I-? (957)
    Pseudomonas syringae H-18, I-? (957)   pv. vesicatoria H-18, I-? (957)
OR: H-19, I-? (592)

Nymphaea tuberosa   (Magnolia water lily)   Nymphaeaceae

A	B	B	C	D	J	J
01	06	09	09	03	09	16

OC: Alternaria sp. H-14, I-1 (837)   Fusarium roseum H-14, I-1 (837)

Ochrocarpus africanus   (African mammee apple)   Clusiaceae

A	B	C	D	J	J	M	M	N	N	O	O	O
01	01	09	01	08	11	01	05	09	10	02	04	14

OC: Attagenus piceus H-4, I-4 (48)   Tineola bisselliella H-4, I-4 (48)

Ocimum americanum   (American basil)   Lamiaceae

C	D	K	M	M	O
09	01	07	05	14	07

OC: Mosquitoes H-5, I-? (87)
OR: H-5, I-? (533)

Ocimum basilicum   (Sweet basil)   Lamiaceae

A	B	C	D	D	J	J	J	J	K	M	M	M	M	N	O	O
03	06	09	01	06	02	04	06	07	06	01	05	15	17	07	10	07	10

OC: Aphis gossypii H-2, I-12 (87)   Leptinotarsa decemlineata H-1, I-7, 8 (105)
    Callosobruchus chinensis H-2, I-7, 12 (519)   Maggots H-2, I-12 (105)
    Dysdercus cingulatus H-2a, I-1 (124)   Mosquitoes H-2, 5, I-1, 12 (78, 80, 105)
    Flies H-1, I-7, 8 (105); H-5, I-10 (1350);   Musca domestica H-1, I-12 (80)
     H-5, I-12 (80)   Oncopeltus fasciatus H-3, I-7, 12 (723)
    Lentinus lepideus H-14, I-? (428)   Polyporus versicolor H-14, I-? (428)
    Lenzites trabea H-14, I-? (428)   Sitophilus oryzae H-2, I-7, 12 (519)

Stegobium paniceum H-2, I-7, 12 (519)   Tribolium castaneum H-2, I-7, 12 (519)
Tetranychus cinnabarinus H-11, I-12 (87)
OR: H-3, I-? (132); H-5, I-10 (1350); H-15, 19, I-? (96, 912)

Ocimum canum   (Hoary basil)   Lamiaceae
A   B   C   D   J   J   M   M   N   O
03   02   09   10   03   08   01   05   07   07
OC: Cyperus rotundus H-24, I-7 (149, 395)   Fleas H-2, I-1 (105); H-5, I-7 (605)
Euproctis fraterna H-2, I-1 (105)
OR: H-19, I-10 (449); H-20, I-7 (504)

Ocimum gratissimum   (Ramtulsi)   Lamiaceae
B   B   C   D   M   O
02   06   09   01   05   07
OC: Mosquitoes H-5, I-? (73, 80)
OR: H-2, I-? (1276); H-32, I-? (1276)

Ocimum sanctum   (Holy basil)   Lamiaceae
A   B   C   D   J   J   J   J   K   K   M   M   M   N   O   O   O
03   06   09   01   02   03   09   10   03   07   01   05   14   07   02   07   10
OC: Alternaria tenius H-14, I-7 (480)   Meloidogyne javanica H-16, I-2, 6, 7
Amaranthus spinosus H-23, I-7 (159)   (100, 616)
Curvularia penniseti H-14, I-7 (480, 481)   Mosquitoes H-1, I-1, 20 (78, 135, 1197);
Dacus correctus H-6, I-7 (473, 696)   H-5, I-6, 7 (117)
Drechslera oryzae H-14, I-7 (174, 1305)   Musca domestica H-1, I-2 (1197)
Dysdercus cingulatus H-3, I-6, 7 (435)   Pericallia ricini H-2, I-? (88, 435);
Exserohilum turcicum H-14, I-15 (827)   H-2b, I-1 (124)
Fleas H-1, I-? (1197)   Pyricularia oryzae H-14, I-7 (174, 1305)
Flies H-5, I-6, 7 (117)   Rhizoctonia solani H-14, I-7 (174)
Helminthosporium sp. H-14, I-7 (480)   Spodoptera litura H-2, I-? (435); H-2b,
Maggots H-1, I-? (105)   I-1 (124)
Meloidogyne incognita H-16, I-2, 6, 7 (100, 616)
OR: H-19, I-7 (427)
AP: Alk=I-16 (1392)

Ocimum suave   (Basil genus)   Lamiaceae
B   C   D   M
06   09   01   14
OC: Mosquitoes H-5, I-? (87)

Ocimum viride   (Mosquito plant)   Lamiaceae
B   C   D   K   K   M   O
06   09   01   03   07   05   07
OC: Mosquitoes H-2, I-7 (105)

Oenanthe biennis   (Not known)   Apiaceae
A   B   C   D
01   06   09   03
OC: Bipolaris sorokiniana H-14, I-1 (944)
OR: H-19, I-8 (1262)
AP: Tan=I-? (1391)

Oenanthe crocata   (Not known)   Apiaceae
A   B   C   D   D   J
01   06   09   10   08   08

OC: Pieris brassicae H-2, I-7 (105)
OR: H-32, I-2 (504)

Oldenlandia corymbosa   (Chay root)   Rubiaceae
    B   B   C   J   M   M   M   O
    06 09 01 03 02 05 16 01
OC: Exserohilum turcicum H-14, I-15 (827)
AP: Alk=I-7 (caffeine)(1392)

Oldfieldia africana   (African oak)   Euphorbiaceae
    A   B   C   D   M   M   O   O
    01 01 09 01 04 05 02 04
OC: Lice H-2, I-4, 7 (105)   Phthirus pubis H-2, I-4, 7 (105)

Olea cuspidata   (Indian olive)   Oleaceae
    A   B   C   D   G   J   J   M
    01 02 09 03 04 04 11 19
OC: Tribolium castaneum H-5, I-7 (179)

Olea europaea   (Olive)   Oleaceae
    A   B   C   D   M   M   M   M   N   O
    01 01 09 06 01 05 16 29 09 12
OC: Sphaerothica humuli H-14, I-? (493, 496, 760)

Omphalea diandra   (Cobnut)   Euphorbiaceae
    A   B   B   C   D   J   J   M   N
    01 01 03 09 01 08 11 01 09
OC: Attagenus piceus H-4, I-9 (48)

Omphalea triandra   (Cobnut)   Euphorbiaceae
    A   B   C   D   J   M   M   N
    01 01 09 01 08 01 20 09
OC: Attagenus piceus H-4, I-10 (48)

Onopordum acanthium   (Cotton thistle)   Asteraceae
    A   B   C   D
    02 07 09 10
OC: Leptinotarsa decemlineata H-4, I-7 (531)   Phytodecta fornicata H-4, I-7 (531)
OR: H-19, I-7 (729)
AP: Alk=I-6, 7, 8 (1372)

Onosmodium occidentale   (Not known)   Boraginaceae
    C
    09
OC: Melanoplus femurrubrum H-3, I-? (87)

Opuntia dillenii   (Prickly pear genus)   Cactaceae
    A   B   C   D   J   J
    01 08 09 04 03 09
OC: Aphids H-1, I-6 (1241)   Rice borers H-1, I-6 (1241)
    Leafhoppers H-1, I-6 (1241)
OR: H-19, I-6 (1241)

Opuntia robusta   (Prickly pear genus)   Cactaceae

    A   B   C   D   J
    01 08 09 09 03

OC: Tobacco mosaic virus H-21, I-7 (184)

Origanum majorana   (Sweet marjoram)   Lamiaceae

    A   B   C   D   D   D   F   J   K   L   M   M   O
    01 06 09 01 05 06 06 06 06 05 05 14 07

OC: Alternaria tenius H-14, I-12 (426)   Fusarium oxysporum
    Botrytis allii H-14, I-12 (185, 426)   f. sp. lycopersici H-14, I-12 (426)
    Ceratocystis ulmi H-14, I-12 (426)   Gibberella fujikuroi H-14, I-12 (426)
    Cladosporium fulvum H-14, I-12 (426)   Mosquitoes H-2, I-12 (105)
    Claviceps purpurea H-14, I-12 (426)   Spodoptera litura H-5, I-7 (126)
    Diplodia maydis H-14, I-12 (426)   Verticillium albo-atrum H-14, I-12 (426)
    Fusarium oxysporum f. sp. conglutinans H-14, I-12 (426)
OR: H-20, I-12 (399)

Orixa japonica   (Not known)   Rutaceae

    A   B   C   D   J
    01 02 09 03 13

OC: Spodoptera litura H-4, I-? (143, 866, 1064)
AP: **Alk**=I-? (dictamnine)(1392); **Alk**=I-2 (kokusaginoline, orixine)(1392); **Alk**=I-2, 9 (koku-
    sagine, kokusaginine, skimmianine)(1392); **Alk**=I-4 (kokusagine)(1392)

Ormocarpum orientale   (Not known)   Fabaceae

    A   B   C   D
    01 01 09 01

OR: H-2, I-? (80)

Ornithogalum umbellatum   (Star-of-Bethlehem)   Liliaceae

    B   C   D   D   J   J   J   L   M   N
    06 09 03 06 03 08 11 05 01 03

OC: Attagenus piceus H-4, I-2 (48)

Orobanche aegyptica   (Not known)   Orobanchaceae

    B   B   C   D   D   J
    06 10 09 02 03 09

OC: Musca domestica H-1, I-1 (785)
OR: H-2, I-? (100)

Orostachys fimbriata   (Not known)   Crassulaceae

    B   C   D   J
    06 09 03 03

OC: Aphids H-1, I-1 (1241)

Oroxylum indicum   (Not known)   Bignoniaceae

    A   B   C
    01 01 09

OC: Flies H-5, I-4 (1345)

Orthocarpus luteus   (Owl's clover)   Scrophulariaceae

    A   B   C   D   D   J
    03 06 09 01 02 03

OC: Attagenus piceus H-4, I-1 (48)

Orthodon grosseserratum　　(Not known)　　Lamiaceae
   C   J
   09 03
OC: Drosophila hydei　H-2, I-6, 7, 8 (101)

Orthosiphum stamineus　　(Java tree)　　Lamiaceae
   C   M   O
   09 05 07
OC: Tobacco mosaic virus　H-21, I-? (1126)
AP: Alk=I-7 (1392)

Oryza sativa　　(Rice)　　Poaceae
   A   B   B   C   D   E   F   G   J   J   M   N
   03 06 09 09 01 10 05 05 03 09 01 10
OC: Meloidogyne incognita　H-16, I-7 (115)　　　Tobacco mosaic virus　H-21, I-2, 7, 8, 10
   Meloidogyne sp.　H-16, I-? (525)　　　　　(1126, 1315, 1317)
   Rotylenchulus sp.　H-16, I-? (525)
AP: Alk=I-7 (1392)

Osmorhiza aristata　　(Not known)　　Apiaceae
   A   B   C   J
   01 06 09 03
OC: Drosophila sp.　H-2, I-2, 7 (101)

Osmunda regalis var. spectabilis　　(Royal fern)　　Osmundaceae
   A   C   D   J
   01 06 03 09
OC: Fusarium oxysporum f. sp. lycopersici　H-14, I-15 (944)
OR: H-7, I-16 (944)

Ostodes paniculata　　(Not known)　　Euphorbiaceae
   A   B   C   D
   01 01 09 01
OC: Musca domestica　H-1, I-16 (785)

Ostostegia integrifolia　　(Not known)　　Lamiaceae
   A   C   K
   01 09 07
OR: H-2, I-4, 6, 7, 20 (1143)

Ougeinia dalbergioides　　(Not known)　　Fabaceae
   A   B   C   D   J   J   M   M
   01 01 09 01 04 08 12 19
OC: Callosobruchus chinensis　H-2, I-4, 6, 7　　　Euproctis fraterna　H-2, I-? (105)
   (105, 605)　　　　　　　　　　　　　　　Plutella xylostella　H-2, I-? (105)
   Caterpillars　H-1, I-4, 7, (1029)　　　　　Spodoptera litura　H-2, I-? (105)
   Crocidolomia binotalis　H-2, I-? (105)
OR: H-32, I-4 (220, 459)
AP: Tri=I-4 (lupeol)(1381)

Ourouparia gambir　　(Gambier)　　Rubiaceae
   C
   09
OC: Popillia japonica　H-5, I-? (1243)
AP: Alk=I-6 (gambinine)(1392)

197

Oxalis anthelmintica　　(Wood sorrel genus)　　Oxalidaceae

   B   C   D
   06 09 01

OC: Mollusks　H-28, I-2 (1143)

OR: H-17, 32, I-2 (1143)

Oxalis corniculata　　(Creeping oxalis)　　Oxalidaceae

   A   B   C   D   J   L   M   M   N   O
   01 06 09 03 03 05 01 05 07 07

OC: Aphis gossypii H-1, I-1, 2 (1241)　　　Weevils　H-1, I-1, 2 (1241)
   Maggots H-1, I-1, 2 (1241)

AP: Alk=I-1 (1365)

Oxalis deppei　　(Lucky clover)　　Oxalidaceae

   A   B   C   D   M
   01 01 09 01 13

OC: Plutella xylostella H-4, I-16 (87)

Pachygone ovata　　(Kadakkodi)　　Menispermaceae

   A   B   C   D   J   J
   01 04 09 01 02 03

OC: Vermin　H-27, I-9 (504)

OR: H-2, 32, I-? (105, 220); H-32, I-9 (504)

AP: Sap=I-? (220)

Pachyrhizus achipa　　(Not known)　　Fabaceae

   B   B   C   D   M
   04 06 06 01 12

OC: Epilachna varivestis H-2, I-? (487)

Pachyrhizus angulatus　　(Yam bean)　　Fabaceae

   B   C   D   M   M   M   N   N
   04 09 01 01 03 12 02 11

OC: Lipaphis erysimi H-2, I-2 (84)

Pachyrhizus erosus　　(Chinese yam bean)　　Fabaceae

   A B C D F J J J J K K L M M M N N N O
   03 04 09 01 07 03 04 08 11 05 06 03 01 05 12 02 03 11 10

OC: Acraea issoria H-1, I-? (1012)　　　　　　Epilachna varivestis H-2, I-2, 7, 10
   Acrosternum hilare H-2, I-10 (101)　　　　　(29, 101, 105, 272, 487)
   Alsophila pometaria H-1, I-2, 7, 10 (272)　Laphygma frugiperda H-2, I-10 (278, 1108)
   Aphis fabae H-2, I-2, 7, 10　　　　　　　　Macrosiphum sonchi H-1, I-? (278)
     (29, 101, 105, 272)　　　　　　　　　　　Myzus persicae H-1, I-2, 7, 10 (272, 278)
   Archips cerasivoranus H-1, I-? (272)　　　Nezara viridula H-1, I-? (1012)
   Ascia monuste H-1, I-? (278)　　　　　　　Oregma lanigera H-1, I-2, 7, 10 (272)
   Attagenus piceus H-4, I-10 (48)　　　　　　Periplaneta americana H-1, I-? (278)
   Blattella germanica H-1, I-2, 7, 10 (278)　Phyllotreta vittata H-1, I-2, 7, 10 (272)
   Bombyx mori H-1, I-9, 10 ((272)　　　　　　Pieris rapae H-1, I-? (272, 870, 1252)
   Caterpillars H-1, I-10 (90)　　　　　　　　Plutella xylostella H-2, I-9, 10
   Cerotoma ruficornis H-2, I-10 (101, 278)　　　(101, 1108)
   Diaphania hyalinata H-2, I-9, 10　　　　　Prenolepis longicornis H-1, I-? (278)
     (101, 278, 1108)　　　　　　　　　　　　Spodoptera frugiperda H-2, I-10 (101)
   Dysdercus megalopygus H-1, I-? (1012)　　Spodoptera litura H-1, I-10 (116)
   Dysdercus sanguinarius H-1, I-9, 10 (1108)

OR: H-2, 32, I-? (80, 87); H-32, 33, I-7, 10 (504, 891, 1349)
AP: Alk=I-6, 7, 9 (1392); Glu=I-7 (pachyrrhizid)(1349)

Pachyrhizus palmatilobus   (Not known)   Fabaceae
   B   C   D   J   K   L   M   M   N
   06 09 01 04 05 03 01 12 02
OC: Diaphania hyalinata H-2b, I-2, 5, 10, 11   Laphygma frugiperda H-1, I-2, 5 (1108)
   (101, 1108)   Pseudaletia unipunctata H-1, I-2, 5
   Dysdercus sp. H-2a, I-10 (101)   (101, 1108)

Pachyrhizus piscipula   (Yam bean genus)   Fabaceae
   B   B   C   D   M
   04 06 06 01 12
OC: Diaphania hyalinata H-1, I-? (1108)   Plutella xylostella H-1, I-? (1108)

Pachyrhizus strigosus   (Yam bean genus)   Fabaceae
   B   C   D   M
   06 09 01 12
OC: Epilachna varivestis H-2, I-? (487)
OR: H-2, I-10 (1345)

Pachyrhizus tuberosus   (Yam bean)   Fabaceae
   B   C   D   J   K   L   M   M   N   N   N
   04 09 01 04 04 03 01 12 02 10 11
OC: Brevicoryne brassicae H-2, I-10 (101)   Vermin H-27, I-? (104)
   Epilachna varivestis H-2, I-10 (101, 487)

Pachyrhizus vernalis   (Yam bean genus)   Fabaceae
   B   C   D
   06 09 01
OR: H-2, I-10 (1345)

Pachyrhizus wrightii   (Yam bean genus)   Fabaceae
   B   C   D   M
   06 09 01 12
OC: Pieris rapae H-1, I-? (1252)

Pachysandra terminalis   (Spurge genus)   Buxaceae
   A   B   C   D   M
   01 06 09 03 12
OC: Agrobacterium tumefaciens H-18, I-? (957)   Xanthomonas campestris
   Erwinia amylovora H-18, I-? (957)   pv. phaseoli H-18, I-? (957)
   Erwinia carotovora H-18, I-? (957)   pv. pruni H-18, I-? (957)
   Pseudomonas syringae H-18, I-? (957)   pv. vesicatoria H-18, I-? (957)
AP: Alk=I-? (1392)

Padina gymnospora   (Algae)   Dictyotales (order)
   A   B   C   D   D
   04 09 02 01 02
OC: Rhizopus oryzae H-14, I-1 (1024)

Paederis scandens   (Not known)   Ranunculaceae
   C   J
   09 03
OC: Aphids H-1, I-1 (1241)

Paeonia brownii   (Peony genus)   Paeoniaceae
   A   B   C   D   J   J
   01 06 09 03 03 08
OC: Attagenus piceus H-4, I-2 (48)   Tineola bisselliella H-4, I-2 (48)
AP: Alk, Tan=I-? (1391)

Paeonia suffruticosa   (Tree peony)   Paeoniaceae
   A   B   C   D   J
   01 02 09 03 14
OC: Ceratitis capitata H-6, I-6, 7 (956)   Dacus cucurbitae H-6, I-6, 7 (956)

Pandanus tectorius   (Thatch screw pine)   Pandanaceae
   A   B   C   D   M   M   M   M   M   M   M   N   N   O
   01 01 09 01 01 04 05 08 17 22 24 09 10 08
OC: Clothes moths H-5, I-8 (105)
AP: Fla=I-? (1377)

Panicum maximum   (Guinea grass)   Poaceae
   A   B   C   D   D   M
   01 06 09 01 03 12
OR: H-24, I-7 (1247); H-33, I-? (220)

Papaver rhoeas   (Flanders poppy)   Papaveraceae
   A   B   C   D   J   J   M
   03 06 09 03 03 04 16
OC: Tylenchulus semipenetrans H-16, I-8 (540)
AP: Alk=I-2, 6, 7, 8 (1371); Alk=I-2, 6, 7, 8, 9 (coptisine, morphine, narcotine, protopine,
   rhoeadine, rhoeagenine, thebaine)(1392); Alk=I-10 (1391)

Papaver somniferum   (Opium)   Papaveraceae
   A   B   C   D   M   M   M   M   M   N   N
   03 06 09 03 01 02 15 20 28 10 12
OC: Cassida nebulosa H-4, I-7 (531)   Phytodecta fornicata H-4, I-7 (531)
   Phyllobius oblongus H-4, I-7 (531)
OR: H-35, I-9 (531)
AP: Alk=I-2, 7, 9 (morphine)(1392); Alk=I-7, 9 (codeine, narcotine)(1392); Alk=I-9
   (aporeine, codamine, crytopine, gnoscopine, hyrocotarnine, lanthopine, laudanidine,
   laudanine, meconidine, narceine, neopine, oxynarcotine, papaveramine, papaverine,
   porphyroxine, protopine, rhoecadine, thebaine, xanthaline)(1392)

Paphiopedilum javanicum   (Lady's slipper genus)   Orchidaceae
   B   C   D
   06 09 01
OC: Flies H-1, I-7 (1241)
AP: Alk=I-7 (1392)

Parabenzoin praecox   (Not known)   Lauraceae
   A   B   C   D   D
   01 02 09 03 08
OC: Spodoptera litura H-4, I-7 (1068)

Parabenzoin trilobum   (Not known)   Lauraceae
   B   C   D   D   J
   02 09 03 08 13

OC: Spodoptera litura H-2, I-7 (143, 382)   Trimeresia miranda H-2, I-7 (382)
OR: H-4, I-? (132)

Paratecoma peroba   (Not known)   Bignoniaceae
    C   J
    09 09
OC: Reticulitermes flavipes H-2, I-? (87)

Paris verticilleata   (Not known)   Liliaceae
    B   C   D   D   L
    06 09 03 08 03
OR: H-2, I-2, 6, 7 (101)

Parthenium alpinum   (Not known)   Asteraceae
    B   D   D
    09 02 04
OC: Heliothis zea H-4, I-7 (1355)   Spodoptera exiqua H-4, I-7 (1355)

Parthenium fruticosum   (Not known)   Asteraceae
    B   D   D
    09 02 04
OC: Heliothis zea H-4, I-7 (1355)   Spodoptera exiqua H-4, I-7 (1355)

Parthenium hysterophorus   (Carrot grass)   Asteraceae
    A   A   B   C   D   F   J   J   J   K   L   M   M   O
    02 03 06 09 01 07 02 03 04 06 05 02 05 02
OC: Crocidolomia binotalis H-4, I-7 (132)   Pyricularia oryzae H-14, I-7 (1297)
    Drechslera oryzae H-14, I-7 (1297)   Rhizoctonia solani H-14, I-7 (1297)
    Dysdercus cingulatus H-2, I-7 (88, 124);   Sclerotium oryzae H-14, I-7 (1297)
     H-3, I-1 (441)   Sclerotium rolfsii H-14, I-7 (1297)
    Melanoplus sanguinipes H-4, I-7 (132)   Spodoptera litura H-2a, I-7 (124)
OR: H-2, 33, I-1 (118); H-24, I-2, 15 (1186, 1286); H-33, I-7 (689)
AP: Alk=I-? (parthenine)(1392); Alk=I-2, 6, 7, 8, 9 (1369, 1372)

Parthenocissus inserta   (Woodbine genus)   Vitaceae
    A   B   C   D   M
    01 03 09 03 13
OC: Leptinotarsa decemlineata H-4, I-7 (531)   Pieris brassicae H-4, I-7 (531)
    Phytodecta formicata H-4, I-7 (531)

Parthenocissus quinquefolia   (Virgina creeper)   Vitaceae
    A   B   C   D   K   M   M   M   O
    01 03 09 03 08 05 16 21 04
OC: Eriosoma lanigerum H-1, I-7 (105)
AP: Alk, Tan=I-? (1391)

Parthenocissus tricuspidata   (Boston ivy)   Vitaceae
    A   B   C   D   M
    01 03 09 03 13
OC: Phyllobius oblongus H-4, I-7 (531)

Passiflora incarnata   (Wild passionflower)   Passifloraceae
    B   D   J   M   M   N   O
    03 03 03 01 05 09 08
OC: Heliothis virescens H-5, I-7 (1240)
AP: Alk=I-2, 6, 7, 9 (passiflorine)(1392); Alk=I-6, 7, 8, 9 (1369)

Passiflora quadrangularis   (Granadilla)   Passifloraceae
     B   C   D   J   J   M   M   N
     03 09 01 08 11 01 13 08
OC: Attagenus piceus H-4, I-7 (48)
AP: Alk=I-? (passiflorine)(1392)

Pastinaca sativa   (Parsnip)   Apiaceae
     A   B   C   D   F   J   M   N
     02 06 06 03 05 03 01 02
OC: Acyrthosiphum pisum H-2, I-2 (388)   Musca domestica H-2, I-2 (388)
    Aedes aegypti H-2, I-2 (388)   Sitophilus granarius H-2, I-2 (388)
    Botrytis cinerea H-14, I-8, 16 (945)   Spodoptera eridania H-2, I-2 (388)
    Ceratitis capitata H-2, I-2 (388)   Tetranychus atlanticus H-11, I-2 (388)
    Drosophila melanogaster H-2, I-2 (388)   Venturia inaequalis H-14, I-? (16)
    Epilachna varivestis H-2, I-2 (388)
OR: H-30, I-2 (87)
AP: Alk=I-16 (1392)

Paullinia pinnata   (Barbasco)   Sapindaceae
     A   B   C   D   L   M   M   O   O   O
     01 01 09 02 03 03 05 02 07 10
OC: Aphids H-1, I-? (681)
OR: H-32, 33, I-2, 6, 10 (504, 1276)

Paulownia tomentosa   (Princess tree)   Bignoniaceae
     A   B   C   D   M   M
     01 01 09 03 06 19
OR: H-5, I-7 (1241)

Pavetta indica   (Gusokan)   Rubiaceae
     A   B   B   C   D   J   M   O
     01 01 02 09 01 09 05 07
OC: Musca domestica H-1, I-16 (785)   Tribolium castaneum H-1, I-16 (785)
AP: Alk=I-7 (1388)

Pavonia zeylanica   (Not known)   Malvaceae
     A   B   B   C   D   D   J   K
     01 01 02 09 01 02 10 06
OC: Dysdercus cingulatus H-2a, I-8 (124)

Peganum harmala   (Harmala shrub)   Zygophyllaceae
     B   C   D   D   J   J   M   M   O   O
     06 09 03 06 04 05 05 16 02 10
OC: Aulacophora foveicollis H-1, I-? (600)   Pediculus humanus capitis H-2, I-2 (105)
    Empoasca devastans H-1, I-? (600)   Tribolium castaneum H-5, I-? (179)
OR: H-2, I-2, 10 (73); H-2, 33, I-? (220); H-19, I-? (844); H-35, I-10 (504)
AP: Alk=I-6, 7 (1374); Alk=I-6, 8 (vasicine)(1392); Alk=I-10 (harmaline, harmalol, harmine,
    peganine, vasicine)(1392)

Pelargonium graveolens   (Rose geranium)   Geraniaceae
     A   B   C   D   K
     01 02 09 03 06
OC: Dysdercus cingulatus H-2, I-? (88); H-2a, I-1 (124)

Pelargonium hortorum　　(House geranium)　　Geraniaceae
　　A　　B　　C　　D　　L　　M
　　01　02　09　02　03　21
OC: Potato virus X　H-21, I-14 (1136)
　　Tobacco mosaic virus　H-21, I-7, 14 (184, 1057, 1131)
OR: H-22, I-8, 14 (667, 969)
AP: Tan=I-? (667, 1057)

Pelargonium odoratissimum　　(Apple geranium)　　Geraniaceae
　　A　　B　　C　　M
　　01　02　09　17
OR: H-2, I-? (105); H-7, I-12 (101)

Pelargonium sp.　　(Geranium)　　Geraniaceae
　　B　　C　　D　　J　　M
　　02　09　02　06　13
OC: Aphids　H-2, I-12 (105)　　　　　　　Plutella xylostella　H-4, I-16 (105)
　　Cochliomyia hominivorax　H-5, I-? (105)　　Popillia japonica　H-2b, 6, I-7, 8 (105, 202)

Pelargonium zonale　　(Geranium genus)　　Geraniaceae
　　A　　B　　C　　D
　　01　02　09　02
OC: Phyllobius oblongus　H-4, I-7 (531)　　　　Pieris brassicae　H-4, I-7 (531)
　　Phytodecta fornicata　H-4, I-7 (531)

Pellionia pulchra　　(Rainbow vine)　　Urticaceae
　　B　　B　　C　　D　　M
　　05　06　09　01　13
OC: Plutella xylostella　H-4, I-? (87)

Pellionia scabra　　(Not known)　　Urticaceae
　　B　　C　　D　　J
　　06　09　03　03
OC: Drosophila hydei　H-2, I-2, 6, 7 (101)

Peltandra virginica　　(Arrow arum)　　Areaceae
　　A　　B　　C　　D　　E　　J　　J　　J　　M　　N　　N
　　01　06　09　03　11　03　08　11　01　03　09
OC: Tineola bisselliella　H-4, I-2 (48)
OR: H-24, I-7 (1215)

Peltophorum suringari　　(Not known)　　Caesalpiniaceae
　　B　　C　　D　　J　　J　　J　　M
　　01　09　01　03　08　11　12
OC: Attagenus piceus　H-4, I-2 (48)　　　　　　Tineola bisselliella　H-4, I-2 (48)

Pentaclethra macroloba　　(Not known)　　Mimosaceae
　　A　　B　　C　　D　　J　　J　　M
　　01　01　09　01　08　11　12
OC: Attagenus piceus　H-4, I-4 (48)

Perezia nana　　(Not known)　　Asteraceae
　　B　　C　　D　　D　　J
　　06　09　02　04　04
OC: Pachyzancla bipunctalis　H-2, I-6, 7 (101)　　Spodoptera eridania　H-1, I-? (1252)

Perezia wrightii   (Not known)   Asteraceae
    B   D
    06   02
OC: Pachyzancla bipunctalis H-2, I-? (1252)   Peridroma saucia H-2, I-? (1252)

Perilla frutescens   (Not known)   Lamiaceae
    A   B   C   D   J   L   M   M
    03 06 09 01 09 05 14 20
OC: Rats H-26, I-1 (1393)

Periploca sepium   (Silver vine genus)   Asclepiadaceae
    A   B   C   D
    01 03 09 03
OC: Pieris rapae H-1, I-2 (1241)   Maggots H-1, I-2 (1241)

Peristrophe bicalyculate   (Not known)   Acanthaceae
    B   B   C   D   M
    02 06 09 01 02
OC: Nematodes H-16, I-? (100)

Petalostemon villosum   (Silky prairie cover)   Fabaceae
    A   B   C   D   J   M
    01 06 09 03 02 12
OC: Aedes aegypti H-2, I-9 (165)

Petiveria alliacea   (Guinea-hen weed)   Phytolaccaceae
    B   C   D   J   M
    02 09 01 11 05
OC: Attagenus piceus H-4, I-2 (48)   Mosquitoes H-2, I-? (101)
    Cimex lectularius H-2, I-? (105)   Musca domestica H-2, I-? (101)
    Clothes moths H-5, I-? (504)
OR: H-2, I-? (1276); H-17, 34, I-? (504)

Petroselinum crispum   (Parsley)   Apiaceae
    A   B   C   D   D   J   K   M   M   M   N   N   O
    03 06 09 03 06 04 05 01 05 14 02 07 09
OC: Cochliomyia americana H-5, I-12 (244)   Ixodes redikorzevi H-13, I-7 (87, 1133)
    Dermacentor marginatus H-13, I-7 (87, 1133) Rhipicephalus rossicus H-13, I-7 (87, 1133)
    Haemaphysalis punctata H-13, I-7 (87, 1133)
AP: Alk=I-2, 7 (1392)

Petunia atkinsiana   (Petunia genus)   Solanaceae
    B   C   D
    06 09 01
OC: Cassida nebulosa H-4, I-7 (531)   Leptinotarsa decemlineata H-4, I-7 (531)
    Hyphantria cunea H-4, I-7 (531)   Pieris brassicae H-4, I-7 (531)

Petunia axillaris   (Large white petunia)   Solanaceae
    B   C   D   D
    06 09 02 03
OC: Manduca sexta H-1, I-7 (327)

Petunia hybrida   (Garden petunia)   Solanaceae
    A   B   C   D   M
    03 06 09 10 13

OC: Manduca sexta H-2b, I-16 (87)
OR: H-1, I-? (1026)

Petunia inflata   (White petunia)   Solanaceae
    A   B   C   D   J   K   M
    03   06   09   03   01   01   13
OC: Manduca sexta H-2b, I-7, 17 (87, 327, 328)

Petunia sp.   (Petunia)   Solanaceae
    A   B   C   D   M
    03   06   09   03   13
OC: Centoptera americana H-2b, I-7, 8 (105)   Meloe violaceus H-2b, I-7, 8 (105)
    Cetonia aurata H-2b, I-7, 8 (105)   Pieris brassicae H-2b, I-7, 8 (105)
    Diloba caeruleocephala H-2b, I-7, 8 (105)   Plutella xylostella H-1, I-? (101)
    Leptinotarsa decemlineata H-1, I-? (101)

Petunia violacea   (Violet-flowered petunia)   Solanaceae
    A   B   C   D   J   K   M
    03   06   09   10   01   01   13
OC: Manduca sexta H-2, I-7, 17 (87, 327, 328)
AP: Alk=I-? (1392)

Phacelia ixodes   (Scorpion weed genus)   Hydrophyllaceae
    B   C   D   D
    06   09   02   04
OR: H-10, I-16 (173)

Phacelia parryi   (Scorpion weed genus)   Hydrophyllaceae
    B   C   D   D
    06   09   02   04
OR: H-10, I-16 (173)

Phaseolus lunatus   (Butter bean)   Fabaceae
    A   B   C   D   J   J   L   M   M   M   M   N
    03   04   09   10   02   03   03   01   02   10   12   10
OC: Maggots H-1, I-7 (163)   Meloidogyne incognita H-16, I-10 (542)
OR: H-32, I-? (968); H-33, I-? (220)

Phaseolus vulgaris   (Kidney bean)   Fabaceae
    A   B   C   D   J   M   M   N
    03   06   09   01   03   01   12   10
OC: Cerotoma ruficornis H-4, I-14 (1136)   Phytodecta fornicata H-4, I-7 (531)
    Diabrotica balteata H-4, I-14 (1136)   Pieris brassicae H-4, I-7 (531)
    Hoplolaimus sp. H-16, I-6, 7 (1271)   Potato virus X H-21, I-14 (1136)
    Leptinotarsa decemlineata H-4, I-7 (531)   Pratylenchus zea H-16, I-? (1274)
    Locusta migratoria H-4, 5, I-7 (596)   Tylenchorhynchus dubius H-16, I-6, 7
    Monilinia fructicola H-14, I-6, 7 (1271)   (1271)
OR: H-24, I-7 (569)

Phellodendron amurense   (Amur cork tree)   Rutaceae
    A   B   C   D   J   M   O
    01   01   09   03   10   05   04
OC: Culex pipiens H-1, I-4, 9 (101)   Mosquitoes H-2, I-9, 18 (96)
    Flies H-1, I-? (105)   Musca domestica H-1, I-? (254); H-2,
    Laspeyresia pomonella H-1, I-? (105, 254);   I-9, 18 (96)
      H-2, I-9, 18 (96)   Spodoptera eridania H-1, I-? (254)

OR: H-20, I-2 (504)
AP: Alk=I-4 (berberine, jatrorrhizine, magnoflorine, palmatine, phellodendrine)(1392)

Phellodendron lavallei   (Cork tree)   Rutaceae
   A   B   C   D
   01 01 09 03
OR: H-2, I-9 (105)
AP: Alk=I-4 (berberine)(1392)

Philadelphus coronarius   (Mock orange genus)   Saxifragaceae
   B   C   D   D
   02 09 01 03
OC: Phyllobius oblongus H-4, I-7 (531)

Philodendron hastatum   (Not known)   Araceae
   B   C   D   J   J   J
   04 09 10 03 08 11
OC: Attagenus piceus H-4, I-7 (48)

Phleum pratense   (Timothy grass)   Poaceae
   A   B   C   D   J
   01 06 09 03 03
OR: H-25, I-7 (1204)

Pholidota protracta   (Not known)   Orchidaceae
   B   B   C   D
   06 10 09 01
OC: Musca domestica H-1, I-1 (785)
OR: H-2, I-? (100)

Pholistoma auritum   (Fiesta flower)   Hydrophyllaceae
   A   B   C   D   K
   03 06 09 06 03
OR: H-24, I-7 (1178)

Phoradendron serotinum   (American mistletoe)   Loranthaceae
   B   B   C   D   J   J
   02 10 09 03 03 08
OC: Culex quinquefasciatus H-2, I-7 (463)
OR: H-19, I-6, 7, 9 (974)
AP: Alk=I-? (tyramine)(1392)

Photinia serrulata   (Not known)   Rosaceae
   A   B   C   D   J
   01 01 09 03 03
OC: Aphids H-1, I-7 (1241)   Phytophthora infestans H-14, I-7 (1241)

Phryma leptostachya   (Not known)   Phrymaceae
   C   D   J   J   J
   09 03 03 08 11
OC: Aedes aegypti H-2, I-2, 6, 7, 8, 9   Attagenus piceus H-4, I-1, 9 (48)
   (77, 843)   Drosophila melanogaster H-2, I-2 (165)

Phryma oblongifolia   (Not known)   Phrymaceae
    C   J
    09   03

OC: Culex pipiens   H-1, I-? (101)   Musca domestica   H-2, I-2 (101)
    Drosophila hydei   H-2, I-7, 8 (101)   Pieris rapae   H-1, I-? (101)

Phyla oatesii   (Frogfruit genus)   Verbenaceae
    A   B   C   D
    01   02   09   01

OC: Mosquitoes   H-1, I-? (87)

Phyllanthus acuminatus   (Berry leaf flower)   Euphorbiaceae
    B   B   C   D   D
    02   06   09   01   02

OC: Andrector ruficornis   H-2a, I-2 (101)   Plutella xylostella   H-2a, I-2 (101)
    Diaphania hyalinata   H-2a, I-2 (101)

Phyllanthus emblica   (Emblica)   Euphorbiaceae
    B   C   D   M   M   N
    02   09   01   01   21   09

OC: Dysdercus cingulatus   H-3, I-? (712)
AP: Tan=I-4 (504)

Phyllanthus maderaspatensis   (Not known)   Euphorbiaceae
    C   D
    06   01

OC: Locusta migratoria   H-4, 5, I-7 (596)

Phyllocladus aspelenifolius   (Celery hop pine)   Podocarpaceae
    A   B   C   D
    01   01   08   02

OR: H-29, I-5 (504)

Phyllostachys nigra   (Black bamboo)   Poaceae
    A   C   D   J   M   N
    01   09   03   14   1a   02

OC: Oncopeltus fasciatus   H-2, I-7 (958)

Physalis alkekengi   (Chinese-lantern plant)   Solanaceae
    A   B   C   D   M   M   N   O
    01   06   09   03   01   05   09   09

OR: H-1, I-1 (1241)
AP: Alk=I-? (1392); Alk=I-10 (1374)

Physalis mollis   (Smooth ground cherry)   Solanaceae
    A   A   B   C   D
    01   03   06   09   03

OC: Flies   H-2, I-6, 7 (96, 101)
AP: Alk=I-6, 7 (1392)

Physalis peruviana   (Strawberry tomato)   Solanaceae
    A   B   C   D   D   D   M   N
    01   06   09   01   04   06   01   09

OC: Epilachna varivestis   H-4, I-7 (349)   Spodoptera littoralis   H-1, I-16 (699)
AP: Alk=I-6, 7, 8, 9 (1371)

Physalis subglabrata   (Husk tomato)   Solanaceae
   A   B   C   D   J
   01 06 09 03 03
OC: Leptinotarsa decemlineata H-5, I-? (548)
OR: H-19, I-1, 14 (704, 732, 1171)

Physianthus albens   (Arauja)   Fabaceae
   C   K   M
   09 01 12
OC: Plusio precationis H-9, I-? (670)

Physocarpus capitatus   (Ninebark genus)   Rosaceae
   A   B   C   D   J
   01 02 09 03 14
OC: Dacus cucurbitae H-6, I-4, 5 (956)

Physostegia parviflora   (Not known)   Lamiaceae
   A   B   C   D   J   J   J
   01 06 09 03 02 03 11
OC: Aedes aegypti H-2, I-7 (165)   Attagenus piceus H-4, I-1 (48)

Physostigma venenosum   (Calabar bean)   Fabaceae
   A   B   C   D   J   K   L   M   M   O
   03 04 09 01 02 06 03 05 12 10
OC: Leptinotarsa decemlineata H-1, I-? (87)
OR: H-2a, I-1 (118); H-24, I-? (118); H-33, I-10 (504, 717)
AP: Alk=I-10 (eseramine, eseridine, geneserine, isophysostigmine, physostigmine, physovenine)
   (1392)

Phytolacca acinosa   (Indian pokeberry)   Phytolaccaceae
   B   C   D   J   K   M   N
   06 09 01 04 05 01 07
OC: Epilachna varivestis H-2, I-2 (29, 101)

Phytolacca americana   (Pokeberry)   Phytolaccaceae
   B   C   D   J   J   M   M   M   O
   06 09 10 03 08 05 13 16 02
OC: Attagenus piceus H-4, I-4 (48)   Tobacco mosaic virus H-21, I-? (1069, 1126)
   Cockroaches H-2, I-2 (104)   Tomato aucuba mosaic virus H-21, I-? (1132)
   Southern mosiac bean virus H-21, I-? (1127) Tomato bushy stunt virus H-21, I-? (1132)
OR: H-15, I-2, 12 (1157); H-19, I-6, 7, 8, 9, 10, 14 (413, 704)
AP: Alk=I-2, 6, 7 (phytolaccine)(1392); Alk=I-6, 7, 9 (1371)

Phytolacca esculenta   (Pokeberry genus)   Phytolaccaceae
   B   C   D   D   J   M   N
   06 09 01 03 03 01 07
OC: Tobacco mosaic virus H-21, I-7 (1306)

Phytolacca rigida   (Pokeberry genus)   Phytolaccaceae
   B   C   D
   06 09 01
OC: Tobacco mosaic virus H-21, I-14 (1074)

Picea glauca   (White spruce)   Pinaceae
    A   B   C   D
    01 01 08 03
OC: Pissodes strobi H-5, I-4 (643)

Picea mariana   (Black spruce)   Pinaceae
    A   B   C   D   M   N
    01 01 08 03 01 06
OC: Pissodes strobi H-5, I-4 (643)

Picea rubens   (Red spruce)   Pinaceae
    A   B   C   D   J
    01 01 08 03 04
OC: Pissodes strobi H-5, I-4 (87, 643)

Picramnia pentandra   (Not known)   Simaroubaceae
    A   B   C   D   D   M   O
    01 01 09 01 02 05 04
OR: H-2, I-? (105, 1276); H-20, I-4 (504)

Picrasma excelsa   (Jamaica quassia)   Simaroubaceae
    A   B   C   D
    01 01 09 01
OC: Aphids H-2, I-5 (81)   Phorodon humuli H-2, I-? (481)
AP: Alk=I-4, 5 (1392)

Picrasma javanica   (Not known)   Simaroubaceae
    A   B   C   D
    01 01 09 01
OR: H-2, I-6, 7 (73)

Picrasma napalensis   (Not known)   Simaroubaceae
    A   B   C   D   F
    01 01 09 03 04
OC: Mosquitoes H-2, I-6, 7 (105, 220)

Picrasma quassioides   (Not known)   Simaroubaceae
    B   C   D   D
    02 09 03 08
OC: Lice H-2, I-4 (104)
OR: H-1, I-5 (504)

Picrorrhiza kurroa   (Not known)   Scrophulariaceae
    A   B   C   D   J   J   M   O   O
    01 06 09 03 04 11 05 02 07
OC: Tribolium castaneum H-5, I-9 (179)
AP: Fla=I-2 (1379)

Pieris floribunda   (Fetter bush)   Ericaceae
    A   B   C   D   E   E   J   J   J
    01 02 09 03 07 11 03 08 11
OC: Attagenus piceus H-4, I-7 (48)

Pieris japonica   (Lily-of-the-valley bush)   Ericaceae
    A   B   B   C   E   E   J
    01 01 02 09 07 11 03
OR: H-2, I-7 (105)

Pilocarpus jaborandii   (Jaborandii)   Rutaceae
    A   B   C   D   M   O
    01 02 09 01 05 07
OC: Culex quinquefasciatus H-2, I-7 (463)
AP: Alk=I-7 (isopilocarpine, pilocarpidine, pilocarpine, pilosine)(1392)

Pilocarpus microphyllus   (Jaborandii)   Rutaceae
    A   B   C   D   M   O
    01 02 09 01 05 07
OC: Culex quinquefasciatus H-2, I-7 (463)
AP: Alk=I-7 (isopilocarpine, pilocarpine, pilosine)(1392)

Pimenta dioica   (Allspice)   Myrtaceae
    A   B   C   D   D   M
    01 01 09 01 02 14
OC: Callosobruchus maculatus H-4, I-? (644)   Popillia japonica H-6, I-7, 12 (313)
AP: Alk=I-9 (1392)

Pimenta racemosa   (Bay rum tree)   Myrtaceae
    A   B   C   D   D   J   M   M   M
    01 01 09 01 02 10 14 15 17
OC: Dacus diversus H-6, I-? (473)   Dacus zonatus H-6, I-? (473)
    Dacus fergugineus H-6, I-? (473)   Mosquitoes H-2, I-12 (99, 105)
OR: H-2, I-12 (101, 1276)

Pimpinella anisum   (Common anise)   Apiaceae
    A   B   C   D   D   J   M   M   N
    03 06 09 04 06 03 01 14 07
OC: Aedes aegypti H-2, I-15 (928)   Drosophila melanogaster H-2, I-15 (928)
    Ants H-5, I-? (105)   Musca domestica H-2b, I-15 (928)
    Attagenus piceus H-4, I-10 (48)   Oncopeltus fasciatus H-4, I-10 (48)
    Chrysomya macellaria H-5, I-12 (199)   Pediculus humanus humanus H-4, I-12
    Cochliomyia americana H-1, I-12 (244)   (105, 244)
    Cochliomyia hominivorax H-2, I-12 (105)   Phytophthora parasitica H-14, I-? (911)
OC: H-19, I-12 (904)

Pimpinella major   (Anise genus)   Apiaceae
    B   C   D
    06 09 03
OC: Dermestres maculatus H-1, I-? (544)   Graphosoma mellonella H-1, I-? (544)
    Dysdercus cingulatus H-1, I-? (544)   Pyrrhocoris apterus H-1, I-? (544)
    Graphosoma italicum H-1, I-? (544)

Pimpinella saxifraga   (Black caraway)   Apiaceae
    B   C   F   J   J   M   M
    06 09 13 03 08 05 14
OC: Phytodecta fornicata H-4, I-7 (531)   Venturia inaequalis H-14, I-? (16)
    Popillio japonica H-5, I-2 (531)
AP: Alk=I-6, 7, 9 (1374)

Pinellia ternata　　(Not known)　　Araceae
　　A　C　J
　　01　09　03
OC: Aphids　H-2, I-2, 7 (1241)　　　　　Rice borers　H-2, I-2, 7 (1241)
　　Pieris rapae　H-2, I-2, 7 (1241)　　　Rondotia menciana　H-2, I-2, 7 (1241)
　　Puccinia rubigavera　H-14, I-2, 7 (1241)　Spider mites　H-11, I-2, 7 (1241)

Pinguicula vulgaris　　(Butterwort)　　Lentibulariaceae
　　B　C　D　E
　　06　09　07　13
OC: Lice　H-2, I-7, 12 (105)
OR: H-9, I-1 (504)

Pinus banksiana　　(Jack pine)　　Pinaceae
　　A　B　C　D　M　M
　　01　01　08　03　04　06
OC: Neodiprion rugifrons　H-4, I-7 (547)　　　Neodiprion swainei　H-4, I-7 (547)
AP: Alk=I-10 (1373)

Pinus densiflora　　(Japanese red pine)　　Pinaceae
　　A　B　C　D
　　01　01　08　03
OC: Monochamus alternatus　H-4, I-7 (633)
OR: H-24, I-7 (1207, 1238)
AP: Alk=I-10 (1373)

Pinus insularis　　(Benguet pine)　　Pinaceae
　　A　B　C　D　M
　　01　01　08　01　20
OC: Rice field insects　H-5, I-6, 7 (137)
AP: Fla=I-6 (1379)

Pinus massoniana　　(Masson pine)　　Pinaceae
　　A　B　C　D　J　J　M　M
　　01　01　08　03　03　3a　04　19
OC: Oulema oryzae　H-1, I-6, 7, 9 (1241)　　　Rice leafhoppers　H-1, I-6, 7, 9 (1241)
　　Phytophthora infestans　H-14, I-6, 7, 9 (1241)

Pinus rigida　　(Pitch pine)　　Pinaceae
　　A　B　C　D　E　J　M
　　01　01　08　03　12　14　06
OC: Oncopeltus fasciatus　H-3, I-6, 7 (958)
AP: Alk=I-10 (1374)

Pinus sp.　　(Not known)　　Pinaceae
　　A　B　C　D
　　01　01　08　03
OC: Botrytis sp.　H-14, I-12 (207)　　　　　Laspeyresia pomonella　H-6, I-12 (314)
　　Chrysomya macellaria　H-5, I-12 (199)　　Monilia sp.　H-14, I-12 (207)
　　Dendroctonus brevicornis　H-2, I-13, 20　Popillia japonica　H-5, I-7 (105, 1243)
　　　(318)　　　　　　　　　　　　　　Pseudococcus gahani　H-2, I-12 (207)
　　Fusarium solani　　　　　　　　　　　Rhizoctonia solani　H-14, I-5 (960)
　　　f. sp. phaseoli　H-14, I-5 (960)　　　Thielaviopsis basicola　H-14, I-5 (960)

Pinus strobus　　(White pine)　　Pinaceae
    A B C D M
    01 01 08 03 19
OC: Manduca sexta H-4, I-7 (669)　　　　　　Panonychus citri H-4, 11, I-7 (669)
AP: Alk=I-10 (1374)

Pinus sylvestris　　(Scotch pine)　　Pinaceae
    A B C D M M M O
    01 01 08 03 04 05 17 12
OC: Dermacentor marginatus H-13, I-? (1133)　　Locusta migratoria H-4, 5, I-7 (596)
    Haemaphysalis punctata H-13, I-? (1133)　　Rhipicephalus rossicus H-13, I-? (1133)
    Ixodes redikorzevi H-13, I-? (1133)
OR: H-19, I-12 (865); H-29, I-1 (101)
AP: Alk=I-10 (1374)

Pinus tabuliformis var. yunnanensis　　(Not known)　　Pinaceae
    A B C D
    01 01 08 03
OC: Aphids H-1, I-7 (1241)　　　　　　Empoasca biguttula H-1, I-7 (1241)

Pinus teada　　(Loblolly pine)　　Pinaceae
    A B C D J M M
    01 01 08 03 10 04 22
OC: Culex quinquefasciatus H-2, I-10 (463)

Pinus virginiana　　(Spruce pine)　　Pinaceae
    A B C D J
    01 01 08 03 10
OC: Culex quinquefasciatus H-2, I-10 (463)

Piper aduncum　　(Pepper genus)　　Piperaceae
    A B B C D M O
    01 02 03 09 01 05 07
OC: Ants H-2, I-? (105, 1276)

Piper betle　　(Betel pepper)　　Piperaceae
    A B C D J J M M N O
    01 03 09 01 03 04 01 05 07 07
OC: Cerotoma trifurcata H-2a, I-7 (101)　　　Pyricularia oryzae H-14, I-7 (174, 1305)
    Cladosporium cucumerinum H-14, I-7 (663)　　Rhizoctonia solani H-14, I-7 (174, 1305)
    Diaphania hyalinata H-2a, I-7 (101)　　　Ustilago hordei H-14, I-7 (181)
    Drechslera oryzae H-14, I-7 (174, 1305)　　Ustilago tritici H-14, I-7 (181)
    Dysdercus flavidus H-2a, I-7 (101)
OR: H-19, I-7, 12 (449, 450); H-20, I-6, 7 (1319)
AP: Alk=I-6, 7 (1319)

Piper cubeb　　(Cubeb pepper)　　Piperaceae
    A B C D J M M O
    01 03 09 01 10 05 14 12
OR: H-2, I-9 (99); H-20, I-9 (504)
AP: Alk=I-9 (piperine)(1392)

Piper guineense   (West African pepper)   Piperaceae
    A   B   C   D   J   M
    01 02 09 01 04 14
OC: Pseudaletia unipuncta H-2, I-9 (101)
AP: Alk=I-9 (piperine)(1392)

Piper kadsura   (Pepper genus)   Piperaceae
    A   B   C   D
    01 03 09 03
OC: Spodoptera litura H-4, I-7 (1068)

Piper longum   (Long pepper)   Piperaecae
    A   B   C   D   J   M   M   O
    01 02 09 08 06 05 09 02
OC: Alternaria tenuis H-14, I-? (1295)   Stored grain pests H-5, I-? (80)
    Drechslera graminea H-14, I-? (1295)
OR: H-19, I-6, 7, 12 (897)
AP: Alk=I-6, 7 (1379); Alk=I-9 (piperine, piplartine)(1384, 1392)

Piper methysticum   (Kava pepper)   Piperaceae
    A   B   C   D   J
    01 02 09 01 02
OC: Alternaria solani H-14, I-2 (945)   Ceratocystis ulmi H-14, I-2 (945)
    Botrytis cinerea H-14, I-2 (945)   Sclerotinia fructicola H-14, I-2 (945)
OR: H-20, 35, I-2 (504)

Piper nigrum   (Black pepper)   Piperaceae
    A   B   C   D   F   J   J   J   J   K   K   K   M   M   O
    01 03 09 01 12 02 04 09 10 05 06 09 05 14 09
OC: Acanthoscelides obtectus H-1, I-? (296);   Mites H-11, I-7 (1320)
      H-1, I-9 (1124)   Mosquitoes H-1, I-9 (99)
    Anthonomus grandis H-2, I-9 (333)   Musca domestica H-1, I-? (80, 90, 466, 699)
    Callosobruchus maculatus H-1, I-? (1096);   Plutella xylostella H-1, I-? (90)
      H-2, I-9 (326, 393, 931, 1124)   Pseudomonas solanacearum H-18, I-? (650)
    Dysdercus cingulatus H-1, I-? (699)   Sitophilus oryzae H-1, I-? (90, 1096);
    Fusarium oxysporum H-14, I-12 (649)   H-2, I-9 (326, 1124)
    Heliothis obsoleta H-5, I-? (301)   Sitophilus zeamais H-1, I-? (699)
    Heliothis zea H-5, I-? (105)   Tribolium castaneum H-5, I-? (695)
OR: H-15, I-7 (1320)
AP: Alk=I-4 (methylpyrroline, piperovatine)(1392); Alk=I-9 (chavicine, piperidine,
    piperine)(1392)

Piper peepuliodes   (Pepper genus)   Piperaceae
    A   C   D   J
    01 09 01 11
OC: Aedes aegypti H-2, I-9 (792)   Musca nebulo H-2, I-9 (792)
    Musca domestica H-2, I-9 (792)

Piper tuberculatum   (Cordoncillo)   Piperaceae
    A   B   C   D   J   K
    01 02 09 01 02 06
OC: Mosquitoes H-2, I-9, 10 (101)
OR: H-2a, I-1 (118)

213

Piper umbellata   (Mano de zopilote)   Piperaceae
    B   C   D   F   F   M   O   O
    06 09 01 08 15 05 02 09
OR: H-5, I-12 (87)

Piptadenia peregrina   (Cohoba tree)   Mimosaceae
    A   B   C   D   J   M   M
    01 01 09 01 04 04 12
OR: H-4, I-? (132); H-35, I-10 (504)

Piqueria trinervia   (Stevia)   Asteraceae
    A   B   C   J
    01 06 09 03
OR: H-24, I-2, 7 (1223)

Piscidia acuminata   (Not known)   Fabaceae
    A   B   C   D   J   K
    01 01 09 01 04 05
OC: Diaphania hyalinata H-2, I-2, 7 (101)   Plutella xylostella H-2, I-7 (101)

Piscidia erythrina   (Jamaican dogwood)   Fabaceae
    A   B   C   D   M   M   M
    01 01 09 01 04 06 12
OC: Diaphania hyalinata H-2, I-? (1252)   Spodoptera eridania H-2, I-? (1252)
OR: H-32, 35, I-2, 4 (504)
AP: Alk=I-? (1392); Alk=I-7, 10 (trigonelline)(1392)

Piscidia grandifolia   (Not known)   Fabaceae
    A   B   C   D   M
    01 02 09 01 12
OC: Cimex lectularius H-5, I-4, 7 (101)

Piscidia piscipula   (Jamaica dogwood)   Fabaceae
    A   B   C   D   J   J   J   J   J   K   L   M   M   M   M
    01 01 09 01 02 03 04 08 11 05 03 04 05 06 12
OC: Aedes aegypti H-2a, I-4 (48)   Pachyzancla bipunctalis H-2, I-7
    Attagenus piceus H-4, I-4 (48)   (101, 190, 304)
    Diaphania hyalinata H-2b, I-2 (101)   Plutella xylostella H-2a, 2b, I-2 (101)
    Dysdercus cingulatus H-2a, I-2 (101)   Spodoptera eridania H-2, I-7 (101)
    Hymenia recurvalis H-2, I-7 (101, 190, 304) Tineola bisselliella H-4, I-4 (48)
OR: H-32, I-2, 4, 5, 6 (105, 211, 504, 941, 1276); H-33, I-9 (620); H-35, I-? (504)

Pistacia chinensis   (Chinese pistachio)   Anacardiaceae
    A   B   C   D   J   M   N
    01 01 09 03 03 01 09
OC: Rice field insects H-1, I-2 (1241)

Pistia stratiotes   (Water lettuce)   Araceae
    A   B   B   D   L   M   M   M   M   M   M   N   O
    01 06 09 01 05 1a 02 05 10 15 27 07 07
OC: Periplaneta americana H-1, I-? (105)   Vermin H-27, I-? (105)
OR: H-2, 5, I-7 (1345); H-15, I-1 (1321); H-17, I-7 (1353)
AP: Sfr=I-7 (1321)

Pisum sativum   (Pea)   Fabaceae
    B   C   J   K   M   M   N
   04 09 11 09 01 12 10
OC: Monilinia fructicola H-14, I-7 (991)   Sitophilus oryzae H-5, I-10 (894, 1124)
    Sclerotinia fructicola H-14, I-11 (753)
AP: Alk=I-7, 10 (trigonelline)(1392)

Pithecellobium trapezifolium   (Not known)   Mimosaceae
    A   B   B   C   D   M
   01 01 02 09 01 12
OC: Bombyx mori H-2, I-7 (78, 284)   Periplaneta americana H-2, I-7 (78)

Pittosporum senacia   (Not known)   Pittosporaceae
    A   B   B   C   D   D   J   J   M
   01 01 02 09 01 02 08 11 13
OC: Attagenus piceus H-4, I-6, 7 (48)

Plagiochila fruticosa   (Liverwort)   Cephaloziaceae
    A   C   D
   04 03 03
OC: Spodoptera exempta H-4, I-1 (990)

Plagiochila hattoriana   (Liverwort)   Cephaloziaceae
    A   C   D
   04 03 03
OC: Spodoptera exempta H-4, I-1 (990)

Plagiochila ovalifolia   (LIverwort)   Cephaloziaceae
    A   C   D
   04 03 03
OC: Spodoptera exempta H-4, I-1 (990)

Plagiochila yokogurensis   (Liverwort)   Cephaloziaceae
    A   C   D
   04 03 03
OC: Spodoptera exempta H-4, I-1 (990)

Plantago lanceolata   (English plantain)   Plantaginaceae
    A   B   C   D   L   M   N
   01 06 09 03 05 01 07
OC: Athalia rosae H-4, I-7 (531)   Phytodecta fornicata H-4, I-7 (531)
    Leptinotarsa decemlineata H-4, I-7 (531)   Pieris brassicae H-4, I-7 (531)
AP: Alk=I-6, 7, 9 (1574)

Plantago major   (Common plantain)   Plantaginaceae
    A   B   C   D   M   M   N   O
   01 06 09 03 01 05 07 07
OC: Cassida nebulosa H-4, I-7 (531)   Phytodecta fornicata H-4, I-7 (531)
    Leptinotarsa decemlineata H-4, I-7 (531)   Pieris brassicae H-4, I-7 (531)
    Locusta migratoria H-4, 5, I-7 (531)   Tanymecus dilaticollis H-4, I-7 (531)
OR: H-19, I-? (172); H-20, I-7 (1320)

Plantago monticola (Plantain genus) Plantaginaceae
  A B C D J
  01 06 09 10 11
OC: Attagenus piceus H-4, I-2 (48)

Plantago rugelli (Plantain genus) Plantaginaceae
  B C
  06 09
OC: Popillia japonica H-4, 5, I-7 (548)

Platanus occidentalis (Eastern sycamore) Platanaceae
  A B C M M M
  01 01 09 04 13 19
OC: Ambrosia psilostachya H-24, I-7 (575)   Panicum scribnerianum H-24, I-7 (575)
    Andropogon virginicus H-24, I-7 (575)   Panicum virgatum H-24, I-7 (575)
    Cynodon dactylon H-24, I-7 (575)   Poa pratensis H-24, I-7 (575)
    Leptinotarsa decemlineata H-4, 5, I-7 (548) Setaria viridis H-24, I-7 (575)
    Lolium multiflorum H-24, I-7 (575)   Tridens flavus H-24, I-7 (575)
AP: Alk=I-10 (1372)

Platonia insignis (Bacury) Clusiaceae
  A B C D J J J M M M N N
  01 01 09 01 03 08 11 01 06 15 09 12
OC: Attagenus piceus H-4, I-9 (48)

Platycarya strobilacea (Not known) Juglandaceae
  A B C D J M
  01 01 09 03 03 16
OC: Agrotis sp. H-1, I-7 (1241)   Pieris rapae H-1, I-7 (1241)
    Aphis gossypii H-1, I-7 (1241)   Pseudoperonospora cubensis H-14, I-7 (1241)
    Mosquitoes H-5, I-7 (1241)   Pyricularia oryzae H-14, I-7 (1241)
    Oulema oryzae H-1, I-7 (1241)   Spider mites H-11, I-7 (1241)

Platycladus orientalis (White cedar) Cupressaceae
  A B C D J J
  01 01 09 03 3a 08
OC: Tobacco mosaic virus H-21, I-7 (184)

Plectranthus rugosus (Swedish ivy) Lamiaceae
  B D K
  09 01 03
OC: Mosquitoes H-5, I-16 (105)

Plumbago auriculata (Cape leadwort) Plumbaginaceae
  A B C D J L
  01 02 09 01 03 04
OC: Drechslera oryzae H-14, I-7 (113)
OR: H-33, I-14 (504)

Plumbago indica (Laurel) Plumbaginaceae
  A B C D J L M M O O
  01 02 09 01 03 04 05 13 02 04
OC: Drechslera oryzae H-14, I-7 (113)
OR: H-33, I-14 (504)

Plumbago zeylanica   (Ceylon leadwort)   Plumbaginaceae

    A   B   C   D   J   J   L   M   M   O   O

   01 06 09 01 07 08 04 05 21 02 07

OC: Caterpillars H-1, I-2, 4 (1029)   Euproctis fraterna H-2, I-2, 4 (91, 105)

    Dactynotus carthani H-1, I-2 (91)   Lipaphis erysimi H-1, I-2 (91)

    Dysdercus koenigii H-1, I-2 (91)   Mites H-11, I-2 (1350)

    Epilachna varivestis H-2, I-2, 4 (105)

OR: H-15, I-2 (411, 1350); H-19, I-2 (411, 776); H-20, I-2, 7 (504); H-33, I-14 (504)

AP: Alk, Tan=I-6 (1360)

Plumeria multiflora   (Plumeria genus)   Apocynaceae

    A   B   C   D   J

   01 01 09 01 08

OC: Alternaria sp. H-14, I-2 (562)   Fusarium oxysporum H-14, I-2 (562)

    Ceratocystis ulmi H-14, I-2 (562)   Monilinia sp. H-14, I-2 (562)

Poa annua   (Meadowgrass)   Poaceae

    A   B   C   D   D   M

   03 06 09 02 08 02

OC: Leptinotarsa decemlineata H-5, I-? (87)

Podocarpus hallii   (White pine genus)   Podocarpaceae

    A   B   C   D   M

   01 01 08 03 04

OC: Musca domestica H-2, I-? (683)

Podocarpus macrophyllus   (Buddhist pine)   Podocarpaceae

    A   B   C   D

   01 01 08 10

OC: Chilo suppressalis H-1, I-? (1003)   Pieris rapae H-1, I-? (1003)

    Culex pipiens molestus H-1, I-? (1003)   Plutella xylostella H-1, I-? (1003)

    Musca domestica H-1, I-? (1003)

AP: Alk=I-7 (1392)

Podocarpus nakaii   (White pine genus)   Podocarpaceae

    A   B   C   D

   01 01 08 03

OC: Chilo suppressalis H-1, I-? (1003)   Pieris rapae H-1, I-? (1003)

    Culex pipiens molestus H-1, I-? (1003)   Plutella xylostella H-1, I-? (1003)

    Musca domestica H-1, I-? (1003)

Podocarpus nivalis   (White pine genus)   Podocarpaceae

    A   B   C   D   J

   01 02 08 03 12

OC: Musca domestica H-2, I-7 (683)

Podophyllum peltatum   (Mandrake)   Berberidaceae

    A   B   C   D   J   J   J   J   L   M   N

   01 06 09 03 03 08 09 11 03 01 09

OC: Attagenus piceus H-4, I-2, 7 (48)   Rodents H-25a, I-1 (944)

    Popillia japonica H-5, I-1 (1243)

Pogogyne parviflora   (Not known)   Lamiaceae
   B   C   J   M
   06 09 11 14
OC: Attagenus piceus H-4, I-8 (48)   Fleas H-5, I-? (104)

Pogostemon patchouli   (Patchouli)   Lamiaceae
   A   B   C   D   J   M   M
   01 06 09 01 06 15 17
OC: Ants H-2, I-7, 15 (90, 163)   Leeches H-28, I-7, 15 (101, 163)
   Aphids H-5, I-12 (119)   Mosquitoes H-5, I-12 (105)
   Callosobruchus chinensis H-1, I-10, 12   Moths H-1, I-7, 15 (90, 101, 163)
    (476)   Pericallia ricini H-1, I-12 (943)
   Clothes moths H-5, I-7, 12 (119, 504) Pseudaletia unipuncta H-5, I-12 (119)
   Cockroaches H-2, I-7, 15 (90, 101, 163) Sitophilus oryzae H-1, I-10, 12 (476)
   Crocidolomia binotalis H-2, I-2, 6, 7 Spodoptera litura H-2, I-2, 6, 7 (105)
    (105)   Stegobium paniceum H-1, I-10, 12 (476)
   Dysdercus koenigii H-3, I-1, 12 (795) Tribolium castaneum H-1, I-10, 12 (476)
OR: H-1, I-7 (220); H-5, I-7 (80, 119, 220, 504, 605); H-19, I-? (904)

Polemonium caeruleum   (Jacob's ladder)   Polemoniaceae
   A   B   C   D   J   M   O   O
   01 06 09 03 03 05 02 07
OR: H-2, I-2 (101); H-31, I-7 (504)

Polyalthia longifolia   (Not known)   Annonaceae
   A   B   C   D   J   J
   01 01 09 01 03 06
OC: Aspergillus oryzae H-14, I-12 (649)   Pyricularia oryzae H-14, I-7 (1297)
   Drechslera oryzae H-14, I-12 (649)   Rhizoctonia solani H-14, I-7 (1297)
   Fusarium oxysporum H-14, I-12 (649)   Sclerotium oryzae H-14, I-7 (1297)
   Pseudomonas solanacearum H-18, I-? (650) Sclerotium rolfsii H-14, I-7 (1297)
OR: H-19, I-7, 12 (900)

Polygala sanguinea   (Milkwort genus)   Polygalaceae
   B   C   J
   06 09 09
OC: Rodents H-25, I-1 (946)

Polygonatum japonicum   (Amotokoto)   Liliaceae
   A   B   C   D   J
   01 06 09 03 03
OC: Drosophila hydei H-2, I-7 (101)

Polygonatum sibiricum   (Solomon's seal genus)   Liliaceae
   A   B   C   D   J
   01 06 09 01 03
OC: Ceratocystis fimbriata H-14, I-2 (1241)   Puccinia rubigavera H-14, I-2 (1241)
   Puccinia graminis tritici H-14, I-2 (1241) Rice borers H-1, I-2 (1241)

Polygonum aubertii   (Fleece vine)   Polygonaceae
   B   C   D   J   L
   04 09 03 01 01
OC: Popillia japonica H-2, I-7 (101)

218

Polygonum aviculare   (Prostrate knotweed)   Polygonaceae
   B   C   D   J   J   J   M   O
   06 09 03 07 09 12 05 07
OC: Phytodecta fornicata H-4, I-7 (531)
OR: H-2, 32, I-? (220); H-24, I-2, 15 (1224, 1225, 1226)

Polygonum baldschuanicum   (Bukhara fleece flower)   Polygonaceae
   A   B   C   D
   01 06 09 03
OC: Phytodecta fornicata H-4, I-7 (531)

Polygonum coccineum   (Water smartweed)   Polygonaceae
   A   B   B   C   D   J   K
   01 06 09 09 03 03 03
OC: Aedes aegypti H-2, I-6, 7, 8 (77, 156, 843) Culex tarsalis H-2, I-6, 7, 8 (156)

Polygonum convolulus   (Knotweed genus)   Polygonaceae
   B   D
   06 03
OC: Phytodecta fornicata H-4, I-7 (531)

Polygonum cuspidatum   (Japanese fleece flower)   Polygonaceae
   A   B   C   D   J   L   M   M   O
   01 07 09 03 09 05 05 16 02
OC: Botrytis cinerea H-14, I-1 (944)

Polygonum flaccidum   (Knotweed genus)   Polygonaceae
   B   C   D
   06 09 01
OC: Vermin H-27, I-? (73, 105)
OR: H-2, I-? (220); H-32, I-? (105, 220)

Polygonum gladrum   (Knotweed genus)   Polygonaceae
   B   C   J   J
   06 09 03 09
OC: Ustilago hordei H-14, I-7 (181)   Ustilago tritici H-14, I-7 (181)
OR: H-22, I-1 (1395)

Polygonum hydropiperoides   (Water pepper)   Polygonaceae
   A   B   C   D   D   E   F   G   J   J   J   K   M   M   N   N
   03 06 09 01 06 11 02 05 01 03 07 03 01 04 01 02
OC: Diacrisia obliqua H-3, I-1, 2 (161)   Rats H-25, I-2 (799)
   Flies H-5, I-? (87, 105, 161)   Stored grain pests H-1, I-? (155)
OR: H-2, 32, I-? (220); H-4, I-? (132); H-5, I-? (73); H-24, I-7 (1325)

Polygonum lapathifolium   (Knotweed genus)   Polygonaceae
   B   C
   06 09
OC: Athalia rosae H-4, I-7 (531)   Phytodecta fornicata H-4, I-7 (531)
   Cassida nebulosa H-4, I-7 (531)   Pieris brassicae H-4, I-7 (531)
   Leptinotarsa decemlineata H-4, I-7 (531)
OR: H-24, I-7, 14 (569)

Polygonum newberryi   (Knotweed genus)   Polygonaceae
   B   C   J
   06 09 14
OC: Dacus cucurbitae H-6, I-6, 7, 8 (956)   Dacus dorsalis H-6, I-6, 7, 8 (956)

Polygonum nodosum   (Knotweed genus)   Polygonaceae
   B   C
   06 09
OC: Cnaphalocrocis medinalis H-1, I-1 (1241)   Rice leafhoppers H-1, I-1 (1241)
   Euproctis pseudoconspersa H-1, I-1 (1241)   Spider mites H-11, I-1 (1241)
   Maggots H-1, I-1 (1241)

Polygonum orientale   (Princess feather)   Polygonaceae
   A   B   C   D   J   J   M
   03 06 09 02 03 13 21
OC: Spodoptera litura H-4, I-7 (1064)
OR: H-2, 32, I-? (220); H-24, I-2, 7, 8 (1248); H-32, I-7 (867)
AP: Alk=I-7 (1392); Tan=I-9, 10 (1391)

Polygonum persicae   (Knotweed genus)   Polygonaceae
   B   C
   06 09
OR: H-2, 32, I-? (220)

Polygonum punctatum   (Knotweed genus)   Polygonaceae
   B   C   D
   06 09 01
OR: H-2, I-? (105); H-32, I-1 (401)

Polygonum tomentosum   (Knotweed genus)   Polygonaceae
   B   C   D   M   N
   06 09 01 01 07
OR: H-2, 32, I-? (220)

Polypodium aureum   (Rabbit's foot fern)   Polypodiaceae
   A   C   D   M   O
   01 06 01 05 02
OC: Xanthomonas campestris pv. phaseoli-sojensis H-18, I-7 (854)

Polypodium punctatum   (Climbing bird's-nest fern)   Polypodiaceae
   A   B   C   D
   01 10 06 01
OC: Erwinia carotovora H-18, I-7 (854)
   Xanthomonas campestris pv. phaseoli-sojensis H-18, I-7 (854)

Polypodium walkarae   (Polypody genus)   Polypodiaceae
   A   B   C   D   K
   01 06 06 01 06
OC: Dysdercus cingulatus H-2, I-1 (124)
OR: H-2, I-? (88)

Polyscias guilfoylei   (Geranium-leaf aralia)   Araliaceae
   A   B   D   J   K   M   M
   01 02 01 07 06 13 17

OC: Dysdercus cingulatus H-2, I-? (88); H-2b, I-4, 7 (124); H-3, I-6, 7 (441)
AP: Alk=I-6, 7 (1370)

Polystichum tsus-simense   (Shield fern genus)   Polypodiaceae
    C D
    06 01

OC: Xanthomonas campestris pv. phaseoli-sojensis H-18, I-7 (854)

Pongamia pinnata   (Poonga oil tree)   Fabaceae

A	B	C	D	F	G	J	J	J	J	K	K	M	M	M	M	M	O	O	O	O
01	01	09	01	06	04	03	04	08	15	05	06	02	03	04	05	06	04	08	10	12
												09	10	12	13	15				

OC: Aphids H-1, I-2, 10 (105)   Opatriodes frater H-2, I-18 (87, 433)
    Callosobruchus chinensis H-1, I-10   Rhizopertha dominica H-4, I-7 (88, 124, 126)
      (494, 516)   Rotylenchulus reniformis H-16, I-10 (1269)
    Lipaphis erysimi H-2, I-10 (91)   Seleron latipes H-4, I-7, 18 (87, 433)
    Meloidogyne javanica H-16, I-18 (925, 1107) Sitophilus oryzae H-2, I-8 (87, 598)
    Mesomorphus villiger H-2, I-18   Sitotroga cerealella H-4, I-7 (88, 124, 126)
      (87, 433, 1326)   Spodoptera litura H-4, I-? (93, 126)
    Nephotettix virescens H-1, I-10, 12 (1326)
OR: H-1, I-10, 12 (1326); H-1, 32, I-? (220); H-4, I-10 (124); H-5, I-9 (162); H-20, I-10, 12
    (504, 1349); H-32, I-2, 4, 10 (220, 459, 983, 1349)
AP: Alk=I-2, 4, 9 (1392)

Populus balsamifera   (Balsam poplar)   Saliacaceae

A	B	C	D	J	M	M	M	O
01	01	09	03	03	04	05	22	15

OR: H-19, I-? (592); H-24, I-7 (1216)

Populus canescens   (Gray poplar)   Saliacaceae

A	B	C	D	M	M	M
01	01	09	10	07	13	22

OR: H-20, I-19 (851); H-29, I-1 (87)

Populus deltoides   (Cottonwood)   Saliacaceae

A	B	C	D
01	06	09	03

OC: Flies H-5, I-7 (1345)
OR: H-19, I-19 (851)

Porophyllum punctatum   (Pioja)   Lamiaceae

A	B	C	D	F	J	K
01	02	09	01	14	03	03

OR: H-2a, I-9 (118)

Portulaca foliosa   (Purslane genus)   Portulacaceae

B	C	D
06	09	01

OC: Locusta migratoria H-4, 5, I-7 (596)

Portulaca oleracea   (Kitchen-garden purslane)   Portulacaceae

A	B	D	D	J	L	M	M	M	N	O
03	06	05	10	03	05	01	02	05	07	07

OC: Athalia rosae H-4, I-7 (531)   Pieris brassicae H-4, I-7 (531)
    Drechslera oryzae H-14, I-7 (113, 432)   Vermin H-27, I-? (1353)
    Meloidogyne incognita H-16, I-7 (115, 1138)
OR: H-22, I-16 (977)
AP: Alk=I-16 (1392)

Potamogeton amplifolius   (Pondweed genus)   Potamogetonaceae
    B   B   C   J   J
    06   09   09   03   16
OC: Alternaria sp. H-14, I-1 (837)   Fusarium roseum H-14, I-1 (837)
OR: H-19, I-1 (837)

Potamogeton nutans   (Pondweed genus)   Potamogetonaceae
    A   B   C   D   D   J   J   J   M   N
    01   06   09   02   03   03   09   16   01   02
OC: Alternaria sp. H-14, I-1 (837)   Fusarium roseum H-14, I-1 (837)
OR: H-19, I-1 (837)

Potamogeton pectinatus   (Pondweed genus)   Potamogetonaceae
    A   B   C   D   J   J   J
    01   06   09   10   03   09   16
OC: Alternaria sp. H-14, I-1 (837)   Fusarium roseum H-14, I-1 (837)
OR: H-19, I-1 (837)

Potamogeton richardsonii   (Pondweed genus)   Potamogetonaceae
    B   C   J   J
    06   09   03   16
OC: Alternaria sp. H-14, I-1 (837)   Fusarium roseum H-14, I-1 (837)
OR: H-19, I-1 (837)

Potamogeton zosteriformis   (Pondweed genus)   Potamogetonaceae
    B   C   J   J
    06   09   03   16
OC: Alternaria sp. H-14, I-1 (837)   Fusarium roseum H-14, I-1 (837)
OR: H-19, I-1 (837)

Potentilla fructicosa   (Golden hardhack)   Rosaceae
    A   B   C   D   J   M   N
    01   02   09   03   09   01   07
OC: Ceratocystis ulmi H-14, I-2 (946)

Potentilla reptans   (Five-finger)   Rosaceae
    A   B   C   D   M   O   O
    01   06   09   03   05   02   07
OC: Phytodecta fornicata H-4, I-7 (531)

Poterium sanguisorba   (Garden burnet)   Rosaceae
    B   C   D   J   M   N
    06   09   03   11   01   07
OC: Attagenus piceus H-4, I-1 (48)

Pouzolzia pentandra   (Not known)   Urticaceae
    B   C   D   J   M
    07   09   01   04   03
OC: Maggots H-1, I-7 (80)

Prangos pabularia   (Prangos)   Apiaceae
    B   C   D   D   J   M   M   O   O
    06   09   03   08   03   02   05   02   09
OC: Mites H-11, I-4 (101)   Snails H-28, I-4 (105)
AP: Alk=I-? (prangosine)(1392)

Premna microphylla   (Not known)   Verbenaceae
    A   C   D
    01   09   02
OC: Maggots H-1, I-2, 7 (1241)   Puccinia rubigavera H-14, I-2, 7 (1241)
    Phytophthora infestans H-14, I-2, 7 (1241)   Rhizoctonia solani H-14, I-2, 7 (1241)
    Puccinia graminis triciti H-14, I-2, 7   Verticillium albo-atrum H-14, I-2, 7 (1241)
      (1241)

Premna odorata   (Alagau)   Verbenaceae
    A   B   C   D   M   O   O
    01   01   09   01   05   07   08
OC: Cimex lectularius H-5, I-7 (430)
OR: H-19, I-4 (982)
AP: Alk, Tan=I-6, 7 (1319); Sap=I-6 (1319)

Prenanthes alba   (Rattlesnake root genus)   Asteraceae
    A   B   C   D   J
    01   06   09   03   09
OC: Ceratocystis ulmi H-14, I-1 (946)

Primula vulgaris   (English primrose)   Primulaceae
    A   B   C   D
    01   06   09   03
OC: Venturia inaequalis H-14, I-? (16)

Pristimera celastroides   (Not known)   Hippocrateaceae
    C   J
    09   04
OC: Pachyzancla bipunctalis H-2, I-6, 7 (101)   Pediculus humanus humanus H-2, I-10 (101)

Proboscidea louisianica   (Unicorn plant)   Martyniaceae
    A   B   C   D   J   M   M   N
    03   05   09   03   04   01   13   09
OC: Diaphania hyalinata H-2, I-6 (101)

Prosopis africana   (African ironwood)   Mimosaceae
    A   B   C   D   J   M   M   M   M   M   M   N   O
    01   01   09   01   04   01   04   05   12   19   21   10   07
OC: Termites H-1, I-? (1353)
OR: H-32, I-11 (504)
AP: Tan=I-? (1353)

Prosopis juliflora   (Mesquite)   Mimosaceae
    A   B   C   D   M   M   M   M   M   M   N   N
    01   02   09   01   01   02   04   06   18   21   10   11
OC: Schistocerca gregaria H-4, 5, I-? (611)
OR: H-24, I-7 (554, 559)
AP: Alk=I-6, 7 (1373, 1392)

Prosopis spicigera   (Mesquite genus)   Mimosaceae
    A   C   D   J   M   M   M   N
    01   09   03   03   01   04   06   01
OC: Alternaria tenuis H-14, I-7 (480)   Helminthosporium sp. H-14, I-7 (480)
    Curvularia oryzae H-14, I-7 (480)
AP: Alk, Sap, Tan=I-6, 7 (1360); Sap, Tan=I-9 (1360)

Protium copal   (Copalillo)   Burseraceae
    A   B   C   F   J
    01   01   09   14   03
OR: H-2, I-? (504); H-2a, I-7 (118)

Prunella vulgaris   (Self-heal)   Lamiaceae
    A   B   C   D   D   M   O
    01   06   09   01   06   05   07
OC: Pieris brassicae H-4, I-7 (531)

Prunus americana   (American wild plum)   Rosaceae
    A   B   C   D   M   N
    01   01   09   03   01   09
OR: H-2, I-7, 8 (101)

Prunus armeniaca   (Apricot)   Rosaceae
    A   B   C   D   M   N   N
    01   01   09   03   01   07   10
OC: Leptinotarsa decemlineata H-4, I-7 (531)   Phytodecta fornicata H-4, I-7 (531)
OR: H-33, I-10 (220)

Prunus buergerana   (Not known)   Rosaceae
    A   B   C   D   J   M   N
    01   01   09   03   03   01   09
OC: Drosophila hydei H-2, I-6, 7 (101)

Prunus caroliniana   (Wild orange)   Rosaceae
    A   B   C   D   M
    01   01   09   03   19
OR: H-2, I-2, 6, 7, 9 (87)

Prunus dulcis   (Almond tree)   Rosaceae
    A   B   C   D   L   M   M   M   M   N   O
    01   01   09   06   03   01   05   10   13   10   12
OC: Blatta orientalis H-5, I-10, 12 (105)   Pediculus humanus capitis H-5, I-6 (105)
    Flies H-5, I-6 (105)

Prunus dulcis var. amara   (Bitter almond)   Rosaceae
    A   B   C   D   M   O   O
    01   01   09   01   05   10   12
OC: Cochliomyia americana H-2, I-10, 12 (244)   Pediculus humanus capitis H-5, I-6 (87)
    Musca domestica H-5, I-6 (87)
OR: H-33, I-10 (220)

Prunus grayana   (Gray's artichoke)   Rosaceae
```
   A   B   C   D   J   M   N
   01  01  09  03  03  01  08
```
OC: Drosophila sp. H-2, I-7 (101)

Prunus japonica   (Japanese plum)   Rosaceae
```
   A   B   C   D   J   M   N
   01  01  09  03  03  01  09
```
OC: Drosophila hydei H-2, I-7 (101)

Prunus laurocerasus   (Cherry laurel)   Rosaceae
```
   A   B   B   C   D   J   M   M   O
   01  01  02  09  03  04  05  13  07
```
OC: Cimex lectularius H-2, I-7 (101)   Ixodes ricinus H-13, I-? (101)
    Claviceps purpurea H-14, I-12 (185)   Mosquitoes H-2, I-7 (101)
    Diplodia maydis H-14, I-12 (185)   Musca domestica H-2, I-7 (101)
    Fleas H-1, I-? (101)

Prunus maackii   (Not known)   Rosaceae
```
   A   B   C   D   K
   01  01  09  03  04
```
OC: Aedes punctor H-2, I-4, 8 (101)   Cimex lectularius H-2, I-4, 8 (101)

Prunus padus   (Bird cherry)   Rosaceae
```
   A   B   C   D   F   J   J   K   M   M   N
   01  01  09  03  03  03  04  05  01  19  09
```
OC: Drosophila sp. H-2, I-4, 7, 15 (101)   Mosquitoes H-2, I-4, 7, 15 (101)
    Flies H-2, I-1 (101)   Musca domestica H-2, I-4, 7, 15 (101)
    Lice H-2, I-1 (101)   Ticks H-13, I-4, 7, 15 (101)
    Midges H-2, I-1 (101)

Prunus persica   (Peach)   Rosaceae
```
   A   B   C   D   F   L   M   N
   01  01  09  03  07  03  01  09
```
OC: Bombyx mori H-2, I-7 (105)   Sphaerothica humuli H-14, I-10, 12
    Cochliomyia hominivorax H-5, I-10, 12 (105)   (493, 496)
OR: H-2, I-7, 8 (101); H-32, I-4, 7 (401); H-33, I-10 (220)

Prunus racemosus   (European bird cherry)   Rosaceae
```
   A   B   C   D   J
   01  01  09  03  04
```
OC: Aedes sp. H-2, I-4 (87)   Ixodes ricinus H-13, I-1 (87)
    Anopheles sp. H-2, I-4, 15 (87)   Midges H-2, I-15 (87)
    Culex sp. H-2, I-4 (87)   Musca domestica H-2, I-15 (87)

Pseudocalymna alliaceum   (Garlic vine)   Bignoniaceae
```
   A   B   C   D   F   G   J
   01  03  09  01  13  04  03
```
OC: Drechslera oryzae H-14, I-7 (113)   Pyricularia oryzae H-14, I-7 (114)

Pseudoelephantopus spicatus   (Not known)   Bignoniaceae
```
   A   B   C   D   F   G   J   K
   03  06  09  01  06  01  12  06
```
OC: Dysdercus cingulatus H-2a, I-7 (125)

225

Pseudolarix kaempferi   (Golden larch)   Pinaceae
    A   B   C
    01   01   08
OC: Aphis gossypii H-1, I-6 (1241)   Spider mites H-11, I-6 (1241)
    Pieris rapae H-1, I-6 (1241)

Psidium guajava   (Guava)   Myrtaceae
    A   B   B   C   D   J   J   M   M   M   N   O   O
    01   01   02   09   01   03   3a   01   05   21   09   07   08
OC: Drechslera oryzae H-14, I-4, 6, 7   Ustilago hordei H-14, I-7 (181)
      (113, 432)   Ustilago tritici H-14, I-7 (181)
    Dysdercus cingulatus H-3, I-? (712)   Vermin H-27, I-7 (1350)
    Tobacco mosaic virus H-21, I-7 (184)
OR: H-19, I-7, 8 (172, 1076); H-24, I-7 (569)
AP: Tan=I-? (1350)

Psoralea corylifolia   (Scurfy pea genus)   Fabaceae
    B   C   D   D   J   M   M   M   O
    06   09   01   02   08   05   12   17   07
OC: Dsydercus koenigii H-3, I-10 (371)
OR: H-3, I-? (132); H-17, I-10 (910); H-20, I-7 (504)

Psoralea glandulosa   (Scurfy pea genus)   Fabaceae
    B   C   D   D   J   M   M   N
    02   09   01   02   11   01   12   07
OC: Attagenus piceus H-4, I-7, 15 (48)
AP: Alk=I-7 (1392)

Psorospernum baumii   (Not known)   Clusiaceae
    A   B   C   D   J
    01   02   09   01   03
OC: Lice H-1, I-2 (87)

Psorospernum febrifugum   (Not known)   Clusiaceae
    A   B   C   D   J   J   J   M   O
    01   02   01   01   03   08   11   05   04
OC: Tineola bisselliella H-4, I-4 (48)

Ptaeroxylon utile   (Cape mahogany)   Meliaceae
    A   B   C   D
    01   01   09   02
OC: Clothes moths H-5, I-5 (504)

Pteridium aquilinum   (Eagle fern)   Polypodiaceae
    B   C   M   M   N   N   O
    06   06   01   05   02   06   07   02
OC: Clothes moths H-5, I-7 (105)   Trichoplusia ni H-4, I-7 (1170)
    Erwinia carotovora H-18, I-7 (854)   Vermin H-27, I-7 (104)
    Locusta migratoria H-4, 5, I-7 (596)
OR: H-17, I-2 (504); H-24, I-7 (1220)
AP: Str, Tri=I-? (1376)

Pteris tremula   (Australian bracken)   Polypodiaceae
    A   C   D
    01   06   01
OC: Erwinia carotovora H-18, I-7 (854)   Pseudomonas solanacearum H-18, I-7 (854)

Pteris vittata   (Ladder braken)   Polypodiaceae
    A   C   D
    01   06   01
OC: Pseudomonas solanacearum H-18, I-7 (854)
    Xanthomonas campestris pv. phaseoli-sojensis H-18, I-7 (854)

Pterocarpus marsupium   (Gum king tree)   Fabaceae
    A   B   C   D   M   M   M   M   O
    01   01   09   01   04   05   12   21   13
OC: Dysdercus cingulatus H-3, I-6 (690)
AP: Alk=I-5 (1392); Alk=I-9, 10 (1360); Sap, Tan=I-4, 7 (1360); Tan=I-13 (1352)

Pterocarya stenoptera   (Chinese wingnut)   Juglandaceae
    A   B   C   D   J   J   K   M   O   O
    01   01   09   03   03   04   05   05   04   07
OC: Epilachna varivestis H-2, I-7 (29)
OR: H-17, I-4, 7 (504)

Pterospermum acerifolium   (Kanak champa)   Byttneriaceae
    A   B   C   D   M
    01   01   09   01   19
OC: Clothes moths H-5, I-8 (105)
OR: H-20, I-8 (105)

Pueraria lobata   (Kudzu vine)   Fabaceae
    A   B   C   D   L   M   M   M   M
    01   03   09   03   05   02   09   12   13
OC: Agrotis sp. H-1, I-2, 7 (1241)   Pieris rapae H-1, I-2, 7 (1241)
    Aphids H-1, I-2, 7 (1241)   Rice field insects H-1, I-2, 7 (1241)
    Maggots H-1, I-2, 7 (1241)   Spider mites H-11, I-2, 7 (1241)

Pueraria phaseoloides   (Tropical kudzu)   Fabaceae
    A   B   C   D   L   M
    01   03   09   03   05   12
OC: Fusarium oxysporum f. sp. saccharum H-14, I-? (952)
AP: Alk=I-10 (1392)

Pueraria yunnanensis   (Not known)   Fabaceae
    A   B   C   J   M
    01   03   09   04   12
OC: Aphis fabae H-2, I-2 (101)
    Epilachna varivestis H-2, I-2, 6 (29, 78, 101)

Punica granatum   (Pomegranate)   Punicaceae
    A   B   C   D   D   J   J   K   M   M   M   M   M   N   O   O   O   O
    01   01   09   06   08   09   10   02   01   05   16   21   22   09   04   07   09   10
OC: Agrobacterium tumefaciens H-18, I-? (957)   Erwinia amylovora H-18, I-? (957)
    Colletotrichum falcatum H-14, I-9 (1047)   Erwinia carotovora H-18, I-? (957)

Mollusks H-28, I-7 (1143)   Xanthomonas campestris
Plutella xylostella H-4, I-7 (87)   pv. phaseoli H-18, I-? (957)
Pseudomonas syringae H-18, I-? (957)   pv. pruni H-18, I-? (957)
Pyricularia oryzae H-14, I-14 (1047)   pv. vesicatoria H-18, I-? (957)
Stored grain pests H-5, I-9 (1345)
OR: H-2, I-? (100); H-17, I-2, 4, 9 (504, 1319, 1350); H-17, 28, 32, I-7 (1143); H-19,
    I-7, 9 (172, 1076); H-24, I-7 (569)
AP: Alk=I-2, 4 (methylpelletierne)(1392); Alk=I-4 (pelletierine)(1350, 1392); Alk=I-4
    (isopelletierine)(1392); Alk, Tan=I-7 (1319); Tan=I-? (1047); Tan=I-9 (504)

Pupalia lappacea   (Not known)   Amaranthaceae
    C D
    09 01
OR: H-25, I-11 (1345)

Purshia tridentata   (Not known)   Rosaceae
    A B C D J
    01 02 09 03 03
OC: Attagenus piceus H-4, I-7 (48)   Leptinotarsa decemlineata H-4, I-7 (1118)
OR: H-32, I-7 (401)

Pycanthus kombo   (Not known)   Myristicaceae
    A C D J J M M
    01 09 01 08 11 17 28
OC: Attagenus piceus H-4, I-10 (48)

Pycnanthemum pilosum   (Mountain mint genus)   Lamiaceae
    A B C D
    01 06 09 03
OC: Botrytis cinerea H-14, I-8 (945)

Pycnanthemum rigidus   (Wild savory)   Lamiaceae
    A B C D J
    01 06 09 03 04
OC: Aphids H-2, I-7 (101)   European cabbage worms H-2, I-7 (101)
    Epilachna varivestis H-2, I-7 (101)

Pycnarrhena manillensis   (Ambal)   Menispermaceae
    B C D M M O
    03 09 01 02 05 02
OR: H-2, I-2 (139)
AP: Alk=I-? (ambalinine, pycnamine, pycnarrhenamine, pycnarrhenine, pycharrhine)(1392);
    Alk=I-2 (ambaline)(1392)

Pyrrosia drakeana   (Felt fern genus)   Polypodiaceae
    A B C D J
    01 10 06 01 03
OC: Aphis gossypii H-1, I-1, 2 (1241)   Tetranychus cinnabarinus H-11, I-1, 2 (1241)
    Caterpillars H-1, I-1, 2 (1241)

Pyrrosia lingua   (Japanese felt fern)   Polypodiaceae
    A B C D
    01 10 06 03
OC: Xanthomonas campestris pv. phaseoli-sojensis H-18, I-7 (854)

Pyrularia pubera   (Oil nut)   Santalaceae
    A   B   C   D   J   J   J   M   N
    01 02 09 03 08 09 11 01 09
OC: Attagenus piceus H-4, I-2, 4, 6, 7 (48)   Botrytis cinerea H-14, I-9 (946)

Pyrus communis   (Pear)   Rosaceae
    A   B   C   D
    01 01 09 01
OC: Pieris brassicae H-4, I-7 (531)
OR: H-15, I-9 (652)

Pyrus serrulata   (Pear genus)   Rosaceae
    A   B   C   D   J
    01 01 09 03 04
OC: Venturia inaequalis H-14, I-7, 8 (16)

Pyrus ussuriensis   (Chinese pear)   Rosaceae
    A   B   C   D   J
    01 01 09 03 04
OC: Venturia inaequalis H-14, I-7, 8 (16)

Quassia africana   (Quassia genus)   Simaroubaceae
    A   B   C   D
    01 01 09 01
OR: H-25, I-10 (1345)

Quassia amara   (West Indian quassia)   Simaroubaceae
    A   B   B   C   D   J   J   J   K   K   M   M   M   O
    01 01 02 09 01 03 04 08 02 06 05 13 21 05
OC: Acyrthosiphum pisum H-2a, I-5 (481)   Hoplocampa minuta H-2a, I-5 (1032)
    Aphids H-2a, I-5 (81)   Macrosiphum ambrosiae H-2a, I-5 (481)
    Aphis fabae H-2a, I-5 (481)   Macrosiphum liriodendri H-2a, I-5 (481)
    Attagenus piceus H-4, I-2 (48)   Macrosiphum rosae H-2a, I-5 (481)
    Bombyx mori H-2a, I-5 (1032)   Phyllaphis fagi H-2a, I-5 (481)
    Chaitophorus populicola H-2a, I-5 (481) Phymatocera aterrima H-2, I-5 (81, 97)
    Diaphania hyalinata H-2a, I-2 (101, 1108) Porosagrotis orthogonia H-2, I-5 (196)
    Hoplocampa flava H-2a, I-5 (1032)
OR: H-2, I-5, 8 (80, 1345)
AP: Alk=I-5 (1392); Tan=I-7 (1319)

Quassia cedron   (Not known)   Simaroubaceae
    A   B   C   D   J
    01 01 09 01 02
OR: H-2, I-10 (1345)

Quassia indica   (Niepa bark tree)   Simaroubaceae
    A   B   C   J
    01 01 09 03
OC: Termites H-2, I-7 (504)

Quercus alba   (White oak)   Fagaceae
    A   B   C   D   M   M   M   O
    01 01 09 03 04 05 21 04

OC: Leptinotarsa decemlineata H-4, I-7 (548)
AP: Tan=I-4, 10 (504, 1391)

Quercus cerris   (Turkey oak)   Fagaceae
  A   B   C   D   M   M
  01 01 09 06 01 12
OC: Venturia inaequalis H-14, I-7 (16)

Quercus eugeniaefolia   (Oak genus)   Fagaceae
  A   B   C   D
  01 01 09 03
OR: H-24, I-2, 7 (1183)

Quercus glandulifera   (Konara oak)   Fagaceae
  A   B   C   D
  01 01 09 03
OC: Agrobacterium tumefaciens H-18, I-? (957)   Monilinia fructicola H-14, I-? (957)
   Colletotrichum lindemuthianum H-14, I-?   Pseudomonas syringae H-18, I-? (957)
    (957)   Xanthomonas campestris
   Endothia parasitica H-14, I-? (957)   pv. phaseoli H-18, I-? (957)
   Erwinia amylovora H-18, I-? (957)   pv. pruni H-18, I-? (957)
   Erwinia carotovora H-18, I-? (957)   pv. vesicatoria H-18, I-? (957)

Quercus robur   (English oak)   Fagaceae
  A   B   C   D   D   M   M   M   N
  01 01 09 03 06 01 06 21 10
OC: Operophtera brumata H-1, I-7 (636)
OR: H-19, I-7 (1042); H-21, I-5 (1351)
AP: Tan=I-4, 5, 7 (504, 636)

Quercus velutina   (Black oak)   Fagaceae
  A   B   D   M   M   M
  01 01 09 12 16 21
OC: Popillia japonica H-5, I-7 (105, 1243)
AP: Tan=I-4 (504)

Quillaja brasiliensis   (Not known)   Rosaceae
  A   B   C   D   M
  01 01 09 01 15
OR: H-2, I-? (105)

Quillaja saponaria   (Soapbark tree)   Rosaceae
  A   B   C   D   M   M   O
  01 01 09 02 05 15 04
OC: Caterpillars H-2, I-5 (105)

Quisqualis indica   (Rangoon creeper)   Combretaceae
  A   B   C   D   M
  01 03 09 01 13
OC: Aphids H-1, I-6 (124)   Erwinia carotovora H-18, I-6 (124)
OR: H-17, I-2, 9 (504)
AP: Alk=I-10 (1392)

Randia nilotica   (Not known)   Rubiaceae
    A   B   C   D   J   M   O
    01 01 09 01 04 05 02
OC: Plutella xylostella H-2, I-2 (101)

Ranunculus abortivus   (Buttercup genus)   Ranunculaceae
    B   C   J
    06 09 03
OC: Erwinia carotovora H-18, I-? (585)
OR: H-19, I-1 (1101)

Ranunculus acris   (Common buttercup)   Ranunculaceae
    A   B   C   D
    01 06 09 03
OC: Athalia rosae H-4, I-7 (531)   Locusta migratoria H-4, 5, I-7 (596)
    Leptinotarsa decemlineata H-4, I-7 (531)

Ranunculus ficaria   (Pilewort)   Ranunculaceae
    A   B   C   D   F   M   N   N
    01 06 09 03 11 01 06 07
OC: Venturia inaequalis H-14, I-2 (16)

Ranunculus flagelliformis   (Buttercup genus)   Ranunculaceae
    B   C   D   D   J
    06 09 08 10 03
OC: Drosophila hydei H-2, I-2, 6, 7 (101)

Ranunculus illyricus   (Buttercup genus)   Ranunculaceae
    A   B   C   D   D   J
    01 06 09 08 10 04
OC: Dermacentor marginatus H-13, I-8 (87, 1133) Ixodes redikorzevi H-13, I-8 (87, 1133)
    Haemaphysalis punctata H-13, I-8 (87, 1133) Rhipicephalus rossicus H-13, I-8 (87, 1133)

Ranunculus macounii   (Buttercup genus)   Ranunculaceae
    B   C   D   D
    06 09 08 10
OC: Aedes aegypti H-2, I-7, 8, 9 (77, 165, 843)

Ranunculus mirissimus   (Buttercup genus)   Ranunculaceae
    B   C   D   D
    06 09 08 10
OC: Drosophila hydei H-2, I-2, 6, 7 (101)

Ranunculus sceleratus   (Blister buttercup)   Ranunculaceae
    B   C   D   J   J   L
    06 09 03 03 06 03
OC: Alternaria tenuis H-14, I-6, 7   Drosophila hydei H-2, I-2, 6, 7 (101)
      (182, 418, 1295)   Exserohilum turcicum H-14, I-15 (827)
    Curvularia lunata H-14, I-6, 7   Fusarium nivale H-14, I-6, 7
      (182, 418, 1295)   (182, 418, 1295)
    Drechslera graminea H-14, I-6, 7   Ustilago hordei H-14, I-7 (181)
      (182, 418, 774, 1295)   Ustilago tritici H-14, I-7 (181)

Ranunculus vernyii   (Buttercup genus)   Ranunculaceae
    B   C   D   D   J
    06 09 08 10 03
OC: Drosophila hydei H-2, I-7 (101)

Ranunculus zuccarini   (Buttercup genus)   Ranunculaceae
    B   C   D   D   J
    06 09 06 10 03
OC: Drosophila hydei H-2, I-2, 7, 8 (101)

Raphanus raphinastrum   (Wild mustard)   Brassicaceae
    B   C
    06 09
OC: Bipolaris sorokiniana H-14, I-? (1240)   Heliothis virescens H-5, I-? (1240)

Raphanus sativus   (Radish)   Brassicaceae
    A   B   C   D   J   K   M   M   N   O   O
    03 06 09 03 03 01 01 05 02 07 14
OC: Cephalosporium sacchari H-14, I-1, 14   Meloidogyne incognita H-16, I-2 (147)
       (1048)   Meloidogyne sp. H-16, I-? (525)
    Drechslera oryzae H-14, I-10 (432)   Musca domestica H-6, I-2 (390, 932)
    Drosophila melanogaster H-1, I-? (390)   Pyricularia oryzae H-14, I-10 (432)
    Fusarium nivale H-14, I-1, 14 (1048)   Rotylenchulus sp. H-16, I-? (525)
    Leptinotarsa decemlineata H-4, I-7 (531)   Venturia inaequalis H-14, I-? (16)
OR: H-15, I-10 (711); H-17, I-2 (432, 1321); H-19, I-10 (654, 750)
AP: Alk, Tan=I-10 (1391)

Ratibida columnifera   (Prairie cornflower)   Asteraceae
    A   B   C   D
    01 06 09 03
OC: Melanoplus femurrubrum H-1, I-? (87)

Rauvolfia serpentina   (Java devil pepper)   Apocynaceae
    A   B   C   D   M   M   O
    01 01 09 01 05 13 02
OC: Musca domestica H-7, I-? (87)   Tribolium confusum H-7, I-? (87)
    Tobacco mosaic virus H-21, I-? (1126)
OR: H-2, I-2 (73); H-32, I-? (220); H-34, I-2 (507)
AP: **Alk**=I-2 (ajmalicine, ajmaline, ajmalinine, alloyohimbine, chandrine, isoajmaline,
    isorauhimbine, isoyohimbine, methylreserpate, neoajmaline, papaverine, rauhimbine,
    raupine, rauwolfinine, rauwolscine, rescinnamine, reserpiline, reserpine, reserpinine,
    reserpoxidine, sarpagine, serpentine, serpentinine, serpine, serpinine, thebaine,
    yohimbine)(1392); **Alk**=I-10 (1392)

Rhamnus alnifolia   (Buckthorn genus)   Rhamnaceae
    A   B   C   D   J
    01 02 09 03 16
OC: Lymantria dispar H-4, I-7 (588)   Malacosoma americana H-4, I-7 (588)

Rhamnus crenata   (Buckthorn genus)   Rhamnaceae
    A   B   C   J
    01 02 09 04
OC: Aphis fabae H-2, I-2 (101)

Rheum officinale   (Medicinal rhubarb)   Polygonaceae
    A   B   C   D   M   O
    01   06   09   03   05   02
OC: Aphids   H-1, I-? (1241)   Puccinia rubigavera   H-14, I-? (1241)
    Glomerella gossypii   H-14, I-? (1241)   Rice borers   H-1, I-? (1241)
    Phytophthora infestans   H-14, I-? (1241)   Xanthomonas campestris
    Pieris rapae   H-1, I-? (1241)   pv. malvacearum   H-18, I-? (1241)
    Puccinia graminis tritici   H-14, I-? (1241)

Rheum rhaponticum   (Rhubarb genus)   Polygonaceae
    A   B   C   D   J   L   M   M   N   O
    01   06   09   03   03   03   01   05   06   06
OC: Mosquitoes   H-2, I-2 (98, 105)
OR: H-2, I-7 (504); H-19, I-? (497); H-33, I-7 (504)

Rheum undulatum   (Bucharian rhubarb)   Polygonaceae
    A   B   C   M   N
    01   06   09   01   06
OC: Leptinotarsa decemlineata   H-4, I-7 (531)   Phytodecta fornicata   H-4, I-7 (531)

Rhinacanthus communus   (Twi)   Acanthaceae
    B   C   D   J   M   O   O
    02   09   02   11   05   02   07
OC: Attagenus piceus   H-4, I-2 (48)
OR: H-20, I-2, 7, 10 (504)

Rhinacanthus nasuta   (Tagak-tagak)   Acanthaceae
    A   B   C   D   J
    01   02   09   01   14
OC: Oncopeltus fasciatus   H-2, I-6 (958)
OR: H-20, I-2, 7, 10 (504)

Rhinanthus crista-galli   (Rattlebox)   Scrophulariaceae
    A   B   C   D
    03   06   09   03
OR: H-2, I-? (104)

Rhipsalis leucorhaphis   (Wickerware cactus)   Cactaceae
    A   B   C   D   J
    01   08   09   03   11
OC: Attagenus piceus   H-4, I-6 (48)

Rhodea japonica   (Not known)   Liliaceae
    B   C   D   M   O
    06   09   03   05   01
OC: Aphids   H-1, I-2, 7 (1241)   Pieris rapae   H-1, I-2, 7 (1241)
AP: Alk=I-1 (1392)

Rhododendron albiflorum   (Azalea genus)   Ericaceae
    A   C   D
    01   09   03
OR: H-24, I-7 (1186)

Rhododendron hunnewellianum   (Nao-yang-wha)   Ericaceae
    A   B   C   D   J
    01   02   09   03   02
OC: Aphis fabae H-2b, I-8 (97)
OR: H-2, I-8 (105, 1345)

Rhododendron japonicum   (Japanese azalea)   Ericaceae
    A   B   C   D   J
    01   02   09   03   04
OC: Rondotia menciana H-2, I-8 (105)

Rhododendron maximum   (Great laurel)   Ericaceae
    A   B   C   D   J   M   M   O
    01   02   09   03   09   04   05   19   07
OC: Fusarium oxysporum f. sp. lycopersici H-14, I-7, 9 (944)
    Glomerella cingulata H-14, I-7, 9 (944)

Rhododendron molle   (Yellow azalea)   Ericaceae
    A   B   C   D   J   J   J   J   J   K   L
    01   02   09   01   03   04   09   10   15   05   05
OC: Bean plataspids H-4, I-8 (101, 1012)   Maggots H-1, I-8 (105)
    "Bugs" H-1, I-8 (105)   Oregma lanigera H-1, I-8 (1012)
    Colaphellus bowringi H-4, I-8 (105)   Phyllotreta vittata H-1, I-7 (1012)
    Epilachna varivestis H-2, I-2, 7 (29)   Tryporyza incertulas H-2, I-? (162)

Rhodomyrtus tomentosa   (Hill guava)   Myrtaceae
    A   B   C   D   J   M   N
    01   02   09   01   09   01   09
OC: Rice borers H-1, I-2, 7 (1241)
OR: H-22, I-16 (1394)
AP: Tri=I-7 (1390)

Rhus chinensis   (Nutgall tree)   Anacardiaceae
    A   B   C   D   M   M
    01   01   09   03   13   21
OC: Aphis gossypii H-1, I-2, 7 (1241)   Rice field insects H-1, I-2, 7 (1241)
    Cnaphalocrocis medinalis H-1, I-2, 7 (1241)
AP: Tan=I-7 (504)

Rhus coriaria   (Sicilian sumac)   Anacardiaceae
    A   B   C   D   D   J   K   K   M
    01   02   09   01   06   02   04   07   21
OC: Eriosoma lanigerum H-2, I-7 (105)   Phylloxera vitifoliae H-2, I-7 (104)
OR: H-19, 22, I-4 (1052)
AP: Tan=I-7 (504, 507)

Rhus glabra   (Vinegar tree)   Anacardiaceae
    A   B   C   D   J   M   M   M   M   N   O
    01   02   09   03   09   01   05   13   16   09   02
OC: Pseudomonas solanacearum H-18, I-1 (944)   Rodents H-25, I-? (944)
OR: H-22, I-? (975)
AP: Tan=I-10 (1391)

Rhus radicans   (Poison ivy)   Anacardiaceae
   A   B   C   D   L
   01 04 09 10 03
OC: Manduca sexta H-4, I-7 (669)   Panonychus citri H-4, 11, I-7 (669)
OR: H-33, I-7 (507)

Rhus typhina   (Staghorn sumac)   Anacardiaceae
   A   B   C   D   J   J   M   M   N
   01 02 09 03 09 14 01 13 09
OC: Fusarium sp. H-14, I-6, 7 (623)   Rodents H-25, I-1, 9 (946)
   Pseudomonas solanacearum H-18, I-? (946)
OR: H-15, 19, I-6, 7 (623)
AP: Alk=I-7 (1369); Tan=I-9, 10 (1391)

Rhus verniciflua   (Lacquer tree)   Anacardiaceae
   A   B   C   D   J   M   M
   01 01 09 03 03 06 20
OC: Aphis gossypii H-1, I-2, 7 (1241)   Cnaphalocrocis medinalis H-1, I-2, 7 (1241)
OR: H-33, I-? (504)

Rhus viminalis   (Sumac genus)   Anacardiaceae
   A   B   C   D   M
   01 01 09 02 13
OC: Manduca sexta H-1, I-? (669)
OR: H-29, I-5 (87)

Ribes sanguineum   (Gooseberry genus)   Saxifragaceae
   A   B   C   D   M
   01 02 09 03 13
OC: Locusta migratoria H-4, 5, I-7 (596)

Ricinus communis   (Castor bean)   Euphorbiaceae
   A   B   C   D   D   E   F   G   J   J   J   J   J   K   K   K   L   M   M   M   O   O
   01 02 09 01 04 07 06 02 03 04 06 09 12 04 05 06 03 05 10 15 07 10
                                                                        20 21 22 12

OC: Aleyrodes vaporariorum H-5, I-7 (886)   Laspeyresia pomonella H-2, I-7 (275)
   Aphelenchus avenae H-16, I-7 (100)   Leaf-cutting ants H-6, I-? (87)
   Aphids H-5, I-7 (886)   Lice H-1, I-7 (101)
   Callosobruchus chinensis H-3, I-12 (494)   Locusta migratoria H-5, I-7 (596)
   Colletotrichum atramentarium H-14, I-18   Meloidogyne incognita H-16, I-7 (100, 115,
      (923)   539, 541, 617); H-16, I-18 (923)
   Ditylenchus cypei H-16, I-? (100)   Meloidogyne javanica H-16, I-? (992); H-16,
   Fleas H-1, I-7 (101)   I-18 (421, 962)
   Flies H-5, I-10, 12 (220, 1353)   Meloidogyne sp. H-16, I-18 (953)
   Fusarium oxysporum   Mites H-1, I-7 (101); H-5, 25, I-7 (886)
      f. sp. lycopersici H-14, I-? (541)   Mosquitoes H-5, I-7 (886)
   Fusarium sp. H-14, I-18 (923)   Musca domestica H-2, I-7 (275)
   Helicotylenchus erythrinae H-16, I-?   Oncopeltus fasciatus H-2, I-9 (958)
      (541); H-16, I-18 (923)   Popillia japonica H-2, I-7 (218)
   Heterodera rostochiensis H-16, I-18 (421)   Pratylenchus delattrei H-16, I-18 (126)
   Heterodera schachtii H-16, I-18 (421)   Pythium aphanidermatum H-14, I-? (875)
   Hoplolaimus indicus H-16, I-18   Rhabdoscelus obscurus H-2a, I-16 (153)
      (100, 539, 541, 612, 617, 923)   Rhizoctonia solani H-14, I-? (541)

Rotylenchulus reniformis H-16, I-18   Tylenchorhynchus brassicae H-16, I-18
  (539, 923)   (100, 539, 541, 617, 923)
Sphaerothica humuli H-14, I-? (493, 760)   Tylenchulus semipenetrans H-16, I-18 (423)
OR: H-1, I-10, 18 (73, 80, 82, 87, 220); H-2a, I-1, 7, 10 (111, 116, 504); H-5, I-12 (1124);
  H-15, I-10, 12 (1320); H-17, I-7 (1353); H-22, I-7, 9 (881, 969, 1287); H-33, I-10 (220)
AP: Alk=I-7, 10 (ricinine)(1392); Alk=I-9 (1369); Tan=I-6, 7 (1320)

Rivinia humilis   (Bloodberry)   Phytolaccaceae
  B   C   D   D   J   J   M
  06 09 01 02 03 11 16
OC: Attagenus piceus H-4, I-1 (48)

Robinia pseudo-acacia   (Black locust)   Fabaceae
  A   B   C   D   J   J   L   M   M   N
  01 01 09 03 03 04 03 1a 12 10
OC: Cassida nebulosa H-4, I-7 (531)   Malacosoma disstria H-1, I-16 (87)
  Flies H-2, I-7 (87)   Phyllobius oblongus H-4, I-7 (531)
  Leptinotarsa decemlineata H-4, I-7 (531)
OR: H-24, I-? (559, 1207); H-33, I-1 (504)
AP: Alk=I-6 (1373)

Rosa arkansana   (Rose genus)   Rosaceae
  A   B   C   D
  01 02 09 03
OC: Melanoplus femurrubrum H-4, I-7 (87, 546)

Rosa carolina   (Pasture rose)   Rosaceae
  A   B   C   D   J
  01 02 09 03 09
OC: Pseudomonas solanacearum H-18, I-1 (946)
AP: Alk=I-10 (1391)

Rosa centifolia   (Cabbage rose)   Rosaceae
  A   B   C   D   D   M
  01 02 09 03 09 17
OC: Leptinotarsa decemlineata H-4, I-7 (531)

Rosa chinensis   (China rose)   Rosaceae
  A   B   C   D   J
  01 02 09 03 12
OC: Alternaria solani H-14, I-8 (1050)   Fusarium nivale H-14, I-8 (1050)
  Bipolaris sorokiniana H-14, I-8 (1050)   Fusarium oxysporum H-14, I-8 (1050)
  Cephalosporium sacchari H-14, I-8 (1050)   Pythium aphanidermatum H-14, I-8 (1050)
  Curvularia lunata H-14, I-8 (1050)   Rhizopus nigricans H-14, I-8 (1050)
  Drechslera oryzae H-14, I-8 (1050)

Rosmarinus officinalis   (Rosemary)   Lamiaceae
  B   C   D   D   J   J   J   M   M   M   M   O   O
  02 09 01 06 03 06 08 05 14 17 18 07 08
OC: Attagenus piceus H-4, I-7 (48, 1243)   Mosquitoes H-2, I-12 (105)
  Cockroaches H-5, I-? (105)   Popillia japonica H-5, I-7 (105, 1243)
OR: H-1, I-? (87, 101)

Rubia cordifolia   (Indian madder)   Rubiaceae

   A   C   D   D   J   M   N
  01 09 06 10 03 01 07

OC: Aphids H-1, I-1 (1241)
OR: H-18, I-2 (735)

Rubia tenuifolia   (Not known)   Rubiaceae

   B   C   J
  06 09 14

OC: Ceratitis capitata H-6, I-6, 7, 9 (956)

Rubus idaeus   (European red raspberry)   Rosaceae

   A   B   C   D   M   N   N
  01 02 09 03 01 07 09

OC: Cassida nebulosa H-4, I-7 (531)   Pieris brassicae H-4, I-7 (531)
   Phytodecta fornicata H-4, I-7 (531)

Rubus idaeus var. strigosus   (American red raspberry)   Rosaceae

   A   B   C   D   J   M   N   N
  01 02 09 03 14 01 07 09

OC: Ceratitis capitata H-1, I-6, 7 (956)

Rubus japonica   (Bramble genus)   Rosaceae

   A   B   C   J   M   N
  01 02 09 03 01 09

OC: Drosophilia hydei H-2, I-2, 6, 7 (101)

Rudbeckia hirta   (Black-eyed Susan)   Asteraceae

   A   A   B   C   D   D   M
  02 03 06 09 02 03 13

OC: Popillia japonica H-5, I-1 (105)
OR: H-19, I-? (1171)

Rumex acetosa   (Garden sorrel)   Polygonaceae

   A   B   C   D   M   N
  01 06 09 03 01 07

OC: Leptinotarsa decemlineata H-4, I-7 (531)   Locusta migratoria H-4, I-7 (531)

Rumex crispulus   (Yellow dock)   Polygonaceae

   A   B   C   D   M   M   O
  01 07 09 03 05 07 02

OC: Parasites (human) H-1, I-? (87)

Rumex dentatus   (Dock genus)   Polygonaceae

   A   B   C   D
  01 06 09 03

OC: Aphids H-1, I-1 (1241)   Spider mites H-11, I-1 (1241)
   Pieris rapae H-1, I-1 (1241)

Rumex obtusifolius   (Bitter dock)   Polygonaceae

   A   B   C   D
  01 06 09 03

OC: Athalia rosae H-4, I-7 (531)   Phytodecta fornicata H-4, I-7 (531)
    Cassida nebulosa H-4, I-7 (531)
AP: Alk=I-2 (picoline)(1392); Alk=I-2, 6, 7, 8 (1368)

Ruscus hypoglossum   (Not known)   Liliaceae
    A   B   C   D   D   M
    01 02 09 03 06 13
OC: Spodoptera littoralis H-4, I-7 (344)
OR: H-2, I-? (132)
AP: Alk=I-16 (1392)

Ruta graveolens var. angustifolia   (Rue)   Rutaceae
    A   B   C   D   J   M   M   N   O
    01 02 09 06 03 01 05 07 07
OC: Agrobacterium tumefaciens H-18, I-? (957)   Pseudomonas syringae H-18, I-? (957)
    Culex quinquefasciatus H-1, I-1 (463)   Xanthomonas campestris
    Erwinia amylovora H-18, I-? (957)   pv. phaseoli H-18, I-? (957)
    Erwinia carotovora H-18, I-? (957)   pv. pruni H-18, I-? (957)
    Popillia japonica H-5, I-1 (105, 1243)   pv. vesicatoria H-18, I-? (957)
OR: H-1, I-7 (105); H-5, I-2, 7, 10 (73); H-32, I-? (220)
AP: Alk=I-1 (1366); Alk=I-6, 7, 8, 9 (1371); Alk=I-9 (kokusaginine, skimmianine)(1392)

Ryania angustifolia   (Not known)   Flacourtiaceae
    A   B   C   D   J   J
    01 01 09 01 03 08
OC: Attagenus piceus H-4, I-2 (48)   Tineola bisselliella H-4, I-2 (48)
    Blattella germanica H-4, I-2 (48)

Ryania speciosa   (Not known)   Flacourtiaceae
    A   B   C   D   F   J   J   J   K   K
    01 01 09 01 03 03 04 12 05 06
OC: Argyria stricticraspis H-2, I-? (81, 126)   Mineola vaccinii H-1, I-? (305)
    Attagenus piceus H-2, I-2, 4, 6 (48, 97)   Ostrinia nubilalis H-2, I-2, 4 (30, 81, 101)
    Blatella germanica H-4, I-2, 4, 6 (48)   Pseudaletia unipuncta H-1, I-2, 4 (101)
    Culex sp. H-1, I-? (572)   Pyrausta nubilalis H-1, I-2 (283)
    Cylas formicarius H-1, I-? (126)   Pyrausta salentialis H-1, I-? (524)
    Dacus dorsalis H-2, I-2, 6 (101)   Scirtothrips dorsalis H-1, I-? (126)
    Diaphania hyalinata H-2, I-2, 4, 6   Spodoptera litura H-1, I-? (126)
      (97, 101)   Thrips H-1, I-? (93)
    Heliothis armigera H-1, I-? (524)   Tineola bisselliella H-4, I-2, 4, 6 (48)
    Laspeyresia pomonella H-1, I-? (97)
AP: Alk=I-2, 6 (ryanodine)(1392)

Sabal mexicana   (Texas palmetto)   Aracaceae
    A   B   C   D   F   J   M
    01 01 09 01 08 03 04
OR: H-5, I-6 (118)

Saccharum officinarum   (Sugarcane)   Poaceae
    A   B   C   D   J   M   N
    01 06 09 01 03 01 06
OC: Bipolaris sorokiniana H-14, I-? (593)
    Fusarium oxysporum f. sp. cubense H-14, I-? (593)
OR: H-19, I-7 (982)

Saccharum spontaneum　　(Thatch grass)　　Poaceae
　　A　 B　 C　 D　 J　 M　 M　 M　 M　 M　 M　 M　 N
　　01　06　09　01　09　01　02　03　07　08　22　24　15
OC: Musca domestica H-1, I-1 (785)　　　　　　Tribolium castaneum H-1, I-1 (785)

Sageretia oppositifolia　　(Not known)　　Rhamnaceae
　　B　 C　 D
　　02　09　03
OR: H-2, I-? (100)

Sagittaria calycinus　　(Swamp potato genus)　　Alismataceae
　　B　 B　 C　 K　 M　 N
　　06　09　09　03　01　02
OC: Oncopeltus fasciatus H-2, I-1 (48)

Sagittaria latifolia　　(Duck potato)　　Alismataceae
　　A　 B　 C　 D　 J　 J　 M　 N
　　01　06　09　10　09　16　01　02
OC: Fusarium roseum H-14, I-? (837)
OR: H-19, I-? (837)

Salix alba var. calva　　(White willow)　　Saliacaceae
　　A　 B　 C　 D
　　01　01　09　03
OR: H-29, I-1 (87)

Salix babylonica　　(Weeping willow)　　Saliacaceae
　　A　 B　 C　 D　 J
　　01　01　09　03　09
OC: Locusta migratoria H-4, 5, I-7 (596)
OR: H-19, I-16, 19 (851, 1395)

Salix fragilis　　(Brittle willow)　　Saliacaceae
　　A　 B　 C　 D
　　01　01　09　03
OC: Venturia inaequalis H-14, I-7 (16)

Salix purpurea　　(Basket willow)　　Saliacaceae
　　A　 B　 C　 D　 D
　　01　02　09　03　06
OC: Venturia inaequalis H-14, I-7 (16)

Salmea scandens　　(Not known)　　Asteraceae
　　A　 B　 C　 D　 J　 M　 O
　　01　02　09　01　10　05　02
OC: Mosquitoes H-2, I-6 (101)
OR: H-32, 34, I-2 (504)

Salsola kali　　(Russian thistle)　　Chenopodiaceae
　　C　 M
　　09　01
OC: Pseudomonas solanacearum H-18, I-8 (945)
OR: H-21, I-1, 8 (945); H-24, I-7 (1218)
AP: Alk=I-? (salsolidine, salsoline)(1392)

Salvadora persica   (Mustard tree)   Salvadoraceae
    A   B   C   M   M   M   M   N   O   O
    01 02 09 01 02 05 06 07 04 09
OC: Schistocerca gregaria H-4, 5, I-? (611)

Salvia apiana   (White sage)   Lamiaceae
    A   B   C   D   D   M   M   N
    01 02 09 02 03 01 18 10
OC: Avena fatua H-24, I-7 (733, 1154)

Salvia dorii   (Grey ball sage)   Lamiaceae
    A   B   C   D
    01 02 09 03
OC: Leptinotarsa decemlineata H-4, I-4 (1118)

Salvia leucophylla   (Purple sage)   Lamiaceae
    B   C   D
    02 09 06
OC: Avena fatua H-24, I-7 (578, 733, 1173)   Bromus rubens H-24, I-7 (578, 1173)
    Bromus mollis H-24, I-7 (548, 1173)   Erodium cicutarium H-24, I-7 (578, 1173)
    Bromus rigidus H-24, I-7   Festuca megalura H-24, I-7 (578, 1173)
      (578, 583, 733, 1173)   Stipa pulchra H-24, I-7 (1173)
OR: H-24, I-7 (586, 989, 1154)
AP: Ter=I-7 (camphene, camphore, cineole, pinene)(1154)

Salvia mellifera   (Black sage)   Lamiaceae
    A   B   C   D   D   J   M
    01 07 09 04 06 11 18
OC: Attagenus piceus H-4, I-1 (48)
OR: H-24, I-7 (988, 989, 1154)
AP: Ter=I-7 (camphene, camphore, cineole, dipentene, pinene)(1154)

Salvia officinalis   (Common sage)   Lamiaceae
    A   B   C   D   J   M   M   N
    01 02 09 06 10 01 14 07
OC: Mosquitoes H-2, I-7 (98, 105)   Pieris napi H-5, I-? (1002)
    Pieris brassicae H-5, I-? (1002)   Pieris rapae H-5, I-? (1002)
OR: H-19, I-19 (851)

Salvia plebeia   (Sage genus)   Lamiaceae
    A   B   C   D   M
    01 07 09 01 10
OC: Aphis gossypii H-2, I-12 (105)   Vermin H-27, I-10 (105)
AP: Alk=I-6, 7 (1392)

Salvia pratensis   (Meadow sage)   Lamiaceae
    A   B   C   D   M   O
    01 06 09 06 05 09
OC: Phyllobius oblongus H-4, I-7 (531)

Salvia sclarea   (Clary sage)   Lamiaceae
    A   A   C   D   D   E   J   M   M
    01 02 09 04 06 09 06 14 17
OC: Aphis gossypii H-2, I-12 (105)

Salvia splendens   (Scarlet sage)   Lamiaceae

  A   B   C   D   M

  01 02 09 01 13

OC: Leptinotarsa decemlineata H-4, I-7 (531)   Pieris brassicae H-4, I-7 (531)

Salvia tiliifolia   (Lindenleaf sage)   Lamiaceae

  A   B   C   D   J

  03 07 09 01 03

OC: Lice H-2, I-7 (504)   Spodoptera frugiperda H-3, I-1 (176)

Salvia verticillata   (Lilac sage)   Lamiaceae

  A   B   C   D   D   J

  01 01 09 02 06 04

OC: Dermacentor marginatus H-13, I-7 (87, 1133) Leptinotarsa decemlineata H-4, I-7 (531)
    Haemaphysalis punctata H-13, I-7 (87, 1133) Rhipicephalus rossicus H-13, I-7 (87, 1133)
    Ixodes redikorzevi H-13, I-7 (87, 1133)

Samadera indica   (Niepa bark tree)   Simaroubaceae

  A   B   C   D   J   J   J   M   M   O   O

  01 01 09 01 03 08 11 05 20 04 05

OC: Attagenus piceus H-4, I-2 (48)   Termites H-2, I-7 (504)

OR: H-2, I-? (104); H-2, I-4 (101)

Samanea saman   (Monkey pod)   Mimosaceae

  A   B   C   J   J   M   M   M   M   M   M   N   O   O

  01 01 09 03 09 01 02 05 12 13 14 19 11 04 07

OC: Drechslera oryzae H-14, I-7 (113)   Termites H-1, I-? (101)
    Rats H-26, I-16 (1393)

OR: H-29, I-1 (87)

AP: **Alk**=I-2, 6, 7, 10 (1318); **Alk**=I-4 (pithecolobine)(1392); **Alk**=I-6, 7 (1373, 1375);
    Alk=I-7 (1392)

Sambucus canadensis   (Canadian elderberry)   Caprifoliaceae

  A   B   C   D   J   M   M   M   N   O

  01 01 09 03 10 01 05 21 09 08

OC: Culex quinquefasciatus H-2, I-2 (463)   Mosquitoes H-2, I-? (99)

OR: H-19, I-7 (1171)

AP: **Tan**=I-9, 10 (1391)

Sambucus nigra   (European elderberry)   Caprifoliaceae

  A   B   C   D   J   J   J   K   M   N   N

  01 02 09 03 03 07 10 06 01 08 09

OC: Botrytis cinera H-14, I-7 (739)   Phyllobius oblongus H-4, I-7 (531)
    Caterpillars H-2, I-7 (105)   Pieris brassicae H-5, I-6, 7 (1002)
    Culex quinquefasciatus H-2, I-2 (463)   Pieris napi H-5, I-6, 7 (1002)
    Cydnus bicolor H-1, I-7 (105)   Pieris rapae H-5, I-6, 7 (1002)
    Mosquitoes H-1, I-? (99)

AP: **Alk**=I-4, 7, 8 (sambucine)(1392)

Sambucus williamsii   (Elderberry genus)   Caprifoliaceae

  A   C   D   J   M

  01 09 03 03 13

OC: Aphids H-1, I-6, 7 (1241)

Sanguinaria canadensis   (Canadian blood root)   Papaveraceae
   A   B   C   D   M
   01 07 09 03 16
OC: Aedes aegypti H-2, I-2, 7   Culex tarsalis H-3, I-5 (156)
   (77, 156, 165, 843)   Drosophila melanogaster H-3, I-2 (165)
OR: H-2, I-? (155); H-19, I-2 (1102); H-24, I-? (156)
AP: Alk=I-2 (allocryptopine, chelerythrine, oxysanguinarine, protopine, sanguinarine)(1392)

Sanguisorba officinalis   (Bloodwort)   Rosaceae
   A   C   D   M   N
   01 09 03 01 07
OC: Aphis gossypii H-1, I-1 (1241)   Sylepta derogata H-1, I-1 (1241)
   Spider mites H-11, I-1 (1241)

Sansevieria anthispica   (Bowstring hemp genus)   Agavaceae
   A   B   C   D   J   M
   01 06 09 05 3a 13
OC: Tobacco mosaic virus H-21, I-7 (184)

Sansevieria hyacinthoides   (African bowstring hemp)   Agavaceae
   A   B   C   D   E   J   M   M   M   O
   01 07 09 05 08 08 03 05 17 02
OC: Attagenus piceus H-4, I-2 (48)
OR: H-17, I-2 (504)

Santalum album   (White sandalwood)   Santalaceae
   A   B   C   D   J   M   M   M   O
   01 01 09 01 06 05 17 19 12
OC: Blatta orientalis H-5, I-12 (105)   Termites H-1, I-? (1352)
   Mosquitoes H-5, I-12 (105)
OR: H-2, I-12 (73, 80); H-5, I-? (80); H-19, I-12 (820)

Santolina chamaecyparissus   (Lavender cotton)   Asteraceae
   A   B   C   D   M   M   M   O
   01 02 09 06 05 14 17 08
OC: Moths H-5, I-6, 8 (504)
OR: H-2, I-? (105); H-15, I-8 (504)

Sapindus marginatus   (Florida soapberry)   Sapindaceae
   A   B   C   D   D   E   E   J   K   L   L   M   M
   01 01 09 01 04 07 09 04 09 03 05 13 15
OC: Rhizopertha dominica H-4, I-10 (88, 124)   Stored grain pests H-5, I-9 (105)
   Sitotroga cerealella H-4, I-10 (88, 124)   Weevils H-5, I-9 (105, 1124)
OR: H-4, I-? (132)

Sapindus mukorossi   (Chinese soapberry)   Sapindaceae
   A   B   C   D   D   E   E   J   J   M   M   M   M
   01 01 09 02 03 07 09 03 04 05 13 15 23
OC: Myzus persicae H-2, I-9 (105)
OR: H-1, I-9 (92); H-32, I-9 (73)
AP: Alk=I-9 (1392)

Sapindus rarak (Soapberry genus) Sapindaceae
  A   B   C   D   D   E   E   L   L   M
  01 01 09 01 04 07 13 07 09 15
OR: H-2, I-? (80); H-2, I-2, 32, I-9 (504)

Sapindus saponaria (Southern soapberry) Sapindaceae
  A   B   C   D   D   E   K   K   M   M   N
  01 01 09 01 02 07 03 09 01 15 09
OC: Mosquitoes H-2a, I-4 (101)
OR: H-2, I-4, 9, 10 (101); H-32, I-4, 9, 10 (983)

Sapindus trifoliatus (Soapberry genus) Sapindaceaè
  A   B   C   D   D   E   E   F   J   K   M   M   M   M   O
  01 01 09 01 04 07 09 06 03 06 04 05 15 21 09
OC: Chirida bipunctata H-1, I-10 (966)   Spodoptera litura H-5, I-? (126)
    Rats H-25, I-10 (799)   Tribolium castaneum H-5, I-? (517)
    Sitophilus oryzae H-5, I-? (517)
OR: H-32, I-4, 9 (459, 504)
AP: **Alk**=I-? (1391); **Alk**=I-? (sanguinarine)(1392); **Sap**=I-6, 7 (1360); **Sap**=I-9 (1352);
    Tan=I-7 (1360)

Sapindus utilis (Soapberry genus) Sapindaceae
  A   C   D
  01 09 06
OC: Parlatoria oleae H-2, I-9 (105)

Sapium ellipticum (Not known) Euphorbiaceae
  A   B   B   C   D   D   J
  01 01 02 09 01 04 04
OC: Maggots H-2, I-4 (87); H-2, I-6 (105)

Sapium indicum (Not known) Euphorbiaceae
  A   B   C   D   L
  01 01 09 01 03
OR: H-2, 32, I-10 (105, 1115); H-32, I-9, 10 (73, 504); H-32, 33, 35, I-10 (220)

Sapium sebiferum (Chinese tallow tree) Euphorbiaceae
  A   B   C   D   M   M   M   M
  01 01 09 03 06 13 15 16
OC: Grasshoppers H-1, I-7, 10 (1241)   Spider mites H-11, I-7, 10 (1241)

Saponaria officinalis (Bouncing bet) Caryophyllaceae
  A   B   C   D   D   J   J
  01 06 09 03 06 02 03
OC: Aedes aegypti H-2, I-7, 9, 10 (165)   Rodents H-25, I-1 (944)
    Leptinotarsa decemlineata H-4, I-7 (531)
OR: H-19, I-? (1171)
AP: **Alk**=I-2, 6, 7, 8 (1372); **Alk**=I-6, 7 (1370); **Sap**=I-? (105)

Sarcobatus vermiculatus (Greasewood) Chenopodiaceae
  A   B   C   D   D   M   M   M   N
  01 02 09 02 03 01 06 22 10
OR: H-24, I-7 (554, 559)
AP: **Alk**=I-6, 7, 8 (1370)

Sarcostemma acidum   (Not known)   Asclepiadaceae
```
A    B    B    C    D
01   02   03   09   01
```
OR: H-2, I-? (73, 220)

Sargentodoxa cuneata   (Not known)   Sargentodoxaceae
```
A    B    C    D    J    M    O    O
01   02   09   03   03   05   02   06
```
OC: Aphids H-1, I-7 (1241)   Cnaphalocrocis medinalis H-1, I-7 (1241)

Sarracenia flava   (Yellow pitcher plant)   Sarraceniaceae
```
A    B    C    D    D    E    M    O
01   03   09   02   03   11   05   07
```
OR: H-2, I-? (143)
AP: Alk=I-2 (veratrine)(1392); Alk=I-7 (1392)

Sassafras albidum   (Sassafras)   Lauraceae
```
A    B    C    D    J    J    M    M    M    N    O
01   01   09   03   03   06   01   05   14   08   02
```
OC: Chrysomya macellaria H-5, I-12 (199)   Cockroaches H-6, I-? (105)
Cochliomyia americana H-2, I-12 (244)   Oncopeltus fasciatus H-3, I-2, 4 (958)
Cochliomyia hominivorax H-2, I-12 (105)
OR: H-20, I-2, 12 (504); H-24, I-2, 7 (1181)
AP: Alk=I-9, 10 (1391)

Satureja chandleri   (Savory genus)   Lamiaceae
```
B    C    D    D
06   09   02   03
```
OC: Oncopeltus fasciatus H-2, I-1 (78)   Periplaneta americana H-2, I-1 (78)

Satureja douglasii   (Yerba buena)   Lamiaceae
```
A    B    C    D    D    J    J    M    N
01   06   09   02   03   08   11   01   07
```
OC: Attagenus piceus H-4, I-1 (48)   Periplaneta americana H-2, I-1 (78)
Oncopeltus fasciatus H-2, I-1 (78)

Satureja hortensis   (Summer savory)   Lamiaceae
```
A    B    C    D    J    M    M    M    N
03   07   09   06   06   01   14   21   07
```
OC: Aphis gossypii H-5, I-12 (105)
AP: Tan=I-9, 10 (1391)

Sauromatum guttatum   (Voodoo lily)   Araceae
```
A    B    C    D    J    K    L
01   06   09   01   01   01   03
```
OC: Flies H-1, 6, I-8 (105)
OR: H-33, I-2 (220)

Saussurea lappa   (Costus root)   Asteraceae
```
A    B    C    D    F    J    J    K    M    M    M    O    O
03   02   09   03   06   04   06   05   05   14   16   02   14
```
OC: Chrotogonus trachypterus H-2, I-2 (116)   Trogoderma granarium H-1, I-? (458)
OR: H-1, I-2 (220); H-2, 5, I-2, 12 (1276, 1352); H-5, I-2, 7, 12 (73, 80, 105); H-15, 19,
I-2 (411); H-20, I-2, 12 (504)
AP: Alk=I-2 (1369); Alk=I-2 (saussurine)(1392)

Saxifraga bronchialis   (Not known)   Saxifragaceae
    A   B   C   D   J
    01 06 09 03 11
OC: Attagenus piceus H-4, I-1 (48)

Schefflera capitata   (Umbrella-tree genus)   Araliaceae
    A   B   C   D   J
    01 01 09 01 09
OC: Rats H-25, I-16 (1393)

Schefflera octophylla   (Rubber tree)   Araliaceae
    A   B   C   D   D
    01 01 09 02 03
OC: Aphids H-1, I-7 (1241)

Schima wallichii   (Not known)   Theaceae
    A   B   C   D   E   M
    01 01 09 01 11 04
OC: Cockroaches H-1, I-4 (1241)
OR: H-15, I-1, 16 (882)

Schinus molle   (Australian pepper)   Anacardiaceae
    A   B   C   D   D   J   M   M   M   N   O
    01 01 09 01 08 3a 01 05 14 09 04
OC: Tobacco mosaic virus H-21, I-7 (184)

Schinus terebinthifolius   (Christmas berry tree)   Anacardiaceae
    A   B   C   C   C   J   L   M
    01 02 01 02 09 3a 05 13
OC: Tobacco mosaic virus H-21, I-7 (184)
AP: Alk=I-4, 6, 7 (184)

Schizonepeta tenuifolia   (Not known)   Lamiaceae
    C
    09
OC: Ceratocystis fimbriata H-14, I-2 (1241)   Puccinia graminis tritici H-14, I-2 (1241)
   Heliothis armigera H-1, I-2 (1241)   Puccinia rubigavera H-14, I-2 (1241)
   Malacosoma neustria H-1, I-2 (1241)   Verticillium albo-atrum H-14, I-2 (1241)
   Phytophthora infestans H-14, I-2 (1241)

Schkuria pinnata   (Not known)   Asteraceae
    C   D   J   J
    09 03 03 11
OC: Attagenus piceus H-4, I-1 (48)   Spodoptera exempta H-4, I-? (143)
   Epilachna varivestis H-4, I-? (143)

Schleichera oleosa   (Ceylon oak)   Sapindaceae
    A   B   C   D   M   M   M   M   M   M   M   M   N   N
    01 01 09 01 01 04 06 15 16 20 21 28 07 09
OR: H-2, I-10 (73, 105, 220)
AP: Tan=I-4 (1352)

Schoenocaulon drummondii   (Texas green lily)   Liliaceae
    B   C   D   L
    06 09 01 03

OC: Musca domestica H-2, I-10 (101, 280)
OR: H-2, I-10, 12 (97); H-33, I-10 (504)

Schoenocaulon officinale   (Sabadilla)   Liliaceae
   A   B   C   D   G   J   J   J   J   K   K   M   O
   01   06   09   01   01   04   08   11   15   05   07   05   10
OC: Acalymma vittata H-1, I-? (302)   Lice H-2, I-10 (87)
    Acrosternum hilare H-1, I-? (293)   Loxostege similalis H-1, I-? (302)
    Anasa tristis H-1, I-? (288, 302, 311);   Lygus elisus H-2, I-10 (96)
      H-1, I-10 (285)   Lygus lineolaris H-1, I-? (287)
    Ancysta perseae H-1, I-? (981)   Melanoplus femurrubrum H-2, I-?
    Aphrophora saratogenesis H-1, I-? (294)   (101, 291, 310)
    Attagenus piceus H-4, I-10 (48)   Meloidae sp. H-2, I-? (101)
    Autographa brassicae H-2, I-10 (282, 311)   Murgantia histrionica H-2, I-10 (96, 292)
    Blissus leucopterus H-1, I-10   Musca domestica H-1, I-10 (280, 627)
      (96, 288, 302, 311)   Nezara viridula H-1, I-? (981)
    Bombyx mori H-1, I-10 (311)   Oncopeltus fasciatus H-1, I-? (291, 310);
    Cockroaches H-2, I-10 (104)   H-2, I-? (101)
    Corythuca cydoniae H-1, I-? (302)   Ostrinia nubilalis H-2, I-? (101)
    Corythuca gossypii H-1, I-? (981)   Philaenus leucophthalmus H-1, I-? (300)
    Culex quinquefasciatus H-2, I-10 (463)   Pieris rapae H-2, I-10 (282, 311)
    Diabrotica duodecimpunctata H-1, I-? (302)   Platythpena scabra H-1, I-? (292)
    Diaphania hyalinata H-2, I-? (101, 1252)   Pseudaletia unipuncta H-2, I-? (101)
    Empoasca fabae H-1, I-? (87, 287)   Spodoptera eridania H-2, I-? (101, 1252)
    Epilachna varivestis H-1, I-? (292)   Thrips tabaci H-2, I-10 (87)
    Grasshoppers H-1, I-10 (287, 311)   Tineola bisseliella H-4, I-10 (48)
    Heliothis armigera H-1, I-10 (292)   Trichoplusia ni H-2, I-? (101)
    Laspeyresia pomonella H-2, I-? (101)   Vermin H-27, I-10 (504)
AP: Alk=I-10 (cevacine, cevadilline, cevadine, cevine, dehydrocevagenine, hydroalkamine, neo-
    sabadine, protocevine, sabadine, sabatine, sabine, vanilloylveracevine, veracevine, vera-
    cevine, veragermine, veratridine)(1392)

Scirpus americanus   (Not known)   Cyperaceae
   B   C
   06   09
OC: Pseudomonas solanacearum H-18, I-1, 9 (945)

Scirpus atrovirens   (Not known)   Cyperaceae
   B   C
   06   09
OC: Pseudomonas solanacearum H-18, I-1, 9 (945)

Scirpus lacustris   (Great bulrush)   Cyperaceae
   B   C   D   M   N   N
   06   09   10   01   02   15
OC: Pseudomonas solanacearum H-18, I-1, 9 (945)

Scleria pergracilis   (Not known)   Cyperaceae
   A   B
   01   06
OC: Mosquitoes H-5, I-7, 9 (73)

Sclerocarya caffra   (Kaffir maruda nut)   Anacardiaceae
   A   C   D   M   N
   01   09   02   01   09

OC: Ticks H-13, I-9 (87)
AP: Alk=I-4 (1392)

Scopolia japonica   (Not known)   Solanaceae
    A   B   C   D
    01   01   09   01
OC: Tobacco mosaic virus H-21, I-? (1126)
AP: Alk=I-? (scopolamine, solanidine)(1392); Alk=I-7 (hyoscyamine, norhyoscyamine)(1392)

Scrophularia lanceolata   (Figwort genus)   Scrophulariaceae
    A   A   B   C   D
    01   02   06   09   10
OC: Aedes aegypti H-2, I-7 (77, 843)
AP: Alk=I-2, 6 (1371)

Scutellaria baicalensis   (Baical skullcap)   Lamiaceae
    A   B   C   D
    01   06   09   03
OC: Mosquitoes H-5, I-6, 7 (1241)

Sebastiana pavoniana   (Arrowwood)   Euphorbiaceae
    B   C   D   L
    02   09   01   03
OR: H-33, I-9, 14 (504)

Secale cereale   (Rye)   Poaceae
    A   B   C   D   M   N
    03   06   09   03   01   10
OC: Ambrosia artemisiifolia H-24, I-? (109)   Meloidogyne incognita H-16, I-? (1240)
   Cerastium vulgatum H-24, I-? (109)   Polygonum persicaria H-24, I-? (109)
   Chenopodium album H-24, I-? (109)   Pratylenchus penetrans H-16, I-? (358)
   Digitaria sanguinalis H-24, I-? (109)   Setaria viridis H-24, I-? (109)
   Heliothis virescens H-5, I-? (1240)
OR: H-24, I-1, 20 (1227, 1229, 1231)

Securinega virosa   (Not known)   Euphorbiaceae
    A   B   C   D   D
    01   02   09   02   03
OR: H-1, I-6, 7 (1241)
AP: Alk, Cou=I-7 (1384); Tri=I-4 (1381)

Sedum dendroideum   (Stonecrop genus)   Crassulaceae
    A   B   C   D   D   M   O
    01   02   09   01   08   05   14
OC: Tobacco mosaic virus H-21, I-7 (184)

Sedum nussbaumerianum   (Stonecrop genus)   Crassulaceae
    A   B   C   D   D   J
    01   02   09   01   08   3a
OC: Tobacco mosaic virus H-21, I-7 (184)

Sedum ternatum   (Stonecrop genus)   Crassulaceae
    B   C
    07   09
OC: Botrytis cinerea H-14, I-1, 9 (184)

Selaginella scandens (Club moss genus) Selaginellaceae

  B   C   D   D   K   M
  06 09 01 02 07 13

OC: Ticks H-5, 13, I-7 (1345)

Selenicereus grandiflora (Not known) Cactaceae

  A   B   C   D   J
  01 08 09 01 11

OC: Attagenus piceus H-4, I-6 (48)

Selinum tenuifolium (Not known) Apiaceae

  B   D
  06 03

OC: Drechslera oryzae H-14, I-? (1360)
AP: **Alk**=I-7, 8 (1379); **Fla**=I-2 (1379)

Semecarpus anacardium (Varnish tree) Anacardiaceae

  A   B   C   D   J   J   L   M   M   M   M   N
  01 01 09 01 04 06 03 01 10 16 20 09

OC: Caterpillars H-2a, I-10 (87)   Termites H-1, I-10, 12 (504, 1352)
    Lice H-2a, I-10 (87)

Semiaquilegia adoxoides (Not known) Ranunculaceae

  A   B   C   D
  01 06 09 03

OC: Aphids H-1, I-2, 10 (1241)   Spider mites H-11, I-2, 10 (1241)

Senecio canus (Not known) Asteraceae

  A   B   C   D   D   J   J
  01 06 09 05 08 02 03

OC: Aedes aegypti H-2, I-7, 8 (165)   Drosophila melanogaster H-2, I-7, 8 (165)

Senecio cineraria (Dusty miller) Asteraceae

  A   B   C   D   M   O
  01 06 09 06 05 14

OC: Plutella xylostella H-4, I-16 (87)
AP: **Alk**=I-10 (jacobine, jacodine, senecionine)(1392)

Senecio ehrenbergianus (Not known) Asteraceae

  B   J
  06 11

OC: Ants H-1, I-2 (78)   Attagenus piceus H-4, I-1 (48)
AP: **Alk**=I-6, 7 (1372)

Senecio jacobaea (Ragwort) Asteraceae

  C   D   M   O
  09 03 05 07

OC: Locusta migratoria H-4, 5, I-7 (596)
OR: H-17, I-7 (504)

Senecio scandens (Groundsel genus) Asteraceae

  A   B   C   D   J   M   O
  01 02 09 10 03 05 07

OC: Aphids H-1, I-7 (1241)   Maggots H-1, I-7 (1241)

Senecio totucans   (Rabanillo)   Asteraceae
    B   C   D   J
    06   09   10   03
OC: Spodoptera frugiperda H-1, I-1, 14 (176)

Senecio vulgaris   (Groundsel genus)   Asteraceae
    B   C   D   M   O
    06   09   03   05   07
OC: Athalia rosae H-4, I-7 (531)   Phytodecta fornicata H-4, I-7 (531)
    Leptinotarsa decemlineata H-4, I-7 (531)   Pieris brassicae H-4, I-7 (531)
AP: Alk=I-16 (condoline, fuchsisenecionine, jacobine, othasenine, platyphylline, retrosine,
    senecifolidine, senecifoline, senecine, senecionine, seneciphylline, silvasenecine)(1392)

Sequoia sempervirens   (Redwood)   Taxodiaceae
    A   B   C   D   J   M
    01   01   09   03   3a   04
OC: Tobacco mosaic virus H-21, I-7 (184)

Sesamum indicum   (Sesame)   Pedaliaceae
    A   B   C   D   D   J   M   N
    03   06   09   01   02   03   01   10
OC: Anasa tristis H-1, I-? (101)   Pediculus humanus capitis H-1, I-14 (1353)
    Autographa oo H-1, I-? (101)   Peronospora tabacina H-14, I-? (482)
    Diaphania hyalinata H-1, I-? (101)   Rhizoctonia solani H-14, I-? (422)
    Hoplolaimus indicus H-16, I-? (612)   Sphaerothica humuli H-14, I-?
    Meloidae sp. H-1, I-? (101)   (493, 496, 760)
    Musca domestica H-1, I-10, 12 (101, 812) Spodoptera eridania H-1, I-? (101)
OR: H-2, I-? (80, 504); H-30, I-12 (101)
AP: Alk=I-10 (1373)

Sesbania aculeata   (Not known)   Fabaceae
    B   B   C   D   D   M   M   M   M
    02   06   09   01   02   02   03   12   22
OC: Glossina sp. H-5, I-7 (105)   Meloidogyne javanica H-16, I-7 (922)
AP: Alk=I-6, 7, 9 (1392)

Sesbania punctata   (Alamba)   Fabaceae
    B   B   C   D   J   M   M
    02   06   09   03   03   02   12
OC: Glossina sp. H-5, I-7 (105)

Sesbania sesban   (Sesban)   Fabaceae
    A   B   B   C   M   M   M   M
    01   01   02   09   02   03   10   12
OR: H-5, I-7 (504); H-22, I-15 (882)

Sesuvium portulacastrum   (Sea purslane)   Aizoaceae
    B   C   D   D   J   J   J   M   N   N
    06   09   01   02   03   08   11   01   06   07
OC: Attagenus piceus H-4, I-1 (48)

Setaria faberii   (Giant foxtail)   Poaceae
    B   C
    06   09
OR: H-24, I-2 (363, 1207)

Setaria glauca (Yellow foxtail) Poaceae

   B   C   D   M   M   N
   06 09 01 01 02 10

OR: H-24, I-2, 7 (1141, 1142)

Setaria italica (Italian millet) Poaceae

   A   B   C   D   D   D   M   N
   03 06 09 01 02 04 01 10

OC: Pratylenchus brachyurus H-16, I-2 (1274)

Shibataea kumasaca (Bamboo) Poaceae

   A   B   C   D   J
   01 07 09 03 03

OC: Drosophila hydei H-2, I-6, 7 (101)

Sidalcea oregana (Not known) Malvaceae

   A   B   C   D   J
   01 06 09 03 14

OC: Dacus cucurbitae H-6, I-2, 6, 7, 9 (956)

Sideroxylon borbonicum (Not known) Sapotaceae

   A   B   C   D
   01 01 09 01

OR: H-2, I-? (104)

Silene alba (White campion) Caryophyllaceae

   A   A   C   D   J   L
   01 02 09 03 03 03

OC: Aedes aegypti H-2, I-2, 6, 7, 9 (77, 843)   Phytodecta fornicata H-4, I-7 (531)

OR: H-24, I-16 (1254)

Silene cserei (Not known) Caryophyllaceae

   A   B   C   D   D   J   J
   01 06 09 06 03 02 03

OC: Aedes aegypti H-2, I-2, 7, 8, 9, 10 (165)

Simarouba versicolor (Not known) Simaroubaceae

   A   B   C   D   J
   01 01 09 01 04

OR: H-2, I-4 (105)

Similax glabra (Not known) Liliaceae

   C   J
   09 03

OC: Aphids H-1, I-1, 2 (1241)

Simmondsia chinensis (Jojoba) Buxaceae

   A   B   C   D   J
   01 02 09 06 3a

OC: Tobacco mosaic virus H-21, I-7 (184)

AP: Alk=I-? (1392)

Sinobamboosa kunishii   (Not known)   Bambucaceae
   C
   09
OR: H-24, I-7 (1247)

Sium suave   (Water parsnip)   Apiaceae
   A B C D M
   01 06 09 03 17
OC: Aedes aegypti H-2, I-2, 7, 8, 9 (843); H-2, I-2, 8 (77)
OR: H-22, I-? (909, 944)

Skimmia laureala   (Nera)   Rutaceae
   A B B C D F J M M
   01 01 02 09 04 03 04 13 17
OC: Tribolium castaneum H-5, I-? (112)
OR: H-5, I-7 (111); H-33, I-16 (220)
AP: **Alk**=I-6, 7 (1378); **Alk**=I-6, 7 (skimmianine)(1392)

Smilax sieboldi   (Greenbrier genus)   Liliaceae
   B C J
   04 09 03
OC: Drosophila hydei H-2, I-10 (101)

Solanum acaule   (Not known)   Solanaceae
   B C
   06 09
OC: Leptinotarsa decemlineata H-4, I-7 (87)
OR: H-4, I-? (132)
AP: **Alk**=I-? (solanidine)(1392)

Solanum auriculatum   (Fumabravo)   Solanaceae
   B C J K
   06 09 03 01
OC: Leptinotarsa decemlineata H-1, 6, I-? (87)
OR: H-2, I-9 (87, 104)
AP: **Alk**=I-9 (solasodine, solaurieidine)(1392)

Solanum aviculare   (Kangaroo apple)   Solanaceae
   A B C D M N
   01 02 09 01 01 09
OC: Pieris brassicae H-4, I-7 (531)
AP: **Alk**=I-4, 5, 7 (1392); **Alk**=I-6, 7, 9 (1368); **Alk**=I-7 (solanidine)(1392); **Alk**=I-7, 9
   (saponin)(1363); **Alk**=I-7, 9 (solasodine)(1392); **Alk, Sap**=I-9 (1363)

Solanum ballsii   (Not known)   Solanaceae
   B C D
   06 09 10
OC: Heterodera rostochiensis H-16, I-2 (101, 718)

Solanum brachystachys   (Not known)   Solanaceae
   B C J
   06 09 14
OC: Ceratitis capitata H-6, I-6, 8, 9 (956)
AP: **Alk**=I-2 (1369)

Solanum calvescens   (Not known)   Solanaceae
   B   C
   06 09
OC: Leptinotarsa decemlineata H-4, I-? (87)
OR: H-4, I-? (132)

Solanum carolinense   (Carolina horse-nettle)   Solanaceae
   D
   03
OC: Flies H-2, I-7, 8 (1345)
OR: H-15, 19, I-9 (413, 501); H-19, I-1, 14 (704); H-34, I-8 (504)
AP: Alk=I-2, 4, 7, 9 (solanidine)(1392); Alk=I-6, 7, 9 (1371)

Solanum chacoense   (Not known)   Solanaceae
   B   C
   06 09
OC: Leptinotarsa decemlineata H-4, I-7 (87, 535)
OR: H-4, I-? (132)
AP: Alk=I-? (solanidine)(1392)

Solanum demissum   (Not known)   Solanaceae
   B   C
   06 09
OC: Leptinotarsa decemlineata H-4, I-7 (87, 132, 535)
OR: H-2, I-? (143)
AP: Alk=I-? (demissidine)(1392)

Solanum dulcamara   (Bittersweet)   Solanaceae
   A   B   C   D   L   L   M   O
   01 03 09 03 03 05 05 06
OC: Venturia inaequalis H-14, I-7 (16)
OR: H-4, I-? (132); H-33, I-1 (220, 507); H-34, I-6 (504)
AP: Alk=I-7, 9 (soladulcindine)(1392); Alk=I-10 (atropine)(1392)

Solanum hyporhodium   (Not known)   Solanaceae
   B   C
   06 09
OC: Meloidogyne sp. H-16, I-? (479)

Solanum integrifolium   (Tomato-fruited eggplant)   Solanaceae
   A   B   C
   03 06 09
OC: Potato virus X H-21, I-14 (1136)

Solanum jamesii   (Not known)   Solanaceae
   B   C   D   M   N
   06 09 02 01 02
OC: Leptinotarsa decemlineata H-2b, I-7 (105)
OR: H-4, I-? (132)
AP: Alk=I-? (demissidine)(1392)

Solanum luteum   (Not known)   Solanaceae
   B   C
   06 09
OC: Leptinotarsa decemlineata H-4, I-? (87)

Solanum mammosum   (Love apple)   Solanaceae
    A   B   C   D   D   L   M   M   O
    01   02   09   01   02   03   05   13   07
OC: Cockroaches H-2, I-9 (1345)   Mosquitoes H-2, I-7 (101)
AP: Alk=I-9 (solanidine)(1392)

Solanum melongena   (Eggplant)   Solanaceae
    A   B   C   D   J   M   M   N   O   O
    01   02   09   10   03   01   05   09   07   09
OC: Heliothis virescens H-4, I-? (1240)   Tobacco mosaic virus H-21, I-7 (557, 794)
    Pieris brassicae H-4, I-7 (531)   Tobacco ring spot virus H-21, I-7 (557, 794)
OR: H-22, I-9 (969); H-24, I-7, 14 (569)
AP: Alk=I-6, 7 (1392); Alk=I-9 (solanidine, trigonelline)(1392); Alk=I-10 (1369)

Solanum nigrum   (Black nightshade)   Solanaceae
    A   B   B   C   J   K   L   M   N   N   N
    03   03   06   09   03   06   03   01   07   09   15
OC: Athalia rosae H-4, I-? (531)   Phytodecta fornicata H-4, I-? (531)
    Cassida nebulosa H-4, I-? (531)   Pieris brassicae H-4, I-? (531)
    Leptinotarsa decemlineata H-5, I-? (548)   Woolly aphids H-2, I-1 (105)
OR: H-2, I-9 (105, 1077); H-4, I-? (132); H-15, I-9 (1320); H-33, I-? (220, 507)
AP: Alk=I-6, 7, 8, 9 (1374); Alk=I-7 (1390); Alk=I-9 (solamargine, solanidine, solasodine)
    (1392)

Solanum pennellii   (Not known)   Solanaceae
    B   C
    06   09
OC: Tetranychus cinnabarinus H-10, 11, I-17 (939)
    Tetranychus urticae H-10, 11, I-17 (939)

Solanum polyadenium   (Not known)   Solanaceae
    A   B   C
    03   06   09
OC: Leptinotarsa decemlineata H-4, I-7 (87, 535)

Solanum pseudocapsicum   (Jerusalem cherry)   Solanaceae
    A   B   C   D   D   L   M
    01   02   09   01   02   03   13
OC: Heliothis virescens H-5, I-? (1240)
OR: H-19, I-? (413); H-19, 20, I-7, 9 (391, 832); H-33, I-9 (507)
AP: Alk=I-2, 7, 9 (1392); Alk=I-6, 7 (1369); Alk=I-9 (solanacopsine, solanidine, solano-
    capsidine)(1392)

Solanum sucrense   (Not known)   Solanaceae
    B   C
    06   09
OC: Heterodera rostochiensis H-16, I-2 (101, 718)

Solanum surattense   (Not known)   Solanaceae
    B   C   J
    06   07   03
OC: Argemone mexicana H-24, I-15 (1185)
AP: Alk=I-1 (1379)

Solanum trifolium   (Not known)   Solanaceae
    B   D   J   M   N
    06 04 03 01 09
OC: Aedes aegypti H-2, I-7 (165)

Solanum tuberosum   (Irish potato)   Solanaceae
    A   B   C   D   M   N
    03 06 09 03 01 03
OC: Drechslera graminea H-14, I-2 (1295)   Potato virus X H-21, I-14 (1136)
    Hyphantria cunea H-4, I-? (531)   Pseudomonas solanacearum H-18, I-7 (158)
    Phyllobius oblongus H-4, I-7 (531)   Tanymecus dilaticollis H-4, I-? (531)
    Phytophthora infestans H-14, I-7 (664)   Venturia inaequalis H-14, I-7 (16)
    Pieris brassicae H-4, I-7 (531)
OR: H-4, I-? (132)
AP: Alk=2, 7, 9 (narcotine, solaridine, trigonelline)(1392)

Solanum xanthocarpum   (Yellow-berried nightshade)   Solanaceae
    A   B   C   D   K   M   M   N   N
    01 06 09 01 06 01 05 09 10
OC: Dysdercus cingulatus H-2a, I-7 (98)   Spodoptera litura H-2, 4, I-? (88, 124)
OR: H-22, I-9 (881)
AP: Alk=I-2, 6, 7 (1392)

Solidago altissima   (Goldenrod genus)   Asteraceae
    A   B   C   D   J   J   M
    01 06 09 01 03 12 26
OC: Drosophila melanogaster H-2, I-2 (374)
OR: H-17, 24, I-2 (1217); H-19, I-2 (1253); H-24, I-7 (1204)
AP: Alk=I-6, 7 (1371)

Solidago bicolor   (White goldenrod)   Asteraceae
    A   B   C   D   J
    01 06 09 01 09
OC: Fusarium oxysporum f. sp. lycopersici H-14, I-1 (944)

Solidago caesia   (Blue-stem goldenrod)   Asteraceae
    A   B   C   D   J
    01 06 09 03 09
OC: Rodents H-25, I-1 (946)

Solidago canadensis   (Canadian goldenrod)   Asteraceae
    A   B   C   D   J   J   M   N
    01 06 09 03 03 09 01 10
OC: Fusarium oxysporum f. sp. lycopersici H-14, I-1 (944)
    Mice H-25, I-1 (407)

Solidago flexicaulis   (Goldenrod genus)   Asteraceae
    A   B   C   D   J   J
    01 06 09 03 03 09
OC: Mice H-25, I-1 (407)

Solidago hispida   (Goldenrod genus)   Asteraceae
    A   B   C   D   J   J
    01 06 09 03 03 09
OC: Mice H-25, I-1 (407)

Solidago juncea   (Goldenrod genus)   Asteraceae
    A   B   C   D   J   J
    01 06 09 03 03 09
OC: Mice H-25, I-1 (407)

Solidago microcephala   (Goldenrod genus)   Asteraceae
    A   B   C   D
    01 06 09 03
OC: Pseudomonas solanacearum H-18, I-1 (945)

Solidago microglossa   (Goldenrod genus)   Asteraceae
    A   B   C   D   J   L
    01 06 09 03 11 05
OC: Attagenus piceus H-4, I-1 (48)

Solidago missouriensis   (Goldenrod genus)   Asteraceae
    A   B   C   D
    01 06 09 03
OC: Melanoplus femurrubrum H-3, I-7 (87, 546)

Solidago occidentalis   (Not known)   Asteraceae
    A   B   C   J
    01 06 09 11
OC: Attagenus piceus H-4, I-8, 16 (48)

Solidago odora   (Sweet goldenrod)   Asteraceae
    A   B   C   D   J   J   J   M
    01 06 09 03 03 08 11 01
OC: Attagenus piceus H-4, I-1 (48)

Solidago rugosa   (Goldenrod genus)   Asteraceae
    A   B   C   D   J   J
    01 06 09 03 03 09
OC: Alternaria solani H-14, I-1 (944)   Mice H-25, I-1 (407)

Solidago serotina   (Goldenrod genus)   Asteraceae
    A   B   C   D   J   J
    01 06 09 03 03 09
OC: Mice H-25, I-1 (407)

Solidago squarrosa   (Goldenrod genus)   Asteraceae
    A   B   C   D   J   J
    01 06 09 03 03 09
OC: Mice H-25, I-1 (407)

Solidago uliginosa   (Goldenrod genus)   Asteraceae
    A   B   C   D   J   J
    01 06 09 03 03 09
OC: Mice H-25, I-1 (407)

Solidago ulmifolia   (Goldenrod genus)   Asteraceae
    B   C   J
    06 09 09
OC: Botrytis cinerea H-14, I-1 (946)

Solidago virgaurea   (European goldenrod)   Asteraceae
    B   C   D   M   O   O
    06 09 03 05 02 07
OC: Athalia rosae H-4, I-7 (531)   Phyllobius oblongus H-4, I-7 (531)
    Cassida nebulosa H-4, I-7 (531)   Phytodecta fornicata H-4, I-7 (531)
    Hyphantria cunea H-4, I-7 (531)   Pieris brassicae H-4, I-7 (531)
    Leptinotarsa decemlineata H-4, I-7 (531) Tanymecus dilaticollis H-4, I-7 (531)
AP: Alk=I-7 (1392)

Sonchus arvensis   (Not known)   Asteraceae
    B   C   D   J
    06 09 03 03
OC: Aphids H-1, I-1 (1241)

Sonchus oleraceus   (Sow thistle)   Asteraceae
    A   B   C   D   M   M   M   N   O
    03 06 09 03 01 02 05 07 14
OC: Cassida nebulosa H-4, I-7 (531)   Pieris brassicae H-4, I-7 (531)

Sophora flavescens   (Not known)   Fabaceae
    A   B   C   D   J   M   M   M   O
    01 01 09 10 03 05 12 13 02
OC: Aphis fabae H-2, I-2 (101)   Cucumber beetles H-2, I-2 (101)
    Aulacophora cattigarensis H-1, I-2 (101, 1012)
OR: H-2, I-6, 7 (104)
AP: Alk=I-2 (anagyrine, baptifoline, hydroxymatrine, matrine, matrime, N-methylcytisine)(1392)

Sophora griffithii   (Not known)   Fabaceae
    A   B   C   D   J   J   M   M
    01 01 09 10 03 04 12 13
OC: Pediculus humanus capitis H-2, I-10 (104)
AP: Alk=I-7, 10 (cystisine, pachycarpine)(1392); Alk=I-10 (cytisine)(1392)

Sophora japonica   (Japanese pagoda tree)   Fabaceae
    A   B   C   D   M   M   M
    01 01 09 03 12 13 16
OC: Aphids H-1, I-2, 7, 8, 9 (1241)   Phyllobius oblongus H-4, I-7 (531)
    Caterpillars H-1, I-2, 7, 8, 9 (1241)   Pieris rapae H-1, I-2, 7, 8, 9 (1241)
    Diaporthe nomurai H-14, I-15 (917)   Puccinia rubigavera H-14, I-2, 7, 8, 9
    Monilinia fructicola H-14, I-? (991)   (1241)
AP: Alk=I-6, 7 (cytisine)(1392); Alk, Tan=I-10 (1391)

Sophora mollis   (Not known)   Fabaceae
    A   B   C   D   D   M   M
    01 02 09 03 08 12 13
OR: H-2, I-10 (73, 105, 220)

Sophora pachycarpa   (Not known)   Fabaceae
    A   B   B   C   M   M
    01 01 02 09 12 13
OC: Aphids H-2, I-? (105)
AP: Alk=I-7 (pachycarpine, sparteine)(1392); Alk=I-10 (matrine, sophocarpine)(1392); Alk=I-16
    (pachycarpidine, sophoramine)(1392)

Sophora secundiflora   (Coral bean)   Fabaceae

   A   B   C   D   D   J   M   M

   01 01 09 02 04 04 12 23

OC: Pseudaletia unipuncta H-2, I-6, 10 (78, 101)

AP: **Alk**=I-6, 7 (1392); **Alk**=I-10 (cytisine)(1392)

Sophora sericea   (Silly sophora)   Fabaceae

   A   B   C   D   J   M   M   M   N

   01 01 09 03 03 01 12 13 02

OC: Attagenus piceus H-4, I-15 (48)

AP: **Alk**=I-10 (cytisine)(1392)

Sophora tetraptera   (Four-wing sophora)   Fabaceae

   A   B   C   D   M   M

   01 01 09 03 12 19

OC: Monilinia fructicola H-14, I-? (991)

AP: **Alk**=I-6, 7, 8, 9 (1368); **Alk**=I-10 (matrine, N-methylcytisine, sophochrysine)(1392)

Sophora tinctoria   (Not known)   Fabaceae

   A   B   C   M

   01 01 09 12

OC: Flies H-5, I-6 (1345)

Sophora tomentosa   (Silverbush)   Fabaceae

   A   B   B   C   D   M   M   M   O   O   O

   01 01 02 09 01 05 12 13 02 07 10

OC: Monilinia fructicola H-14, I-? (991)

OR: H-2, I-? (89); H-32, I-14 (1276)

AP: **Alk**=I-7, 9 (1392); **Alk**=I-10 (cytisine)(1392)

Sorbus arbutifolia   (Mountain ash)   Rosaceae

   A   B   C   D   J   M

   01 01 09 03 14 12

OC: Ceratitis capitata H-6, I-6, 7 (956)   Dacus cucurbitae H-6, I-6, 7 (956)

OR: H-19, I-7 (728)

Sorghum bicolor   (Sorghum)   Poaceae

   A   B   C   D   D   J   K   K   M   M   N

   03 06 09 02 04 03 01 06 01 02 10

OC: Ambrosia artemisiifolia H-24, I-? (109)   Lepidium sativum H-24, I-? (691)

    Cerastium vulgatum H-24, I-? (109)   Polygonum persicaria H-24, I-? (109)

    Chenopodium album H-24, I-? (109)   Rhizoctonia solani H-14, I-7 (422)

    Diabrotica virgifera H-2, I-4 (87, 323)   Schizaphis graminum H-4, I-1 (1019)

    Digitaria sanguinalis H-24, I-? (109)   Setaria viridis H-24, I-? (109)

    Fusarium solani f. sp. phaseoli H-14, I-7 (422)

OR: H-24, I-1, 7 (152, 1228, 1229, 1237)

AP: **Alk**=I-? (hordinine)(1392)

Sorghum bicolor var. saccharatum   (Sweet sorghum)   Poaceae

   A   B   C   D   D

   03 06 09 02 03

OC: Fusarium oxysporum f. sp. cubense H-14, I-? (952)

Sorghum dochna   (Snowden)   Poaceae
   B   C   D   D
   06 09 01 02
OR: H-2, I-? (87)

Sorghum halepense   (Johnson grass)   Poaceae
   A   C   D   J   M
   01 09 03 03 02
OC: Amaranthus retroflexus H-24, I-2, 7 (577)   Bromus tectorum H-24, I-2, 7 (577)
    Aristida oligantha H-24, I-2, 7 (577)   Digitaria sanguinalis H-24, I-2, 7 (577)
    Bromus japonicus H-24, I-2, 7 (577)   Setaria viridis H-24, I-2, 7 (577)

Sorghum sudanense   (Sudan grass)   Poaceae
   B   C   D
   06 09 01
OC: Ambrosia artemisiifolia H-24, I-1 (109)   Polygonum persicaria H-24, I-1 (109)
    Cerastium vulgatum H-24, I-1 (109)   Rhizoctonia solani H-14, I-? (353)
    Chenopodium album H-24, I-1 (109)   Setaria viridis H-24, I-1 (109)
    Digitaria sanguinalis H-24, I-1 (109)

Spathobolus roxburghii   (Not known)   Fabaceae
   B   C   D   F   J   K   M
   02 09 01 06 04 01 12
OC: Ceratitis capitata H-2a, I-2 (111)
OR: H-2, I-? (82); H-32, I-2 (767, 1276)

Spergularia marina   (Sand spurry)   Caryophyllaceae
   B   C   E   J
   06 09 03 11
OC: Attagenus piceus H-4, I-1 (48)

Sphaeralcea angustifolia   (Narrowleaf globe mallow)   Malvaceae
   A   B   C   D   J   J
   01 06 09 03 02 03
OC: Spodoptera frugiperda H-3, I-1 (176)

Sphaeranthus indicus   (Not known)   Asteraceae
   B   C   M   M   N   N   O
   06 09 1a 05 07 10 10
OR: H-16, I-1 (73); H-17, I-7 (504); H-32, I-2 (220)
AP: Alk=I-? (sphaeranthine)(1392)

Sphaeropteris cooperi   (Australian tree fern)   Cyatheaceae
   A   C   D
   01 06 01
OC: Xanthomonas campestris pv. phaseoli-sojensis H-18, I-7 (854)

Spigelia anthelmia   (Worm grass genus)   Loganiaceae
   B   C   M   M
   06 09 03 05
OC: Mosquitoes H-2, I-? (101)   Musca domestica H-2, I-? (101)
OR: H-17, 33, I-? (504)
AP: Alk=I-7 (spigeline)(1392)

Spigelia humboltiana   (Worm grass genus)   Loganiaceae
　　B　C　D　J
　　06　09　02　03
OC: Attagenus piceus H-4, I-2 (48)   Tineola bisselliella H-4, I-2 (48)

Spigelia marilandica   (Indian pink)   Loganiaceae
　　A　B　C　J　J　M　O
　　01　06　09　03　11　05　02
OC: Attagenus piceus H-4, I-2 (48)   Tribolium castaneum H-2, I-? (1124)
　　Tineola bisselliella H-4, I-2 (48)
OR: H-17, I-2 (504)
AP: Alk=I-? (spigeline)(1392)

Spilanthes acmella   (Brazil cress)   Asteraceae
　　A　B　C　D　J　M　M　M　N　O
　　03　06　09　01　03　01　02　05　07　02
OC: Anopheles sp. H-1, I-? (762)   Mosquitoes H-2, I-8 (78, 101, 504, 805)
　　Culex sp. H-1, I-? (762)
OR: H-2, I-? (105); H-2, 32, I-9 (1276); H-32, I-7 (504); H-34, I-8 (504, 1353)
AP: Alk=I-16 (1392)

Spilanthes mauritiana   (Not known)   Asteraceae
　　A　B　B　C　D　G　J
　　03　05　06　09　01　07　07
OC: Anopheles sp. H-2, I-8 (87)
OR: H-33, I-9 (87)

Spilanthes oleracea   (Not known)   Asteraceae
　　B　C　D
　　06　09　01
OC: Anopheles sp. H-1, I-8 (762)   Culex pipiens H-1, I-? (971)
AP: Alk=I-? (spilanthine)(1392)

Spinacia oleracea   (Spinach)   Chenopodiaceae
　　A　B　C　D　D　M　N
　　03　06　09　01　02　01　07
OC: Cucumber mosaic virus H-21, I-? (1073)   Tobacco mosaic virus H-21, I-? (1073)
　　Latent potato ring spot virus H-21, I-?   Tobacco necrotic virus H-21, I-? (1073)
　　　(1073)   Tobacco ring spot virus H-21, I-? (1073)
　　Potato virus X H-21, I-14 (1136)

Spiraea nipponica   (Tosa spirea)   Rosaceae
　　A　B　C　D　J
　　01　02　03　09　03
OC: Drosophila hydei H-2, I-7 (101)

Spondianthus preussi   (Ebai)   Euphorbiaceae
　　A　B　C　D　K
　　01　01　09　01　02
OC: Rats H-25, I-4, 10 (504, 1345)

Spondianthus ungandensis   (Twianga)   Euphorbiaceae
　　A　B　C　D　K
　　01　01　09　01　02
OC: Rats H-25, I-4 (504)

Sprekelia formosissima   (Jacobean lily)   Amaryllidaceae
    B   C   D   J   J   J
    06 09 01 03 08 11
OC: Attagenus piceus H-4, I-3 (48)
AP: Alk=I-3 (haemanthamine, haemanthidine, lycorine, tazettine)(1392)

Spyridium filamentosum   (Not known)   Rhamnaceae
    A   B   C   D
    01 02 09 01
OC: Mucor racemosus H-14, I-? (1024)

Stachys niederi   (Nettle genus)   Lamiaceae
    B   C   D
    06 09 01
OC: Drosophila hydei H-2, I-2, 6, 7 (78)

Stachys officinalis   (Common betony)   Lamiaceae
    A   B   C   D
    01 02 09 10
OC: Popillia japonica H-5, I-1 (1243)
AP: Alk=I-? (betonicine, stachydrine, turicine)(1392)

Stachytarpheta jamaicensis   (Devil's coachwhip)   Verbenaceae
    A   A   B   C   D   J   M   M   O
    01 03 06 09 01 03 02 05 07
OC: Drechslera oryzae H-14, I-7 (113)   Vermin H-27, I-? (1353)
    Pyricularia oryzae H-14, I-7 (114)
OR: H-17, I-7 (504)
AP: Alk=I-6, 7 (1365, 1369); Sap=I-6, 7 (1319)

Stachytarpheta mutabilis   (Not known)   Verbenaceae
    B   C   D
    06 09 01
OC: Locusta migratoria H-4, I-7 (706)   Spodoptera littoralis H-4, I-7 (706)
    Schistocerca gregaria H-4, I-7 (706)
AP: Alk=I-6, 7 (1392)

Staphylea pinnata   (European bladdernut)   Staphyleaceae
    A   B   C   D
    01 01 09 03
OC: Cassida nebulosa H-4, I-7 (531)   Phytodecta fornicata H-4, I-7 (531)
    Phyllobius oblongus H-4, I-7 (531)

Stavia pratensis   (Not known)   Lamiaceae
    C
    09
OC: Leptinotarsa decemlineata H-4, I-7 (531)   Pieris brassicae H-4, I-7 (531)

Stavia splendens   (Not known)   Lamiaceae
    C
    09
OC: Leptinotarsa decemlineata H-4, I-7 (531)   Pieris brassicae H-4, I-7 (531)

260

Stavia verticillata　　(Not known)　　Lamiaceae
　　C
　　09
OC: Leptinotarsa decemlineata H-4, I-7 (531)

Stellaria media　　(Chickweed)　　Caryophyllaceae
　　A　　B　　C　　D　　M　　N
　　03　06　09　06　01　07
OC: Athalia rosae H-4, I-7 (531)

Stellera chamaejasme　　(Lang-tu)　　Thymelaeaceae
　　B　　C　　D　　J
　　06　09　03　03
OC: Aphids H-1, I-2 (1241)　　　　　　　　Malacosoma neustria H-2, I-2 (1012)
　　Colaphellus bowringi H-1, I-2 (1241)　　Pieris rapae H-1, I-2 (1241)
　　Flies H-1, I-2 (1241)　　　　　　　　　Puccinia graminis tritici H-14, I-2 (1241)
　　Maggots H-1, I-2 (1241)　　　　　　　Puccinia rubigavera H-14, I-2 (1241)

Stemona burkillii　　(Nohn-tai-yahk)　　Rosaceae
　　B　　C　　D　　J　　L
　　02　09　01　04　03
OC: Maggots H-2, I-2 (84)
OR: H-2, I-2 (504)

Stemona collinsae　　(Nohn-tai-yahk)　　Rosaceae
　　B　　C　　D　　J
　　02　09　01　04
OC: Maggots H-2, I-2 (84)
OR: H-2, I-2 (504)

Stemona curtisil　　(Nohn-tai-yahk)　　Rosaceae
　　B　　C　　D　　J
　　02　09　01　04
OC: Maggots H-2, I-2 (84)
OR: H-2, I-2 (504)

Stemona japonica　　(Not known)　　Stemonaceae
　　B　　C　　D　　J
　　02　09　03　03
OC: Aphids H-1, I-2 (1241)　　　　　　　　Glomerella gossypii H-14, I-2 (1241)
　　Cimex lectularius H-1, I-2 (1241)　　　Maggots H-1, I-2 (1241)
　　Erwinia carotovora H-18, I-2 (1241)　　Rhizoctonia solani H-14, I-2 (1241)
　　Flies H-1, I-2 (1241)　　　　　　　　　Trichodetes sp. H-1, I-2 (1241)

Stemona sessilifolia　　(Not known)　　Stemonaceae
　　A　　B　　C　　D　　J
　　01　02　09　01　03
OC: Aphids H-1, I-2 (1241)　　　　　　　　Glomerella gossypii H-14, I-2 (1241)
　　Cimex lectularius H-1, I-2 (1241)　　　Maggots H-1, I-2 (1241)
　　Erwinia carotovora H-18, I-2 (1241)　　Rhizoctonia solani H-14, I-2 (1241)
　　Flies H-1, I-2 (1241)　　　　　　　　　Trichodetes sp. H-1, I-2 (1241)
OR: H-2, I-2 (504); H-21, I-2 (733)
AP: Alk=I-2 (hodorine, protestemonine, stemonine)(1392)

Stemona tuberosa   (Nohn-tai-yahk)   Stemonaceae

    A   B   C   D   J   J   J   K   M   O
    01 02 09 01 02 03 04 06 01 02

OC: Aphids H-2, I-2 (1241)   Fleas H-2, I-2 (105, 1241)
    Calandra oryzae H-1, I-2 (1031)   Glomerella gossypii H-14, I-2 (1241)
    Caterpillars H-2, I-? (84)   Hypomeces squamosus H-1, I-2 (1031)
    Catopsilia crocale H-1, I-2 (1031) Lepidoptera sp. H-1, I-2 (105)
    Cimex lecturalis H-2, I-2 (1241)   Lice H-2, I-2 (105)
    Crickets H-2, I-2 (105)   Maggots H-2, I-2 (84, 1241)
    Dysdercus cingulatus H-1, I-2 (1031) Malacosoma neustria H-1, I-? (1012)
    Erwinia carotovora H-18, I-2 (1241) Rhizoctonia solani H-14, I-2 (1241)
OR: H-19, I-2 (733)
AP: Alk=I-2 (hypotuberostemonine, isotuberostemonine, oxotuberostemonine, stemonine, tubero-
    stemonine)(1392)

Stenochlaena tenuifolia   (Not known)   Polypodiaceae

    A   B   C   D
    01 10 06 01

OC: Xanthomonas campestris pv. phaseoli-sojensis H-18, I-7 (854)

Sterculia foetida   (Indian almond)   Sterculiaceae

    A   B   C   D   J   M   M   M   M   M   N   N   O
    01 01 09 01 03 01 03 05 06 28 09 10 04

OC: Drechslera oryzae H-14, I-7 (113)
    Musca domestica H-7, I-4, 7, 10, 12 (87, 322)
OR: H-5, 7, I-4, 7 (1345)
AP: Alk=I-10 (1392); Tan=I-9 (1360)

Stereospermum suaveolens   (Not known)   Bignoniaceae

    A   B   C   D
    01 01 09 01

OC: Reticulitermes flavipes H-2, 5, I-5 (87)   Reticulitermes lucifugens H-2, 5, I-5 (87)
OR: H-22, I-2 (881)

Stillingia sylvatica   (Queen's-delight)   Euphorbiaceae

    A   B   C   D   J   K   M   O
    01 01 09 03 02 03 05 02

OC: Fleas H-5, I-1 (1345)

Streblus asper   (Not known)   Moraceae

    A   B   C   D   M   M   M   O   O
    01 01 09 01 05 22 27 04 14

OR: H-2, I-? (100)

Strobilanthes ixocephalus   (Mexican petunia genus)   Acanthaceae

    A   B   C   D
    01 06 09 01

OC: Dysdercus koenigii H-3, I-1, 12 (795)

Strophanthus divaricatus   (Not known)   Apocynaceae

    A   B   C   D   J   K
    01 02 09 03 03 02

OC: Aphids H-1, I-6, 7, 9 (1241)   Rats H-25, I-6, 7, 9 (1241)
    Pacydiplosis oryzae H-1, I-6, 7, 9 (1241) Rice borers H-1, I-6, 7, 9 (1241)

Strophanthus hispidus   (Not known)   Apocynaceae
    B   C   D   J   M   O
    02   09   01   04   05   02
OC: Pediculus humanus capitis H-2, I-10 (87)

Strychnos ignatii   (St. Ignatius bean)   Loganiaceae
    A   B   B   C   D   J   L   M   O
    01   01   03   09   01   04   03   05   04
OC: Vermin H-27, I-10 (97)
OR: H-33, I-2, 4, 10, 14 (103, 504, 849)
AP: Alk=I-2, 4, 5, 6, 7, 10 (1389); Alk=I-10 (brucine, strychnine)(1392)

Strychnos nux-vomica   (Strychnine tree)   Loganiaceae
    A   B   B   C   D   L   M   M   M   M   M   M   N   N   O
    01   01   03   09   01   03   01   04   05   07   16   19   20   07   09   10
OC: Rodents H-25, I-10 (220, 1098)   Termites H-8, I-7 (1352)
OR: H-2, I-? (1098); H-2, 32, I-4 (105, 1276); H-32, I-10 (73, 220); H-33, I-4, 10 (102, 220,
    504, 507, 983)
AP: Alk=I-? (novacine)(1392); Alk=I-2, 6, 7, 10 (1320); Alk=I-7 (strychnicine)(1392); Alk=I-10
    (B-colubrine, brucine, struxine, strychnine)(1392); Sap=I-10 (1320)

Stylosanthes scabra   (Not known)   Fabaceae
    B   C   D   D   J   K   M
    06   09   01   02   01   01   12
OC: Boophilus microplus H-10, 13, I-17 (177)

Stylosanthes viscosa   (Not known)   Fabaceae
    B   C   D   D   J   K   M
    06   09   01   02   01   01   12
OC: Boophilus microplus H-10, 13, I-17 (177)

Swartzia madagascariensis   (Not known)   Caesalpiniaceae
    A   B   B   C   D   J   K   L   M   M
    01   01   02   09   01   04   09   04   13   19
OC: Monilinia fructicola H-14, I-? (991, 1353)   Termites H-5, I-9 (87)
    Stored grain pests H-5, I-9 (87)
OR: H-2, I-9, 10 (87); H-32, I-9, 10 (101, 1353)

Swertia chinensis   (Not known)   Gentianaceae
    B   C   D
    06   09   03
OC: Aphis gossypii H-1, I-1 (1241)   Puccinia graminis tritici H-14, I-1 (1241)
    Phytophthora infestans H-14, I-1 (1241)   Spider mites H-11, I-1 (1241)
    Pieris rapae H-1, I-1 (1241)
AP: Alk=I-6, 7 (1373)

Swertia chirata   (Chirata)   Gentianaceae
    A   A   B   C   D   J
    01   03   06   09   08   08
OC: Popillia japonica H-2, I-? (105); H-5, I-1 (1243)
OR: H-5, I-? (100)

Swietenia mahogani   (West Indies mahogany)   Meliaceae
    A   B   C   D   J   M   O
    01 01 09 01 11 05 04
OC: Cryptotermes brevis H-8, I-5 (936)

Symphoricarpos albus   (Waxberry)   Caprifoliaceae
    A   B   C   D   M
    01 02 09 03 13
OC: Venturia inaequalis H-14, I-7 (16)
OR: H-22, I-? (667)

Symphytum officinale   (Common comfrey)   Boraginaceae
    A   B   C   D   F   G   J   M   M   M   N   N
    01 06 09 10 06 03 03 01 02 21 06 07
OC: Drechslera oryzae H-14, I-7 (113)   Pyricularia oryzae H-14, I-7 (114)

Symplocos celastrinea   (Sweetleaf genus)   Symplocaceae
    A   B   B   C   D   D
    01 01 02 09 01 02
OR: H-4, I-? (132)

Symplocos gardneriana   (Sweetleaf genus)   Symplocaceae
    A   B   C   D   J
    01 02 09 02 09
OC: Rats H-26, I-16 (1393)

Symplocos paniculata   (Asiatic sweetleaf)   Symplocaceae
    A   B   C   D   J   J
    01 01 09 03 03 09
OC: Fusarium nivale H-14, I-? (1295)
OR: H-19, I-6 (881); H-21, I-7 (881, 1040); H-22, I-? (881)

Syngonium auritum   (Five-fingers)   Araceae
    A   B   C   D   J   J
    01 03 09 01 08 11
OC: Aedes egyptii H-2, I-6, 7 (48)   Tineola bisselliella H-4, I-6, 7 (48)
    Attagenus piceus H-4, I-2 (48)

Synsepalum dulcificum   (Miraculous fruit)   Sapotaceae
    B   C   D   M   N
    02 09 01 01 09
OR: H-4, I-? (132)

Syringa oblicata   (Lilac genus)   Oleaceae
    A   B   B   C   J   J
    01 01 02 09 03 11
OC: Attagenus piceus H-4, I-1 (48)

Syringa vulgaris   (Common lilac)   Oleaceae
    A   B   C   D   J   J   J   M   M   O
    01 02 09 06 03 10 16 05 13 08
OC: Leptinotarsa decemlineata H-4, I-7 (531)   Phyllobius oblongus H-4, I-7 (531)
OR: H-19, I-? (592); H-24, I-4 (906)
AP: Alk=I-7 (1373)

Syzygium aromaticum　　(Clove)　　Myrtaceae

　　A　　B　　C　　D　　M　　M　　M
　　01　01　09　01　13　14　17

OC: Attagenus piceus H-1, I-? (105)　　　　　Dacus zonatus H-5, I-? (473)
　　Callosobruchus maculatus H-4, I-? (644)　　Fusicladium effusum H-14, I-12 (352)
　　Chrysomya macellaria H-5, I-? (199)　　　　Lice H-2, I-12 (105)
　　Clothes moths H-5, I-12 (105)　　　　　　　Mealy bugs H-2, I-12 (45)
　　Cochliomyia hominivorax H-5, I-12 (105)　　Pediculus humanus humanus H-2, I-12 (105)
　　Ctenocephalides canis H-2, I-12 (105)　　　Phytophthora parasitica H-14, I-? (911)
OR: H-1, I-6 (80); H-15, 19, I-? (399, 905, 907); H-20, I-3, 12 (504)

Syzygium cumini　　(Black plum)　　Myrtaceae

　　A　　B　　C　　D　　M　　M　　M　　M　　M　　M　　N　　O　　O　　O
　　01　01　09　01　01　02　04　05　06　21　09　04　07　10

OC: Colletotrichum falcatum H-14, I-4 (1047)　　Dacus diversus H-6, I-? (473)
　　Dacus caudatus H-6, I-? (473)　　　　　　　Pyricularia oryzae H-14, I-4 (1047)
AP: Alk=I-? (jambosine)(1392); Alk=I-4, 6, 7 (1363); Sfr=I-7 (1319); Tan=I-4 (1047, 1352);
　　Tan=I-6, 7 (1319)

Syzygium montanum　　(Not known)　　Myrtaceae

　　A　　B　　C　　D
　　01　01　09　01

OC: Rats H-26, I-16 (1393)

Syzygium paniculatum　　(Australian brushcherry)　　Myrtaceae

　　B　　C　　D
　　01　09　01

OC: Tobacco mosiac virus H-21, I-7 (184)

Tabebuia avellanedae　　(Trumpet tree genus)　　Bignoniaceae

　　A　　B　　C　　D　　M
　　01　01　09　01　13

OC: Reticulitermes flavipes H-2, I-5 (87)　　　Reticulitermes lucifugens H-5, I-5 (87)

Tabebuia flavescens　　(Trumpet tree genus)　　Bignoniaceae

　　A　　B　　C　　D　　D　　M
　　01　01　09　01　02　13

OC: Reticulitermes flavipes H-2, I-5 (87)　　　Reticulitermes lucifugens H-5, I-5 (87)

Tabebuia riparia　　(White wood)　　Bignoniaceae

　　A　　B　　B　　C　　M
　　01　01　02　09　13

OR: H-29, I-1 (87)

Tabebuia rosea　　(Pink poui)　　Bignoniaceae

　　A　　B　　C　　D　　J　　J　　M
　　01　01　09　01　13　15　13

OC: Sitophilus oryzae H-2, I-8 (87, 598)

Tabernaemontana dichotoma　　(Red bay genus)　　Apocynaceae

　　A　　C　　D
　　01　09　01

OC: Dysdercus cingulatus H-3, I-? (712)
OR: H-33, I-10 (220)
AP: Alk=I-4 (1392)

Tabernaemontana divaricata   (Flowers-of-love)   Apocynaceae
   A   B   C   D
  01 02 09 01
OC: Helicotylenchus indicus H-16, I-15 (1330)   Tylenchorhynchus brassicae H-16, I-15
   Hoplolaimus indicus H-16, I-15 (1330)   (1330)
   Rotylenchulus reniformis H-16, I-15 (1330)   Tylenchus filiformis H-16, I-15 (1330)
AP: **Alk**=I-4 (coronarine, tabernaemontanine)(1392); **Alk**=I-6, 7 (1392); **Alk**=I-7, 10 (1388);
   Str=I-6 (B-sitosterol)(1381, 1383)

Tabernaemontana pandacaqui   (Red bay genus)   Apocynaceae
   A   B   C   D   J
  01 01 09 01 03
OC: Pyricularia oryzae H-14, I-7 (114)
OR: H-18, I-7 (982)
AP: **Alk**=I-6, 7 (1392)

Tacca pinnatifida   (Polynesian arrowroot)   Taccaceae
   B   E   M   M   N   O
  06 07 1a 05 02 02
OC: Stored grain pests H-5, I-7 (533)
AP: **Alk**=I-2 (1363); **Alk**=I-3, 7, 10 (1392)

Tagetes erecta   (African marigold)   Asteraceae
  A B B C D D F G J J J K K K M M M M O O
  03 02 06 09 01 02 06 03 01 03 04 01 03 06 05 13 16 27 02 07
OC: Aphis craccivora H-2a, I-2 (1005)   Plutella xylostella H-2a, I-2 (1005)
   Drechslera oryzae H-14, I-7, 8 (113, 432)   Pratylenchus penetrans H-16, I-2 (924)
   Dysdercus cingulatus H-3, I-? (88, 441)   Pratylenchus sp. H-16, I-2 (924)
   Helicotylenchus indicus H-16, I-2 (793)   Pratylenchus zea H-16, I-? (384)
   Helicotylenchus sp. H-16, I-2 (924)   Pyricularia oryzae H-14, I-8 (114, 432)
   Hoplolaimus indicus H-16, I-2 (538, 793)   Radopholus similis H-16, I-2 (949)
   Hoplolaimus sp. H-16, I-2 (924)   Rotylenchulus reniformis H-16, I-2
   Meloidogyne arenaria H-16, I-2 (796)   (538, 793)
   Meloidogyne incognita H-16, I-2, 7   Rotylenchulus sp. H-16, I-2 (924)
     (115, 147, 351, 542, 793, 1005)   Tylenchorhynchus brassicae H-16, I-2
   Meloidogyne javanica H-16, I-2 (796)   (538, 793)
   Musca domestica H-2a, I-2 (1005)   Tylenchorhynchus sp. H-16, I-2 (924)
   Nephotettix virescens H-2a, I-2 (1005)   Tylenchus filiformis H-16, I-2 (538, 793)
   Ostrinia furnacalis H-2a, I-2 (1005)   Uramyces phaseoli H-14, I-2 (147)
OR: H-17, I-7 (504, 1320); H-18, I-2 (132); H-20, I-19 (1148); H-24, I-? (1005)
AP: **Alk**=I-9, 10 (1391); **Sfr**=I-2 (1320)

Tagetes lucida   (Sweet-scented marigold)   Asteraceae
   A   B   C   D
  01 06 09 01
OC: Nematodes H-16, I-? (100)

Tagetes minuta   (Mexican marigold)   Asteraceae
   A   B   C   J   J   J   J   M   O
  03 02 09 03 04 09 17 05 07
OC: Aedes aegypti H-1, I-7, 8 (94, 850)   Blow flies H-5, I-12 (105)
   Ants H-1, I-? (94)   Ceratitis capitata H-6, I-6, 7, 8 (956)
   Belonolaimus longicaudatus H-16, I-? (955)   Criconemoides ornatum H-16, I-? (955)

Dacus cucurbitae H-6, I-6, 7, 8 (956)   Musca domestica H-1, I-1 (785)
Dacus dorsalis H-6, I-6, 7, 8 (956)   Nematodes H-16, I-? (100)
Dysdercus koenigii H-3, I-1, 12 (787)   Oncopeltus fasciatus H-1, I-? (958)
Maggots H-1, I-7 (87)   Tribolium castaneum H-1, I-1 (785)
Meloidogyne arenaris H-16, I-? (955)   Trichodorus christiei H-16, I-? (955)
Meloidogyne incognita H-16, I-? (955) Xiphinema americanum H-16, I-? (955)
Meloidogyne javanica H-16, I-? (955)

OR: H-2, I-? (132); H-5, I-7, 12 (504, 507); H-16, I-? (100)
AP: Alk=I-4, 6 (1392)

Tagetes nana   (Marigold genus)   Asteraceae
B   C   D
06 09 03
OC: Anguina tritici H-16, I-? (868)   Heterodera rostochiensis H-16, I-? (868)
Ditylenchus dipsaci H-16, I-? (868)   Pratylenchus penetrans H-16, I-? (868)

Tagetes patula   (French marigold)   Asteraceae
A   B   B   C   D   D   J   K   K   M
03 02 06 09 01 02 09 01 06 13
OC: Aphis craccivora H-5, I-2 (1005)   Ostrinia furnacalis H-1, I-2 (1005)
Epilachna varivestis H-2, I-2 (474)   Pieris rapae H-1, I-? (474)
Meloidogyne incognita H-16, I-2   Plutella xylostella H-1, I-2 (90, 1005)
  (529, 999, 1005)   Pratylenchus alleni H-16, I-? (999)
Musca domestica H-1, I-2 (1005)   Pratylenchus penetrans H-16, I-? (963)
Nematodes H-16, I-? (100)   Rotylenchulus sp. H-16, I-? (529)
Nephotettix virescens H-1, I-2 (1005) Tylenchorhynchus claytoni H-16, I-? (963)
Nilaparvata lugens H-1, I-? (90)   Uramyces phaseoli H-14, I-? (147)

Tagetes tenuifolia   (Signet marigold)   Asteraceae
A   B   C   D
03 06 09 01
OC: Nematodes H-16, I-? (100)

Tamarindus indica   (Indian tamarind)   Caesalpiniaceae
A B C D J J J M M M M M M M N N O O O O
01 01 09 01 02 03 09 01 04 05 12 13 16 19 21 09 11 04 07 10 11
OC: Dysdercus cingulatus H-3, I-? (712)   Ustilago tritici H-14, I-7 (181)
Meloidogyne incognita H-16, I-? (542) Xanthomonas campestris
Ustilago hordei H-14, I-7 (181)   pv. campestris H-18, I-11 (1357)
OR: H-1, 8, I-5 (1353); H-15, 19, I-9 (411); H-17, I-7 (1319); H-19, I-9, 14 (172); H-21,
I-8 (881)
AP: Fla=I-4, 6 (1379); Tan=I-4, 6, 7 (1319, 1352)

Tamus communis   (Yam)   Dioscoreaceae
B   C   D   J   M   N   N
04 09 06 04 01 07 15
OC: Pediculus humanus capitis H-2, I-2 (104)   Venturia inaequalis H-14, I-? (16)
AP: Alk=I-2 (1392)

Tanacetum vulgare   (Tansy)   Asteraceae
A   B   C   D   J   J   J   M   M   O
01 03 09 03 02 03 06 05 17 07
OC: Ants H-5, I-12 (105)
Aphids H-5, I-? (74, 75)   Musca domestica H-2, I-? (87)
                           Pieris rapae H-6, I-? (75)

Tanghinia venenifera   (Not known)   Apocynaceae
```
    A   B   C   D   J   J   J   L
    01  01  09  01  03  08  11  03
```
OC: Attagenus piceus H-4, I-10 (48)
OR: H-33, I-? (504)
AP: Alk=I-? (tanghinine)(1392)

Taraxacum officinale   (Dandelion)   Asteraceae
```
    A   B   C   D   L   M   M   N   N   O
    01  06  09  03  05  01  05  07  08  02
```
OC: Cassida nebulosa H-4, I-7 (531)   Locusta migratoria H-4, 5, I-7 (596)
    Leptinotarsa decemlineata H-5, I-? (87)   Phytodecta fornicata H-4, I-7 (531)

Taxus baccata   (English yew)   Taxaceae
```
    A   B   C   D   D   L   M   M   N
    01  01  09  03  06  03  01  19  09
```
OC: Beetles H-2a, I-7 (1025)   Lymantria dispar H-1a, I-7 (1025)
    Flies H-2a, I-7 (1025)   Lymantria monacha H-1a, I-7 (1025)
    Livestock pests H-2a, I-7 (504)   Malacosoma neustria H-1a, I-7 (1025)
OR: H-1, I-7, 9, 15 (73); H-15, I-? (905); H-32, I-7, 9 (220); H-33, I-7 (220, 507)
AP: Alk=I-6, 7, 9 (ephidrine, taxine)(1392); Fla=I-4 (1378)

Taxus brevifolia   (Western yew)   Taxaceae
```
    A   B   C   D   M
    01  01  09  03  04
```
OC: Pyrrhocoris apterus H-3, I-5 (684)

Taxus canadensis   (American yew)   Taxaceae
```
    A   B   C   D   L   M   M   N
    01  02  09  03  03  01  13  09
```
OC: Erwinia carotovora H-18, I-7, 8 (585)   Leptinotarsa decemlineata H-4, I-7, 8 (585)
OR: H-19, I-9 (413); H-33, I-10 (504)
AP: Alk=I-7 (taxinine)(1392); Alk=I-10 (1392)

Taxus cuspidata   (Japanese yew)   Taxaceae
```
    A   B   C   D   M   M   M
    01  01  09  03  04  16  19
```
OC: Chilo suppressalis H-1, I-? (1003)   Pieris rapae H-1, I-? (1003)
    Culex pipiens molestus H-1, I-? (1003)   Plutella xylostella H-1, I-? (1003)
    Musca domestica H-1, I-? (1003)
AP: Alk=I-7 (1392)

Tecoma indica   (Yellowbells genus)   Bignoniaceae
```
    A   B   B   C   J
    01  01  02  09  15
```
OC: Sitophilus oryzae H-2, I-8 (87, 400, 1124)

Tectaria cicutaria   (Button fern)   Polypodiaceae
```
    A   C   D
    01  06  01
```
OC: Pseudomonas solanacearum H-18, I-7 (854)   Xanthomonas campestris
    Rats H-26, I-16 (1393)   pv. phaseoli-sojensis H-18, I-7 (854)

Tectaria incisa   (Button fern genus)   Polypodiaceae
    A   C   D
    01 06 01
OC: Pseudomonas solanacearum H-18, I-7 (854)
    Xanthomonas campestris pv. phaseoli-sojensis H-18, I-7 (854)

Tectona grandis   (Teak)   Verbenaceae
    A   B   C   D   J   J   M   M   M   M   M   M   M   O   O
    01 01 09 01 08 10 04 05 13 16 19 20 21 05 07
OC: Cryptotermes brevis H-1, I-5 (741)   Reticulitermes flavipes H-1, I-5 (741)
    Dysdercus cingulatus H-3, I-6 (471, 690)   Termites H-5, I-5 (105)
    Maggots H-5, I-5 (105)
OR: H-19, I-7 (982)
AP: Alk=I-6 (1360); Alk=I-9 (1374); Sap, Tan, Ter=I-4 (1359); Tan=I-7 (1350)

Tephrosia candida   (White tephrosia)   Fabaceae
    A   B   C   D   J   M   M   M
    01 02 09 01 08 07 10 12
OC: Aphids H-2, I-10 (605)   Euproctis fraterna H-2, I-10 (605)
    Aphis fabae H-2, I-2, 6, 10 (105)   Pericallia ricini H-2, I-10 (605)
    Coccus viridis H-1, I-10 (1029)   Plutella xylostella H-2, I-10 (605)
    Crocidolomia binotalis H-2, I-10 (605)   Toxoptera aurantii H-1, I-? (681)
    Epacromia tamulus H-2, I-10 (605)
    Epilachna viginti-octopunctata H-2, I-10 (605)
OR: H-2, I-4, 7 (73); H-2, 32, I-? (220); H-32, I-2, 7, 10, 11 (459, 983)
AP: Alk=I-6, 7 (1392)

Tephrosia diffusa   (Not known)   Fabaceae
    B   C   D   D   M   M   N
    06 09 01 02 05 12 02
OC: Pediculus humanus capitis H-2, I-2 (105)

Tephrosia grandiflora   (Not known)   Fabaceae
    A   B   C   D   M
    01 02 09 02 12
OR: H-2, I-2 (105)

Tephrosia heckmannia   (Not known)   Fabaceae
    B   C   D   D   M
    06 09 01 02 12
OC: Cimex lectularius H-2, I-7 (105)   Papaipema nebris H-2, I-7 (105)

Tephrosia latidens   (Not known)   Fabaceae
    B   C   D   D   M
    06 09 01 02 12
OC: Musca domestica H-2, I-2 (105, 1087)

Tephrosia lindheimeri   (Not known)   Fabeceae
    B   C   D   D   M
    06 09 01 02 12
OC: Musca domestica H-2, I-10 (105)
AP: Alk=I-6, 7, 8, 9 (1373)

Tephrosia macropoda   (Hoary pea genus)   Fabaceae
   B   C   D   D   J   M
   06 09 01 02 08 12
OC: Aphids H-2, I-2, 6 (1124)   Orgyia antiqua H-1, I-? (676)
    Aphis fabae H-1, I-2, 6 (34, 676)   Pediculus humanus capitis H-1, I-? (105)
    Cheimatobia brumata H-1, I-? (44)   Selenia tetralunaria H-2, I-2 (34, 676)
    Moths H-2, I-2, 6 (1124)
AP: Alk=I-6, 7, 9 (1392)

Tephrosia nyikensis   (Not known)   Fabaceae
   B   C   D   D   M
   06 09 01 02 12
OC: Toxoptera aurantii H-1, I-? (681)
AP: Alk=I-6, 7, 8, 9 (1373)

Tephrosia purpurea   (Purple tephrosia)   Fabaceae
   A   B   C   D   D   J   J   L   M   M   M   O
   01 01 09 01 02 03 07 05 05 09 10 12 02
OC: Caterpillars H-2, I-? (89)   Spodoptera litura H-5, I-7 (126)
OR: H-2, 32, I-? (220); H-2a, I-2, 7 (118); H-5, I-2, 7 (73, 124, 126); H-24, I-1 (1212);
    H-32, I-2, 10 (459, 504)
AP: Alk=I-6, 7 (1392)

Tephrosia repentina   (Dan-ratchasi)   Fabaceae
   B   C   D   D   M
   06 09 01 02 12
OR: H-2, I-? (80)

Tephrosia toxicaria   (Fishdeath tephrosia)   Fabaceae
   B   C   J   J   J   M
   06 09 03 08 11 12
OC: Aphids H-2, I-? (38)   Oncopeltus fasciatus H-2a, I-2 (48)
    Attagenus piceus H-4, I-2 (48)   Tineola bisselliella H-4, I-2 (48)
    Cheimatobia brumata H-2, I-? (44)   Toxoptera aurantii H-1, I-? (681)
    Musca domestica H-1, I-2 (238)
OR: H-32, I-2 (504)

Tephrosia villosa   (Not known)   Fabaceae
   A   B   C   D   D   F   J   K   L   M
   03 06 09 01 02 06 11 06 05 12
OC: Callosobruchus chinensis H-2, I-10 (605)   Dysdercus koenigii H-2, I-10 (91)
    Coccus viridis H-1, I-7, 10 (1029)   Lipaphis erysimi H-2, I-10 (91)
    Crocidolomia binotalis H-2, I-10   Spodoptera litura H-2, I-10 (605, 1029);
     (605, 1029)   H-5, I-2 (126)
    Dactynotus carthani H-2, I-10 (91)

Tephrosia virginiana   (Devil's shoestring)   Fabaceae
   A   B   C   D   D   E   J   M   M   O
   01 06 09 03 04 07 03 05 12 02
OC: Aphis gossypii H-2, I-2 (57, 208)   Culex quinquefasciatus H-1, I-2 (463)
    Brevicoryne brassicae H-2, I-2 (208, 263) Datana ministra H-2, I-2 (57, 208)
    Busseola fusca H-2, I-2 (208, 256)   Echidnophaga gallinaceae H-2, I-2 (208)
    Ctenocephalus felis H-2, I-2 (208)   Epilachna varivestis H-1, I-7 (284)

Flies H-1, I-? (1087)
Goniocotes gigas H-2, I-2 (208)
Goniocotes hologaster H-2, I-2 (208)
Holopsyllus affinis H-2, I-2 (208)
Hypoderma lineata H-2, I-2 (208)
Leptinotarsa decemlineata H-2, I-2
    (57, 208)
Lipaphis erysimi H-2, I-2 (57, 208)

Malacosoma americana H-2, I-2 (57, 208)
Musca domestica H-2, I-2 (208, 261)
Myzus rosarum H-1, I-? (263)
Phthirus pubis H-2, I-2 (208)
Pulex irritans H-2, I-2 (208)
Rhopalosiphum rufomaculata H-1, I-? (263)
Trichodectus canis H-2, I-2 (208)

OR: H-17, 32, I-2 (504)
AP: **Alk**=I-6, 7 (1392); **Alk**=I-6, 8 (1368)

Tephrosia vogelii   (Vogel tephrosia)   Fabaceae
   A   B   C   D   F   G   J   K   L   M   M   M   M
   01 02 09 01 06 04 03 06 03 09 10 12 13
OC: Aphis citri H-2, I-7 (157)
   Aphis fabae H-1, I-7 (105)
   Cheimatobia brumata H-2, I-? (44)
   Crocidolomia binotalis H-2, I-7 (157)
   Epilachna varivestis H-1, I-? (284)
   Henosepilachna sparsa H-2, I-7 (157)
   Orgyia antiqua H-2, I-? (43)

   Pediculus humanus capitis H-2, I-7 (1353)
   Plutella xylostella H-2, I-7 (157)
   Rats H-25, I-7 (1353)
   Selenia tetralunaria H-2, I-? (44)
   Thrips H-2, I-7, 11 (1353)
   Toxoptera aurantii H-2, I-7, 10, 11 (681)

OR: H-2, I-? (38, 39, 264, 404); H-32, I-? (405); H-32, 33, I-1, 7, 11 (192, 1353)
AP: **Alk**=I-6 (1373)

Tephrosia wallichii   (Cha-kram)   Fabaceae
   B   C   D   D   M
   06 09 01 02 12
OR: H-2, I-? (80)

Terminalia arjuna   (Arjuna tree)   Combretaceae
   A   B   C   D   M   M   O
   01 01 09 01 04 05 04
OC: Colletotrichum flacatum H-14, I-4 (1047)   Rats H-26, I-16 (1393)
   Pyricularia oryzae H-14, I-4 (1047)
OR: H-19, I-? (411)
AP: **Tan**=I-4, 9 (1047, 1352, 1360); **Tri**=I-4 (friedelin)(1381)

Terminalia bellirica   (Myrobalan)   Combretaceae
   A   B   C   D   J   M   M   M   N
   01 01 09 01 03 01 16 21 09
OC: Alternaria tenius H-14, I-? (1295)   Fusarium nivale H-14, I-? (1295)
   Curvularia lunata H-14, I-? (1295)   Pyricularia oryzae H-14, I-9 (1047)
   Drechslera graminea H-14, I-? (1295)
OR: H-15, 19, I-? (411); H-22, I-9 (881); H-32, I-9 (73); H-32, 33, I-10 (220)
AP: **Alk, Tan**=I-9 (1360); **Sap, Ter**=I-7 (1359); **Tan**=I-9 (507, 1352)

Terminalia catappa   (Indian almond)   Combretaceae
   A   B   C   D   M   M   M   M   M   N   N   O   O
   01 01 09 01 01 04 05 16 21 09 10 04 07
OC: Mites H-11, I-7 (105)   Popillia japonica H-5, I-? (105, 1243)
OR: H-17, 20, I-7 (1318); H-19, I-6, 7 (172, 1076)
AP: **Glu**=I-6, 7 (1318); **Sap, Tan**=I-6, 7 (1318); **Tan**=I-2, 4, 7, 9 (504, 507, 1352)

Terminalia chebula   (Myrobalan)   Combretaceae
```
   A   B   C   D   M
   01  01  09  01  04
```
OC: Pyricularia oryzae H-14, I-? (1047)
AP: Tan=I-9 (507)

Terminalia paniculata   (Tropical almond genus)   Combretaceae
```
   A   B   C   D   J
   01  01  09  01  10
```
OC: Dysdercus cingulatus H-3, I-6 (690)
AP: Sap, Tan, Ter=I-7 (1359)

Tetragonia tetragonioides   (New Zealand spinach)   Tetragoniaceae
```
   A   B   C   D   J   J    M   M   N   O
   03  06  09  10  03  3a   01  05  07  07
```
OC: Drosophila hydei H-2, I-2, 6 (101)   Tobacco ring spot virus H-21, I-14 (1071)
AP: Alk=I-2, 6, 7, 9 (1392); Alk=I-6, 7 (1372); Alk, Sap=I-7 (1365)

Tetrapterys acutifolia   (Not known)   Malpighiaceae
```
   B   C   D   K
   04  09  01  05
```
OC: Flies H-2, I-? (101)   Mosquitoes H-2, I-? (101)

Teucrium canadense   (American germander)   Lamiaceae
```
   B   C   D
   06  09  10
```
OC: Botrytis cinerea H-14, I-1, 8, 9 (945)

Thalictrum polygamum   (King-of-the-meadow)   Ranunculaceae
```
   A   B   C   D   J
   01  06  09  03  09
```
OC: Botrytis cinerea H-14, I-1 (944)
AP: Alk=I-2, 7, 8 (1392)

Thamnosma africana   (Not known)   Rutaceae
```
   A   B   C   D
   01  06  09  02
```
OC: Ants H-2, I-1 (87)   Fleas H-2, I-1 (87)

Thamnosma montana   (Turpentine broom)   Rutaceae
```
   B   D
   02  04
```
OR: H-24, I-2, 7 (554, 556)

Thelypteris dentata   (Downy wood fern)   Polypodiaceae
```
   A   C   D
   01  06  01
```
OC: Pseudomonas solanacearum H-18, I-7 (854)
   Xanthomonas campestris pv. phaseoli-sojensis H-18, I-7 (854)
AP: Str, Tri=I-? (1376)

Thelypteris setigera   (Beech fern genus)   Polypodiaceae
```
   A   C   D
   01  06  01
```

OC: Erwinia carotovora H-18, I-7 (854)
   Xanthomonas campestris pv. phaseoli-sojensis H-18, I-7 (854)

Theobroma cacao   (Cocoa)   Byttneriaceae
   A   B   C   D   M   M   M   N   O   O
   01 01 09 01 01 05 21 10 02 10
OC: Fleas H-5, I-10 (105)
AP: **Alk**=I-2, 6, 7 (caffeine, theobromine)(1392); **Alk, Sap, Tan**=I-6, 7 (1318)

Thespesia populnea   (Portia tree)   Malvaceae
   A   B   C   D   J   M   M   M   M   O
   01 01 09 01 09 03 04 05 06 19 07
OC: Rats H-26, I-? (697)
OR: H-22, I-9 (881); H-32, I-? (983)

Thevetia gaumeri   (Not known)   Apocynaceae
   A   B   B   C   D   E   J   J   J
   01 01 02 09 01 07 03 08 11
OC: Attagenus piceus H-4, I-2, 6, 7 (48)

Thevetia ovata   (Not known)   Apocynaceae
   A   B   B   C   D   E   J   J   J
   01 01 02 09 01 07 03 08 11
OC: Attagenus piceus H-4, I-9 (48)

Thevetia peruviana   (Yellow oleander)   Apocynaceae
   A   B   C   D   E   J   J   K   L   M   M   M   O   O
   01 01 09 01 07 03 11 06 03 05 09 13 04 14
OC: Aphids H-2a, I-10 (101, 871)   Diacrisia obliqua H-3, I-? (161)
   Aphis craccivora H-6, I-? (90)   Mealy bugs H-1, I-10 (871)
   Attagenus piceus H-4, I-7, 9 (48)   Tobacco mosaic virus H-21, I-? (1126)
   Callosobruchus chinensis H-5, I-? (599)
OR: H-17, I-4 (1353); H-19, I-7 (776); H-32, I-? (73); H-32, I-4, 7 (983); H-32, 33, I-4,
   10, 14 (220, 1350); H-33, I-2, 14 (1353)
AP: **Alk**=I-6, 7 (1392); **Glu**=I-10 (thevetin)(1353)

Thuja occidentalis   (White cedar)   Cupressaceae
   A   B   C   D   J   M   O
   01 01 08 03 04 05 12
OC: Musca domestica H-2, I-7 (101)
OR: H-20, I-12 (504)
AP: **Alk**=I-6, 7 (berberine)(1392)

Thuja plicata   (Giant cedar)   Cupressaceae
   A   B   C   D   J   J   M
   01 01 08 03 08 09 14
OC: Fomes annosus H-14, I-5 (770, 954)   Lenzites saepiaria H-14, I-5 (954)
   Lentinus lepideus H-14, I-5 (770, 887, 954) Pissodes strobi H-5, I-6 (1020)
OR: H-5, I-5 (101); H-15, I-5 (658); H-19, I-5 (658, 665, 737, 986)

Thymus pannonicus   (Thyme)   Lamiaceae
   A   B   C   D   J   M   M
   01 06 09 03 04 14 17
OC: Dermacentor marginatus H-13, I-7 (87, 1133) Ixodes redikorzevi H-13, I-7 (87, 1133)
   Haemaphysalis punctata H-13, I-7 (87, 1133) Rhipicephalus rossicus H-13, I-7 (87, 1133)

Thymus vulgaris (Thyme) Lamiaceae

A B C D J M
01 02 09 06 06 14

OC: Alternaria tenuis H-14, I-12 (426) Gibberella fujikuroi H-14, I-12 (426)
Botrytis allii H-14, I-12 (426) Lentinus lapideus H-14, I-12 (428)
Ceratocystis ulmi H-14, I-12 (426) Lenzites trabea H-14, I-12 (428)
Cladosporium fulvum H-14, I-12 (426) Pieris brassicae H-5, I-? (1002)
Claviceps purpurea H-14, I-12 (426) Pieris napi H-5, I-? (1002)
Diplodia maydis H-14, I-12 (426) Pieris rapae H-5, I-? (1002)
Fusarium oxysporum Polyporus versicolor H-14, I-12 (428)
  f. sp. conglutinans H-14, I-12 (426) Ustilago avenae H-14, I-12 (426)
  f. sp. lycopersici H-14, I-12 (426) Verticillium albo-atrum H-14, I-12 (426)
Fusicladium effusum H-14, I-12 (352)
OR: H-15, I-12 (399)

Tilia cordata (Small-leaved European linden) Tiliaceae

A B C D M M M M M O
01 01 09 03 03 05 06 18 19 08
OC: Phytodecta fornicata H-4, I-7 (531)

Tinospora rumphii (Makabuhay) Menispermaceae

B C D J J M O
04 09 01 03 08 05 07
OC: Musca domestica H-2, I-? (90) Plutella xylostella H-2, I-? (90)
OR: H-33, I-1 (983)
AP: Alk=I-6, 7 (berberine)(1392)

Tinospora tuberculata (Galancha tinospora) Menispermaceae

A B C D J M O O O
01 03 09 03 09 05 02 06 07
OR: H-1, I-? (871); H-19, I-12 (449); H-21, I-1 (881)
AP: Alk=I-2, 6 (1360)

Tithonia diversifolia (Wild sunflower) Asteraceae

A B C D F G K L M
01 02 09 01 06 04 06 05 05
OC: Dysdercus cingulatus H-2a, I-7 Sitophilus zeamais H-2a, I-7 (125)
   (90, 125, 699) Spodoptera exempta H-2a, I-7 (125, 699)
Musca domestica H-2, I-2 (90) Tribolium castaneum H-2a, I-7 (125)
Plutella xylostella H-2, I-7 (90, 125, 699)
OR: H-2, I-7 (117)

Tortula ruralis (True moss) Bryales (order)

A C
04 04
OC: Agropyron spicatum H-24, I-7 (1174) Stipa thurberiana H-24, I-7 (1174)
Sitanion hystrix H-24, I-7 (1174)

Tournefortia hirsutissima (Not known) Boraginaceae

B C D D J
02 09 01 02 03
OC: Attagenus piceus H-4, I-2 (48)
OR: H-4, I-? (96, 105); H-5, I-? (1276)

Tournefortia volubilis   (Not known)   Boraginaceae
    B   C   D   D   J   J   J   J
    02 09 01 02 03 04 08 11
OC: Attagenus piceus H-4, I-1 (48)   Ticks H-13, I-2 (105)
AP: Alk=I-6, 7, 9 (1370)

Trachyspermum ammi   (Not known)   Apiaceae
    B   C   D
    06 09 01
OC: Alternaria brassicae H-14, I-12 (417)   Curvularia lunata H-14, I-12 (383)
    Alternaria solani H-14, I-12 (417)   Drechslera oryzae H-14, I-12 (383, 417)
    Alternaria tenuis H-14, I-12 (417)   Fusarium moniliforme H-14, I-12 (417)
    Alternaria tenuissima H-14, I-12 (383)   Fusarium poae H-14, I-12 (383)
    Colletotrichum falcatum H-14, I-12 (417) Fusarium solani H-14, I-12 (417)
    Colletotrichum lindemuthianum H-14, I-12 Rhizoctonia solani H-14, I-12 (383, 417)
      (417)

Tradescantia virginiana   (Common spiderwort)   Commelinaceae
    A   B   C   D   J   M   N
    01 06 09 03 09 01 16
OC: Pieris brassicae H-4, I-7 (531)
OR: H-22, I-1 (946)

Tribulus terrestris   (Puncture vine)   Zygophyllaceae
    A   A   C   D   D   K   M   M   O
    01 03 09 01 02 06 05 09 09
OC: Dysdercus cingulatus H-2a, I-8 (124)   Spodoptera litura H-2, 4, I-? (88, 126)
OR: H-2, I-? (220)
AP: Alk=I-6, 7, 8 (1392)

Trichilia cuneata   (Not known)   Meliaceae
    A   B   C   D
    01 01 09 01
OC: Mites H-11, I-7, 9 (105)

Trichilia roka   (Not known)   Meliaceae
    A   B   C   D   J   J
    01 01 09 01 07 11
OC: Choristoneura fumiferana H-2, I-? (94)   Spodoptera eridania H-2, I-2, 4 (94, 1036)
    Epilachna varivestis H-2, I-2, 4 (94, 1036)

Trichilia trifoliata   (Not known)   Meliaceae
    A   B   C   D
    01 01 09 01
OC: Vermin H-27, I-9 (105)
OR: H-32, I-4 (73)

Trichodesma sedgevickeanum   (Not known)   Boraginaceae
    C   J   J
    09 02 03
OR: H-24, I-1 (1212)

Trichosanthes anguina   (Serpent cucumber)   Cucurbitaceae
    A   B   C   D   J   M   N
    03 04 09 01 03 01 09

OC: Meloidogyne incognita H-16, I-10 (542)
AP: Str=I-1 (1382)

Tridax procumbens   (Coat buttons)   Asteraceae
    A   B   C   D   G   J   J   J   J   K   K
    03 06 09 01 04 03 11 13 15 04 06
OC: Fusarium nivale H-14, I-1, 14 (1048)   Sitophilus zeamais H-1, I-? (125)
    Oncopeltus fasciatus H-2a, I-8 (48)   Tribolium castaneum H-1, I-? (125)
    Sitophilus oryzae H-2, I-8 (87, 598)
OR: H-24, I-? (559)
AP: Alk=I-6, 7 (1392)

Trifolium pratense   (Red clover)   Fabaceae
    A   B   C   D   J   M   M   M   M   O
    01 06 09 03 12 02 05 12 16 08
OC: Athalia rosae H-4, I-7 (531)   Phyllobius oblongus H-4, I-7 (531)
    Cassida nebulosa H-4, I-7 (531)   Phytodecta fornicata H-4, I-7 (531)
    Fusarium nivale H-14, I-6, 7 (666)   Pieris brassicae H-4, I-7 (531)
    Leptinotarsa decemlineata H-4, I-7 (531)   Potato virus X H-21, I-14 (1136)
    Meloidogyne sp. H-16, I-20 (424)   Red clover vein mosaic virus H-21, I-? (1128)
    Monilinia fructicola H-14, I-? (991)   Sclerotinia trifoliorum H-14, I-6, 7 (666)
OR: H-24, I-6, 7 (1250)

Trifolium procumbens   (Irish shamrock)   Fabaceae
    B   C   D   M
    06 09 03 12
OC: Botrytis cinerea H-14, I-1 (946)

Trifolium repens   (White clover)   Fabaceae
    B   C   D   L   M   M   M   O
    06 09 03 02 05 12 18 08
OC: Delia brassicae H-1, I-? (498)   Manduca sexta H-4, I-2 (642)
    Heliothis virescens H-5, I-? (1240)   Panonychus citri H-4, 11, I-2 (642)
    Heteronychus arator H-4, I-2 (642)   Pratylenchus zea H-16, I-? (1274)

Triglochin maritima   (Shore podgrass)   Juncaginaceae
    B   C   D   E   J   M   N
    06 09 03 11 03 01 07
OC: Drosophila melanogaster H-1, I-6, 7 (165)

Trigonella foenum-graecum   (Fenugreek)   Fabaceae
    A   B   C   D   D   D   J   J   J   M   M   M   M   M   N   N   O
    03 06 09 01 04 06 04 09 12 01 02 05 12 14 16 07 10 10
OC: Rats H-26, I-10 (1393)   Stored grain pests H-5, I-12 (87, 1391)
    Sitophilus granarius H-5, I-7 (113)
OR: H-5, I-12 (73)
AP: Alk=I-10 (trigonelline)(1353, 1391, 1392)

Trilisa odoratissima   (Deer's tongue)   Asteraceae
    C   D   M
    09 03 17
OC: Clothes moths H-5, I-7 (104)

Trillium erectum (Squaw root) Liliaceae

  A B C D D J J J J

  01 06 09 03 08 02 03 08 11

OC: Attagenus piceus H-5, I-2 (48)   Popillia japonica H-4, I-2 (105, 1243)

Trillium grandiflorum (White wake-robin) Liliaceae

  A B C D

  01 06 09 03

OC: Colletotrichum lindemuthianum H-14, I-?   Endothia parasitica H-14, I-? (957)
  (957)   Monilinia fructicola H-14, I-? (957)

Trillium sessile var. giganteum (Wake-robin) Liliaceae

  A B C D J J

  01 06 09 03 03 11

OC: Attagenus piceus H-4, I-2 (48)

Tripleurospermum maritimum (Scentless false chamomile) Asteraceae

  A B C D J

  01 06 09 03 04

OC: Musca autumnalis H-2, I-8 (101)

Tripterygium forrestii (Three-winged nut) Celastraceae

  A B C D J J

  01 03 09 03 04 08

OC: Acraea issoria H-1, I-2 (1012)   Laspeyresia pomonella H-2, I-2 (29, 101)
  Aphis fabae H-4, I-2 (101)   Oides decimpunctata H-1, I-2 (1012)
  Archips cerasivoranus H-4, I-2 (101) Oregma lanigera H-1, I-2 (101, 1012)
  Aulacophora cattigarensis H-1, I-2 (1012) Paleacrita vernata H-2, I-2 (29, 101)
  Bean plataspids H-4, I-2 (101)   Phaedon brassicae H-2, I-2 (101)
  Colaphellus bowringi H-1, I-2 (289, 1012); Phyllotreta vittata H-2, I-2 (101, 289)
    H-4, I-2 (101)   Pseudaletia unipuncta H-2, I-2 (101)
  Coptosoma cribraria var. punctatissima   Rhaphidopalpa chinensis H-1, I-2 (289)
    H-1, I-2 (1012)   Spodoptera litura H-2, I-2 (101)
  Epilachna varivestis H-4, I-2 (101)   Trichoplusia ni H-4, I-2 (101)
  Hymenia recurvalis H-2, I-2 (101)

Tripterygium hypoglaucum (Not known) Celastraceae

  A B C D

  01 02 09 03

OC: Caterpillars H-1, I-2, 7 (1241)   Maggots H-1, I-2, 7 (1241)
  Colaphellus bowringi H-1, I-2, 7 (1241) Oulema oryzae H-1, I-2, 7 (1241)
  Euproctis pseudoconspersa H-1, I-2, 7,9 Pieris rapae H-1, I-2, 7 (1241)
    (1241)

Tripterygium wilfordii (Thunder-god vine) Celastraceae

  A B C D F J K

  01 03 09 03 09 04 05

OC: Anthonomus eugenii H-2, I-1 (87)   Gastrophysa cyanea H-2, I-2 (87)
  Bombyx mori H-4, I-2 (46)   Hymenia recurvalis H-2, I-2 (87)
  Caterpillars H-2a, I-2 (97)   Laspeyresia pomonella H-2, I-2
  Colaphellus bowringi H-2, I-2 (105)   (46, 101)
  Epilachna varivestis H-2, I-2 (101)   Leptinotarsa decemlineata H-4, I-2 (46, 101)
  Evergestis rimosalis H-2, I-2 (101)   Malacosoma americana H-4, I-2 (46)

Ostrinia nubilalis H-1, I-2 (101)   Pseudaletia unipuncta H-2, I-2 (87)
Periplaneta americana H-1, I-2 (101)   Pyrausta nubilalis H-1, I-? (309)
Phaedon brassicae H-2, I-2 (105)   Tryporyza incertulas H-2, I-2 (162)
Pieris rapae H-2, I-2 (46)   Udea rubigalis H-2, I-2 (101)
Plutella xylostella H-2, I-2 (46, 101)
OR: 1P H-19, I-6 (735)
AP: Alk=I-2 (wilfodeine, wilforgine, wilforine, wilfortrine, wilforzine)(1392)

Triticum aestivum   (Wheat)   Poaceae
   A   B   C   D   J   K   K   M   M   N
   03   06   09   03   02   01   06   01   02   10
OC: Ambrosia artemisiifolia H-24, I-1 (109)   Meloidogyne sp. H-16, I-20 (424)
   Bipolaris sorokiniana H-14, I-? (593)   Polygonum persicaria H-24, I-1 (109)
   Cerastium vulgatum H-24, I-1 (109)   Rhizoctonia solani H-14, I-? (960)
   Chenopodium album H-24, I-1 (109)   Setaria viridis H-24, I-1 (109)
   Digitaria sanguinalis H-24, I-1 (109)   Thielaviopis basicola H-14, I-? (960)
   Fusarium solani   Verticillium albo-atrum H-14, I-? (361)
      f. sp. phaseoli H-14, I-? (960)
OR: H-24, I-1 (951, 1228, 1229, 1230)

Tropaeolum majus   (Nasturtium)   Tropaeolaceae
   A   B   C   D   J   J   K   M   M   M   M   M   N
   03   04   09   01   01   03   01   01   13   14   17   27   07
OC: Colletotrichum lindemuthianum H-14, I-14   Exserohilum turcicum H-14, I-15 (827)
   (1105)   Monilinia fructicola H-14, I-14 (1105)
   Eriosoma lanigerum H-2, I-2 (104)
OR: H-15, I-10, 12 (1157); H-19, I-7 (1116); H-20, I-? (504)
AP: Alk=I-6, 7, 8 (1371); Sfr=I-7 (1116)

Tsuga canadensis   (Canadian hemlock)   Pinaceae
   A   B   C   D   M   M   M   N   N
   01   01   08   03   01   04   21   04   07
OC: Manduca sexta H-4, I-7 (669)   Pyrrhocoris apterus H-3, I-4 (684)
   Panonychus citri H-4, 11, I-7 (669)
AP: Tan=I-5 (504)

Tussilago farfara   (Colts-foot)   Asteraceae
   A   B   C   D   J   M   M   N   O   O
   01   06   09   03   03   01   05   07   07   08
OC: Mosquitoes H-2, I-2 (98, 105)

Tylophora fasciculata   (Not known)   Asclepiadaceae
   B   C   D
   02   09   01
OC: Rats H-25, I-2 (104, 105, 220)   Vermin H-27, I-2, 7 (104)
AP: Alk=I-? (tylophorine)(1392)

Tylophora ovata   (Not known)   Asclepiadaceae
   B   C   D   K
   06   09   01   02
OC: Lice H-1, I-1, 2 (1241)   Rats H-25, I-? (1241)
   Maggots H-1, I-1, 2 (1241)

Typha angustifolia   (Narrow-leaf cattail)   Typhaceae
   B   C   D   J   J   M   N
   06   09   10   03   16   01   15

OC: Alternaria sp. H-14, I-1 (837)   Fusarium roseum H-14, I-1 (837)
OR: H-19, I-1 (837)

Typha latifolia   (Bulrush)   Typhaceae
   A   B   C   D   E   M   N
   01 06 09 03 11 01 02
OR: H-24, I-7 (1215)

Typhonium giganteum   (Not known)   Araceae
   A   B   D
   01 06 01
OC: Aphids H-2, I-2, 6 (1241)   Puccinia graminis tritici H-14, I-2, 6
   Maggots H-2, I-2, 6 (1241)   (1241)
   Phytophthora infestans H-14, I-2, 6 (1241) Puccinia rubigavera H-14, I-2, 6 (1241)

Udotea flabellum   (Algae)   Udoteaceae
   A   B   C
   04 09 02
OC: Rhizopus oryzae H-14, I-1 (1024)

Ulex europaeus   (Furze)   Fabaceae
   B   C   D   E   E   J   L   M   M
   02 09 03 07 12 08 02 09 12
OC: Aphis fabae H-2, I-10 (105)
OR: H-33, I-6 (504)
AP: Alk=I-4, 6, 8, 10 (cytisine)(1392); Alk=I-8 (anagyrine)(1392); Alk=I-10 (1391)

Ulmus americana   (American white elm)   Ulmaceae
   A   B   C   D   M
   01 01 09 03 04
OC: Manduca sexta H-4, I-7 (669)   Panonychus citri H-4, 11, I-7 (669)
AP: Alk=I-10 (1374)

Ulmus parvifolia   (Chinese elm)   Ulmaceae
   A   B   C   D
   01 01 09 03
OC: Fusarium lateritium f. sp. mori H-14, I-15 (917)
   Fusarium roseum H-14, I-15 (917)
   Fusarium solani f. sp. mori H-14, I-15 (917)

Umbellularia californica   (California laurel)   Lauraceae
   A   B   C   D   M   M   M
   01 01 09 03 04 13 19
OC: Fleas H-5, I-7 (104)
OR: H-29, I-1 (104)
AP: Alk=I-6, 7, 9 (1368); Alk=I-10 (1373)

Urginea maritima   (Sea onion)   Liliaceae
   A   B   B   C   D   D   J   J
   01 01 06 09 02 06 01 04
OC: Caterpillars H-1, I-? (105)   Rats H-25, I-3 (504, 507)
   Mosquitoes H-5, I-3 (105)   Rodents H-25, I-3 (1085, 1098)
   Moths H-1, I-? (105)   Spodoptera littoralis H-4, I-7 (344)
OR: H-2, I-? (132)
AP: Alk=I-? (caffeine)(1392)

Urtica breweri   (Not known)   Urticaceae
   B   C   D   J   M
   06 06 01 11 03
OC: Attagenus piceus H-4, I-1 (48)

Urtica dioica   (Stinging nettle)   Urticaceae
   A   B   C   D   M   M   M   N   N   O   O
   01 06 09 03 01 03 05 07 15 07 10
OC: Athalia rosae H-4, I-7 (531)   Locusta migratoria H-4, 5, I-7 (596)

Urtica urens   (Dog nettle)   Urticaceae
   A   B   C   D   J   J   J   M   M   N   O   O
   01 06 09 10 02 03 04 01 05 07 07 08
OC: Phytodecta fornicata H-4, I-7 (531)   Tylenchulus semipenetrans H-16, I-7 (540)
   Pieris brassicae H-4, I-7 (531)
AP: Alk=I-2, 6, 7, 9 (1374); Alk=I-17 (5-hydroxytryptamine)(1392)

Vaccinium leschenaultii   (Blueberry genus)   Ericaceae
   A   B   C   M   N
   01 02 09 01 09
OC: Rats H-25, I-16 (1393)

Vaccinium membranaceum   (Mountain blueberry)   Ericaceae
   A   B   C   D   J
   01 06 09 03 14
OC: Oncopeltus fasciatus H-1, I-2 (958)

Vaccinium oxycoccos   (Small cranberry)   Ericaceae
   A   B   C   D   E   E   J
   01 02 09 03 02 07 11 03
OC: Drosophila hydei H-2, I-2, 6, 7 (101)

Valeriana clematitis   (Hierpa del sapo)   Valerianaceae
   A   B   C   D
   01 06 09 01
OC: Spodoptera frugiperda H-1, I-18 (176)

Valeriana officinalis   (Garden heliotrope)   Valerianaceae
   A   B   C   D   F   J   J   K   M   M   M   M   O
   01 06 09 04 03 04 09 01 05 13 14 17 02
OC: Trogoderma granarium H-5, I-2 (112)
AP: Alk=I-? (1391); Alk=I-? (chatinine, valerine)(1392); Alk=I-6, 7 (1374)

Valeriana wallichii   (Valeriana genus)   Valerianaceae
   B   C   D   M
   06 09 08 17
OC: Tribolium castaneum H-5, I-2 (179)

Vallisneria americana   (Wild celery)   Hydrocharitaceae
   B   B   C   D   J   J   J   M
   06 09 09 03 03 09 16 02
OC: Fusarium roseum H-14, I-1 (837)
OR: H-19, I-1 (837)

Vateria chinensis　　(Not known)　　Dipterocarpaceae

　　A　B　C　D　J
　　01　01　09　01　09

OC: Rats　H-26, I-16 (1393)

Vateria indica　　(Manila copal)　　Dipterocarpaceae

　　A　B　C　D　J　K　M　M　M　M　M　N　O
　　01　01　09　01　09　03　01　05　06　15　20　10　13

OC: Ants　H-1, I-? (105)　　　　　　　　　Rats　H-26, I-16 (1393)

OR: H-19, I-12 (897)

AP: **Alk**=I-4 (bergenin, coumarin, isocoumarin)(1386)

Veratrum album　　(European white hellebore)　　Liliaceae

　　A　B　C　D　J　J　J　J　K　L　M　O
　　01　06　09　03　03　04　08　11　05　03　05　02

OC: Attagenus piceus　H-4, I-1 (48)　　　　　Lymantria monacha　H-1, I-6 (1025)
　　Beetles　H-1, I-6 (1025)　　　　　　　　Malacosoma neustria　H-1, I-6 (1025)
　　Caterpillars　H-2, I-1 (87)　　　　　　　Musca domestica　H-2, I-2 (80); H-2b,
　　Cockroaches　H-2, I-2 (80)　　　　　　　　I-7 (635)
　　Diaphania hyalinata　H-2, I-2 (101)　　　Popillia japonica　H-5, I-2 (1243)
　　Flies　H-2, I-2 (80)　　　　　　　　　　Rats　H-25, I-3 (1345)
　　Grasshoppers　H-2, I-2 (104)　　　　　　Taeniothrips gladioli　H-1, I-? (490)
　　Lice　H-2, I-2 (80)　　　　　　　　　　Thrips tabaci　H-1, I-? (490)
　　Lymantria dispar　H-1, I-6 (1025)

OR: H-34, I-2 (504)

AP: **Alk**=I-? (angeloylzgadenine, deacetylgermitetrine, deacetylneoprotoveratrine, neogerm-
　　budine, protoveratridine, veratetrine, veratramine, veratridine)(1392); **Alk**=I-2 (504);
　　Alk=I-2 (geralbine, germerine, germine, germitetrine, isorubijervine, jervine, proto-
　　veratrine, rubijervine, rubiverine, synaine, veralbidine, veratrobasine, veratroylzy-
　　gadenine, verine)(1392); **Alk**=I-7 (protoveratrine)(1392)

Veratrum californicum　　(Corn lily)　　Liliaceae

　　A　B　C　D　L
　　01　06　09　03　01

OC: Ants　H-1, I-8 (219)　　　　　　　　　Flies　H-1, I-8 (219)
　　Beetles　H-1, I-8 (219)

OR: H-26, I-2 (504)

AP: **Alk**=I-7 (1370)

Veratrum dahuricum　　(Not known)　　Liliaceae

　　A　B　C　D　J　J
　　01　06　09　03　03　08

OC: Cimex lectularius　H-2, I-8 (101)　　　Lice　H-2, I-8 (101)
　　Flies　H-2, I-8 (101)　　　　　　　　　Mosquitoes　H-2, I-8 (101)

Veratrum grandiflorum　　(False hellebore genus)　　Liliaceae

　　A　B　C　D　J
　　01　06　09　03　03

OC: Drosophila hydei　H-2, I-6, 7 (101)

AP: **Alk**=I-2 (jervine, veratramine)(1392); **Alk**=I-2, 7 (1373)

Veratrum lobelianum　　(Not known)　　Liliaceae

　　A　B　C　D
　　01　06　09　03

OC: Musca domestica H-2, I-2 (87)
AP: Alk=I-2, 3 (jervine)(1392); Alk=I-2, 6, 7 (protorerotrine)(1392)

Veratrum maackii var. japonicum   (False hellebore genus)   Liliaceae

    A   B   C   D
    01 06 09 03

OC: Flies H-1, I-2 (1241)

Veratrum maximowicz   (Not known)   Liliaceae

    A   B   C   D   J
    01 06 09 03 03

OC: Drosophila hydei H-2, I-7 (101)

Veratrum nigrum   (Black hellebore)   Liliaceae

    A   B   C   D   J   J   L   M
    01 06 09 03 02 03 03 04

OC: "Bugs" H-2, I-2 (101)   Malacosoma neustria H-1, I-4, 6 (1012, 1026)
    Drosophila hydei H-2, I-2 (101)   Mosquitoes H-2, I-2 (101)
    Flies H-2, I-2 (101)   Rondotia menciana H-2, I-? (105)
    Lice H-2, I-2 (101)
AP: Alk=I-? (germerine, jervine, rubijervine, veratroylzygadenine)(1392); Alk=I-10 (1392)

Veratrum viride   (American false hellebore)   Liliaceae

    A   B   C   D   F   J   K
    01 06 09 03 01 04 05

OC: Aphis fabae H-2, I-2 (531)   Musca domestica H-2, I-2 (531)
    Blattella germanica H-2, I-2 (531)   Myzus persicae H-2, I-2 (531)
    Diaphania hyalinata H-2, I-?   Oncopeltus fasciatus H-2, I-2 (531)
      (101, 246, 1252)   Ostrinia nubilalis H-2, I-2 (101)
    Hymenia recurvalis H-2, I-? (101, 1252) Periplaneta americana H-2, I-2 (269, 531)
OR: H-2a, 2b, I-2 (105, 504)
AP: Alk=I-2 (cevadine, deacetylneoprotoveratrine, gembudine, germerine, germidine, germine,
    germitrine, isogermidine, jervine, neogermbudine)(1392)

Verbascum blattaria   (Moth mullein)   Scrophulariaceae

    A   B   C   D
    02 06 09 01

OC: Phytodecta fornicata H-4, I-7 (531)
OR: H-19, I-? (1171)

Verbascum lychnitis   (White mullein)   Scrophulariaceae

    A   B   C   D   D
    02 06 09 02 03

OC: Mice H-5, I-8 (504)

Verbascum phlomoides   (Clasping mullein)   Scrophulariaceae

    A   B   C   D   J
    02 06 09 03 03

OC: Cassida nebulosa H-4, I-7 (531)   Mice H-5, I-? (504)
    Leptinotarsa decemlineata H-4, I-7 (531)   Phytodecta fornicata H-4, I-7 (531)
OR: H-32, I-? (504)
AP: Alk=I-6, 7, 8, 9 (1374)

Verbascum thapsiforme   (Wool mullein)   Scrophulariaceae

    A   B   C
    02 06 03

OC: Musca domestica H-2, I-1 (101)
OR: H-2, I-1 (87)

Verbascum thapsus   (Common mullein)   Scrophulariaceae
    A   D   D   J   M   M   M   O   O   O
    02   03   04   03   05   28   30   07   08   10
OC: Agrobacterium tumefaciens H-18, I-1 (585)   Erwinia carotovora H-18, I-1 (585)
OR: H-19, I-6, 7 (729, 1262); H-32, I-10 (73, 220, 504); H-35, I-10 (220)
AP: Alk=I-2, 7 (1371); Alk=I-8 (1369)

Verbena bonariensis   (Vervain genus)   Verbenaceae
    A   B   C   D   J
    01   06   09   01   09
OC: Rats H-26, I-1 (1393)

Verbena hastata   (Blue vervain)   Verbenaceae
    A   B   C   D   J
    01   06   09   01   09
OC: Pseudomonas solanacearum H-18, I-1 (946)

Verbena officinalis   (Holy wort)   Verbenaceae
    A   B   C   D   D   M   M   O
    03   06   09   03   06   05   13   07
OC: Aphids H-1, I-1 (1241)   Puccinia graminis tritici H-14, I-1 (1241)
    Erwinia carotovora H-18, I-1 (1241)   Puccinia rubigavera H-14, I-1 (1241)
    Pieris rapae H-1, I-1 (1241)   Rhizoctonia solani H-14, I-1 (1241)
OR: H-20, I-7 (504)

Verbena urticifolia   (White vervain)   Verbenaceae
    A   B   D   J   M
    01   06   03   03   13
OC: Aedes aegypti H-2, I-10 (165)

Verbesina encelioides var. exauriculata   (Butter daisy)   Asteraceae
    A   B   C   J   J   J
    03   02   09   03   08   11
OC: Attagenus piceus H-4, I-1 (48)

Verbesina greenmannii   (Crown beard genus)   Asteraceae
    C   D   J
    09   03   14
OC: Dacus cucurbitae H-6, I-4 (956)

Vernonia amygdalinal   (Bitter leaf)   Asteraceae
    A   B   C   D
    01   02   09   01
OR: H-5, I-7 (1345)

Vernonia anthelmintica   (Kinka oil ironweed)   Asteraceae
    A   B   C   D   J   M   O
    01   06   09   01   03   05   07
OC: Meloidogyne incognita H-16, I-10 (100, 616) Meloidogyne javanica H-16, I-10 (100, 616)
OR: H-2, I-10 (73); H-2, 5, I-? (105); H-20, I-7 (504)

Vernonia colorata   (Ironweed genus)   Asteraceae
    A   C   K
    01 09 08
OC: Rodents H-25, I-7 (945)
OR: H-5, I-7 (1345)

Vernonia fasiculea   (Not known)   Asteraceae
    A   C   J   J
    01 09 02 03
OC: Aedes aegypti H-2, I-8 (165)

Vernonia gigantea   (Not known)   Asteraceae
    A   C   D
    01 09 01
OC: Spodoptera eridania H-2, I-5 (125)   Spodoptera frugiperda H-2, I-5 (125)
OR: H-4, I-7 (1177)

Vernonia glauca   (Not known)   Asteraceae
    A   C   D
    01 09 01
OC: Spodoptera eridania H-4, I-? (125)   Spodoptera frugiperda H-4, I-? (125)
OR: H-4, I-? (1177)

Vernonia hymenolepis   (Ironweed genus)   Asteraceae
    C
    09
OR: H-24, I-? (1176)

Vernonia shirensis   (Not known)   Asteraceae
    A   C   D
    01 09 02
OC: Pediculus humanus capitis H-2, I-? (87)

Veronica agrestis   (Speedwell genus)   Scrophulariaceae
    B   C   D   M
    06 09 03 13
OC: Leptinotarsa decemlineata H-4, I-7 (531)   Pieris brassicae H-4, I-7 (531)

Veronica cinerea   (Speedwell genus)   Scrophulariaceae
    B   C   D   M
    06 09 03 13
OC: Alternaria tenuis H-14, I-7 (480)   Helminthosporium sp. H-14, I-7 (480)
    Curvularia penniseti H-14, I-7 (480)
OR: H-19, I-7 (776)

Veronica polita   (Speedwell genus)   Scrophulariaceae
    C
    09
OC: Athalia rosae H-4, I-7 (531)   Phytodecta fornicata H-4, I-7 (531)

Veronicastrum axillare   (Not known)   Scrophulariaceae
    A   B   C   D
    01 06 09 03
OC: Aphids H-1, I-1 (1241)   Spider mites H-11, I-1 (1241)
    Rice borers H-1, I-1 (1241)

Vetiveria nigritana   (Vetiver genus)   Poaceae
   A   B   C   D   D   M   M
   01 06 09 01 05 02 17
OC: Locusta migratoria H-4, 5, I-7 (596)

Vetiveria zizanioides   (Vetiver genus)   Poaceae
   A   B   C   D   J   J   J   J   K   M   M   M   M   M
   01 06 09 01 02 03 06 11 06 02 04 17 22 24
OC: Attagenus piceus H-4, I-15 (48)   Latheticus oryzae H-2, I-7 (88)
    Cimex lectularius H-2, I-2 (105)   Lice H-2, I-2 (105)
    Clothes moths H-2, I-2 (105, 1098)   Oncopeltus fasciatus H-2, I-2 (48)
OR: H-2, 5, I-2 (1276); H-2a, I-8 (1241)

Viburnum atrocyaneum   (Arrowwood genus)   Caprifoliaceae
   A   B   B   C   D   D   M
   01 01 02 09 02 03 13
OR: H-2, I-? (100)

Viburnum lantana   (Wayfaring tree)   Caprifoliaceae
   A   B   C   D   M
   01 01 09 03 13
OC: Venturia inaequalis H-14, I-9 (16)

Vicia cracca   (Bird vetch)   Fabaceae
   A   B   C   J   M   M   N
   01 06 09 03 01 12 10
OC: Epilachna varivestis H-2, I-7 (101)
AP: **Alk**=I-6, 7, 8 (1371)

Vicia faba   (Spring vetch)   Fabaceae
   C   D   M   M   M   N
   09 03 01 02 12 10
OC: Alternaria solani H-14, I-2, 6 (738)   Potato virus X H-21, I-14 (1136)
    Monilia fructigena H-14, I-2, 6 (738)
OR: H-15, I-? (170); H-24, I-? (951)
AP: **Alk**=I-10 (convincine)(1392)

Vicia villosa   (Hairy vetch)   Fabaceae
   A   B   C   D   M
   01 06 09 03 12
OC: Pratylenchus zea H-16, I-? (1274)

Viguiera reticulatus   (Not known)   Asteraceae
   A   C   J
   01 09 03
OR: H-24, I-7 (554, 559)

Vinca minor   (Lesser periwinkle)   Apocynaceae
   A   B   C   D   J   M   O
   01 02 09 03 03 05 07
OC: Spodoptera frugiperda H-2, I-1 (176)
AP: **Alk**=I-6 (vincaminorine, vinine)(1392); **Alk**=I-7 (isovincamine, minorine, perivincine, pubescine, vincamine, vincaminorine, vinine)(1392)

Viola phalacrocarpoides   (Not known)   Violaceae
   B   C   D   D   J
   06 09 03 08 03
OC: Drosophila hydei H-2, I-2, 6, 7 (101)

Viola septentrionalis   (Northern blue violet)   Violaceae
   B   C
   06 09
OC: Manduca sexta H-4, I-7 (669)   Panonychus citri H-4, 11, I-7 (669)

Viola takedana var. variegata   (Violet genus)   Violaceae
   B   C   D   D   J
   06 09 03 08 03
OC: Drosophila hydei H-2, I-2, 6, 7 (101)

Viscum album   (Mistletoe)   Loranthaceae
   B   B   C   D
   02 10 09 03
OR: H-10, I-14 (87); H-24, I-7, 14 (569)
AP: Alk=I-7 (hyramine, phenethylamine)(1392)

Vitex agnus-castus   (Monk's pepper tree)   Verbenaceae
   A   B   C   D   D   D   M   M   M
   01 01 09 01 02 06 14 17 24
OC: Flies H-5, I-? (105)   Stored grain pests H-5, I-6 (1345)

Vitex cannabifolia   (Not known)   Verbenaceae
   A   B   C   D
   01 01 09 02
OC: Agrotis sp. H-1, I-1 (1241)   Maggots H-1, I-1 (1241)
    Aphids H-1, I-1 (1241)   Spider mites H-11, I-1 (1241)

Vitex negundo   (Indian privet)   Verbenaceae
   A   B   C   D   F   J   J   J   K   K   M   O   O
   01 01 09 01 05 03 04 11 06 09 05 09 10
OC: Achaea janata H-2, I-7 (105)   Pyricularia oryzae H-14, I-7 (114)
    Callosobruchus chinensis H-2, I-7 (95, 475) Rice field insects H-5, I-6, 7 (137)
    Clothes moths H-5, I-7 (605)   Sitophilus oryzae H-2, I-? (161)
    Euproctis fraterna H-2, I-7 (105)   Sitotroga cerealella H-2, I-7 (121, 1124)
    Latheticus oryzae H-2a, I-7 (124)   Spodoptera litura H-2, I-7 (105)
    Musca domestica H-2a, I-? (87)   Stored grain pests H-5, I-7, 10 (73)
    Pericallia ricini H-2, I-7 (105)   Tryporyza incertulas H-2, I-? (95)
    Plutella xylostella H-2, I-7 (105)
OR: H-2, I-7 (88, 90, 101, 1351); H-5, I-6, 7 (1352); H-17, I-9 (1318); H-24, I-7 (188)
AP: Alk=I-? (bicucine, corlumidine)(1124); Alk=I-2 (allocryptopine, bicuculline, bulbocapnine, corybulbine, corycavine, corydaline, corydine, corytuberine, crytopine)(1392); Alk=I-6, 7 (1318); Alk=I-7 (nishindine)(1392)

Vitis labrusca   (Fox grape)   Vitaceae
   A   B   C   D   M   M   M   N
   01 03 09 03 01 02 21 09
OC: Fusarium roseum H-14, I-16 (917)   Lymantria dispar H-5, I-? (340)
AP: Tan=I-10 (504)

Vitis vinifera   (Wine grape)   Vitaceae
  A   B   C   D   M   N
  01 03 09 03 01 09
OC: Leptinotarsa decemlineata H-4, I-7 (531)

Vittadinia australis   (Not known)   Asteraceae
  A   B   C   D   J
  01 06 09 03 09
OC: Musca domestica H-2, I-1 (785)
OR: H-2, I-? (100); H-22, I-1 (1395)

Vriesea heliconioides   (Not known)   Bromeliaceae
  C
  09
OC: Dacus dorsalis H-6, I-8 (473)

Walsura piscidia   (Not known)   Meliaceae
  A   C
  01 09
OR: H-1, I-? (220); H-32, I-4, 9 (220, 459, 1276)

Warburgia ugandensis   (Not known)   Canellaceae
  A   C   D
  01 09 01
OC: Spodoptera exempta H-2, 4, I-? (132, 143)

Washingtonia filifera   (Desert fan palm)   Arecaceae
  A   B   C   D   D   J   M   M   N   N
  01 01 09 05 06 02 01 03 09 15
OR: H-24, I-9 (1242)

Wedelia natalensis   (Not known)   Asteraceae
  B   C   D   D
  06 09 01 02
OC: Vermin H-27, I-2, 7 (87)

Wedelia prostrata   (Not known)   Asteraceae
  A   B   C   D   D   F   G   K   L
  03 06 09 01 02 06 04 06 04
OC: Dysdercus cingulatus H-2a, I-8 (125)
OR: H-33, I-7 (125)

Wendlandia wallichii   (Not known)   Rubiaceae
  A   C   D
  01 09 01
OC: Rats H-26, I-16 (1393)

Wikstroemia chamaedaphne   (Not known)   Thymelaeaceae
  A   B   C   D   J
  01 02 09 01 04
OC: Oides decimpunctata H-1, I-? (1241)

Wikstroemia indica   (Not known)   Thymelaeaceae
  A   B   C   D
  01 02 09 01

OC: Rice borers H-1, I-1 (1241)
OR: H-32, I-? (504)
AP: Alk=I-2, 7 (37)

Wikstroemia nutans   (Not known)   Thymelaeaceae
    B   C   D
    06   09   02
OR: H-1, I-? (101)

Wikstroemia sandwicensis   (Not known)   Thymelaeaceae
    B   C   D   J
    06   09   01   11
OC: Attagenus piceus H-4, I-? (48)   Tineola bisselliella H-4, I-? (48)

Willardia mexicana   (Nesco)   Fabaceae
    A   B   C   D   J   J   J   J   K   M
    01   01   09   04   03   04   08   11   05   12
OC: Attagenus piceus H-4, I-7 (48)   Oncopeltus fasciatus H-4, I-7 (48)
    Diaphania hyalinata H-2, I-4 (83, 101)   Pediculus humanus humanus H-2, I-4 (101)
    Evergestis rimosalis H-2, I-4 (101)   Spodoptera eridania H-2, I-4 (101)
    Hymenia recurvalis H-2, I-4 (101)   Tineola bisselliella H-4, I-7 (48)
    Livestock pests H-1, I-4 (504)
OR: H-4, I-15 (507)

Withania somnifera   (Not known)   Solanaceae
    B   C   D   D   J   M   M   O
    02   09   01   06   09   05   15   07
OC: Aulacophora foveicollis H-1, I-? (600)   Lice H-2, I-2, 7 (1353)
    Epilachna varivestis H-1, I-7 (349)   Lipaphis erysimi H-1, I-? (600)
OR: H-2, I-? (220); H-19, I-? (411); H-22, I-1 (881, 1041); H-35, I-? (504)
AP: Alk=I-? (nicotine, somniferine, somniferinine, somnine, withananine, withananinine,
    withania) (1392); Alk=I-2 (1368); Alk=I-2, 6, 7 (1357); Alk=I-6 (1371, 1379)

Woodfordia floribunda   (Not known)   Lythraceae
    A   B   C   D
    01   02   09   01
OC: Schistocerca gregaria H-4, I-? (615)

Woodwardia japonica   (Chain fern genus)   Polypodiaceae
    A   C   J   J   K   K   M
    01   06   03   04   05   06   13
OC: Aphids H-1, I-2, 7 (1241)   Spider mites H-11, I-2, 7 (1241)

Xanthium canadense   (Burweed)   Asteraceae
    B   C   D   J   J
    06   09   03   11   12
OC: Attagenus piceus H-4, I-1 (48)   Drosophila melanogaster H-3, I-2 (373, 713)

Xanthium pennsylvanicum   (Not known)   Asteraceae
    B   C   D   J
    06   09   10   13
OC: Ceratocystis ulmi H-14, I-7 (568)   Fusarium oxysporum H-14, I-7 (568)
OR: H-19, I-? (1171)

Xanthium sibiricum   (Not known)   Asteraceae
    B   C   D   J
    06   09   01   03
OC: Aphids H-1, I-1, 10 (1241)   Spider mites H-11, I-1, 10 (1241)
    Pieris rapae H-1, I-1, 10 (1241)

Xanthium spinosum   (Cotweed)   Asteraceae
    C   D   J   J   M   O
    09   03   08   11   05   07
OC: Attagenus piceus H-4, I-6, 7 (48)   Oncopeltus fasciatus H-1, I-6, 7 (78)
    Mosquitoes H-1, I-6, 7 (78)
AP: **Alk**=I-? (1392)

Xanthium strumarium   (Burweed)   Asteraceae
    B   C   D   J   J   L   M   M   N   N   O
    06   09   10   02   03   02   01   05   07   15   07
OC: Tylenchulus semipenetrans H-16, I-9 (540)   Ustilago tritici H-14, I-7 (181)
    Ustilago hordei H-14, I-7 (181)
AP: **Alk**=I-9 (1370)

Xeromphis spinosa   (Not known)   Rubiaceae
    A   B   B   C   D   J   J   J   J   J   K   K   M   N
    01   01   02   09   01   03   04   08   09   11   02   09   1a   09
OC: Aphids H-2, I-9 (105)   Euproctis fraterna H-2, I-2 (105)
    Attagenus piceus H-4, I-2, 9 (48)   Plutella xylostella H-2, I-2 (101)
    Coccus viridis H-1, I-2 (87)   Stored grain pests H-5, I-2, 9 (87, 105)
    Epacromia tamulus H-2, I-9 (105)
OR: H-2, I-2, 9 (220, 1276); H-22, I-9 (881); H-32, I-4, 6, 7, 9 (459, 605, 968, 1276, 1351)
AP: **Alk**=I-10 (1392); **Fla**=I-4, 7 (1377); **Fla, Sap**=I-2 (1380); **Tan**=I-4, 7, 9 (1377)

Xylocarpus granatum   (Not known)   Meliaceae
    A   C   J
    01   09   16
OC: Epilachna varivestis H-4, I-4 (1145)   Spodoptera littoralis H-4, I-4 (1145)
    Spodoptera exempta H-4, I-4 (1145)
AP: **Alk**=I-7 (1392); **Fla**=I-6, 7 (1377)

Xylocarpus molluscens   (Not known)   Meliaceae
    C
    09
OC: Spodoptera exempta H-4, I-9 (763)
AP: **Alk**=I-10 (1392)

Xysmalobium stellatum   (Not known)   Asclepiadaceae
    B   C   D   D
    06   09   01   02
OR: H-1, I-? (504)

Xysmalobium undulatum   (Not known)   Asclepiadaceae
    B   C   D   D
    06   09   01   02
OC: Maggots H-1, I-? (87)

Yucca aloifolia   (Spanish-bayonet)   Agavaceae
   A   B   C   D   J
   01 01 09 01 11
OC: Attagenus piceus H-4, I-7 (48)

Yucca schidigera   (Dagger plant)   Agavaceae
   A   B   C   D   D   J
   01 02 09 02 05 04
OC: Diaphania hyalinata H-2, I-7 (101)   Udea rubigalis H-2, I-7 (101)
    Hymenia recurvalis H-2, I-7 (101, 1252)   Urbanus proteus H-2, I-7 (101)
    Laspeyresia pomonella H-2, I-7 (101)

Zaluzania augusta   (Not known)   Asteraceae
   C
   09
OC: Spodoptera frugiperda H-3, I-7 (176)

Zamia loddigesii   (Not known)   Zamiaceae
   A   B   C   D   D
   01 01 07 01 02
OC: Rats H-25, I-2 (1345)

Zantedeschia aethiopica   (Trumpet lily)   Araceae
   B   C   D
   06 09 02
OR: H-1, I-2, 6 (1241)
AP: Alk=I-6, 7, 8 (etiopine)(1392)

Zanthoxylum alatum   (Wingleaf prickly ash)   Rutaceae
   A   B   C   D   J   J   K   M   M   M   N   O
   01 01 09 10 03 04 05 01 05 17 09 10
OC: Aspergillus oryzae H-14, I-12 (649)   Musca domestica H-2, I-9, 12 (87)
    Fusarium oxysporum H-14, I-12 (649)   Pseudomonas solanacearum H-18, I-? (650)
    Locusta migratoria H-2, I-4, 9, 12 (87, 658)
OR: H-1, I-? (220); H-2, I-? (80); H-19, I-9 (429); H-32, I-4, 9 (73, 220, 459)
AP: Alk=I-? (1391); Alk=I-4 (berberine)(1392)

Zanthoxylum americanum   (Toothache tree)   Rutaceae
   A   B   C   D   J   J   J   J   M   M   O
   01 01 09 03 02 03 08 10 05 17 04
OC: Lymantria dispar H-5, I-? (340)   Popillia japonica H-1, I-? (773); H-2, I-2
    Mosquitoes H-4, I-9 (105)   (105); H-5, I-4 (1243)
OR: H-34, I-4 (504)
AP: Alk=I-4 (berberine, N-methylcinnamide, O-methyltyramine)(1392)

Zanthoxylum armatum   (Prickly ash genus)   Rutaceae
   A   B   C   D
   01 01 09 03
OC: Tribolium castaneum H-5, I-4, 6, 10 (179)

Zanthoxylum chalybea   (Prickly ash genus)   Rutaceae
   A   B   C   D   J   M   M   M   M   N   O
   01 01 09 01 16 01 05 14 17 07 02

OC: Epilachna varivestis H-4, I-4 (1145)   Spodoptera littoralis H-4, I-4 (1145)
    Spodoptera exempta H-4, I-4 (1145)
AP: **Alk**=I-2 (1374)

Zanthoxylum clava-herculis   (Southern prickly ash)   Rutaceae
    A   B   B   C   D   J   J   J   K   M   M   O   O
    01   01   02   09   03   04   10   11   05   05   17   04   09
OC: Caterpillars H-5, I-4, 7 (97, 105, 773)   Musca domestica H-2, I-4 (96, 773)
    Diaphania hyalinata H-2, I-4 (101)   Pseudaletia unipuncta H-2, I-4 (101)
    Flies H-1, I-4 (813)   Urbanus proteus H-2, I-4 (101)
    Mosquitoes H-2, I-4 (99)
OR: H-34, I-4 (504)
AP: **Alk**=I-2, 4, 5 (1389); **Alk**=I-5 (773, 1371); **Alk**=I-6, 7, 9 (1369)

Zanthoxylum dimorphophyllum   (Prickly ash genus)   Rutaceae
    A   B   C   D
    01   01   09   03
OC: Pyricularia oryzae H-14, I-7, 10 (1241)   Rice borers H-1, I-7, 10 (1241)

Zanthoxylum hamiltonianum   (Not known)   Rutaceae
    A   B   C   D   D   G   J
    01   02   09   02   03   03   03
OC: Mosquitoes H-2, I-2 (73, 105, 504, 773)
AP: **Sap**=I-? (105)

Zanthoxylum holstii   (Prickly ash genus)   Rutaceae
    A   B   C
    01   01   09
OC: Epilachna varivestis H-4, I-? (1145)   Spodoptera littoralis H-4, I-? (1145)
    Spodoptera exempta H-4, I-? (1145)

Zanthoxylum malcomia   (Not known)   Rutaceae
    A   B   C   D   D   M
    01   01   09   02   03   17
OR: H-2, I-? (100)

Zanthoxylum monophyllum   (Prickly ash genus)   Rutaceae
    A   B   C   J
    01   01   09   09
OC: Hemileuca oliviae H-4, I-6 (632)   Melanoplus sanguinipes H-4, I-6 (632)
    Hypera postica H-4, I-6 (632)   Schizaphis graminum H-4, I-6 (632)

Zanthoxylum piperitum   (Japanese pepper)   Rutaceae
    A   B   C   D
    01   01   09   03
OC: Culex pipiens H-1, I-? (105)
OR: H-22, I-? (971)
AP: **Alk**=I-? (samshoamide)(1392); **Alk**=I-2, 10 (magnoflorine)(1392)

Zanthoxylum planispinum   (Wing-leaved prickly ash)   Rutaceae
    A   C   D   J
    01   09   03   02
OC: Pyricularia oryzae H-14, I-7, 10 (1241)   Rice borers H-1, I-7, 10 (1241)

Zanthoxylum simulans   (Prickly ash genus)   Rutaceae
   A   B   C   D
   01 02 09 03
OC: Aphis gossypii H-1, I-7, 9 (1241)

Zanthoxylum xanthoxyloides   (Prickly ash genus)   Rutaceae
   A   B   C   D   M   M   O   O
   01 01 09 01 05 14 04 10
OC: Termites H-8, I-5 (1353)   Vermin H-27, I-2 (1353)
OR: H-32, I-4 (1353); H-34, I-2 (804, 1353)
AP: Alk=I-2 (artarine, fagaramine)(1392); Alk=I-4 (fagaridine, skimmianine)(1392)

Zea mays   (Corn)   Poaceae
   A   B   C   D   M   N
   03 06 09 10 01 10
OC: Chionaspis salicis-nigrae H-1, I-12 (1086)   Panonychus citri H-4, 11, I-7 (669)
    Colletotrichum lindemuthianum H-14, I-14   Peronospora tabacina H-14, I-? (482)
      (1105)   Phenacoccus gossypii H-1, I-12 (101, 1086)
    Fusarium solani   Phytodecta fornicata H-4, I-7 (531)
      f. sp. phaseoli H-14, I-20 (422, 960)   Pieris brassicae H-4, I-7 (531)
    Lepidosaphes ulmi H-1, I-12 (101, 1086)   Pratylenchus penetrans H-16, I-12 (1054)
    Leptinotarsa decemlineata H-4, I-7 (531)   Rhizoctonia solani H-14, I-20 (353, 422)
    Manduca sexta H-4, I-7 (699)   Rotylenchulus sp. H-16, I-? (525)
    Meloidogyne sp. H-16, I-? (525)   Sphaerothica humuli H-14, I-? (496)
    Monilinia fructicola H-14, I-14 (1105)   Thielaviopis basicola H-14, I-20 (960)
    Ostrinia nubilalis H-4, I-? (101)   Zabrotes subfasciatus H-2, I-12 (334)
OR: H-24, I-7 (569, 1227, 1228)
AP: Alk=I-7 (hordinine)(1392)

Zebrina pendula   (Wandering jew)   Commelinaceae
   A   B   C   D   D   J   M
   01 06 09 01 04 03 09
OC: Drechslera oryzae H-14, I-7 (113)

Zephyranthes drummondii   (Not known)   Amaryllidaceae
   A   B   C   D   J   J   J
   03 06 09 10 03 08 11
OC: Attagenus piceus H-4, I-3 (48)
AP: Alk=I-? (lycorine)(1392)

Zieria smithii   (Not known)   Rutaceae
   C   J
   09 06
OC: Aedes sp. H-5, I-7, 12 (101)   Dacus cacuminatus H-6, I-7 (335, 473)
    Blow flies H-5, I-7, 12 (101)
OR: H-2, 3, I-7, 12 (101); H-19, I-12 (775); H-30, I-12 (101)
AP: Alk=I-2, 6, 7 (1392)

Zigadenus glaucus   (White camas)   Liliaceae
   A   B   C   D   J
   01 06 09 03 03
OC: Attagenus piceus H-4, I-3 (48)

Zigadenus paniculatus   (Sand corn)   Liliaceae

A B C D J
01 06 09 03 03

OC: Attagenus piceus H-4, I-3 (48)   Oncopeltus fasciatus H-2a, I-3 (48)

Zingiber officinale   (Common ginger)   Zingiberaceae

B C D D J M M O
06 09 01 02 03 05 14 07

OC: Drechslera oryzae H-14, I-2 (113)   Sclerotium rolfsii H-14, I-2 (1297)
    Rhizoctonia solani H-14, I-2 (1297)   Tribolium castaneum H-5, I-? (179)
    Sclerotium oryzae H-14, I-2 (1297)
OR: H-34, I-? (1353)

Zinnia elegans   (Common zinnia)   Asteraceae

A B C D
03 06 09 01

OC: Leptinotarsa decemlineata H-4, I-7 (531)
AP: Alk=I-2 (nicotine)(1392); Alk=I-7 (anabasine, nicotine, nornicotine)(1392)

Zizania aquatica   (Wild rice)   Poaceae

A B B C D J M N
01 06 09 09 03 16 01 10

OC: Fusarium roseum H-14, I-1 (837)
OR: H-19, I-1 (837)

Zizyphus jujuba   (Chinese date)   Rhamnaceae

A B C D D J M M M N O
01 01 09 03 06 03 01 02 05 21 09 07

OC: Alternaria tenuis H-14, I-7 (480)   Helminthosporium sp. H-14, I-7 (480)
    Curvularia penniseti H-14, I-7 (480)
OR: H-32, I-9 (983, 1353)
AP: Tan=I-4 (1353)

ROTENONE

J K
04 05

OC: Anasa tristis H-1, I-2 (14)   Leptopha minor H-1, I-2 (298)
    Anuraphis roseus H-1, I-2 (14)   Lipaphis erysimi H-1, I-2 (206)
    Aonidiella aurantii H-1, I-2 (295)   Macrosiphum liriodendri H-1, I-2 (14)
    Aphis fabae H-1, I-2 (14)   Menopon pallidum H-1, I-2 (14)
    Aphis persicae-niger H-1, I-2 (14)   Mineola vaccinii H-1, I-2 (305)
    Aphis pomi H-1, I-2 (14, 213, 794)   Musca domestica H-1, I-2 (233)
    Aphis pseudobrassicae H-1, I-2 (216) Myzus persicae H-1, I-2 (14)
    Aphis sorbi H-1, I-2 (213, 216)   Philaenus leucophthalmus H-1, I-2 (300)
    Aphis spiraecola H-1, I-2 (206)   Pieris rapae H-1, I-2 (14)
    Aulocaspis rosae H-1, I-2 (216)   Popilia japonica H-1, I-2 (14)
    Blattella germanica H-1, I-2 (14)   Pseudococcus citri H-1, I-2 (216)
    Chionaspis euonymi H-1, I-2 (216)   Psylla pyricola H-1, I-2 (303)
    Culicine sp. H-1, I-2 (14)   Pteronidea ribesii H-1, I-2 (216)
    Diabrotica duodecimpuntata H-1, I-2 (206) Tetranychus cinnabarinus H-11, I-2 (206)
    Doryphora 10-lineata H-1, I-2 (14)   Thermobia domestica H-1, I-2 (281)
    Empoasca fabae H-1, I-2 (367)   Thrips tabaci H-1, I-2 (14)
    Epilachna varivestis H-1, I-2 (14, 206) Typhlocyba comes H-1, I-2 (14)
    Eriosoma lanigerum H-1, I-2 (216, 794)
OR: H-2, I-2 (549); H-25, I-2 (656)

BACTERIA

Agrobacterium tumefaciens
(Crown gall)
Allium cernuum
Allium sativum
Allium tricoccum
Artabotrys hexapetalus
Barbarea vulgaris
Berberis thunbergii
Convolvulus arvensis
Davidia involucrata
Evodia daniellii
Fontanesiana editorum
Lepidium draba
Magnolia glandiflora
Nymphaea odorata
Pachysandra terminalis
Punica granatum
Quercus glandulifera
Ruta graveolens
   var. angustifolia
Verbascum thapsus

Corynebacterium flaccumfaciens
(Vascular wilt of bean)
Allium sativum
Artabotrys hexapetalus

Corynebacterium michiganense
(Bacterial canker of tomato)
Allium cepa
Allium sativum
Artabotrys hexapetalus
Cannabis sativa
Nuphar luteum

Erwinia amylovora
(Fire blight of pears, apples)
Davidia involucrata
Evodia daniellii
Fontanesiana editorum
Magnolia grandiflora
Nymphaea odorata
Pachysandra terminalis
Punica granatum
Quercus glandulifera
Ruta graveolens
   var. angustifolia

Erwinia aroideae
(Soft rot of tomato)
Allium sativum

Erwinia carotovora
(Soft rot of potato)
Adiantum capillus-veneris
Allium cernuum
Allium sativum
Allium tricoccum
Artabotrys hexapetalus
Asplenium nidus
Barbarea vulgaris
Berberis thunbergii
Carpinus tschonoskii
Celastrus scandens
Cibotium chamissoi
Cibotium schieldei
Convolvulus arvensis
Cymbopogon nardus
Davallia pentaphylla
Davidia involucrata
Doodia media
Drynaria quercifolia
Erythronium americanum
Evodia daniellii
Fontanesiana editorum
Ilex decidua
Madhuca indica
Magnolia grandiflora
Medeola virginiana
Microlepia strigosa
Nephrolepis biserrata
Nymphaea odorata
Pachysandra terminalis
Polypodium punctatum
Pteridium aquilinum
Pteris tremula
Punica granatum
Quercus glandulifera
Quisqualis indica
Ranunculus abortivus
Ruta graveolens
   var. angustifolia
Stemona japonica
Stemona sessilifolia
Stemona tuberosa

Erwinia carotovora
   (cont'd.)
Taxus canadensis
Thelypteris setigera
Verbascum thapsus
Verbena officinalis

Pseudomonas lachrymans
(Angular leaf spot of cucumber)
Allium sativum

Pseudomonas phaseolicola
(Bacterial blight of bean)
Allium sativum
Artabotrys hexapetalus

Pseudomonas solanacearum
(Tomato wilt)
Acanthospermum hispidum
Adiantum trapeziforme
Allium sativum
Angiopteris evecta
Anogeissus leiocarpus
Artabotrys hexapetalus
Aster macrophyllus
Baccharis halimifolia
Bixa orellana
Blechnum spicant
Blumea eriantha
Cibotium chamissoi
Cibotium glaucum
Cibotium schieldei
Cyrtomium falcatum
Davallia pentaphylla
Dennstaedtia punctilobula
Dryopteris filix-mas
Echinacea pallida
Eupatorium purpureum
Geranium maculatum
Helianthus annuus
Heuchera americana
Hibiscus moscheutos
Hieracium aurantiacum
Kosteletzky virginica
Limonium carolinianum
Lycopersicon lycopersicum
Lygodium japonicum
Microlepia strigosa
Nuphar advena
Piper nigrum
Polyalthia longifolia
Pteris tremula
Pteris vittata
Rhus glabra

Pseudomonas solanacearum
   (cont'd)
Rhus typhina
Rosa carolina
Salsola kali
Scirpus americanus
Scirpus atrovirens
Scirpus lacustris
Solanum tuberosum
Solidago microcephala
Tectaria cicutaria
Tectaria incisa
Thelypteris dentata
Verbena hastata
Zanthoxylum alatum

Pseudomonas syringae
(Bacterial canker of stone fruit)
Carpinus tschonoskii
Davidia involucrata
Evodia daniellii
Fontanesiana editorum
Magnolia grandiflora
Nymphaea odorata
Pachysandra terminalis
Punica granatum
Quercus glandulifera
Ruta graveolens
   var. angustifolia

Streptomyces scabies
(Potato scab)
Glycine max

Xanthomonas campestris
   pv. campestris
(Black vein of cabbage)
Allium sativum
Artabotrys hexapetalus
Didymocarpus pedicellata
Elaeis guineensis
Eucalyptus botryoides
Hydnocarpus kurzii
Tamarindus indica

Xanthomonas campestris pv. citri
(Citrus canker)
Elaeis guineensis
Hydnocarpus kurzii

Xanthomonas campestris
   pv. malvacearum
(Bacterial blight of cotton)
Rheum officinale

295

Xanthomonas campestris pv. oryzae
(Bacterial blight of rice)
Allium sativum
Artabotrys hexapetalus

Xanthomonas campestris pv. phaseoli
(Bacterial blight of bean)
Artabotrys hexapetalus
Aster macrophyllus
Davidia involucrata
Dryopteris filix-mas
Evodia daniellii
Fontanesiana editorum
Magnolia grandiflora
Nymphaea odorata
Pachysandra terminalis
Punica granatum
Quercus glandulifera
Ruta graveolens
    var. angustifolia

Xanthomonas campestris
  pv. phaseoli-sojensis
(Bacterial blight of soybeans)
Adiantum capillus-veneris
Adiantum trapeziforme
Aglaomorpha meyenianum
Angiopteris evecta
Asplenium nidus
Blechnum spicant
Cibotium chamissoi
Cibotium glaucum
Cibotium schieldei
Cyrtomium falcatum
Dennstaedtia punctilobula
Doodia media
Drynaria quercifolia
Lygodium japonicum
Microlepia strigosa
Nephrolepis biserrata
Nephrolepis exaltata
Polypodium aureum
Polypodium punctatum
Polystichum tsus-simense
Pteris vittata
Pyrrosia lingua

Xanthomonas campestris
  pv. phaseoli-sojensis
  (cont'd)
Sphaeropteris cooperi
Stenochlaena tenuifolia
Tectaria cicutaria
Tectaria incisa
Thelypteris dentata
Thelypteris setigera

Xanthomonas campestris pv. pruni
(Bacterial spot of stone fruits)
Davidia involucrata
Evodia daniellii
Fontanesiana editorum
Magnolia grandiflora
Nymphaea odorata
Pachysandra terminalis
Punica grantum
Quercus glandulifera
Ruta graveolens
    var. angustifolia

Xanthomonas campestris
  pv. vesicatoria
(Bacterial spot of tomato)
Allium sativum
Artabotrys hexapetalus
Davidia involucrata
Evodia daniellii
Fontanesiana editorum
Magnolia grandiflora
Nymphaea odorata
Pachysandra terminalis
Punica granatum
Quercus glandulifera
Ruta graveolens
    var. angustifolia

Xanthomonas campestris pv. vitians
(Bacterial leaf spot of lettuce)
Artabotrys hexapetalus

Xanthomonas sp.
  Allium sativum

FUNGI

Alternaria brassicae
(Black leaf spot of crucifers)
Aegopodium podagraria
Cestrum diurnum
Cymbopogon martinii
Cymbopogon oliverii
Trachyspermum ammi

Alternaria citri
(Black rot of citrus fruit)
Humulus lupulus

Alternaria solani
(Early blight of potato/tomato)
Avena sativa
Chenopodium album
Cinnamomum zeylanicum
Cymbopogon martinii
Cymbopogon oliverii
Erigeron linifolius
Lawsonia inermis
Moringa pterygosperma
Piper methysticum
Rosa chinensis
Solidago rugosa
Trachyspermum ammi
Vicia faba

Alternaria sp.
Catalpa speciosa
Didymocarpus pedicellata
Lemna minor
Maughania chappar
Nymphaea tuberosa
Plumeria multiflora
Potamogeton amplifolius
Potamogeton nutans
Potamogeton pectinatus
Potamogeton richardsonii
Potamogeton zosteriformis
Typha angustifolia

Alternaria tenuis
(Leaf spot of cotton, onion)
Acacia loculata
Allamanda cathartica
Allium cepa
Allium sativum
Amaranthus viridis
Argemone mexicana
Artabotrys hexapetalus

Alternaria tenuis
(cont'd)
Azadirachta indica
Beta vulgaris
Betula lenta
Cestrum diurnum
Clematis gouriana
Cucumis prophetarum
Curcuma amada
Curcuma zedoaria
Cymbopogon citratus
Datura stramonium
Eruca vesicaria
Ficus religiosa
Gualtheria procumbens
Laurus nobilis
Lawsonia inermis
Lycopersicon lycopersicum
Madhuca indica
Melia azedarach
Mentha piperita
Mentha pulegium
Ocimum sanctum
Origanum majorana
Piper longum
Prosopis spicigera
Ranunculus sceleratus
Terminalia bellirica
Thymus vulgaris
Trachyspermum ammi
Veronica cinerea
Zizyphus jujuba

Alternaria tenuissima
(Brown spot of onion)
Anethum graveolens
Cuminum cyminum
Trachyspermum ammi

Aspergillus alliaceus
(Stem rot of cereus, opuntia)
Brassica nigra

Aspergillus niger
(Onion bulb rot)
Allium cepa
Allium sativum
Artemisia absinthium
Brassica nigra
Curcuma amada
Curcuma angustifolia

Aspergillu niger
(cont'd)
Curcuma aromatica
Curcuma zedoaria
Cymbopogon nardus
Hydnocarpus kurzii
Moringa pterygosperma
Nigella sativa

Aspergillus oryzae
(Rot of stored rice)
Acanthospermum hispidum
Beta vulgaris
Blumea eriantha
Brassica oleracea
  var. gongyloides
Brassica rapa
  var. rapifera
Humulus lupulus
Polyalthia longifolia
Zanthoxylum alatum

Bipolaris maydis
(Southern corn leaf blight)
Anagallis arvensis
Coffea arabica
Erigeron linifolius

Bipolaris sorokiniana
(Leaf spot of grass)
Arctium minus
Avena sativa
Brassica napus
Brassica sp.
Impatiens capensis
Medicago sativa
Melilotus officinalis
Oenanthe biehnis
Raphanus raphinastrum
Rosa chinensis
Saccharum officinarum
Triticum aestivum

Botrytis allii
(Neck rot of onion)
Allium cepa
Allium sativum
Betula lenta
Brassica nigra
Cymbopogon citratus
Laurus nobilis
Mentha piperita
Origanum majorana
Thymus vulgaris

Botrytis cinerea
(Leaf blight of lettuce)
Amaranthus hybridus
Arisaema triphyllum
Borago officinalis
Campanula trachelium
Chelidonium majus
Desmodium paniculatum
Eleocharis obtusa
Erythrophleum suaveolens
Eupatorium maculatum
Helianthus giganteus
Impatiens pallida
Ligustrum vulgare
Lonicera japonica
Medicago sativa
Nasturtium officinale
Pastinaca sativa
Piper methysticum
Polygonum cuspidatum
Pycnanthemum pilosum
Pyrularia pubera
Sambucus nigra
Sedum ternatum
Solidago ulmifolia
Teucrium canadense
Thalictrium polygamum
Trifolium procumbens

Botryis sp.
Pinus sp.

Cephalosporium sacchari
(Wilt of sugarcane)
Allium sativum
Brassica oleracea
  var. capitata
Brassica rapa
Raphanus sativus
Rosa chinensis

Ceratocystis fimbriata
Artemisia annua
Camellia oleifera
Polygonatum sibiricum
Schizonepeta tenuifolia

Ceratocystis ulmi
(Dutch elm disease)
Allium cepa
Allium sativum
Aster macrophyllus
Avena sativa
Betula lenta

Ceratocystis ulmi
  (cont'd)
Betula papyrifera
Chelidonium majus
Cymbopogon citratus
Dryopteris austriaca
Erigeron philadelphicus
Euphorbia dentata
Laurus nobilis
Lycopodium obscurum
  var. dendroideum
Maclura pomifera
Origanum majorana
Piper methysticum
Plumeria multiflora
Potentilla fructicosa
Prenanthes alba
Thymus vulgaris
Xanthium pennsylvanicum

Cercospora cruenta
(Leafspot of mungbean)
Allium sativum
Euphorbia pulcherrima
Leucaena leucocephala
Mirabilis jalapa

Cladosporium cucumerinum
(Cucumber scab)
Allium sativum
Piper betle

Cladosporium fulvum
(Leaf mold of tomato)
Allium sativum
Betula lenta
Betula papyrifera
Coriandrum sativum
Cymbopogon citratus
Laurus nobilis
Mentha piperita
Origanum majorana
Thymus vulgaris

Claviceps purpurea
(Ergot of wheat)
Allium cepa
Allium sativum
Betula lenta
Cymbopogon citratus
Laurus nobilis
Origanum majorana
Prunus laurocerasus
Thymus vulgaris

Colletotrichum atramentarium
(Potato anthracnose)
Arachis hypogaea
Azadirachta indica
Madhuca indica
Ricinus communis

Colletotrichum capsici
(Ripe rot of pepper)
Allium sativum
Cestrum diurnum

Colletotrichum circinans
(Onion smudge)
Allium cepa
Allium sativum
Brassica nigra

Colletotrichum falcatum
(Red rot of sugarcane)
Acacia catechu
Acacia nilotica
Anagallis arvensis
Carissa opaca
Cassia fistula
Cassia tora
Cymbopogon martinii
Cymbopogon oliverii
Emblica officinalis
Geranium nepalense
Punica granatum
Syzygium cumini
Terminalia arjuna
Trachyspermum ammi

Colletotrichum gloeosporides
(Wither tip of citrus)
Cestrum diurnum

Colletotrichum lagenarium
(Anthracnose of cucurbits)
Aegopodium podagraria

Colletotrichum lindemuthianum
(Bean anthracnose)
Allium cepa
Allium sativum
Chenopodium album
Cucumis melo
Cymbopogon martinii
Cymbopogon oliverii
Davidia involucrata
Evodia daniellii

Colletotrichum lindemuthianum
  (cont'd)
Fontanesiana editorum
Hosta minor
Impatiens capensis
Liriope spicata
Magnolia grandiflora
Quercus glandulifera
Trachyspermum ammi
Trillium grandiflorum
Tropaeolum majus
Zea mays

Colletotrichum papayae
(Papaya anthracnose)
Anagallis arvensis

Colletotrichum pisi
(Pea anthracnose)
Avena sativa

Colletotrichum trifolii
(Red clover anthracnose)
Allium cepa
Allium sativum

Curvularia lunata
(Leaf spot of gladioli)
Allamanda cathartica
Allium cepa
Allium sativum
Alocasia macrorrhiza
Anethum graveolens
Artabotrys hexapetalus
Cestrum diurnum
Cinnamomum zeylanicum
Clematis gouriana
Cuminum cyminum
Curcuma amada
Lycopersicon lycopersicum
Nigella sativa
Ranunculus sceleratus
Rosa chinensis
Terminalia bellirica
Trachyspermum ammi

Curvularia oryzae
(Leaf spot of rice)
Curcuma angustifolia
Curcuma aromatica
Nigella sativa
Prosopis spicigera

Curvularia penniseti
(Bajra leaf spot)
Acacia loculata
Allium cepa
Allium sativum
Amaranthus viridis
Beta vulgaris
Cucumis prophetarum
Datura stramonium
Eruca vesicaria
Ficus religiosa
Lawsonia inermis
Melia azedarach
Mentha piperita
Ocimum sanctum
Veronica cinerea
Zizyphus jujuba

Curvularia sp.
Curcuma amada

Diaporthe nomurai
Sophora japonica

Diplodia maydis
(Stalk and ear rot of corn)
Allium cepa
Allium sativum
Betula lenta
Cymbopogon citratus
Laurus nobilis
Origanum majorana
Prunus laurocerasus
Thymus vulgaris

Drechslera graminea
(Striped disease of barley)
Allamanda cathartica
Allium cepa
Allium sativum
Artabotrys hexapetalus
Cestrum diurnum
Clematis gouriana
Cyperus rotundus
Lycopersicon lycopersicum
Nigella sativa
Piper longum
Ranunculus sceleratus
Solanum tuberosum
Terminalia bellirica

Drechslera oryzae
(Leaf spot of rice)
Abelmoschus esculentus
Aerva lantana
Allium sativum
Amaranthus spinosus
Amorphophallus campanulatus
Antidesma pentandrum
Averrhoa bilimbi
Basella alba
Blumea balsamifera
Brassica integrifolia
Callicarpa candicans
Carica papaya
Catharanthus roseus
Chrysophyllum oliviforme
Cinnamomum zeylanicum
Corchorus olitorius
Curcuma amada
Curcuma angustifolia
Curcuma aromatica
Curcuma zedoaria
Cymbopogon martinii
Cymbopogon oliverii
Cymbopogon sp.
Dryopteris cochealata
Erigeron linifolius
Euphorbia hirta
Euphorbia pulcherrima
Gardenia jasminoides
Impatiens balsamina
Ipomoea aquatica
Ipomoea batatas
Ixora coccinea
Jatropha gossypifolia
Jatropha podagrica
Leucosyke capitellata
Lipia geminata
Mentha arvensis
Mirabilis jalapa
Nerium oleander
Ocimum sanctum
Parthenium hysterophorus
Piper betle
Plumbago auriculata
Plumbago indica
Polyalthia longifolia
Portulaca oleracea
Pseudocalymna alliaceum
Psidium guajava
Raphanus sativus
Rosa chinensis
Samanea saman
Selinum tenuifolium

Drechslera oryzae
 (cont'd)
Stachytarpheta jamaicensis
Sterculia foetida
Symphytum officinale
Tagetes erecta
Trachyspermum ammi
Zebrina pendula
Zingiber officinale

Endothia parasitica
(Chestnut blight fungi)
Castanea mollissima
Davidia involucrata
Evodia daniellii
Fontanesiana editorum
Hosta minor
Liriope spicata
Magnolia grandiflora
Quercus glandulifera
Trillium grandiflorum

Exserohilum turcicum
(Northern corn leaf blight)
Ammania baccifera
Ammania multiflora
Anagallis arvensis
Callistemon citrinus
Cleome viscosa
Erigeron canadensis
Erigeron linifolius
Lawsonia inermis
Medicago sativa
Mollugo pentaphylla
Ocimum sanctum
Oldenlandia corymbosa
Ranunculus sceleratus
Tropaeolum majus

Fomes annosus
(Heart rot of timber)
Thuja plicata

Fusarium chlamydosporum
Antheum graveolens
Cuminum cyminum

Fusarium culmorum
(Cereal rot)
Aegopodium podograria
Allium cepa
Allium sativum
Eugenia heyneana
Eupatorium ayapana

Fusarium graminearum
(Corn root rot)
Allium cepa
Allium sativum

Fusarium lateritium f. sp. mori
(Twig blight)
Morus bombycis
Ulmus parvifolia

Fusarium moniliforme
(Root rot of corn)
Allium cepa
Allium sativum
Cestrum diurnum
Cymbopogon martinii
Cymbopogon oliverii
Lipia geminata
Trachyspermum ammi

Fusarium nivale
(Pink snow mold of grass)
Abelmoschus esculentus
Allamanda cathartica
Allium cepa
Allium sativum
Artabotrys hexapetalus
Brassica oleracea var. capitata
Cassia auriculata
Clematis gouriana
Cyperus rotundus
Dolichos lablab
Lycopersicon lycopersicum
Ranunculus sceleratus
Raphanus sativus
Rosa chinensis
Symplocos paniculata
Terminalia bellirica
Tridax procumbens
Trifolium pratense

Fusarium oxysporum
(Wilts)
Acanthospermum hispidum
Allium sativum
Beta vulgaris
Blumea eriantha
Brassica oleracea
  var. gongyloides
Brassica rapa
Brassica rapa
  var. rapifera
Carpinus tschonoskii
Cestrum diurnum

Fusarium oxysporum
  (cont'd)
Curcuma zedoaria
Eugenia heyneana
Eupatorium ayapana
Heliopsis helianthoides
Heliopsis longipes
Lepidium virginicum
Lycopersicon lycopersicum
Madhuca indica
Piper nigrum
Plumeria multiflora
Polyalthia longifolia
Rosa chinensis
Xanthium pennsylvanicum
Zanthoxylum alatum

Fusarium oxysporum
  f. sp. conglutinans
(Wilt of cabbage)
Allium cepa
Allium sativum
Betula lenta
Cymbopogon citratus
Laurus nobilis
Lycopersicon lycopersicum
Origanum majorana
Thymus vulgaris

Fusarium oxysporum f. sp. cubense
(Wilt of buckwheat)
Crotalaria anagyroides
Dolichos lablab
Mucuna deeringiana
Saccharum officinarum
Sorghum bicolor
  var. saccharatum

Fusarium oxysporum f. sp. lini
(Flax wilt)
Arachis hypogaea
Brassica rapa
Brassica sp.

Fusarium oxysporum
  f. sp. lycopersici
(Tomato wilt fungi)
Allium cepa
Allium sativum
Arachis hypogaea
Azadirachta indica
Betula lenta
Chelidonium majus
Cymbopogon citratus

Fusarium oxysporum
  f. sp. lycopersici
  (cont'd)
Dryopteris austriaca
Humulus lupulus
Impatiens capensis
Laurus nobilis
Lycopersicon lycopersicum
Madhuca indica
Moringa pterygosperma
Origanum majorana
Osmunda regalis
  var. spectabilis
Rhododendron maximum
Ricinus communis
Solidago bicolor
Solidago canadensis
Thymus vulgaris

Fusarium oxysporum f. sp. pisi
(Pea wilt)
Lycopersicon lycopersicum

Fusarium oxysporum f. sp. saccharum
(Sugarcane wilt)
Pueraria phaseoloides

Fusarium oxysporum f. sp. udum
(Pigeon pea wilt)
Allium sativum
Arachis hypogaea
Brassica rapa
Brassica sp.

Fusarium oxysporum
  f. sp. vasinfectum
(Cotton wilt)
Arachis hypogaea
Brassica rapa
Brassica sp.

Fusarium poae
(Carnation bud rot)
Allium cepa
Allium sativum
Anethum graveolens
Cuminum cyminum
Trachyspermum ammi

Fusarium roseum
(Root and stem rot)
Broussonetia kazinoki
Broussonetia papyrifera
Calla palustris

Fusarium roseum
  (cont'd)
Carex lacustris
Castanea crenata
Ceratophyllum demersum
Chara vulgaris
Cudrania tricuspidata
Diospyros kaki
Elodea canadensis
Lemna minor
Morus bombycis
Myriophyllum spicatum
Nuphar variegatum
Nymphaea tuberosa
Potamogeton amplifolius
Potamogeton nutans
Potamogeton pectinatus
Potamogeton richardsonii
Potamogeton zosteriformis
Sagittaria latifolia
Typha angustifolia
Ulmus parvifolia
Vallisneria americana
Vitis labrusca
Zizania aquatica

Fusarium solani
(Root rot)
Beta vulgaris
Brassica rapa
  var. rapifera
Brassica oleracea
  var. gongyloides
Cestrum diurnum
Cinnamomum zeylanicum
Cymbopogon martinii
Cymbopogon oliverii
Trachyspermum ammi

Fusarium solani f. sp. mori
Castanea crenata
Diospyros kaki
Ulmus parvifolia

Fusarium solani f. sp. phaseoli
(Dry root rot of bean)
Avena sativa
Coffea sp.
Hordeum vulgare
Pinus sp.
Sorghum bicolor
Triticum aestivum
Zea mays

Fusarium sp.
Allium cepa
Allium sativum
Arachis hypogaea
Azadirachta indica
Catalpa speciosa
Didymocarpus pedicellata
Lomatium dissectum
Madhuca indica
Maughania chappar
Rhus typhina
Ricinus communis

Fusicaldium effusum
(Pecan scab fungus)
Syzygium aromaticum
Thymus vulgaris

Gibberella fujikuroi
(Kernel rot of corn)
Allium cepa
Allium sativum
Betula lenta
Cymbopogon citratus
Laurus nobilis
Origanum majorana
Thymus vulgaris

Gloeosporium limetticola
(Wither tip disease of lime)
Citrus aurantiifolia

Glomerella cingulata
(Anthracnose)
Aegopodium podograria
Allium sativum
Erigeron linifolius
Impatiens capensis
Rhododendron maximum

Glomerella gossypii
(Cotton anthracnose)
Camellia oleifera
Daphne genkwa
Firmiana simplex
Rheum officinale
Stemona japonica
Stemona sessilifolia
Stemona tuberosa

Helminthosporium sp.
(Leaf blights)
Acacia loculata
Allium cepa

Helminthosproium sp.
  (cont'd)
Amaranthus viridis
Argemone mexicana
Beta vulgaris
Boswellia serrata
Catalpa speciosa
Cucumis prophetarum
Curcuma domestica
Datura stramonium
Didymocarpus pedicellata
Eruca vesicaria
Eugenia heyneana
Eupatorium ayapana
Ficus religiosa
Lawsonia inermis
Lepidium virginicum
Maughania chappar
Melia azedarach
Mentha piperita
Ocimum sanctum
Prosopis spicigera
Veronica cinerea
Zizyphus jujuba

Lentinus lepideus
(Brown rot of conifers)
Allium sativum
Capsicum annuum
Cassia auriculata
Cinnamomum zeylanicum
Coriandrum sativum
Cymbopogon nardus
Foeniculum vulgare
Laurus nobilis
Lavandula angustifolia
Myristica acuminata
Ocimum basilicum
Thuja plicata
Thymus vulgaris

Lenzites saepiaria
(Timber rot)
Thuga plicata

Lenzites trabea
(Timber rot)
Allium sativum
Capsicum annuum
Cassia auriculata
Cinnamomum zeylanicum
Coriandrum sativum
Cymbopogon nardus
Foeniculum vulgare

Lenzites trabea
   (cont'd)
Laurus nobilis
Lavandula angustifolia
Myristica acuminata
Ocimum basilicum
Thymus vulgaris

Macrophomina phaseolina
(Jute stem rot)
Cymbopogon nardus
Elaeis guineensis
Madhuca indica

Monilia fructigena
(Rot of stone fruits)
Vicia faba

Monilia sp.
(Rot of stone fruits)
Pinus sp.

Monilinia fructicola
(Brown rot of peach/apricot)
Allium sativum
Chenopodium album
Cichorium intybus
Clematis virginiana
Cucumis melo
Davidia involucrata
Evodia daniellii
Ginkgo biloba
Hosta minor
Liriope spicata
Magnolia grandiflora
Phaseolus vulgaris
Pisum sativum
Quercus glandulifera
Sophora japonica
Sophora tetraptera
Sophora tomentosa
Swartzia madagascariensis
Trifolium pratense
Trillium grandiflorum
Tropaeolum majus
Zea mays

Monilinia sp.
(Brown rot of stone fruits)
Plumeria multiflora

Mucor racemosus
(Storage rot of sweet potato)
Amphiroa fragilissima
Dictyosphaerium divaricata

Mucor racemosus
   (cont'd)
Dictyosphaerium favulosa
Dictyosphaerium indica
Hormothamnion entermorphoides
Spyridium filamentosum

Mucor sp.
(Storage rot of root crops)
Curcuma amada

Myrothecium verricaria
(Rice seed fungus)
Allium cepa
Allium sativum
Andropogon schoenanthus

Ophiobolus graminis
(Take-all of wheat)
Avena sativa

Peronospora tabacina
(Blue mold of tobacco)
Aleurites fordii
Arachis hypogaea
Brassica napus
Glycine max
Gossypium hirsutum
Linum usitatissimum
Sesamum indicum
Zea mays

Pestalotia sp.
(Leaf spots)
Allium sativum

Phoma sp.
(Stem canker)
Cestrum diurnum

Phomopsis sp.
(Stem canker)
Allium sativum

Phymatotrichum omnivorum
(Cotton root rot)
Mahonia swaseyi
Mahonia trifoliata

Phytophthora cinnamoni
(Avocado root rot)
Medicago sativa

Phytophthora citrophthora
(Citrus root rot)
Humulus lupulus

Phytophthora drechsleri
(Tuber rot of potato)
Carthamus tinctorius

Phytophthora infestans
(Late blight of potato)
Aconitum kusnezoffii
Acorus tatarinowii
Akebia trifoliata
Allium tuberosum
Anemone chinensis
Anemone hupehensis
Artemisia annua
Artemisia apiacea
Belamcanda chinensis
Carpesium abrotanoides
Clerodendrum bungei
Cnidium monnieri
Coriaria sinica
Dryopteris crassirhizoma
Epimedium sagittatum
Euphorbia helioscopia
Leonurus sibiricus
Lindera strychnifolia
Photinia serrulata
Pinus massoniana
Premna microphylla
Rheum officinale
Schizonepeta tenuifolia
Solanum tuberosum
Swertia chinensis
Typhonium giganteum

Phytophthora parasitica
(Brown rot of citrus)
Cassia fistula
Cuminum cyminum
Curcuma amada
Luvungia scandens
Mentha arvensis
Pimpinella anisum
Syzygium aromaticum

Plasmodiophora brassicae
(Club foot of crucifers)
Brassica sp.

Polyporus versicolor
(Wood rot)
Allium sativum
Capsicum annuum

Polyporus versicolor
   (cont'd)
Cassia auriculata
Cinnamomum zeylanicum
Coriandrum sativum
Cymbopogon nardus
Foeniculum vulgare
Laurus nobilis
Lavandula angustifolia
Myristica acuminata
Ocimum basilicum
Thymus vulgaris

Pseudoperonospora cubensis
(Downy mildew of cucumber)
Ailanthus giralolii
Allium sativum
Dryopteris crassirhizoma
Leonurus sibiricus
Platycarya strobilacea

Puccinia glumarum
(Yellow wheat rust)
Anemone hupehensis
Artemisia annua

Puccinia graminis tritici
(Wheat rust)
Aconitum kusnezoffii
Agrimonia pilosa
Ailanthus giralolii
Anemone chinensis
Angelica dahurica
Arisaema serratum
Arisaema thunbergii
Artemisia annua
Artemisia apiacea
Atractylodes chinensis
Belamcanda chinensis
Camellia oleifera
Carpesium abrotanoides
Clematis aethusifolis
Cnidium monnieri
Cyrtomium fortunei
Daphne genkwa
Dryopteris crassirhizoma
Epimedium sagittatum
Euphorbia esula
Euphorbia helioscopia
Euphorbia pekinensis
Houttuynia cordata
Lindera strychnifolia
Liquidambar taiwaniana
Macleaya cordata
Polygonatum sibiricum

Puccinia graminis tritici
(cont'd)
Premna microphylla
Rheum officinale
Schizonepeta tenuifolia
Stellera chamaejasme
Swertia chinensis
Typhonium giganteum
Verbena officinalis

Puccinia rubigavera
(Rust)
Aconitum kusnezoffii
Agrimonia pilosa
Anemone chinensis
Belamcanda chinensis
Camellia oleifera
Carpesium abrotanoides
Clematis aethusifolis
Cnidium monnieri
Epimedium sagittatum
Euphorbia esula
Euphorbia pekinensis
Leonurus sibiricus
Lindera strychnifolia
Pinellia ternata
Polygonatum sibiricum
Premna microphylla
Rheum officinale
Schizonepeta tenuifolia
Sophora japonica
Stellera chamaejasme
Typhonium giganteum
Verbena officinalis

Pyricularia oryzae
(Rice blast)
Acacia catechu
Acacia nilotica
Allium sativum
Amorphophallus campanulatus
Antidesma pentandrum
Blumea balsamifera
Brassica integrifolia
Carissa opaca
Cassia fistula
Cassia tora
Cestrum diurnum
Clerodendrum indicum
Cnidium monnieri
Curcuma domestica
Cymbopogon sp.
Derris elliptica
Emblica officinalis

Pyricularia oryzae
(cont'd)
Erythroxylum coca
Euphorbia pulcherrima
Geranium nepalense
Impatiens balsamina
Ipomoea batatas
Jatropha gossypifolia
Kalanchoe pinnatum
Leonurus sibiricus
Ocimum sanctum
Parthenium hysterophorus
Piper betle
Platycarya strobilacea
Polyalthia longifolia
Pseudocalymna alliaceum
Punica granatum
Raphanus sativus
Stachytarpheta jamaicensis
Symphytum officinale
Syzygium cumini
Tabernaemontana pandacaqui
Tagetes erecta
Terminalia arjuna
Terminalia bellirica
Terminalia chebula
Vitex negundo
Zanthoxylum dimorphophyllum
Zanthoxylum planispinum

Pythium aphanidermatum
(Damping-off of sugar beets)
Anagallis arvensis
Arachis hypogaea
Boswellia serrata
Erigeron linifolius
Eugenia heyneana
Eupatorium ayapana
Ricinus communis
Rosa chinensis

Pythium irregulare
(Melon root rot)
Avena sativa

Pythium sp.
(Damping off, root rot)
Curcuma amada
Humulus lupulus

Pythium ultimum
(Root rot of sweet potato)
Lepidium virginicum

Rhizoctonia bataticola
(Root and stem rot)
Cestrum diurnum

Rhizoctonia oryzae
(Stem rot of rice)
Lomatium dissectum

Rhizoctonia solani
(Root and stem rot)
Anethum graveolens
Arachis hypogaea
Arisaema serratum
Arisaema thunbergii
Artemisia apiacea
Atractylodes chinensis
Avena sativa
Azadirachta indica
Catalpa speciosa
Cestrum diurnum
Cuminum cyminum
Curcuma domestica
Cymbopogon martinii
Cymbopogon oliverii
Cymbopogon sp.
Fagopyrum esculentum
Glycine max
Hordeum vulgare
Humulus lupulus
Lomatium dissectum
Madhuca indica
Medicago sativa
Ocimum sanctum
Parthenium hysterophorus
Pinus sp.
Piper betle
Polyalthia longifolia
Premna microphylla
Ricinus communis
Sesamum indicum
Sorghum bicolor
Sorghum sudanense
Stemona japonica
Stemona sessilifolia
Stemona tuberosa
Trachyspermum ammi
Triticum aestivum
Verbena officinalis
Zea mays
Zingiber officinale

Rhizoctonia sp.
(Root and crown rot)
Lepidium virginicum

Rhizopus nigricans
(Soft rot of sweet potato)
Anagallis arvensis
Beta vulgaris
Brassica oleracea
  var. gongyloides
Catalpa speciosa
Humulus lupulus
Moringa pterygosperma
Rosa chinensis

Rhizopus oryzae
(Rot of stored rice)
Amphiroa fragilissima
Dictyosphaerium divaricata
Hormothamnion entermorphoides
Padina gymnospora
Udotea flabellum

Sclerotinia fructicola
(Brown rot of stone fruit)
Collinsonia canadensis
Dipsacus fullonum
Helianthus decapetalus
Humulus lupulus
Impatiens capensis
Piper methysticum
Pisum sativum

Sclerotinia trifoliorum
(Stem rot, wilt of clover)
Trifolium pratense

Sclerotium bataticola
(Cotton rot)
Humulus lupulus

Sclerotium oryzae
(Stem rot of rice)
Curcuma domestica
Cymbopogon sp.
Parthenium hysterophorus
Polyalthia longifolia
Zingiber officinale

Sclerotium rolfsii
(Root rot of peanut)
Azadirachta indica
Curcuma domestica
Curcuma zedoaria
Cymbopogon sp.
Erigeron linifolius
Lepidium virginicum
Madhuca indica

Scelerotium rolfsii
   (cont'd)
Parthenium hysterophorus
Polyalthia longifolia
Zingiber officinale

Sclerotium sp.
(Root and stem rot)
Curcuma amada

Septoria nodorum
(Blight of wheat)
Aegopodium podograria

Sphaceloma ampelinum
(Anthracnose of grape)
Allium sativum

Sphaerothica humuli
(Hop powdery mildew)
Arachis hypogaea
Brassica napus
Brassica sp.
Glycine max
Gossypium hirsutum
Linum usitatissimum
Olea europaea
Prunus persica
Ricinus communis
Sesamum indicum
Zea mays

Thielaviopis basicola
(Bean root rot)
Hordeum vulgare
Pinus sp.
Triticum aestivum
Zea mays

Trichoderma viride
(Green mold rot of garlic)
Allium sativum
Curcuma angustifolia
Curcuma aromatica
Eugenia heyneana
Eupatorium ayapana
Murraya paniculata

Uramyces phaseoli
(Bean rust)
Tagetes erecta
Tagetes patula

Uramyces sp. (Rust)
Nicotiana glutinosa

Ustilago avenae
(Black smut of oat)
Allium cepa
Allium sativum
Betula lenta
Cymbopogon citratus
Laurus nobilis
Myrrhis odorata
Thymus vulgaris

Ustilago hordei
(Covered smut of barley)
Allamanda cathartica
Allium sativum
Anagallis arvensis
Artabotrys hexapetalus
Cannabis sativa
Cestrum diurnum
Clematis gouriana
Euphorbia pulcherrima
Kalanchoe pinnatum
Lawsonia inermis
Leucas procumbens
Lycopersicon lycopersicum
Millingtonia hortensis
Piper betle
Polygonum gladrum
Psidium guajava
Ranunculus scelaratus
Tamarindus indica
Xanthium strumarium

Ustilago tritici
(Loose smut of barley)
Allamanda cathartica
Allium sativum
Anagallis arvensis
Artabotrys hexapetalus
Cannabis sativa
Cestrum diurnum
Clematis gouriana
Eriobotrya japonica
Euphorbia pulcherrima
Kalanchoe pinnatum
Lawsonia inermis
Lycopersicon lycopersicum
Manihot esculenta
Millingtonia hortensis
Piper betle
Polygonum glabrum
Psidium guajava
Ranunculus scelaratus
Tamarindus indica
Xanthium strumarium

**Venturia inaequalis
(Apple scab)**
Acer pseudoplatanus
Aesculus hippocastanum
Allium cepa
Anemone nemorosa
Atriplex patula
Bellis perennis
Brassica sinapistrum
Caltha palustris
Castanea sativa
Chrysanthemum segetum
Clinopodium vulgare
Cornus sanguinea
Endymion non-scriptus
Hedera helix
Hieracium boreale
Lonicera periclymenum
Lycopersicon lycopersicum
Medicago lupulina
Pastinaca sativa
Pimpinella saxifraga
Primula vulgaris
Pyrus serrulata
Pyrus ussuriensis
Quercus cerris
Ranunculus ficaria
Raphanus sativus
Salix fragilis
Salix purpurea
Solanum dulcamara

Venturia inaequalis
(cont'd)
Solanum tuberosum
Symphoricarpos albus
Tamus communis
Viburnum lantana

**Verticillium albo-atrum
(Verticillium wilt)**
Allium cepa
Allium sativum
Avena sativa
Betula lenta
Carpesium abrotanoides
Clematis aethusifolis
Cymbopogon citratus
Epimedium sagittatum
Euphorbia pekinensis
Hordeum vulgare
Laurus nobilis
Medicago sativa
Origanum majorana
Premna microphylla
Schizonepeta tenuifolia
Thymus vulgaris
Triticum aestivum

**Verticillium dahliae
(Dahlia wilt)**
Medicago sativa

INSECTS

Abraxas miranda
(Caterpillar)
Cocculus trilobus

Acalymma vittata
(Striped cucumber beetle)
Aleuritis fordii
Amorpha fruticosa
Azadirachta indica
Derris sp.
Juglans nigra
Nicotiana sp.
Schoenocaulon officinale

Acanthoscelides obtectus
(Bean beetle)
Heliopsis longipes
Piper nigrum

Achaea janata
(Croton caterpillar)
Annona reticulata
Azadirachta indica
Diospyros montana
Vitex negundo

Acraea issoria
(China glass butterfly)
Millettia pachycarpa
Pachyrhizus erosus
Tripterygium forrestii

Acrida exatana
(Locust)
Azadirachta indica

Acromyrmex cephalotes
(Leaf-cutting ant)
Citrus maxima
Helianthus annuus
Linium usitatissimum

Acromyrmex octospinosus
(Leaf-cutting ant)
Arachis hypogaea
Brassica napus
Citrus maxima
Helianthus annuus
Linum usitatissimum

Acrosternum hilare
(Green stink bug)
Pachyrhizus erosus
Schoencaulon officinale

Acyrthosiphum pisum
(Pea aphid)
Amorpha fruticosa
Anabasis aphylla
Anacyclus pyrethrum
Annona muricata
Brassica rapa
  var. rapifera
Chrysanthemum cinerariifolium
Derris sp.
Nicotiana sp.
Pastinaca sativa
Quassia amara

Aedes aegypti
(Yellow fever mosquito)
Achillea millefolium var. lanulos
Acorus calamus
Allium sativum
Amianthium muscitoxicum
Amorpha fruticosa
Annona glabra
Annona muricata
Annona squamosa
Apocynum androsaemifolium
Aristolochia bracteata
Asclepias incarnata
Asclepias speciosa
Asclepias syriaca
Astragalus adsurgens
  var. robustior
Astragalus canadensis
Astragalus flexuosus
Astragalus pectinatus
Astragalus racemosus
Astragalus tenellus
Astragalus verilliflexus
Camelina microcarpa
Celastrus scandens
Chrysanthemum cinerariifolium
Clerodendron inerme
Conringia orientalis
Convulvulus arvensis
Corispermum hyssopifolium

Corydalis aurea
Curcuma domestica
Cymbopogon citratus
Delphinium virescens
Dolichos buchani
Duranta repens
Euquisetum fluviatile
Eupatorium maculatum
Gypsophila paniculata
Heliopsis longipes
Laportea canadensis
Lobelia siphilitica
Lupinus argenteus
Lysimachia hybridae
Menispermum canadense
Menispermum cocculus
Nepeta cataria
Nicotiana rustica
Pastinaca sativa
Petalostemom villosum
Phryma leptostachya
Physostegia parviflora
Pimpinella anisum
Piper peepuliodes
Piscidia piscipula
Polygonum coccineum
Ranunculus macounii
Sanguinaria canadensis
Saponaria officinalis
Scrophularia lanceolata
Senecio canus
Silene alba
Silene cserei
Sium suave
Solanum trifolium
Syngonium auritum
Tagetes minuta
Verbena urticifolia
Vernonia fasiculea

Aedes nigromaculis
(Mosquito)
Allium sativum

Aedes punctor
(Mosquito)
Eucalyptus botryoides
Eucalyptus paniculata
Prunus maackii

Aedes sierrensis
(Western tree-hole mosquito)
Allium sativum

Aedes sp.
(Moquito)
Backhousia myrtifolia
Doryphora sassafras
Kirchneriella irregularis
Melaleuca bracteata
Prunus racemosus
Zieria smithii

Aedes triseriatus
(Tree-hole mosquito)
Allium sativum

Agrotis ipsilon
(Black cutworm)
Azadirachta indica

Agrotis sp.
(Cutworm)
Anemone chinensis
Artemisia apiacea
Carpesium abrotanoides
Clematis armandii
Clematis chinensis
Daphne genkwa
Illicium lanceolatum
Lycoris africana
Platycarya strobilacea
Pueraria lobata
Vitex cannabifolia

Ahasverus advena
(Grain beetle)
Anacardium occidentale
Euonymus europaea

Aleurothrixus floccosus
(Woolly white fly)
Aleurites fordii
Azadirachta indica

Aleyrodes vaporariorum
(White fly)
Consolida regalis
Delphinium staphisagria
Ricinus communis

Alsophila pometaria
(Fall cankerworm)
Pachyrhizus erosus

Amsacta moorei
(Red hairy caterpillar)
Azadirachta indica

Anasa tristis
(Squash bug)
Anabasis aphylla
Camellia sinensis
Derris sp.
Erigeron canadensis
Erigeron flagellaris
Haplophyton cimicidum
Heliopsis longipes
Lonchocarpus utilis
Nicotiana sp.
'Rotenone'
Schoenocaulon officinale
Sesamum indicum

Anastrepha ludens
(Mexican fruit fly)
Haplophyton cimicidum

Ancysta perseae
(Avocado lace bug)
Schoenocaulon officinale

Andrector ruficornis
Mammea americana
Phyllanthus acuminatus

Anomala cupripes
(June bug)
Celastrus orbiculatus
Melia azedarach

Anopheles quadrimaculatus
(Malaria mosquito)
Aconitum barbatum
Chrysanthemum cinerariifolium
Dalbergia retusa
Haplophyton cimicidum
Kirchneriella irregularis

Anopheles sp.
(Mosquito)
Backhousia myrtifolia
Prunus racemosus
Spilanthes acmella
Spilanthes mauritiana
Spilanthes oleracea

Antestiopsis lineaticollis
(Variegated coffee bug)
Chrysanthemum cinerariifolium

Antestiopsis orbitalis bechuana
(Coffee shield bug)
Azadirachta indica

Antestiopsis sp.
(Antestia bug)
Azadirachta indica

Anthonomus eugenii
(Pepper weevil)
Tripterygium wilfordii

Anthonomus grandis
(Cotton boll weevil)
Abelmoschus esculentus
Alchornea triplinervia
Aleurites fordii
Enterolobium cyclocarpum
Hibiscus syriacus
Piper nigrum

Anthrenus flavipes
(Furniture carpet beetle)
Azadirachta indica

Antigastra catalaunalis
(Sesame webworm)
Azadirachta indica

Ants
(General category)
Acorus calamus
Anthriscus vulgaris
Baeckea frutescens
Brugmansia arborea
Cassia alata
Chenopodium foetidum
Cissus rheifolia
Citrus limon
Delphinium laxiflorum
Euphorbia dendroides
Ferula assa-foetida
Hydnocarpus wightiana
Litsea quatemalensis
Melinis minutiflora
Momordica foetida
Pimpinella anisum
Piper aduncum
Pogostemon patchouli
Senecio ehrenbergianus
Tagetes minuta

Tanacetum vulgare
Thamnosma africana
Vateria indica
Veratrum californicum

Anuraphis maidiradicis
(Corn root aphid)
Citrus limon

Anuraphis roseus
(Apple aphid)
'Rotenone'

Aonidiella aurantii
(California red scale)
Aleurites fordii
Azadirachta indica
'Rotenone'

Aonidiella citrina
(Yellow scale)
Azadirachta indica

Aphidius phorodontis
(Pulse aphid)
Nicotiana sp.

Aphids
(General category)
Acorus tatarinowii
Agrimonia pilosa
Ailanthus giralolii
Alangium platanifolium
Allium fistulosum
Allium tuberosum
Amianthium muscitoxicum
Anabasis aphylla
Anemone chinensis
Angelica dahurica
Annona muricata
Annona palustris
Annona squamosa
Arisaema serratum
Arisaema thunbergii
Artemisia japonica
Asclepias syriaca
Azadirachta indica
Balanites roxburghii
Bidens bipinnata
Bryonia alba
Cardaria draba
Cassia didymobotiya
Cassia laevigata
Cassia multijuga
Cayratia japonica

Chloranthus japonicus
Chrysanthemum cinerariifolium
Cibotium barometz
Cinnamomum camphora
Citrus limon
Clematis aethusifolis
Clerodendrum bungei
Colchicum autumnale
Consolida regalis
Coriandrum sativum
Coriaria sinica
Croton californicus
Croton klotzschianus
Croton tiglium
Cucurbita moschata
Cynanchum auriculatum
Cyrtomium fortunei
Cytisus laburnum
Cytisus scoparius
Daphne genkwa
Datura stramonium
Delphinium staphisagria
Derris philippinensis
Derris polyantha
Derris sp.
Derris uliginosa
Desmodium caudatum
Dichapetalum ruhlandii
Dichroa febrifuga
Dictamnus dasycarpus
Digitalis grandiflora
Digitalis purpurea
Dioscorea hispida
Dioscorea nipponica
Duboisia hopwoodii
Duchesnea indica
Epimedium sagittatum
Eremocarpus setigerus
Eucalyptus globulus
Euphorbia helioscopia
Euphorbia pekinensis
Euphorbia tirucalli
Evodia rutaecarpa
Firmiana simplex
Gerbera piloselloides
Gliricidia sepium
Glochidion puberum
Gossypium herbaceum
Grindelia humilis
Gynura segetum
Houttuynia cordata
Hyoscyamus niger
Ipomoea nil
Ipomoea purpurea
Ipomoea quamoclit

Juglans mandschurica
Lactuca sativa
Lasiosiphon eriocephalus
Lepidium ruderale
Lindera strychnifolia
Lobelia chinensis
Lycopersicon lycopersicum
Lycoris africana
Melinis minutiflora
Menispermum dauricum
Metaplexis japonica
Myrica rubra
Nepeta cataria
Nicotiana glauca
Nicotiana rustica
Nicotiana sp.
Nicotiana tabacum
Opuntia dillenii
Orostachys fimbriata
Paederis scandens
Paullinia pinnata
Pelargonium sp.
Photinia serrulata
Picrasma excelsa
Pinellia ternata
Pinus tabuliformis
  var. yunnanensis
Pogostemon patchouli
Pongamia pinnata
Pueraria lobata
Pycnathemum rigidus
Quassia amara
Quisqualis indica
Rheum officinale
Rhodea japonica
Ricinus communis
Rubia cordifolia
Rumex dentatus
Sambucus williamsii
Sargentodoxa cuneata
Schefflera octophylla
Semiaquilegia adoxoides
Senecio scandens
Similax glabra
Sonchus arvensis
Sophora japonica
Sophora pachycarpa
Stellera chamaejasme
Stemona japonica
Stemona sessilifolia
Stemona tuberosa

Strophanthus divaricatus
Tanacetum vulgare
Tephrosia candida
Tephrosia macropoda
Tephrosia toxicaria
Thevetia peruviana
Typhonium giganteum
Verbena officinalis
Veronicastrum axillare
Vitex cannabifolia
Woodwardia japonica
Xanthium sibiricum
Xeromphis spinosa

Aphis citri
(Citrus aphid)
Derris malaccensis
Melia azedarach
Tephrosia vogelii

Aphis craccivora
(Cowpea, peanut aphid)
Camellia sinensis
Nicotiana tabacum
Tagetes erecta
Tagetes patula
Thevetia peruviana

Aphis fabae
(Black bean aphid
Aganope gabonica
Amorpha fruticosa
Anabasis aphylla
Annona reticulata
Annona squamosa
Arisaema purpureogaleatum
Buddleia lindleyana
Centella asiatica
Chrysanthemum cinerariifolium
Citrus limon
Consolida regalis
Delphinium staphisagria
Derris elliptica
Derris fordii
Derris malaccensis
  var. sarawakensis
Derris polyantha
Derris sp.
Dryopteris filix-mas
Duboisia hopwoodii
Lantana camara

Lobelia inflata
Lonchocarpus chrysophyllus
Lonchocarpus densiflorus
Millettia pachycarpa
Mundulea suberosa
Neorautanenia fisifolia
Nicotiana sp.
Nicotiana sylvestris
Pachyrhizus erosus
Pueraria yunnanensis
Quassia amara
Rhamnus crenata
Rhododendron hunnewellianum
'Rotenone'
Sophora flavescens
Tephrosia candida
Tephrosia macropoda
Tephrosia vogelii
Tripterygium forrestii
Ulex europaeus
Veratrum viride

Aphis gossypii
(Cotton, melon aphid)
Akebia trifoliata
Aloysia triphylla
Amorpha fruticosa
Anemone hupehensis
Artemisia annua
Artemisia apiacea
Artemisia argyi
Azadirachta indica
Camellia sinensis
Cnidium monnieri
Commelina communis
Coriandrum sativum
Cymbopogon citratus
Datura metel
Dracocephalum moldavica
Ehretia dicksonii
Euphorbia fischeriana
Glechoma longituba
Lavandula angustifolia
Leonurus sibiricus
Litsea cubeba
Lycopersicon hirsutum
Mentha haplocalyx
Mentha longifolia
Momordica cochinchinensis
Platycarya strobilacea
Ocimum basilicum
Oxalis corniculata
Pseudolarix kaempferi
Pyrrosia drakeana

Rhus chinensis
Rhus verniciflua
Salvia plebeia
Salvia sclarea
Sanguisorba officinalis
Satureja hortensis
Swertia chinensis
Tephrosia virginiana
Zanthoxylum simulans

Aphis hederae
(Ivy aphid)
Nicotiana sp.

Aphis maidis
(Corn aphid)
Aconitum ferox
Camellia sinensis

Aphis medicaginis
(Cowpea aphid)
Derris elliptica
Derris philippenensis

Aphis mellifera
(Aphid)
Azadirachta indica

Aphis persicae-niger
(Peach aphid)
Amanita muscaria
'Rotenone'

Aphis pomi
(Green apple aphid)
Anabasis aphylla
Chrysanthemum cinerariifolium
Consolida regalis
Delphinium staphisagria
Nicotiana sp.
'Rotenone'

Aphis populifoliae
(Carolina poplar aphid)
Nicotiana sp.

Aphis pseudobrassicae
(Aphid)
'Rotenone'

Aphis sorbi
(Rosy aphid)
Chrysanthemum cinerariifolium
'Rotenone'

Aphis spiraecola
(Spiraea aphid)
Chrysanthemum cinerariifolium
Hura crepitans
Nicotiana sp.
'Rotenone'

Aphis tavaresi
(Citrus aphid)
Camellia sinensis
Mundulea suberosa
Nicotiana sp.

Aphrophora saratogenesis
(Saratoga spittle beetle)
Schoenocaulon officinale

Archips cerasivoranus
(Ugly nest caterpillar)
Pachyrhizus erosus
Tripterygium forrestii

Argyria stricticraspis
(Sugarcane early shoot borer)
Ryania speciosa

Argyrotaenia velutinana
(Red-banded leafroller)
Aleurites fordii
Azadirachta indica

Ascia monuste
(Great Southern white butterfly)
Mammea americana
Pachyrhizus erosus

Ascotis selenaria
(Giant looper)
Ailanthus giralolii
Clematis chinensis
Euphorbia helioscopia
Gossypium herbaceum

Asphondylia sp.
(Gall midge)
Chrysanthemum cinarariifolium

Aspidiotus perniciosa
(San Jose scale)
Consolida regalis
Delphinium staphisagria

Athalia proxima
(Mustard sawfly)
Aconitum ferox
Acorus calamus
Aloe barbadensis
Crinum bulbispemum
Crinum defixum
Euphorbia royleana
Lantana camara

Athalia rosae
(Saw fly)
Amaranthus retroflexus
Anagallis arvensis
Arctium tomentosum
Artemisia vulgaris
Chelidonium majus
Chenopodium album
Convolvulus arvensis
Erigeron canadensis
Euphorbia helioscopia
Fragaria ananassa
Galinsoga parviflora
Glechoma hederacea
Glycine max
Hordeum vulgare
Impatiens parviflora
Juglans regia
Mercurialis annua
Momordica charantia
Nicandra physalodes
Plantago lanceolatum
Polygonum lapathifolium
Portulaca oleracea
Ranunculus acris
Rumex obtusifolius
Senecio vulgaris
Solanum nigrum
Solidago virgaurea
Stellaria media
Trifolium pratense
Urtica dioica
Veronica polita

Atherigona soccata
(Sorghum shoot fly)
Azadirachta indica

Atta cephalotes
(Leaf-cutting ant)
Hymenaea courbaril

Attagenus piceus
(Black carpet beetle)
Abutilon theophrasti
Actaea arguta
Adonis vernalis
Afrormosia laxiflora
Agelaea pentagyna
Akebia quinata
Allionia incarnata
Amanita muscaria
Amianthium muscitoxicum
Amoreuxia wrightii
Amorpha glabra
Ancistrocladus barterii
Anemia mexicana
Anenome nuttalliana
Annona cherimola
Annona glabra
Annona muricata
Annona squamosa
Apium graveolen
Apocynum cannabinum
Apocynum subiricum
Aralia elata
Aralia humilis
Aralia spinosa
Ardisia escallonioides
Ardisia picardae
Arenaria peploides
Aristolochia serpentaria
Artocarpus altilis
Artocarpus heterophyllus
Asclepias eriocarpa
Asclepias kansana
Asclepias labriformis
Asclepias speciosa
Asclepias syriaca
Asclepiodora viridis
Astelia cunninghamii
Baccharis coridifolia
Baccharis floribunda
Balanites aegyptica
Barringtonia asiatica
Barringtonia racemosa
Bellis perennis
Bersama paulinioides
Bidens pilosa
Caesalpinia coriaria
Calotropis procera
Canavalia ensiformis
Canna flaccida
Carapa guianensis
Careya arborea
Cassine xylocarpum
Castela texana

Caulophyllum thalicitroides
Cayoponia ficifolia
Cerototheca sesamoides
Chaenactis douglasii
Chenopodium ambrosioides
Clibodium arboreum
Clitoria arborescens
Cocculus indicus
Collinsonia anisata
Colubrina asiatica
Combretum caoucia
Convallaria majalis
Corispermum hyssopifolium
Crepis bursifolia
Crossosoma bigelovii
Crotalaria sagittalis
Croton texensis
Cryptostegia grandiflora
Cucubita foetidissima
Cuscuta americana
Cuscuta racemosa
Cynoglossum officinale
Cyrilla racemiflora
Dacryodes hexandra
Daphne cneorum
Deeringia celosioides
Delphinium staphisagria
Dichapetalum cymosum
Dichapetalum toxicaria
Didymopanax morototoni
Didymopanax tremulum
Dieffenbachia sequine
Dioscorea bulbifera
Dolichos buchani
Dolichos lablab
Drymaria pachyphylla
Duranta repens
Eclipta alba
Enterolobium contortisiliquum
Enrerolobium cyclocarpum
Eragrostis cilianenis
Erigeron annuus
Erigeron glabellus
Erodium cicutarium
Erythrophleum couminga
Erythrophleum ivorense
Erythrophleum suaveolans
Ethulia conyzoides
Euonymus atropurpurea
Euonymus europaea
Eupatorium aromaticum
Evodia hupehensis
Fevillea cordifolia
Funastrum gracile
Gelsemium sempervirens

Gleditsia amorphoides
Gleditsia aquatica
Gleditsia sinensis
Grindelia perennis
Grindelia squarrosa
Gustavia augusta
Gymnocladus dioica
Gynocardia odorata
Halesia carolina
Haplophyton cimicidum
Harpullia arborea
Hedera helix
Helenium quadridentalum
Heliopsis longipes
Hiptage benghalensis
Hybanthus yucatansis
Ilex verticillata
Ipomopsis aggregata
Jacquinia barbasco
Jatropha angustidens
Jatropha macrorhiza
Joannesia princeps
Kleinhovia hospita
Laetia calophylla
Lantana horrida
Lemaireocereus gummosus
Leonotis leonurus
Leucothoe axillaris
Lewisia rediviva
Lophopetalum toxicum
Luffa acutangula
Luina hypoleuca
Lupinus mutabilis
Lycopodium clavatum
Lycoris radiata
Machaeranthera varians
Maesa indica
Maesa rufescens
Maillardia bordonica
Malouetia obtusiloba
Malouetia tamaquarina
Mammea americana
Mandevilla mollissima
Marsdenia clausa
Melia azedarach
Melicope erythrococca
Menabea venenata
Mertensia lanceolata
Momordica charantia
Monotropa uniflora
Nerium oleander
Ochrocarpus africanus
Omphalea diandra
Omphalea triandra
Ornithogalum umbellatum

Orthocarpus luteus
Pachyrhizus erosus
Paeonia brownii
Passiflora quadrangularis
Peltophorum suringari
Pentaclethra macroloba
Petiveria alliacea
Philodendron hastatum
Phryma leptostachya
Physostegia parviflora
Phytolacca americana
Pieris floribunda
Pimpinella anisum
Piscidia piscipula
Pittosporum senacia
Plantago monticola
Platonia insignis
Podophyllum peltatum
Pogogyne parviflora
Poterium sanguisorba
Psoralea glandulosa
Purshia tridentata
Pycanthus kombo
Pyrularia pubera
Quassia amara
Rhinacanthus communus
Rhipsalis leucorhaphis
Rivinia humilis
Rosmarinus officinalis
Ryania angustifolia
Ryania speciosa
Salvia mellifera
Samadera indica
Sansevieria hyacinthoides
Satureja douglasii
Saxifraga bronchialis
Schkuria pinnata
Schoenocaulon officinale
Selenicereus grandiflora
Senecio ehrenbergianus
Sesuvium portulacastrum
Solidago microglossa
Solidago occidentalis
Solidago odora
Sophora sericea
Spergularia marina
Spigelia humboltiana
Spigelia marilandica
Sprekelia formosissima
Syngonium auritum
Syringa oblicata
Syzygium aromaticum
Tanghinia venenifera
Tephrosia toxicaria
Thevetia gaumeri

Thevetia ovata
Thevetia peruviana
Tournefortia hirsutissima
Tournefortia volubilis
Trillium erectum
Trillium sessile var. giganteum
Urtica breweri
Veratrum album
Verbesina encelioides
   var. exauriculata
Vetiveria zizanioides
Wikstroemia sandwicensis
Willardia mexicana
Xanthium canadense
Xanthium spinosum
Xeromphis spinosa
Yucca aloifolia
Zephyranthes drummondii
Zigadenus glaucus
Zigadenus paniculatus

Aulacophora abdominalis
(Cucurbit beetle)
Datura stramonium

Aulacophora cattigarensis
(Black cucurbit beetle)
Millettia pachycarpa
Sophora flavescens
Tripterygium forrestii

Aulacophora foveicollis
(Red pumpkin beetle)
Aconitum ferox
Acorus calamus
Annona squamosa
Azadirachta indica
Calotropis procera
Jatropha curcus
Melia azedarach
Peganum harmala
Withania somnifera

Aulacophora hilaris
(Pumpkin beetle)
Annona squamosa

Aulocaspis rosae
(Rose scale)
'Rotenone'

Autographa brassicae
(Cabbage looper)
Cassia tora
Chrysanthemum cinerariifolium
Derris elliptica
Derris malaccensis
Derris sp.
Lonchocarpus utilis
Nicotiana sp.
Schoenocaulon officinale

Autographa sp.
(Caterpillar)
Erigeron canadensis
Erigeron filagellaris
Haplophyton cimicidum
Sesamum indicum

Autoserica sp.
(Mulberry scarab beetle)
Millettia pachycarpa

Bagrada cruciferarum
(Painted bug)
Argemone mexicana
Melia azedarach

Bagrada picta
(Harlequin bug)
Acorus calamus
Argemone mexicana
Calotropis procera
Melia azedarach
Nerium oleander

Bean plataspids
(General category)
Rhododendron molle
Tripterygium forrestii

Beetles
Adina cordifolia
Agauria salicifolia
Clerodendrum glabrum
Colchicum autumnale
Cymbopogon nardus
Daphne mezereum
Delonix regia
Erythrina senegalensis
Mundulea suberosa
Taxus baccata
Veratrum album
Veratrum californicum

Bemisia tabaci
(Tobacco white fly)
Nicotiana rustica

Blattella germanica
(German cockroach)
Asclepias tuberosa
Baccharis glutinosa
Brassica rapa var. rapifer
Ceropegia dichotoma
Citrus limon
Crossosoma bigelovii
Cuscuta americana
Daemia tomentosa
Delphinium virescens
Drymaria pachyphylla
Flourensia cernua
Gleditsia amorphoides
Haplophyton cimicidum
Lathyrus sylvestris
Madia glomerata
Mammea americana
Mandevilla foliosa
Melia azedarach
Pachyrhizus erosus
'Rotenone'
Ryania angustifolia
Ryania speciosa
Veratrum viride

Blatta orientalis
(Oriental cockroach)
Ananus sativus
Armoracia rusticana
Gaultheria procumbens
Hedeoma pulegioide
Matricaria recutita
Prunus dulcis
Santalum album

Blissus leucopterus
(Chinch bug)
Amorpha fruticosa
Schoenocaulon officinale

Blow flies
(General category)
Aconitum napellus
Conium maculatum
Dacrydium franklinii
Dracunculus vulgaris
Tagetes minuta
Zieria smithii

Boarmia selenaria
(Giant looper)
Azadirachta indica

Bombyx mori
(Silkworm)
Acorus calamus
Ajuga remota
Amianthium muscitoxicum
Annona squamosa
Chrysanthemum cinerariifolium
Cinnamomum zeylanicum
Clematis dioica
Clibodium surinamense
Croton tiglium
Derris elliptica
Derris uliginosa
Melia azedarach
Millettia pachycarpa
Myristica fragrans
Nicotiana sp.
Pachyrhiyzus erosus
Pithecellobium trapezifolium
Prunus persica
Quassia amara
Schoenocaulon officinale
Tripterygium wilfordii

Brevicoryne brassicae
(Cabbage aphid)
Annona squamosa
Chrysanthemum cinerariifolium
Derris sp.
Latuca sp.
Lonchocarpus utilis
Melia azedarach
Nicotiana sp.
Pachyrhizus tuberosus
Tephrosia virginiana

Brithys pancratii
(Caterpillar)
Mundulea suberosa
Nicotiana sp.

Bruchid sp.
Mundulea suberosa
Nigella sativa

Bruchus pisorum
(Pea beetle)
Derris sp.

'Bugs' (General category)
Aconitum anthora
Aconitum baicalense
Aconitum barbatum
Cimicifuga foetida
Euphorbia dendroides
Ipomoea muricata
Myrica gale
Rhododendron molle
Veratrum nigrum

Busseola fusca
(Maize stem borer)
Derris elliptica
Derris malaccensis
Tephrosia virginiana

Cacoecia argyrospila
(Fruit tree leafroller)
Arachis hypogaea

Caenurgin crassiuscula
(Clover looper)
Araujia sericifera

Caenurgin erechtea
(Looper)
Aralia sericifera

Calandra oryzae
(Rice billbug)
Stemona tuberosa

Callosobruchus analis
(Bean beetle)
Acorus calamus
Allium cepa
Carum copticum

Callosobruchus chinensis
(Azuki bean beetle)
Acorus calamus
Allium sativum
Annona reticulata
Annona squamosa
Aristolochia bracteata
Azadirachta indica
Brassica latifolia
Brassica sp.
Cassia fistula
Citrullus colocynthis
Cleistanthus collinus
Cocculus trilobus
Coffea arabica

Gendarusa vulgaris
Gossypium hirsutum
Ipomoea cornea
Justicia gendarussa
Madhuca latifolia
Maranta arundinaceae
Mundulea suberosa
Nigella sativa
Ocimum basilicum
Ougeinia dalbergioides
Pogostemon patchouli
Pongamia pinnata
Ricinus communis
Tephrosia villosa
Thevetia peruviana
Vitex negundo

Callosobruchus maculatus
(Cowpea beetle)
Annona reticulata
Arachis hypogaea
Azadirachta indica
Brassica nigra
Brassica sp.
Capsicum annuum
Cinnamomum zeylanicum
Citrus aurantiifolia
Citrus limon
Citrus maxima
Citrus reticulata
Curcuma domestica
Erythrina flabelliformis
Myristica acuminata
Myristica fragrans
Pimenta dioica
Piper nigrum
Syzygium aromaticum

Callosobruchus phaseoli
(Bean beetle)
Citrus aurantium

Carpophilus hemipterus
(Dried fruit beetle)
Azadiracta indica

Caryedon serratus
(Brucid beetle)
Hyptis spicigera

Cassida nebulosa
(Tortoise beetle)
Artemisa vulgaris
Asclepias syriaca

322

Brassica oleracea var. botrytis
Cichorium intybus
Cirsium arvens
Clematis vitalba
Convolvulus arvensis
Fragaria ananassa
Galinsoga parviflora
Lepidium draba
Malva neglecta
Nicandra physalodes
Papaver somniferum
Petunia atkinsiana
Plantago major
Polygonum lapathifolium
Robinia pseudo-acacia
Rubus idaeus
Rumex obtusifolius
Solanum nigrum
Solidago virgaurea
Sonchus oleraceus
Staphylea pinnata
Taraxacum officinale
Trifolium pratense
Verbascum phlomoides

Castelytra zealandica
(Grass grub)
Lotus pedunculatus
Medicago sativa

Caterpillars
(General category)
Acalypha indica
Aconitum carmichaelii
Aconitum kusnezoffii
Amianthium muscitoxicum
Anemone chinensis
Annona squamosa
Arisaema serratum
Arisaema thunbergii
Artemisia absinthium
Artemisia argyi
Brassica latifolia
Celastrus angulatus
Citrullus colocynthis
Clematis armandii
Datura stramonium
Delonix regia
Emilia tuberosa
Houttuynia cordata
Illicium lanceolatum
Ipomoea purpurea
Juglans nigra
Liquidambar taiwaniana
Lycopersicon lycopersicum

Nerium oleander
Ougeinia dalbergioides
Pachyrhizus erosus
Plumbago zeylanica
Pyrrosia drakeana
Quillaja saponaria
Sambucus nigra
Semecarpus anacardium
Sophora japonica
Stemona tuberosa
Tephrosia purpurea
Tripterygium hypoglaucum
Tripterygium wilfordii
Urginea maritima
Veratrum album
Zanthoxylum clava-herculis

Catopsilia crocale
(Caterpillar)
Stemona tuberosa

Centoptera americana
Petunia sp.

Centrococcus insolitus
(Mealy bug)
Nicotiana rustica

Ceratitis capitata
(Mediterranean fruit fly)
Acorus calamus
Alangium salviifolium
Angelica archangelica
Bersama abyssinica
Brongniantia benthamiana
Chlorogalum pomeridianum
Clematis virginiana
Coffea arabica
Coffea robusta
Derris elliptica
D. malaccensis var. sarawakensis
Diospyros discolor
Dorycnium rectum
Iva axillaris
Juglans nigra
Kallstroemia maxima
Lechea maritina
Licania salicifolia
Paeonia suffruticosa
Pastinaca sativa
Rubia tenuifolia
Rubus idaeus var. strigosus
Solanum brachystachys
Sorbus arbutifolia
Spathobolus roxburghii

Tagetes minuta

Ceratitis rosa
(Natah fruit fly)
Copaifera officinalis

Ceratomia catalpae
(Catalpa sphinx)
Derris sp.
Lonchocarpus utilis

Cerotoma ruficornis
(Bean beetle)
Mammea americana
Pachyrhizus erosus
Phaseolus vulgaris

Cerotoma trifurcata
(Bean leaf beetle)
Amorpha fruticosa
Leonotis nepetifolia
Piper betle

Cetonia aurata
Petunia sp.

Chaitophorus populicola
(Aphid)
Quassia amara

Cheimatobia brumata
(Winter moth)
Lonchocarpus chrysophyllus
Lonchocarpus densiflorus
Nicotiana rustica
Tephrosia macropoda
Tephrosia toxicaria
Tephrosia vogelii

Chilo suppressalis
(Asiatic rice borer)
Achyranthes radix
Cyathula capitata
Podocarpus macrophyllus
Podocarpus nakaii
Taxus cuspidata

Chionaspis euonymi
(Euonymous scale)
'Rotenone'

Chionaspis salicis-nigrae
(Willow scurfy scale)
Gossypium hirsutum
Zea mays

Chirida bipunctata
(Tortoise beetle)
Acacia concinna
Azadirachta indica
Justicia adhatoda
Sapindus trifoliatus

Choristoneura fumiferana
(Spruce budworm)
Trichilia roka

Chrotogonus trachypterus
(Surface grasshopper)
Azadirachta indica
Melia azedarach
Saussurea lappa

Chrotoicetes terminifera
Azadirachta indica

Chrysochus auritus
(Dogbane beetle)
Derris sp.
Lochocarpus utilis

Chrysomphalus sp.
(Red scale)
Consolida regalis
Delphinium staphisagria

Chrysomya macellaria
(Screwworm)
Brassica sp.
Cinnamomum camphora
Cymbopogon nardus
Foeniculum vulgare
Pimpinella anisum
Pinus sp.
Sassafras albidum
Syzygium aromaticum

Cicadella viridis
(Large leafhopper)
Clerodendrum trichotomum
Millettia pachycarpa

Cicindela trifasciata
(Tiger beetle)
Nepeta cataria

Cimex lectularus
(Bed bug)
Acorus calamus
Anthemis cotula
Boenninghausenia albiflora

Cannabis sativa
Consolida ambigua
Delphinum formosum
Eugenia haitiensis
Hyptis suaveolens
Inula helenium
Ledum palustre
Melaleuca leucadendron
Petiveria alliacea
Piscidia grandifolia
Premna odorata
Prunus laurocerasus
Prunus maackii
Stemona japonica
Stemona sessilifolia
Stemona tuberosa
Tephrosia heckmannia
Veratrum dahuricum
Vitiveria zizanioides

Citrus psylla
(Citrus psyllid)
Mundulea suberosa
Nicotiana sp.

Cladius pectinicornis
(Rose slug)
Chrysanthemum cinerariifolium

Clothes moth
(General category)
Acorus calamus
Carum bulbocastanum
Cinchona officinalis
Cinnamomum camphora
Cleome brachycarpa
Hedychium spicatum
Inula helenium
Juniperus sabina
Juniperus virginiana
Justicia gendarussa
Nigella sativa
Pandanus tectorius
Petiveria alliacea
Pogostemon patchouli
Ptaeroxylon utile
Pteridium aquilinum
Pterospermum acerifolium
Syzygium aromaticum
Trilisa odoratissima
Vetiveria zizanioides
Vitex negundo

Cnaphalocrocis medinalis
(Rice leafroller)
Actinidia chinensis
Azadirachta indica
Baeckea frutescens
Brucea javanica
Datura stramonium
Euphorbia fischeriana
Mahonia bealei
Polygonum nodosum
Rhus chinensis
Rhus verniciflua
Sargentodoxa cuneata

Coccids
(General category)
Arachis hypogaea
Gliricidia sepium

Coccus viridis
(Green scale)
Annona reticulata
Annona squamosa
Aristolochia indica
Citrullus colocynthis
Derris elliptica
Dillenia indica
Gendarussa vulgaris
Mundulea suberosa
Tephrosia candida
Tephrosia villosa
Xeromphis spinosa

Cochliomyia americana
(American screwworm)
Betula lenta
Brassica sp.
Carum carvi
Cassia auriculata
Cuminum cyminum
Foeniculum vulgare
Juniperus sabina
Mentha pulegium
Petroselinum crispum
Pimpinella anisum
Prunus dulcis var. amara
Sassafras albidum

Cochliomyia hominivorax
(Screwworm)
Chenopodium ambrosioides
Cinnamomum camphora

Cinnamomum cassia
Cinnamomum zeylanicum
Citrus aurantium
Colchicum autumnale
Copaifera lansdorfii
Coriandrum sativum
Cymbopogon nadus
Erythronium americanum
Euphorbia marginata
Foeniculum vulgare
Guaiacum officinale
Hedeoma pulegioides
Humulus lupulus
Juniperus oxycedrus
Juniperus virginiana
Mentha spicata
Myristica fragans
Pelargonium sp.
Pimpinella anisum
Prunus persica
Sassafras albidum
Syzygium aromaticum

Cockroaches
(General category)
Aconitum barbatum
Amianthium muscitoxicum
Artemesia vulgaris
Azadirachta indica
Chrysanthemum cinerariifolium
Chrysanthemum parthenium
Cinnamomum cassia
Cucumis sativus
Cymbopogon nardus
Delhpinium cheilantum
Delphinium dictyocarpum
Delphinium grandiflorum
Delphinium laxiflorum
Delhpinium retropilosum
Derris uliginosa
Haplophyton cimicidum
Lavandula angustifolia
Lycopersicon lycopersicum
Macrosiphona hypoleuca
Melia azedarach
Mundulea suberosa
Nelumbo lutea
Phytolacca americana
Pogostemon patchouli
Rosmarinus officinalis
Sassafras albidum
Schima wallichii
Schoenocaulon officinale
Solanum mammosum
Veratrum album

Colaphellus bowringi
(Cabbage leaf beetle)
Aconitum kusnezoffii
Actinidia chinensis
Celastrus angulatus
Datura stramonium
Euphorbia fischeriana
Millettia pachycarpa
Millettia reticulata
Rhododendron molle
Stellera chamaejasme
Tripterygium forrestii
Tripterygium hypoglaucum
Tripterygium wilfordii

Conotrachelus nenuphar
(Plum curculio)
Aleurites fordii
Azadirachta indica
Colchicum autumnale

Coptosoma cribraria var.
punctatissima (Bean plataspid)
Millettia pachycarpa
Tripterygium forrestii

Corcyra cephalonica
(Rice moth)
Azadirachta indica

Corythuca cydoniae
(Hawthorn lace bug)
Schoenocaulon officinale

Corythuca gossypii
(Cotton lace bug)
Schoenocaulon officinale

Crickets
(General category)
Melia azedarach
Stemona tuberosa

Crocidolomia binotalis
(Cabbage heart caterpillar
Annona reticulata
Annona squamosa
Azadirachta indica
Brassica juncea
Datura metel
Derris elliptica
Derris malaccensis
Dillenia indica
Madhuca latifolia
Mundulea suberosa

Ougeinia dalbergioides
Parthenium hysterophorus
Pogostemon patchouli
Tephrosia candida
Tephrosia villosa
Tephrosia vogelii

Cryptolestes pusillus
(Flat grain beetle)
Azadirachta indica

Cryptotermes brevis
(West Indian dry wood termite)
Guaiacum officinale
Maclura pomifera
Swietenia mahogani
Tectona grandis

Ctenocephalides canis
(Dog flea)
Boenninghausenia albiflora
Mammea americana
Syzygium aromaticum

Ctenocephalus felis
(Cat flea)
Tephrosia virginiana

Cucumber beetles
(General category)
Sophora flavescens

Culex fatigans
(Mosquito)
Acorus calamus
Anacardium occidentale
Azadirachta indica
Cymbopogan citratus

Culex peus (Mosquito)
Allium sativum

Culex pipiens
(Northern house mosquito)
Aconitum japonicum
Amanita pantherina
Armoracia rusticana
Betula platyphylla
Hyptis suaveolens
Lepiota procera
Macleaya cordata
Phellodendron amurense
Phryma oblongifolia
Spilanthes oleracea
Zanthoxylum piperitum

Culex pipiens molestus
(House mosquito)
Achyranthes radix
Cyathula capitata
Podacarpus macrophyllus
Podacarpus nakaii
Taxus cuspidata

Culex pipiens pallens
(House mosquito)
Juniperus recurva

Culex quinquefasciatus
(Southern house mosquito)
Abelmoschus moschatus
Allium sativum
Anabasis aphylla
Avena sativa
Camellia japonica
Capsicum frutescens
Colchicum autumnale
Dryopteris filix-mas
Duboisia hopwoodii
Kirchneriella irregularis
Lychnis coronaria
Nicotiana sp.
Phoradendron flavescens
Phoradendron serotinum
Pilocarpus jaborandii
Pilocarpus microphyllus
Pinus taeda
Pinus virginiana
Ruta graveolens var. angustifolia
Sambucus canadensis
Sambucus nigra
Schoenocaulon officinalis

Culex sp.
(Mosquito)
Gliricidia sepium
Hura crepitans
Mammea americana
Prunus racemosus
Ryania speciosa
Spilanthes acmella

Culex tarsalis
(House mosquito)
Allium sativum
Amorpha fruticosa
Asclepias syriaca
Eupatorium maculata
Polygonum coccineum
Sanguinaria canadensis

Culicine sp.
(Mosquito)
'Rotenone'

Cydnus bicolor
(Burrowing bug)
Sambucus nigra

Cylas formicarius
(Sweet potato weevil
Ryania speciosa

Dactynotus carthani
(Niger aphid)
Acorus calamus
Croton tiglium
Plumbago zeylanica
Tephrosia villosa

Dacus cacuminatus
(Fruit fly)
Zieria smithii

Dacus caudatus
(Fruit fly)
Euginia jabolasa
Syzygium cumini

Dacus correctus
(Mango fruit fly)
Ocimum sanctum

Dacus cucurbitae
(Melon fly)
Acorus calamus
Alangium salviifolium
Arnica chamissonis var. incana
Asclepias curassavica
Bensama abyssinica
Brongniantia benthamiana
Calystegia soldanella
Coffea robusta
Diospyros discolor
Iva axillaris
Juglans nigra
Paeonia suffruticosa
Physocarpus capitatus
Polygonum newberryi
Sidalcea oregana
Sorbus arbutifolia
Tagetes minuta
Verbesina greenmannii

Dacus diversus
(Fruit fly)
Carica papaya
Cymbopogon nardus
Mangifera indica
Pimenta racemosa
Syzygium cumini

Dacus dorsalis
(Oriental fruit fly)
Acorus calamus
Brexia madagascariensis
Cassia fistula
Cinnamomum merkadoi
Coffea robusta
Couroupita guionensis
Cymbopogon marginatus
Diospyros discolor
Iva axillaris
Licania salicifolia
Polygonum newberryi
Ryania speciosa
Tagetes minuta
Vriesea heliconioides

Dacus fergugineus
(Fruit fly)
Carica papaya
Colocasia esculenta
Pimenta racemosa

Dacus zonatus
(Fruit fly)
Carica papaya
Colocasia esculenta
Cymbopogon nardus
Eugenia esculenta var. aromatica
Pimenta racemosa
Syzygium aromaticum

Datana ministra
(Yellow-necked caterpillar)
Tephrosia virginiana

Delia brassicae
(Cabbage root fly)
Lactuca sativa
Trifolium repens

Dendroctonus brevicornis
(Western pine beetle)
Pinus sp.

Dermestres maculatus
(Hide beetle)
Acorus calamus
Angelica sylvestris
Carlina acaulis
Geranium macrorrhizum
Pimpinella major

Diabrotica balteata
(Banded cucumber beetle)
Phaseolus vulgaris

Diabrotica bivittata
(Striped cucumber beetle)
Mammea americana

Diabrotica duodecimpunctata
(Spotted cucumber beetle)
Amorpha fruticosa
Chrysanthemum cinerariifolium
Nicotiana sp.
'Rotenone'
Schoenocaulon officinale

Diabrotica punctata
(12-spotted cucumber beetle)
Chrysanthemum cinerariifolium

Diabrotica undecimpunctata
(Spotted cucumber beetle)
Aleurites fordii
Azadirachta indica

Diabrotica virgifera
(Western corn rootworm)
Sorghum bicolor

Diacrisia obliqua
(Jute hairy caterpillar)
Azadirachta indica
Calotropis gigantea
Diospyros montana
Polygonum hydropiperoides
Thevetia peruviana

Diacrisia virginica
(Yellow woolly bear)
Gliricidia sepium

Diaphania hyalinata
(Melon worm)
Aeschynomene sensitiva
Albizia stipulata
Balanites aegyptica
Calopogonium coeruleum

Chlorogalum pomeridianum
Cinchona calisaya
Delphinium sp.
Encelia farinosa
Encelia flagellaris var. radians
Erigeron bellidiastrum
Erigeron canadensis
Erigeron flagellaris
Haplophyton cimicidum
Helenium elegans
Heliopsis longipes
Humulus lupulus
Jacquinia aristata
Leonotis nepetifolia
Mammea americana
Pachyrhizus erosus
Pachyrhizus palmatilobus
Pachyrhizua piscipula
Phyllanthus acuminatus
Piper betle
Piscidia acuminata
Piscidia erythrina
Piscidia piscipula
Proboscidea louisianica
Quassia amara
Ryania speciosa
Schoenocaulon officinale
Sesamum indicum
Veratrum album
Veratrum viride
Willardia mexicana
Yucca schidigera
Zanthoxylum clava-herculis

Diaphorina citri
(Asiatic citrus psyllid)
Melia azedarach
Melia toosendan

Diloba caeruleocephala
Petunia sp.

Doryphora 10-lineata
(Potato beetle)
Chrysanthemum cinerariifolium
'Rotenone'

Drosophila hydei
(Fruit fly)
Acer carpinifolium
Aconitum japonicum
Allium nipponicum
Alnus firma
Alnus hirsuta
Amsonia elliptica

329

Anemone coronaria
Anemone raddeana
Anodendron affine
Arenaria leptoclados
Athyrium pterorachis
Betula sollennis
Brassica cernua
Buxus japonica
Canavalia maritima
Cardiocrinium cordatum
Carex clivorum
Carex siderosticata var. glabra
Cephalotaxus drupacea
Cephalotaxus harringtonia
Chrysosplenium flagelliferum
Chrysosplenium yesoense
Clematis terniflora
Cocculus trilobus
Coniogramme japonica
Dryopteris bissetiana
Dryopteris erythrosora
Erythronium dens-canis
Euphorbia adenochlora
Geranium eriostemon var. onoei
Hepatica nobilis var. nipponica
Hieracum japonicum
Hydrocotyle javanica
Isopyrum stoloniferum
Leucothoe keiskei
Libanotis ugoensis
Ligustrum obtusifolium
Lycoris radiata
Lysimachia mauritiana
Macleaya cordata
Nerium oleander
Orthodon grosseserratum
Pellionia scabra
Phryma oblongifolia
Polygonatum japonicum
Prunus buergerana
Prunus japonica
Ranunculus flagelliformis
Ranunculus mirissimus
Ranunculus sceleratus
Ranunculus vernyii
Ranunculus zuccarini
Rubus japonica
Shibataea kumasaca
Similax sieboldi
Spiraea nipponica
Stachys niederi
Tetragonia tetragonioides
Vaccinium oxycoccos
Veratrum grandiflorum
Veratrum maximowicz

Veratrum nigrum
Viola phalacrocarpoides
Viola takedana var. variegata

**Drosophila melanogaster
(Vinegar fly)**
Ageratum conyzoides
Ageratum houstonianum
Anethum graveolens
Apocynum subiricum
Artemisia monosperma
Asclepias incarnata
Asclepias speciosa
Asclepias syriaca
Astragalus canadensis
Astragalus pectinatus
Brassica oleracea
Brassica oleracea var. acephala
Brassica oleracea var. capitata
Brassica oleracea var. gemmifera
Brassica rapa var. rapifera
Conringia orientalis
Coreopis lanceolata
Cynoglossum officinale
Eupatorium japonicum
Eupatorium perfoliatum
Gaura coccinea
Glycyrrhiza lepidota
Gypsophila paniculata
Lupinus argenteus
Menispermum canadense
Pastinaca sativa
Phryma leptostachya
Pimpinella anisum
Raphanus sativus
Sanguinaria canadensis
Senecio canus
Solidago altissima
Triglochin maritima
Xanthium canadense

**Drosophila sp.
(Fruit fly)**
Achihhea sibirica
Eriobotrya japonica
Euonymus japonica
Eupatorium japonicum
Nepeta subsessilis
Osmorhiza aristata
Prunus grayana
Prunus padus

330

Dysdercus cingulatus
(Red cotton stainer)
Acorus calamus
Ageratum conyzoides
Allium sativum
Angelica sylvestris
Annona reticulata
Anthocephalus cadamba
Azadirachta indica
Canna indica
Carlina acaulis
Catharanthus roseus
Citrus aurantium
Crassocephalum crepedioides
Datura stramonium
Erythrina variegata
Geranium macrorrhizum
Helichrysum hookeri
Lantana camara
Leucas aspera
Macaranga peltata
Manihot esculenta
Nephrolepis exaltata
Ocimum basilicum
Ocimum sanctum
Parthenium hysterophorous
Pavonia zeylanica
Pelargonium graveolens
Phyllanthus emblica
Pimpinella major
Piper nigrum
Piscidia piscipula
Polypodium walkarae
Polyscias guilfoylei
Pseudoelephantopus spicatus
Psidium guajava
Pterocarpus marsupium
Solanum xanthocarpum
Stemona tuberosa
Tabernaemontana dichotoma
Tagetes erecta
Tamarindus indica
Tectona grandis
Terminalia paniculata
Tithonia diversifolia
Tribulus terrestris
Wedelia prostrata

Dysdercus flavidus
(Cotton stainer)
Ageratum conyzoides
Leonotis nepetifolia

(Dysdercus koenigii
(Red cotton stainer)
Acorus calamus
Annona squamosa
Argemone mexicana
Aristolochia bracteata
Bambusa arundinacea
Croton tiglium
Dendrocalamus strictus
Elephantopus scaber
Embelia ribes
Lantana camara
Plumbago zeylanica
Pogostemon patchouli
Psoralea corylifolia
Strobiianthes ixocephalus
Tagetes minuta
Tephrosia villosa

Dysdercus megalopygus
(Cotton stainer)
Millettia pachycarpa
Pachyrhizus erosus

Dysdercus sanguinarius
(Cotton stainer)
Albizia stipulata
Pachyrhizus erosus

Dysdercus similis
(Red cotton bug)
Mesua ferrea

Dysdercus sp.
(Cotton stainer)
Pachyrhizus palmatilobus

Dysdercus suturellus
(Cotton stainer)
Azadirachta indica

Earias fabia
(Bhendi fruit borer)
Chrysanthemum cinerariifolium

Earias insulana
(Spiny bollworm)
Azadirachta indica

Echidnophaga gallinaceae
(Sticktite flea)
Tephrosia virginiana

Empoasca biguttula
(Leafhopper)
Nicotiana tabacum
Pinus tabuliformis var. yunnanensis

Empoasca devastans
(Cotton leafhopper)
Acorus calamus
Calotropis procera
Chrysanthemum cinerariifolium
Nerium oleander
Peganum harmala

Empoasca fabae
(Potato leafhopper)
Amorpha fruticosa
Chrysanthemum cinerariifolium
Nicotiana sp.
'Rotenone'
Schoenocaulon officinale

Empoasca maligna
(Apple leafhopper)
Nicotiana sp.

Endelomyia rosae
(Rose slug)
Chrysanthemum cinerariifolium

Epacromia tamulus
(grasshopper)
Annona reticulata
Annona squamosa
Mundulea suberosa
Tephrosia candida
Xeromphis spinosa

Ephestia cautella
(Almond moth)
Azadirachta indica

Ephestia elutella
(Wheat grain storage moth)
 Chrysanthemum cinerariifolium

Epicauta lemniscata
(Three-striped blister beetle)
Amorpha fruticosa

Epicauta pennsylvanica
(Black blister beetle)
Chrysanthemum cinerariifolium
Derris sp.

Epicauta vittata
(Striped blister beetle)
Haplophyton cimicidum

Epilachna sp.
(Beetle)
Citrullus colocynthis
Datura metel
Dillenia indica
Diospyros montana

Epilachna varivestis
(Mexican bean beetle)
Aconitum chinense
Aconitum villosum
Aesculus californica
Amorpha fruticosa
Ardisia crispa var. dielsii
Arisaema consanguineum
Arisaema erubescens
Arisaema purpureogaleatum
Azadirachta indica
Brassica rapa var. rapifera
Canthium euroides
Cedrela ciliata
Centella asiatica
Chrysanthemum cinerariifolium
Delphinium delavayi
Derris elliptica
Derris malaccensis
Derris sp.
Haplophyton cimicidum
Lonchocarpus urucu
Lonchocarpus utilis
Millettia pachycarpa
Nicandra physalodes
Nicotiana sp.
Pachyrhizus achipa
Pachyrhizus erosus
Pachyrhizus strigosus
Pachyrhizus tuberosus
Pastinaca sativa
Physalis peruviana
Phytolacca acinosa
Plumbago zeylanica
Pterocarya stenoptera
Pueraria yunnanensis
Pycnanthemum rigidus
Rhododendron molle
'Rotenone'
Schkuria pinnata
Schoenocaulon officinale
Tagetes patula
Tephrosia virginiana

Tephrosia vogelii
Trichilia roka
Tripterygium forrestii
Tripterygium wilfordii
Vicia cracca
Withania somnifera
Xylocarpus granatum
Zanthoxylum chalybea
Zanthoxylum holstii

Epilachna viginti-octopunctata
(Beetle)
Annona squamosa
Mundulea squamosa
Mundulea suberosa
Tephrosia candida

Eriosoma lanigerum
(Woolly apple aphid)
Anabasis aphylla
Nicotiana sp.
Parthenocissus quinquefolia
Rhus coriaria
'Rotenone'
Tropaeolum majus

Eriosoma tesselatum
(Woolly aphid)
Chrysanthemum cinerariifolium

Erythroneura comes
(Grape leafhopper)
Chrysanthemum cinerariifolium
Derris sp.

Euphydryas chalcedona
(Baltimore caterpillar)
Derris elliptica

Euproctis fraterna
(Hairy caterpillar)
Acalypha indica
Annona reticulata
Annona squamosa
Aristolochia indica
Azadirachta indica
Citrullus colocynthis
Datura metel
Dillenia indica
Justicia gendarussa
Madhuca latifolia
Menispermum cocculus
Mundulea suberosa
Ocimum canum

Plumbago zeylanica
Tephrosia candida
Vitex negundo
Xeromphis spinosa

Euproctis laniata
(Castor hairy caterpillar)
Azadiracta indica

Euproctis lunata
(Caterpillar)
Azadirachta indica

Euproctis pseudoconspersa
(Caterpillar)
Actinidia chinensis
Clerodendrum trichotomum
Macleaya cordata
Myrica rubra
Polygonum nodosum
Tripterygium hypoglaucum

Eupterote mollifera
(Caterpillar)
Camellia sinensis

European cabbage worms
(General category)
Pycnanthemum rigidus

Eutettix tenellus
(Bean leafhopper)
Chrysanthemum cinerariifolium

Evergestis rimosalis
(Crossed-striped cabbage worm)
Delphinium sp.
Derris sp.
Eremocarpus setigerus
Lonchocarpus utilis
Tripteryqium wilfordii
Willardia mexicana

Ferrisiana virgata
(Striped mealy bug)
Camellia sinensis

Flea beetles
(General category)
Digitalis grandiflora
Ipomoea purpurea
Lepidium ruderale
Nicotiana rustica

333

Fleas
(General category)
Acorus calamus
Aloe striata
Anthemis cotula
Aquilaria agallocha
Asclepias curassavica
Blumea lacera
Bumelia retusa
Cannabis sativa
Cimicifuga foetida
Daphne mezereum
Datura metel
Gliricidia sepium
Hedeoma pulegioides
Indigofera suffruticosa
Juniperus virginiana
Justicia adhatoda
Ledum palustre
Mammea americana
Melia azedarach
Myrica cerifera
Nigella sativa
Ocimum canum
Ocimum sanctum
Pogogyne parviflora
Prunus laurocerasus
Ricinus communis
Stemona tuberosa
Stillingia sylvatica
Thamnosma africana
Theobroma cacao
Umbellularia californica

Flies
(General category)
Aconitum anthora
Aconitum baicalense
Aconitum barbatum
Aconitum excelsum
Aconitum lycoctonum
Aconitum volubile
Acorus calamus
Adansonia digitata
Allium ampeloprasum
  var. porrum
Aloe ferox
Amanita muscaria
Amanita pantherina
Amianthium muscitoxicum
Amorphophallus campanulatus
Andrachne ovalis
Anemone altaica
Anthemis cotula

Artemisia annua
Artemisia vulgaris
Aureolaria virginica
Baptisia tinctoria
Boswellia dalzielii
Calotropis gigantea
Carissa carandas
Carya glabra
Chrysanthemum cinerariifolium
Colchicum autumnale
Cucurbita pepo
Cymbopogon nardus
Daphne mezereum
Delphinium cheilantum
Delphinium dictyocarpum
Delphinium grandiflorum
Delphinium laxiflorum
Delphinium retropilosum
Delphinium sp.
Dendrobium aggregatum
Derris uliginosa
Eugenia haitiensis
Funastrum clausum
Gardenia lucida
Gymnocladus dioica
Hedeoma pulegioides
Helianthus annuus
Heliopsis parvifolia
Inula helenium
Jacaranda obtusifolia
  var. rhombifolia
Juglans nigra
Justicia adhatoda
Laurus nobilis
Linaria vulgaris
Lomatia silaifolia
Lycopersicon lycopersicum
Melanthium virginicum
Mirabilis nyctaginea
Mundulea suberosa
Nigella sativa
Ocimum basilicum
Ocimum sanctum
Oroxylum indicum
Paphiopedilum javanicum
Phellodendron amurense
Physalis mollis
Polygonum hydropiperoides
Populus deltoides
Prunus dulcis
Prunus padus
Ricinus communis
Robinia pseudo-acacia
Sauromatum guttatum

Solanum carolinense
Sophora tinctoria
Stellera chamaejasme
Stemona japonica
Stemona sessilifolia
Taxus baccata
Tephrosia virginiana
Tetrapterys acutifolia
Veratrum album
Veratrum californicum
Veratrum dahuricum
Veratrum maackii
  var. japonicum
Veratrum nigrum
Vitex agnus-castus
Zanthoxylum clava-herculis

Fowl lice
(General category)
Acorus calamus
Calotropis procera

Frankliniella fusca
(Tobacco thrips)
Cochlospermum religiosum

Fruit flies
(General category)
Haplophyton cimicidum
Lactuca sativa

Galerucella luteola
(Elm leaf beetle)
Chrysanthemum cinerariifolium
Derris sp.

Galleria mellonella
(Greater wax moth)
Azadirachta indica
Brassica oleracea var. capitata

Gargaphia solani
(Eggplant lace bug)
Chrysanthemum cinerariifolium
Haplophyton cimicidum

Gasterophilus intestinalis
(Horse botfly)
Dichrostachys cinerea

Gastrophysa cyanea
(Green dock beetle)
Tripterygium wilfordii

Glossina sp.
(Tsetse fly)
Amomum melegueta
Cissus producta
Cymbopogon citratus
Cymbopogon flexuosus
Euphorbia balsamifera
Euphorbia tirucalli
Melinis minutiflora
Sesbania aculeata
Sesbania punctata

Gnorimoschema lycopersicella
(Tomato pinworm)
Chrysanthemum cinerariifolium
Derris sp.
Nicotiana sp.

Goniocotes gigas
(Large hen louse)
Tephrosia virginiana

Goniocotes hologaster
(Poultry fluff louse)
Tephrosia virginiana

Graphosoma italicum
(Pentatomid 'bug')
Acorus calamus
Angelica sylvestris
Carlina acaulis
Geranium macrorrhizum
Pimpinella major

Graphosoma mellonella
(Pentatomid 'bug')
Acorus calamus
Angelica sylvestris
Carlina acaulis
Geranium macrorrhizum
Pimpinella major

Grasshoppers
(General category)
Aconitum kusnezoffii
Amianthium muscitoxicum
Andrographis paniculata
Azadirachta indica
Balanites roxburghii
Bothriochloa intermedia
Chrysanthemum cinerariifolium
Cyperus rotundus
Delphinium sp.
Euphorbia tirucalli

Haplophyton cimicidum
Lycium halimifolium
Lycopersicon lycopersicum
Lycoris africana
Melia azedarach
Sapium sebiferum
Schoenocaulon officinale
Veratrum album

Gryllotalpa sp.
(Mole crickets)
Baeckea frutescens
Daphne genkwa

Hadena oleracea
(Tomato moth)
Atractylis gummifera
Derris elliptica

Haematopinus eurysternus
(Short-nosed cattle louse)
Delphinium sp.

Heliothis armigera
(American bollworm)
Artemisia annua
Artemisia japonica
Azadirachta indica
Gliricidia sepium
Ryania speciosa
Schizonepeta tenuifolia
Schoenocaulon officinale

Heliothis obsoleta
(Corn earworm)
Piper nigrum

Heliothis virescens
(Tobacco budworm)
Alchornea triplinervia
Azadirachta indica
Brassica nigra
Callicarpa americana
Cassia obtusifolia
Cnidoscolus texanus
Gossypium hirsutum
Helianthus annuus
Ipomoea quamoclit
Melia azedarach
Passiflora incarnata
Raphanus raphinastrum
Secale cereale
Solanum melongena
Solanum pseudocapsicum
Trifolium repens

Heliothis zea
(Corn earworm)
Araujia sericifera
Bersama abyssinica
Lycopersicon hirsutum
Melia azedarach
Parthenium alpinum
Parthenium fruticosum
Piper nigrum

Hellula rogatalis
(Cabbage webworm)
Azadirachta indica

Hemileuca oliviae
(Range caterpillar)
Zanthoxylum monophyllum

Henosepilachna sparsa
(28-spotted ladybird beetle)
Derris malaccensis
Tephrosia vogelii

Heteronychus arator
(Black beetle)
Lolium perenne
Lotus corniculatus
Lotus pedunculatus
Lupinus angustifolius
Medicago sativa
Trifolium repens

Hieroglyphus nigrorepletus
(Kharif grasshopper)
Aconitum ferox

Holopsyllus affinis
(Flea)
Tephrosia virginiana

Holotrichia consanguinea
(Beetle)
Azadirachta indica

Holotrichia insularis
(Beetle)
Azadirachta indica

Holotrichia ovata
(Beetle)
Melia azedarach

Holotrichia serrata
(Sugarcane root grubs)
Azadirachta indica

Hoplocampa flava
(Plum sawfly)
Quassia amara

Hoplocampa minuta
(Plum sawfly)
Quassia amara

Hydrellia philippina
(Rice whorl maggot)
Andrographis paniculata
Gliricidia sepium

Hymenia recurvalis
(Hawaiian beet webworm)
Celastrus angulatus
Celastrus orbiculatus
Cnidoscolus urens
Marah fabaceus
Piscidia piscipula
Tripterygium forrestii
Tripterygium wilfordii
Veratrum viride
Willardia mexicana
Yucca schidigera

Hypera postica
(Alfalfa weevil)
Zanthoxylum monophyllum

Hyphantria cunea
(Fall webworm)
Amaranthus retroflexus
Brassica olaracea var. gongyloides
Capsicum annuum
Chelidonium majus
Chenopodium vulvaria
Clematis vitalba
Consolida regalis
Glechoma hederacea
Hibiscus syriacus
Lamium maculatum
Lepidium draba
Lycopersicon lycopersicum
Nicandra physalodes
Nicotiana sp.
Petunia atkinsiana
Solanum tuberosum
Solidago virgaurea

Hypoderma lineata
(Cattle grub)
Tephrosia virginiana

Hypomeces squamosus
(Weevil)
Stemona tuberosa

Hypsa ficus
(caterpillar)
Annona reticulata
Mundulea suberosa

Hypsipyla grandella
(Caterpillar)
Cedrela ciliata

Idiocerus sp.
(Mango hopper)
Annona reticulata
Aristolochia indica
Clerodendrum imame
Menispermum cocculus
Mundulea suberosa

Indarbela quadrinotata
(Bark-eating caterpillar)
Azadirachta indica

Inopus rubriceps
(Soldier fly)
Brassica oleracea

Jassid sp.
(Jassids)
Nicotiana rustica

Laphygma frugiperda
(Fall armyworm)
Mammea amaricana
Pachyrhizus erosus
Pachyrhizus palmatilobus

Lasioderma serricorne
(Tobacco beetle)
Azadirachta indica

Lasius americanus
(American corn field ant)
Citrus limon

Laspeyresia molesta
(Peach moth)
Chrysanthemum cinerariifolium

Laspeyresia pomonella
(Codling moth)
Aleurites fordii
Anabasis aphylla

Azadirachta indica
Chrysanthemum cinerariifolium
Derris sp.
Erigeron affinis
Glycine max
Haplohpyton cimicidum
Helenium mexicana
Heliopsis longipes
Nicotiana sp.
Phellodendron amurense
Pinus sp.
Ricinus communis
Ryania speciosa
Schoencaulon officinale
Tripterygium forrestii
Tripterygium wilfordii
Yucca schidigera

Latheticus oryzae
(Long-headed flour beetle)
Acorus calamus
Azadirachta indica
Vetiveria zizanioides
Vitex negundo

Leaf-cutting ants
(General category)
Arachis hypogaea
Citrus sinensis
Glycine max
Helianthus annuus
Linum usitatissimum
Ricinus communis

Leafhoppers
(General category)
Catalpa ovata
Glochidion puberum
Gynura segetum
Nicotiana rustica
Opuntia dillenii

Leanium sp.
(Green bug)
Annona reticulata

Leichenum canoliculatum
Nepeta cataria

Lepidoptera sp.
(Caterpillar)
Stemona tuberosa

Lepidosaphes ulmi
(Oyster-shell scale)
Gossypium hirsutum
Zea mays

Lepisma saccharina
(Silverfish)
Ginkgo biloba

Leptinotarsa decemlineata
(Colorado potato beetle)
Acer saccharium
Ailanthus altissima
Amaranthus ascendens
Amaranthus retroflexus
Amianthium muscitoxicum
Amorpha fruticosa
Antirrhinum majus
Apocynum cannabinum
Arctium minus
Arctium tomentosum
Artemisia rigida
Artemisia tridentata
Artemisia vulgaris
Atropa belladonna
Azadirachta indica
Beta altissima
Beta esculenta
Bothriochloa ischaemum
Brassica kaber
Brassica oleracea var. capitata
Calystegia sepium
Canna indica
Capsicum annuum
Centaurea pannonica
Chenopodium album
Chenopodium hybridum
Chrysanthemum cinerariifolium
Chrysanthemum indicum
Chrysothamnus nauseosus
Cichorium intybus
Clematis vitalba
Colchicum autumnale
Convolvulus arvensis
Coriandrum sativum
Cucumis sativus
Dahlia pinnata
Delphinium glaucum
Derris metaloides
Derris sp.
Dryopteris marginalis
Echinochloa crus-galli

Erigeron sphaerocephalcum
Euphorbia helioscopia
Fragaria ananassa
Glycine max
Gomphrena globosa
Haplophyton cimicidum
Helianthus annuus
Lonchocarpus utilis
Lycium halimifolium
Lycopersicon lycopersicum
Lycopersicon piminellifolium
Maclura aurantiaca
Medicago sativa
Melilotus alba
Mentha piperita
Mentha pulegium
Mundulea suberosa
Nepeta cataria
Nicandra physalodes
Nicotiana tabacum
Nicotiana rustica
Nicotiana sp.
Nicotiana suberosa
Ocimum basilicum
Onopordum acanthium
Parthenocissus inserta
Petunia atkinsiana
Petunia sp.
Phaseolus vulgaris
Physalis subglabrata
Physostigma venenosum
Plantago lanceolatum
Plantago major
Platanus occidentalis
Poa annua
Polygonum lapathifolium
Prunus armeniaca
Purshia tridentata
Quercus alba
Ranunculus acris
Rhaphanus sativus
Rheum undulatum
Robinia pseudo-acacia
Rosa centifolia
Rumex acetosa
Salvia dorii
Salvia splendens
Salvia verticillata
Saponaria officinalis
Senecio vulgaris
Solanum acaule
Solanum auriculatum
Solanum calvescens
Solanum chacoense
Solanum demissum

Solanum jamesii
Solanum luteum
Solanum nigrum
Solanum polyadenium
Solidago virga-aurea
Stavia pratensis
Stavia splendens
Stavia verticilliata
Syringa vulgaris
Taraxacum officinale
Taxus canadensis
Tephrosia virginiana
Trifolium pratense
Tripterygium wilfordii
Verbascum phlomoides
Veronica agrestis
Vitis vinifera
Zea mays
Zinnia elegans

Leptopha minor
(Ash lace bug)
'Rotenone'

Leucania venalba
(White-veined rice armyworm)
Melia azedarach
Melia toosendan

Leucinodes orbonalis
(Eggplant fruit borer)
Azadirachta indica
Chrysanthemum cinerariifolium

Lice (animal)
(General category)
Aconitum anthora
Aconitum baicalense
Aconitum barbatum
Aconitum excelsum
Aconitum lycoctonum
Aconitum volubile
Ajuga bracteosa
Anemone altaica
Annona squamosa
Aquilaria agallocha
Aristolochia rotunda
Artemisia vulgaris
Balsamodendron playfairii
Bandeiraea simplicifolia
Bumelia retusa
Calopogonium vellutium
Centaurium erythraea
Citrus limon
Consolida regalis

Datura metel
Delphinium cheilantum
Delphinium dictyocarpum
Delphinium formosum
Delphinium grandiflorum
Delphinium laxiflorum
Delphinium retropilosum
Delphinium sp.
Delphinium staphisagria
Dioscorea deltoidea
Dioscorea prazeri
Eriosema psoraleoides
Erythroxylum coca
Euonymus europaea
Ficus carica
Gloriosa simplex
Hedera helix
Helenium quadridentalum
Indigofera tinctoria
Inula helenium
Ledum groenlandicum
Ledum palustre
Lycopodium complanatum
Mammea americana
Mussaenda kajewskii
Myrica gale
Oldfieldia arficana
Picrasma quassioides
Pinguicula vulgaris
Prunus padus
Psorospermum baumii
Ricinus communis
Salvia tiliifolia
Schoenocaulon officinale
Semecarpus anacardium
Stemona tuberosa
Syzygium aromaticum
Tylophora ovata
Veratrum album
Veratrum dahuricum
Veratrum nigrum
Vetiveria zizanioides
Withania somnifera

Lipaphis erysimi
(False cabbage aphid)
Aconitum ferox
Acorus calamus
Annona squamosa
Argemone mexicana
Barbarea aristata
Chrysanthemum cinerariifolium
Croton tiglium
Datura metel
Derris sp.

Embelia ribes
Helianthus annus
Jatropha curcas
Lantana camara
Lonchocarpus utilis
Melia azedarach
Nicotiana sp.
Pachyrhizus angulatus
Plumbago zeylanica
Pongamia pinnata
'Rotenone'
Tephrosia villosa
Tephrosia virginiana
Withania somnifera

Liriomyza sativae
(Vegetable leafminer)
Azadirachta indica

Liriomyza trifolii
(Chrysanthemum leafminer)
Azadirachta indica

Livestock pests
(General category)
Aesculus pavia
Amorpha fruticosa
Dracunculus vulgaris
Laurus nobilis
Millettia auriculata
Taxus baccata
Willardia mexicana

Locusta migratoria
(Migratory locust)
Acacia albida
Acacia hockii
Acacia tortilis
Ageratum conyzoides
Ageratum houstonianum
Albizia glaberima
Arachis hypogaea
Aristolochia ringens
Arum maculatum
Asclepias longifolia
Atriplex nummularia
Azadirachta indica
Bellis perennis
Boscia angustifolia
Brassica hirta
Brassica oleracea
Celastrus angulatus
Centaurea nigra
Citrus aurantium
Cleome monophylla

Combretum nigricans
Corchorus olitorius
Dioscorea balcana
Epilobium angustifolium
Eucalyptus dalrympleana
Eupatorium odoratum
Euphorbia peplis
Ficus carica
Gossypium hirsutum
Heliotropium subulatum
Ipomoea purpurea
Lavandula angustifolia
Lavandula officinale
Ligustrum vulgare
Lotus corniculatus
Malus pumila
Medicago sativa
Melia azedarach
Morus alba
Phaseolus vulgaris
Phyllanthus maderaspatensis
Pinus sylvestris
Plantago major
Portulaca foliosa
Pteridium aquilinum
Ranunculus acris
Ribes sanguineum
Ricinus communis
Rumex acetosa
Salix babylonica
Senecio jacobaea
Stachytarpheta mutabilis
Taraxacum officinale
Urtica dioica
Veteveria nigritana
Zanthoxylum alatum

(Locusta oleraceae)
(Locust)
Allium oleraceum

Locusts
(General category)
Azadirachta indica
Calotropis procera
Celastrus angulatus
Delphinium sp.
Melia azedarach

Loxostege similalis
(Garden webworm)
Schoenocaulon officinale

Lucanus cervus
(Stag beetle)
Aconitum nepellus

Lygus elisus
(Pale legume bug)
Chrysanthemum cinerariifolium
Schoenocaulon officinale

Lygus hesperus
(Lygus bug)
Chrysanthemum cinerariifolium

Lygus lineolaris
(Tarnished plant bug)
Amorpha fruticosa
Schoenocaulon officinale

Lymantria dispar
(Gypsy moth)
Alnus rugosa
Azadirachta indica
Bursera simaruba
Catalpa speciosa
Fraxinus americana
Kalmia latifolia
Liriodendron tulipifera
Melaleuca leucadendron
Rhamnus alnifolia
Taxus baccata
Veratrum album
Vitis labrusca
Zanthoxylum americanum

Lymantria monacha
(Forest moth)
Taxus baccata
Veratrum album

Macrobasis unicolor
(Blister beetle)
Amorpha fruticosa

Macrosiphoniella sanborni
(Chrysanthemum aphid)
Annona glabra
Annona muricata
Annona palustris
Annona reticulata
Annona squamosa
Chrysanthemum cinerariifolium
Euonymus europaea
Mundulea suberosa

Macrosiphum ambrosiae
(Potato aphid)
Cochlospermum religiosum
Quassia amara

Macrosiphum granarium
(Grain aphid)
Millettia pachycarpa

Macrosiphum liriodendri
(Tulip tree aphid)
Derris elliptica
Derris uliginose
Quassia amara
'Rotenone'

Macrosiphum rosae
(Rose aphid)
Anabasis aphylla
Camellia sinensis
Chrysanthemum cinerariifolium
Consolida regalis
Delphinium staphisagria
Nicotiana sp.
Quassia amara

Macrosiphum solanifolii
(Potato aphid)
Annona reticulata
Annona squamosa
Derris malaccensis

Macrosiphum sonchi
(Potato aphid)
Mammea americana
Pachyrhizus erosus

Macrosteles divisus
(6-spotted leafhopper)
Chrysanthemum cinerariifolium
Derris sp.

Locusts
(General category)
Azadirachta indica
Calotropis procera
Celastrus angulatus
Delphinium sp.
Melia azedarach

Maggots
(General category)
Acacia pinnata
Acalypha indica
Aconitum kusnezoffii

Aconitum subrosulatum
Adina cordifolia
Ailanthus giralolii
Anemone hupehensis
Arisaema thunbergii
Aristolochia bracteata
Artemisia argyi
Asparagus cochinchinensis
Bambusa bambos
Callilepis laureola
Calpurnia intrusa
Calpurnia subdecandra
Camellia oleifera
Cleistanthus collinus
Clerodendrum glabrum
Consolida ambigua
Coriaria sinica
Cynanchum arnottianum
Datura stramonium
Delphinium coeruleum
Delphinium elatum
Delphinium vestitum
Delphinium yunnanense
Desmodium caudatum
Dioscorea cylindrica
Dioscorea hispida
Duchesnea indica
Echinops echinatus
Erythrina variegata
Euphorbia antiquorum
Euphorbia esula
Evodia rutaecarpa
Excoecaria agallocha
Fluggea leucopyrus
Fluggea melanthesioides
Fluggea virosa
Glochidion puberum
Gmelina arborea
Lobelia chinensis
Macleaya cordata
Myrica rubra
Ocimum basilicum
Ocimum sanctum
Oxalis corniculata
Periploca sepium
Phaseolus lunatus
Polygonum nodusum
Pouzolzia pentandra
Premna microphylla
Pueraria lobata
Rhododendron molle
Sapium ellipticum
Senecio scandens
Stellera chamaejasme
Stemona burkillii

Stemona collinsae
Stemona curtisil
Stemona japonica
Stemona sessilifolia
Stemona tuberosa
Tagetes minuta
Tectona grandis
Tripterygium hypoglaucum
Tylophora ovata
Typhonium giganteum
Vitex cannabifolia
Xysmalobium undulatum

Malacosoma americana
(Eastern tent caterpillar)
Chrysanthemum cinerariifolium
Rhamnus alnifolia
Tephrosia virginiana
Tripterygium wilfordii

Malacosoma disstria
(Forest tent caterpillar)
Acer rubrum
Robinia pseudo-acacia

Malacosoma neustria
(Apricot tent caterpillar)
Celastrus angulatus
Derris elliptica
Hyoscymus niger
Millettia pachycarpa
Nicotiana rustica
Schizonepeta tenuifolia
Stellera chamaejasme
Stemona tuberosa
Taxus baccata
Veratrum album
Veratrum nigrum

Mamestra picta
(Zebra caterpillar)
Chrysanthemum cinerariifolium

Manduca sexta
(Tobacco hornworm)
Acer saccharium
Allamanda neriifolia
Annona cherimola
Aristolochia elegans
Caesalpinia pulcherrima
Chrysanthemum coccineum
Conium maculatum
Cycas revoluta
Dioclea megacarpa
Eucalyptus cine

Euphorbia pulcherrima
Fagus grandifolia
Hydrocotyle americana
Lactuca sativa
Lantana camara
Lycopersicon hirsutum
Lycopersicon Lycopersicum
Malus sylvestris
Nicandra physalodes
Nicotiana benthamiana
Nicotiana repanda
Nicotiana stocktoni
Petunia axillaris
Petunia hybrida
Petunia inflata
Petunia violacea
Pinus strobus
Rhus radicans
Rhus viminalis
Schoenocaulon officinale
Trifolium repens
Tsuga canadensis
Ulmus americana
Viola septentrionalis
Zea mays

Mealy bugs
(General category)
Arachis hypogaea
Asclepias syriaca
Duboisia hopwoodii
Syzygium aromaticum
Thevetia peruviana

Melanoplus femurrubrum
(Red-legged grasshopper)
Acer negundo
Chrysopsis villosa
Elymus canadensis
Helianthus petiolaris
Liatris punctata
Onosmodium occidentale
Ratibida columnifera
Rosa arkansas
Solidago missouriensis

Melanoplus mexicana
(Mexican grasshopper)
Medicago sativa

Melanoplus sanguinipes
(Migratory grasshopper)
Parthenium hysterophorus
Zanthoxylum monophyllum

Meloe violaceus
(Blister beetle)
Petunia sp.

Meloidae sp.
(Blister beetle)
Amorpha fruticosa
Schoenocaulon officinale
Sesamum indicum

Menopon biseriatum
(Chicken lice)
Derris elliptica

Menopon pallidum
(Chicken lice)
'Rotenone'

Mesomorphus villiger
(Tobacco beetle)
Pongamia pinnata

Midges
(General category)
Prunus padus
Prunus racemosus

Mineola scitulella
(Prune worm)
Chrysanthemum cinerariifolium

Mineola vaccinii
(Cranberry fruit worm)
'Rotenone'
Ryania speciosa

Mole crickets
(General category)
Euphorbia cyparissias

Monochamus alternatus
(Pine sawfly)
Metasequoia glyptostroboides
Pinus densiflora

Mosquitoes
Abrus precatorius
Aconitum anthora
Aconitum baicalense
Aconitum excelsum
Aconitum lycoctonum
Aconitum volubile
Acorus calamus
Aeschynomene sensitiva
Albizia lebbek

Allium sativum
Allium schoenoprasum
Anacardium occidentale
Anemone altaica
Aristolochia bracteata
Artemisia annua
Artemisia argyi
Artemisia vulgaris
Avena sativa
Azolla sp.
Bambusa arundinacea
Bambusa bambos
Berberis aristata
Bidens cernua
Bixa orellana
Blumea lyrata
Boswellia carteri
Boswellia dalzielii
Brassica geniculata
Brassica nigra
Butea superba
Calotropis gigantea
Canarium schweinfurthii
Cannabis sativa
Capsella bursa-pastoris
Carum carvi
Cassia alata
Cassia hirsuta
Cassia spectabilis
Chara foetida
Chara fragilis
Chenopodium ambrosioides
Chrysanthemum balsamita
Chrysanthemum cinerariifolium
Citrus aurantium
Clausena anisata
Clibodium erosum
Cnicus benedictus
Commiphora abyssinica
Croton eluteria
Croton tiglium
Cucurbita pepo
Cymbopogon nardus
Cymbopogon winterianus
Datura candida
Delphinium cheilantum
Delphinium dictyocarpum
Delphinium glaucum
Delphinium grandiflorum
Delphinium laxiflorum
Delphinium retropilosum
Derris elliptica
Derris philippinensis
Derris uliginosa
Descurainia pinnata

344

Descurainia sophia
Detarium senegalense
Diospyros argentea
Doryphora sassafras
Dryopteris filix-mas
Duranta repens
Echinacea angustifolia
Elsholtzia blanda
Emilia tuberosa
Erigeron affinis
Eucalyptus globulus
Eugenia haitiensis
Euphorbia dendroides
Euphorbia thymifolia
Euphorbia tirucalli
Goniothalamus tapis
Heliopsis longipes
Hura polyandra
Hydrangea arborescens
Hydrastis canadensis
Hyptis spicigera
Inula helenium
Inula viscosa
Jacaranda obtusifolia var. rhombifolia
Jatropha curcas
Justicia adhatoda
Koelreuteria paniculata
Lansium domesticum
Lepidium flavum
Leucas martinicensis
Liriodendron tulipifera
Litsea cubeba
Mangifera indica
Melaleuca leucadendron
Melinis minutiflora
Mentha pulegium
Mirabilis jalapa
Myrica rubra
Nigella sativa
Ocimum americanum
Ocimum basilicum
Ocimum gratissimum
Ocimum sanctum
Ocimum suave
Ocimum viride
Origanum majorana
Petiveria alliacea
Phellodendron amurense
Phyla oatesii
Picramsa napalensis
Pimenta racemosa
Piper nigrum
Piper tuberculatum
Platycarya strobilacea
Plectranthus rugosus

Pogostemon patchouli
Prunus laurocerasus
Prunus padus
Rheum rhaponticum
Ricinus communis
Rosmarinus officinalis
Salmea scandens
Salvia officinalis
Sambucus canadensis
Sambucus nigra
Santalum album
Sapindus saponaria
Scleria pergracilis
Scutellaria baicalensis
Solanum mammosum
Spigelia anthelmia
Spilanthes acmella
Tetrapterys acutifolia
Tussilago fanfana
Urginea maritima
Veratum dahuricum
Veratum nigrum
Xanthium spinosum
Zanthoxylum americanum
Zanthoxylum clava-herculis
Zanthoxylum hamiltonianum

Moths
(General category)
Acorus calamus
Anacardium occidentale
Chenopodium botrys
Cinchona pubescens
Citrullus colocynthis
Cymbopogon marginatus
Gerea viscida
Juniperus virginiana
Ledum palustre
Melilotus officinalis
Momordica foetida
Myrica cerifera
Pogostemon patchouli
Santolina chamaecyparissus
Tephrosia macropoda
Urginea maritima

Murgantia histrionica
(Harlequin bug)
Chrysanthemum cinerariifolium
Derris sp.
Lonchocarpus utilus
Schoenocaulon officinale

Musca autumnalis
(Autumn stable fly)
Tripleurospermum maritimum

Musca domestica
(House fly)
Achillea millefolium
Achyranthes radix
Acorus calamus
Aerva javanica
Aeschynomene sensitiva
Ageratum conyzoides
Albizia lebbeck
Allium sativum
Alpinia afficinarum
Amanita muscaria
Annona squamosa
Anthemis arvensis
Anthemis tinctoria
Artemisia monosperma
Artemisia roxburghiana
Asphodelus tenuifolius
Azadirachta indica
Brassica juncea
Brassica oleracea
Brassica oleracea var. acephala
Brassica oleracea var. botrytis
Brassica oleracea var. capitata
Brassica oleracea var. gemmifera
Brassica oleracea var. gongyloides
Brassica oleracea var. italica
Brassica rapa var. rapifera
Cannabis sativa
Cassia hirsuta
Chrysanthemum cinerariifolium
Colchicum autumnale
Croton caudatus
Croton tiglium
Cyathula capitata
Cymbopogon citratus
Datura candida
Delphinium orientale
Derris sp.
Derris uliginosa
Digitalis grandifolia
Dryopteris filix-mas
Echinacea angustifolia
Emilia tuberosa
Erigeron affinis
Eupatorium glandulosum
Euphorbia cyparissias
Euphorbia dendroides
Fumaria schleicheri
Haplophyton cimicidum
Heliopsis gracilus

Heliopsis helianthoides
Heliopsis longipes
Heliopsis parviflora
Hieracium pilosella
Impatiens longipes
Jatropha curcas
Lantana camara
Launaea nudicaulis
Lepidium sativum
Leptospermum scoparium
Madhuca indica
Mangifera indica
Melaleuca bracteata
Miliusa velutina
Millettia pachycarpa
Myristica fragrans
Nicandra physalodes
Nicotiana sp.
Ocimum basilicum
Ocimum sanctum
Orobanche aegyptica
Ostodes paniculata
Pastinaca sativa
Pavetta indica
Petiveria alliacea
Phellodendron amurense
Pholidota protracta
Phryma oblongifolia
Pimpinella anisum
Piper nigrum
Piper peepuloides
Podocarpus hallii
Podocarpus macrophyllus
Podocarpus nakii
Podocarpus nivalis
Prunus dulcis var. amara
Prunus laurocerasus
Prunus padus
Prunus racemosus
Raphanus sativus
Rauvolfia serpentina
Ricinus communis
'Rotenone'
Saccharum spontaneum
Schoenocaulon drummondii
Schoenocaulon officinale
Sesamum indicum
Spigelia anthelmia
Sterculia foetida
Tagetes erecta
Tagetes minuta
Tagetes patula
Tanacetum vulgare
Taxus cuspidata
Tephrosia latidens

Tephrosia lindheimeri
Tephrosia toxicaria
Tephrosia virginiana
Thuja occidentalis
Tinospora rumphii
Tithonia diversifolia
Veratrum album
Veratrum lobelianum
Veratrum viride
Verbascum thapsiforme
Vitex negundo
Vittadinia australis
Zanthoxylum alatum
Zanthoxylum clava-herculis

Musca nebulo
(House fly)
Acorus calamus
Annona squamosa
Cymbopogon citratus
Piper peepuliodes

Mylabris phalerata
(Yellow-black blister beetle)
Millettia pachycarpa

Myllocerus sp.
(Leaf weevil)
Azadirachta indica

Mythimna separata
(Rice ear cutting caterpillar)
Azadirachta indica

Myzus cerasi
(Black cherry aphid)
Consolida regalis
Delphinium staphisagria

Myzus persicae
(Green peach aphid)
Consolida regalis
Delphinium straphisagria
Derris elliptica
Derris uliginosa
Madhuca indica
Mammea americana
Melia azedarach
Nicotiana sp.
Nicotiana tabacum
Pachyrhizus erosus
'Rotenone'
Sapindus mukorossi
Veratrum viride

Myzus rosarum
(Green chrysanthemum aphid)
Tephrosia virginiana

Neodiprion rugifrons
(Red-headed jackpine sawfly)
Pinus banksiana

Neodiprion swainei
(Swaine jackpine sawfly)
Pinus banksiana

Nephantis serinopa
(Black-headed coconut caterpillar)
Azadirachta indica

Nephotettix bipunctatus
(Rice leafhopper)
Cocculus trilobus

Nephotettix virescens
(Rice green leafhopper)
Annona squamosa
Azadirachta indica
Melia azedarach
Pongamia pinnata
Tagetes erecta
Tagetes patula

Nezara viridula
(Southern green stink bug)
Derris sp.
Lonchocarpus utilis
Pachyrhizus erosus
Schoenocaulon officinale

Nilaparvata lugens
(Brown planthopper)
Aconitum napellus
Annona squamosa
Azadirachta indica
Echinochloa crus-galli
Eclipa alba
Melia azedarach
Melia toosendan
Tagetes patula

Nymphula depunctalis
(Rice caseworm)
Gliricidia sepium

Oides decimpunctata
(10-spotted grape leaf beetle)
Millettia pachycarpa
Tripterygium forrestii

Wikstroemia chamaedaphne

Oncopeltus fasciatus
(Migratory milkweed bug)
Ageratum houstonianum
Anabasis aphylla
Annona glabra
Annona montana
Annona palustris
Annona senegalensis
Annona squamosa
Azadirachta indica
Bidens pilosa
Calodendrum capense
Camptotheca acuminata
Centrosema virginianum
Cephalotaxus fortunii
Clethra alnifolia
Coffea robusta
Crysophila argentea
Cyphostemma kilimandscharia
Delphinium virescens
Derris sp.
Dolichos buchani
Drymaria pachyphylla
Enterolobium cyclocarpum
Epilobium glandulosum
Eucalyptus camaldulensis
Euphorbia maculata
Haplophyton cimicidum
Hypoestes verticillaris
Iris douglasiana
Leonotis leonulrus
Leonotis nepetifolia
Lonchocarpus urucu
Macrosiphonia hypoleuca
Mammea americana
Maytenus senegalensis
Melia azedarach
Menispermum coculus
Mundulea suberosa
Nicotiana sp.
Ocimum basilicum
Phyllostachys nigra
Pimpinella anisum
Pinus rigida
Rhinacanthus nasuta
Ricinus communis
Sagittaria calycinus
Sassafras albidum
Satureja chandleri
Satureja douglasii
Schoenocaulon officinale
Tagetes minuta
Tephrosia toxicaria

Tridax procumbens
Vaccinium membranaceum
Veratrum viride
Vetiveria zizanioides
Willardia mexicana
Xanthium spinosum
Zigadenus paniculatus

Opatroides frater
(Tobacco beetle)
Pongamia pinnata

Operophtera brumata
(Winter moth)
Quercus robur

Ophiomyia reticulipennis
(Stem fly)
Azadirachta indica
Chrysanthemum cinerarifol

Oregma lanigera
(Sugarcane woolly aphid)
Aleurites fordii
Camellia sinensis
Glycine max
Millettia pachycarpa
Pachyrhizus erosus
Rhododendron molle
Tripterygium forrestii

Orgyia antiqua
(Rusty tussock moth)
Lonchocarpus chrysophyllu
Tephrosia macropoda
Tephrosia vogelii

Orseolia oryzae
(Rice gall midge)
Azadirachta indica
Melia toosendan

Orthezia insignis
(Lantana bug)
Mundulea suberosa
Nicotiana sp.

Oryctes rhinoceros
(Rhinoeros beetle)
Acorus calamus
Clerodendrum infortunatum

Oryzaephilus surinamensis
(Saw-toothed grain beetle)
Anacardium occidentale
Annona squamosa
Azadirachta indica
Derris elliptica
Derris fordii
Euonymus europaea
Mundulea suberosa

Ostrinia furnacalis
(Asian corn borer)
Lantana camara
Melia azedarach
Melia toosendan
Tagetes erecta
Tagetes patula

Ostrinia nubilalis
(Europeon corn borer)
Amianthium muscitoxicum
Asclepias labriformis
Delphinium sp.
Erigeron repens
Gingko biloba
Haplophyton cimicidum
Marah fabaceus
Ryania speciosa
Schoenocaulon officinale
Tripterygium wilfordii
Veratrum viride
Zea mays

Oulema oryzae
(Rice leaf beetle)
Actinidia chinensis
Anemone hupehensis
Pinus massoniana
Platycarpa strobilacea
Tripterygium hypoglaucum

Pachyzancla bipunctalis
(Southern beet webworm)
Cnidoscolus urens
Enterolobium cyclocarpum
Erigeron bellidiastrum
Helenium mexicana
Heliopsis longipes
Mammea americana
Perezia nana
Perezia wrightii
Piscidia piscipula
Pristimera celastroides

Pacydiplosis oryzae
(Rice stem gall midge)
Strophanthus divaricatus

Paleacrita vernata
(Spring canker worm)
Tripteryqium forrestii

Panstrongylus megistus
(Assassin bug)
Ageratum conyzoides
Ageratum houstonianum

Papaipema nebris
(Cornstalk borer)
Tephrosia heckmannia

Papilio polyxenes
(Black swallowtail butterfly)
Brassica nigra
Brassica oleracea var. acephala

Paramyelois transitella
Azadirachta indica

Parasa herbifera
(Caterpillar)
Derris sp.
Dioscorea piscatorum

Parasaissetia nigra
(Nigra scale)
Azadirachta indica

Parasites (human)
(General category)
Annona cherimola
Haplophyton cimicidum
Rumex crispulus

Pareva vesta
(China glass butterfly)
Millettia pachycarpa

Parlatoria oleae
(Olive scale)
Sapidus utilis

Pectinophora gossypiella
(Pink cotton bollworm)
Ailanthus giralolii
Ajuga remota
Helianthus annuus
Melia azedarach

Pediculus humanus capitis
(Head lice)
Ageratrum houstonianum
Angelica sylvestris
Annona muricata
Annona reticulata
Cleome gynandra
Consolida regalis
Delphinium sp.
Delphinium staphisagria
Dichapetalum toxicaria
Echinops echinatus
Euonymus atropurpurea
Euphorbia lateriflora
Eurotia lantana
Gloriosa superba
Grewia carpinifolia
Gynandropsis pentaphylla
Hedera helix
Heliopsis longipes
Hibiscus vitifolius
Ipomoea batatas
Khaya nyasica
Maytenus senegalensis
Peganum harmala
Prunus dulcis
Prunus dulcis var. amara
Sesamum indicum
Sophora griffithii
Strophanthus hispidus
Tamus communis
Tephrosia diffusa
Tephrosia macropoda
Tephrosia vogelii
Vernonia shirensis

Pediculus humanus humanus
(Body lice)
Calpurina aurea
Carum carvi
Cassia occidentalis
Citrus aurantiifolia
Consolida ambigua
Heliotropium arborescens
Juniperus sabina
Justicia schrimperiana
Menispermum cocculus
Pimpinella anisum
Pristimera celastroides
Syzygium aromaticum
Willardia mexicana

Pericallia ricini
(Woolly bear)
Acalypha indica

Acorus calamus
Allium sativum
Annona squamosa
Cinnamomum camphora
Cymbopogon nardus
Datura metel
Datura stramonium
Eucalyptus sp.
Gaultheria procumbens
Mundulea suberosa
Ocimum sanctum
Pogostemon patchouli
Tephrosia candida
Vitex negundo

Peridroma saucia
(Variegated cutworm)
Mammea americana
Perezia wrightii

Periphyllus lyropictus
(Norway maple aphid)
Chrysanthemum cinerariifolium
Derris sp.

Periplaneta americana
(American cockroach)
Amorpha glabra
Bidens pilosa
Dennettia tripetala
Enterolobium cyclocarpum
Erythrina fusca
Mammea americana
Pachyrhizus erosus
Pistia stratiotes
Pithecellobium trapezifolium
Satureja chandleri
Satureja douglasii
Tripterygium wilfordii
Veratrum viride

Phaedon brassicae
(Cabbage beetle)
Celastrus angulatus
Leucothoe grayana
Tripterygium forrestii
Tripterygium wilfordii

Phaedon cochleariae
(Mustard beetle)
Atractylis gummifera
Cheiranthus cheiri
Chrysanthemum cinerariifolium
Erysimum cheiranthoides
Erysimum hieraciifolium

Phalera bucephala
(Buff-tip butterfly)
Derris elliptica

Phenacoccus gossypii
(Mexican mealy bug)
Arachis hypogaea
Gossypium hirsutum
Zea mays

Philaenus leucophthalmus
(Spittle bug)
'Rotenone'
Schoenocaulon officinale

Phorodon humuli
(Hop aphid)
Picrasma excelsa

Phthirus pubis
(Crab louse)
Oldfieldia africana
Tephrosia virginiana

Phyllaphis fagi
(Woolly beech aphid)
Quassia amara

Phyllobius oblongus
(Weevil)
Allium cepa
Brassica napus var. arvensis
Celtis occidentalis
Chenopodium album
Clematis vitalba
Convolvulus arvensis
Euonymus europaea
Forsythia viridissima
Fraxinus excelsior
Helianthus annuus
Ligustrum vulgare
Lonicera tatarica
Maclura aurantiaca
Medicago sativa
Melilotus alba
Papaver somniferum
Parthenocissus tricuspidata
Pelargonium zonale
Philadelphus coronarius
Robinia pseudo-acacia
Salvia pratensis
Sambucus nigra
Solanum tuberosum
Solidago virgaurea
Sophora japonica

Staphylea pinnata
Syringa vulgaris
Trifolium pratense

Phyllocnistis citrella
(Citrus leafminer)
Azadirachta indica
Melia azedarach
Melia toosendan
Nicotiana tabacum

Phyllotreta downsei
(Radish flea beetle)
Azadirachta indica

Phyllotreta nemorum
(Flea beetle)
Iberis amara
Iberis umbellata

Phyllotreta tetrastigma
(Flea beetle)
Cheiranthus cheiri
Erysimum cheiranthoides
Erysimum hieraciifolium
Iberis umbellata

Phyllotreta undulata
(Flea beetle)
Cheiranthus cheiri
Erysimum cheiranthoides
Erysimum hieraciifolium
Iberis umbellata

Phyllotreta vittata
(Cabbage flea beetle)
Chrysanthemum cinerariifolium
Derris sp.
Pachyrhizus erosus
Rhododendron molle
Tripterygium forrestii

Phylloxera sp.
Juniperus oxycedrus

Phylloxera vitifoliae
(Grape phylloxera)
Rhus coriaria

Phymatocera aterrima
(Sawfly)
Annona squamosa
Derris elliptica
Quassia amara

351

Phytodecta fornicata
(Luzerne leaf beetle)
Allium cepa
Amaranthus retroflexus
Arctium tomentosum
Artemisia vulgaris
Beta esculenta
Canna indica
Centaurea pannonica
Chenopodium album
Chenopodium hybridum
Cichorium intybus
Cirsium arvense
Clematis vitalba
Convolvulus arvensis
Datura stramonium
Euphorbia helioscopia
Fragaria ananassa
Galinosoga parviflora
Helianthus annuus
Lathyrus tuberosus
Lepidium draba
Melilotus officinalis
Mercurialis annua
Nicandra physalodes
Onopordum acanthium
Papaver somniferum
Parthenocissus inserta
Pelargonium zonale
Phaseolus vulgaris
Pimpinella saxifraga
Plantago lanceolatum
Plantago major
Polygonum aviculare
Polygonum baldschuanicum
Polygonum convolvulus
Polygonum lapathifolium
Potentilla reptans
Prunus armeniaca
Rheum undulatum
Rubus idaeus
Rumex obtusifolius
Senecio vulgaris
Silene alba
Solanum nigrum
Solidago virgaurea
Staphylea pinnata
Taraxacum officinale
Tilia cordata
Trifolium pratense
Urtica urens
Verbascum blattaria
Verbascum phlomoides
Veronica polita
Zea mays

Pieris brassicae
(White cabbage butterfly)
Acorus calamus
Amaranthus retroflexus
Arctium minus
Argemone mexicana
Artemisia abrotanum
Artemisia absinthium
Artemisia vulgaris
Azadirachta indica
Beta esculenta
Calotropis procera
Capsicum annuum
Centaurea pannonica
Chelidonium majus
Chenopodium album
Chenopodium hybridum
Chenopodium vulvaria
Chrysanthemum cinerariifolium
Cichorium intybus
Cirsium arvense
Clematis vitalba
Convolvulus arvensis
Cytisus scoparius
Dahlia pinnata
Delphinium hybridum
Derris elliptica
Echinochloa crus-galli
Erigeron canadensis
Fragaria ananassa
Glycine max
Helianthus annuus
Hibiscus syriacus
Impatiens parviflora
Juglans regia
Lamium maculatum
Lycopersicon lycopersicum
Melia azedarach
Nicotiana sp.
Nicotiana tabacum
Oenanthe crocata
Parthenocissus inserta
Pelargonium zonale
Petunia atkinsiana
Petunia sp.
Phaseolus vulgaris
Plantago lanceolatum
Plantago major
Polygonum lapathifolium
Portulaca oleracea
Prunella vulgaris
Pyrus communis
Rubus idaeus
Salvia officinalis
Salvia splendens

Sambucus nigra
Senecio vulgaris
Solanum aviculare
Solanum melongena
Solanum nigrum
Solanum tuberosum
Solidago virgaurea
Sonchus oleraceus
Stavia pratensis
Stavia splendens
Thymus vulgaris
Tradescantia virginiana
Trifolium pratense
Urtica urens
Verbascum phlomoides
Veronica agrestis
Zea mays

Pieris napi
(Mustard white cabbage butterfly)
Allium cepa
Lycopersicom lycopersicum
Salvia officinalis
Sambucus nigra
Thymus vulgaris

Pieris rapae
(Imported cabbage worm)
Achyranthes radix
Aconitum kusnezoffii
Actinidia chinensis
Ailanthus giralolii
Allium cepa
Ampelopsis japonica
Anemone hupehensis
Angelica dahurica
Arisaema serratum
Arisaema thunbergii
Aristolochia grandiflora
Aristolochia maxima
Artemisia apiacea
Artemisia argyi
Carpesium abrotanoides
Cayratia japonica
Chrysanthemum cinerariifolium
Clematis aethusifolis
Clematis armandii
Clematis chinensis
Coriaria sinica
Cyathula capitata
Datura stramonium
Derris elliptica
Derris malaccensis
Derris sp.
Equisetum arvense

Erigeron canadensis
Erigeron flagellaris
Euphorbia fischeriana
Euphorbia pekinensis
Haplophyton cimicidum
Iris dichotoma
Luffa aegyptiaca
Lycopersicon lycopersicum
Lycoris africana
Mammea americana
Melia azedarach
Melia toosendan
Millettia pachycarpa
Mormordica cochinchinensis
Nepeta cataria
Nicotiana sp.
Pachyrhizus erosus
Pachyrhizus wrightii
Periploca sepium
Phryma oblongifolia
Pinellia ternata
Platycarya strobilacea
Podocarpus macrophyllus
Podocarpus nakaii
Pseudolarix kaempferi
Pueraria lobata
Rheum officinale
Rhodea japonica
'Rotenone'
Rumex dentatus
Salvia officinalis
Sambucus nigra
Schoenocaulon officinale
Sophora japonica
Stellera chamaejasme
Swertia chinensis
Tagetes patula
Tanacetum vulgare
Taxus cuspidata
Thymus vulgaris
Tripterygium hypoglaucum
Tripterygium wilfordii
Verbena officinalis
Xanthium sibiricum

Piesma quadratum
(Beet leaf bug)
Azadirachta indica

Pissodes strobi
(White pine weevil)
Picea glauca
Picea mariana
Picea rubens
Thuja plicata

Plagiodera versicolora
(Willow leaf beetle)
Celastrus angulatus

Planococcus citri
(Citrus mealy bug)
Aleurites fordii
Azadirachta indica

Platythpena scabra
(Green clover worm)
Schoenocaulon officinale

Plodia interpunctella
(Phycitid moth)
Chrysanthemum cinerariifolium

Plusio precationis
(Green looper)
Physianthus albens

Plutella xylostella
(Diamond-back moth)
Acalypha indica
Achyranthes radix
Annona reticulata
Annona squamosa
Azadirachta indica
Balanites aegyptica
Begonia pearcei
Bletilla striata
Buxus sempervirens
Calopogonium coeruleum
Chrysanthemum cinerariifolium
Cinchona calisaya
Cissus rhombifolia
Citrullus colocynthis
Citrus aurantium
Cyathula capitata
Delphinium grandiflorum
Derris elliptica
Derris fordii
Derris malaccensis
Derris sp.
Eranthis hyemalis
Euonymus japonica
Euphorbia lathyris
Hedera helix
Hemerocallis dumortieri
Heuchera sanguinea
Impatiens wallerana
Jacquinia aristata
Lantana camara
Lilium longiflorum
Lycopersicon lycopersicum

Madhuca latifolia
Mammea americana
Matricaria matricarioides
Mundulea suberosa
Nerium oleander
Ougeinia dalbergioides
Oxalis deppei
Pachyrhizus erosus
Pachyrhizus piscipula
Pelargonium sp.
Pellionia pulchra
Petunia sp.
Phyllanthus acuminatus
Piper nigrum
Piscidia acuminata
Piscidia piscipula
Podocarpus macrophyllus
Podocarpus nakaii
Punica granatum
Randia nilotica
Senecio cineraria
Tagetes erecta
Tagetes patula
Taxus cuspidata
Tephrosia candida
Tephrosia vogelii
Tinospora rumphii
Tithonia diversifolia
Tripterygium wilfordii
Vitex negundo
Xeromphis spinosa

Poecilocera picta
(Grasshopper)
Abrus precatorius
Azadirachta indica

Popillia japonica
(Japanese beetle)
Acacia longifolia
Aesculus pavia
Aloe barbadensis
Arctostaphylos uva-ursi
Aster novae-angliae
Aureolaria pedicularia
Azadirachta indica
Buxus sempervirens
Caesalpinia coriaria
Capsella bursa-pastoris
Castanea dentata
Ceanothus americanus
Chelone glabra
Chenopodium ambrosioides
Chlorophora tinctoria
Claytonia virginica

Comandra umbellata
Coreopsis grandiflora
Derris sp.
Erigeron canadensis
Erysimum perofskianum
Eupatorium hyssopifolium
Guarea rusbyi
Gymnocladus dioica
Haematoxylum campechianum
Halesia carolina
Helleborus niger
Helonias bullate
Hieracium pratense
Ilex opaca
Lachnanthes carolina
Magnolia virginiana
Maianthemum canadense
Nigella sativa
Ourouparia gambir
Pelagonium sp.
Pimenta dioica
Pimpinella saxifraga
Pinus sp.
Plantago rugelli
Podophyllum peltatum
Polygonum aubertii
Quercus velutina
Ricinus communis
Rosmarinus officinalis
'Rotenone'
Rudbeckia hirta
Ruta graveolens
Stachys officinalis
Swertia chirata
Terminalia catappa
Tkillium erectum
Veratrum album
Zanthoxylum americanum

Porosagrotis orthogonia
(Pale western cutworm)
Quassia amara

Powderpost beetles
(General category)
Anacardium occidentale

Prenolepis longicornis
(Ant)
Mammea americana
Pachyrhizus erosus

Pseudaletia unipuncta
(Armyworm)
Annona muricata

Capparis horride
Chlorogalum pomeridianum
Cinnamomum camphora
Dryopteris filix-mas
Heliopsis helianthoides
Heliopsis scabra
Mammea americana
Pachyrhizus palmatilobus
Piper guineense
Pogostemon patchouli
Ryania speciosa
Schoenocaulon officinale
Sophora secundiflora
Tripterygium forrestii
Tripterygium wilfordii
Zanthoxylum clava-herculis

Pseudococcus citri
(Greenhouse mealy bug)
'Rotenone'

Pseudococcus gahani
(Citrophilus mealy bug)
Pinus sp.

Psylla mali
(Apple sucker psyllid)
Nicotiana rustica

Psylla pyricola
(Pear psyllid)
Cocculus trilobus
Nicotiana sp.
'Rotenone'

Pteronidea ribesii
(Imported currant worm)
'Rotenone'

Pteronus ribesii
(Gooseberry sawfly)
Chrysanthemum cinerariifolium
Derris elliptica
Hyoscyamus niger

Pulex irritans
(Human flea)
Tephrosia virginiana

Pyrausta nubilalis
(European corn borer)
Brassica oleracea var. capitata
Chrysanthemum cinerariifolium
Derris elliptica
Derris malaccensis

Gingko biloba
Nicotiana sp.
Ryania speciosa
Tripterygium wilfordii

Pyrausta salentialis
(Cornstalk borer)
Ryania speciosa

Pyrilla perpusilla
(Sugarcane leafhopper)
Melinis minutiflora

Pyrrhocoris apterus
(Tobacco leaf bug)
Abies balsamea
Acorus calamus
Angelica sylvestris
Carlina acaulis
Larix laricina
Pimpinella major
Taxus brevifolia
Tsuga canadensis

Rephidopalpa foveiollis
(Red pumpkin beetle)
Melia azedarach

Reticulitermes flavipes
(Eastern subterranean termite)
Aesculus hippocastanum
Cassia fistula
Convallaria majalis
Digitalis lanata
Digitalis purpurea
Glycyrrhiza glabra
Guaiacum officinale
Markhamia stipulata
Nicotiana sp.
Paratecoma peroba
Stereospermum suaveolens
Tabebuia avellanedae
Tabebuia glavenscens
Tectona grandis

Reticulitermes lucifugens
(Termite)
Stereospermum suaveolens
Tabebuia avellanedae
Tabebuia flavescens

Rhabdoscelus obscurus
(New Guinea sugarcane weevil)
Euphorbia hirta
Ricinus communis

Rhaphidopalpa chinensis
(Yellow melon leaf beetle)
Tripterygium forrestii

Rhizopertha dominica
(Lesser grain borer)
Acacia concinna
Acorus calamus
Annona squamosa
Azadirachta indica
Curcuma domestica
Justicia adhatoda
Maranta arundinaceae
Melia azedarach
Pongamia pinnata
Sapindus marginatus

Rhodnius prolixus
Ageratum conyzoides
Ageratum houstonianum

Rhopalosiphum nympheae
(Water-lily aphid)
Azadirachta indica

Rhopalosiphum persicae
(Greek peach aphid)
Derris elliptica
Nicotiana sp.

Rhopalosiphum rufomaculata
(Aphid)
Anabasis aphylla
Nicotiana sp.
Tephrosia virginiana

Rice borers
(General category)
Aconitum kusnezoffii
Actinidia chinensis
Artemisia annua
Artemisia apiacea
Brucea javanica
Camellia oleifera
Catalpa ovata
Coriaria sinica
Datura stramonium
Desmodium caudatum
Dichroa febrifuga
Euscaphis japonica
Gynura segetum
Litsea cubeba
Menispermum dauricum
Opuntia dillenii
Pinellia ternata

Polygonatum sibiricum
Rheum officinale
Rhodomyrtus tomentosa
Strophanthus divaricatus
Veronicastrum axillare
Wikstroemia indica
Zanthoxylum dimorphophyllum
Zanthoxylum planispinum

Rice field insects
(General category)
Aegle glutinosa
Alocasia macrorrhiza
Amorphophallus campanulatus
Annona reticulata
Aphanamixis polystachya
Bambusa vulgaris
Buddleia officinalis
Camellia oleifera
Corchorus capsularis
Cordyline roxburghiana
Corypha elata
Enterolobium saman
Kleinhovia hospita
Pinus insularis
Pistacia chinensis
Pueraria lobata
Rhus chinensis
Vitex negundo

Rice leafhoppers
(General category)
Ailanthus giralolii
Brucea javanica
Clausena excavata
Datura stramonium
Derris trifoliata
Pinus massoniana
Polygonum nodosum

Rondotia menciana
(Mulberry white caterpillar)
Croton tiglium
Datura stramonium
Houttuynia cordata
Lycoris africana
Pinellia ternata
Rhododendron japonicum
Veratrum nigrum

Saissetia nigra
(Black scale)
Azadirachta indica

Scale insects
(General category)
Aloe succotrina
Ipomoea purpurea

Schistocerca gregaria
(Desert locust)
Ageratum houstonianum
Allium cepa
Anethum sowa
Arundo donax
Azadirachta indica
Calotropis gigantea
Calotropis procera
Canna indica
Cassia auriculata
Crinum bulbispermum
Cyamopsis tetragonolobus
Hymenocallis littoralis
Lotus corniculatus
Malvastrum tricumpisdatum
Melia azedarach
Prosopis juliflora
Salvadora persica
Stachytarpheta mutabilis
Woodfordia floribunda

Schizaphis graminum
(Wheat aphid)
Grindelia humilis
Hordeum vulgare
Sorghum bicolor
Zanthoxylum monophyllum

Scirtothrips citri
(Citrus thrips)
Anabasis aphylla

Scirtothrips dorsalis
(Chili thrips)
Ryania speciosa

Scolytus multistriatus
(Smaller European elm bark beetle)
Carya ovata
Juglans nigra

Selenia tetralunaria
(Purple thorn moth)
Lonchocarpus chrysophyllus
Lonchocarpus densiflorus
Tephrosia macropoda
Tephrosia vogelii

Seleron latipes
(Beetle)
Pongamia pinnata

Siphocoryne indobrassicae
(Mustard aphid)
Aconitum ferox

Sitona cylindricollis
(Sweet clover weevil)
Meliotus infesta
Meliotus officinalis

Sitophilus granarius
(Granary weevil)
Anacardium occidentale
Angelica sylvestris
Artemisia absinthium
Pastinaca sativa
Trigonella foenum-graecum

Sitophilus oryzae
(Rice weevil)
Acacia concinna
Acorus calamus
Agave americana
Annona squamosa
Argemone mexicana
Atalantia monophylla
Azadirachta indica
Bougainvillea sp.
Caesalpinia pulcherrima
Callistephus chinensis
Calotropis gigantae
Canna indica
Capsicum frutescens
Chrysanthemum cinerariifolium
Citrus limon
Cocos nucifera
Crotalaria juncea
Curcuma domestica
Delonix regia
Euphorbia pulcherrima
Ganoderma lucidum
Hibiscus rosa-sinensis
Kaempferia galanga
Lantana camara
Leucothoe grayana
Luffa acutangula
Mammea americana
Maranta arundinaceae
Mentha spicata
Nerium oleander
Nigella sativa
Ocimum basilicum

Piper nigrum
Pisum sativum
Pogostemon patchouli
Pongamia pinnata
Sapindus trifoliatus
Tabebuia rosea
Tecoma indica
Tridax procumbens
Vitex negundo

Sitophilus zeamays
(Corn weevil)
Ageratum conyzoides
Piper nigrum
Tithonia diversifolia
Tridax procumbens

Sitotroga cerealella
(Angoumois grain moth)
Acacia concinna
Acorus calamus
Aphanamixis polystachya
Artemisia absinthium
Atalantia monophylla
Azadirachta indica
Carum copticum
Chrysanthemum cinerariifolium
Clerodendron infortunatum
Justicia adhatoda
Melia azedarach
Nicotiana tabacum
Pongamia pinnata
Sapindus marginatus
Vitex negundo

Sogatella furcifera
(White-backed planthopper)
Annona squamosa
Azadirachta indica
Melia azadarach

Spodoptera abyssina
(Rice noctuid)
Melia azedarach
Melia toosendan

Spodoptera eridania
(Southern armyworm)
Acacia chiapensis
Acacia farnesiana
Annona muricata
Artemisia ludoviciana
Celastrus angulatus
Derris sp.
Gliricidia sepium

Gymnocladus dioica
Haplophyton cimicidum
Humulus lupulus
Hura crepitans
Mammea americana
Pastinaca sativa
Perezia nana
Phellodendron amurense
Piscidia erythrina
Piscidia piscipula
Schoenocaulon officinale
Sesamum indicum
Trichilia roka
Vernonia gigantea
Vernonia glauca
Willardia mexicana

Spodoptera exempta
(African armyworm)
Ajuga remota
Clausena anisata
Clerodendrum myricoides
Plagiochila fruticosa
Plagiochila hattoriana
Plagiochila ovalifolia
Plagiochila yokogurensis
Schkuria pinnata
Tithonia diversifolia
Warburgia ugandensis
Xylocarpus granatum
Xylocarpus molluscens
Zanthoxylum chalybea
Zanthoxylum holstii

Spodoptera exigua
(Beet armyworm)
Croton tiglium
Parthenium alpinum
Parthenium fruticosum

Spodoptera frugiperda
(Fall armyworm)
Ajuga remota
Alchemilla procumbens
Arctostaphylos pungens
Azadirachta indica
Baccharis ramulosa
Calopogonium coeruleum
Cirsium ehrenbergii
Erodium cicutarium
Haplopappus heterophylles
Mammea americana
Melia azedarach
Mentha pulegium
Montanoa grandiflora

Pachyrhizus erosus
Salvia tiliifolia
Senecio totucans
Sphaeralcea angustifolia
Valeriana clematitis
Vernonia gigantea
Vernonia glauca
Vinca minor
Zaluzania augusta

Spodoptera littoralis
(Egyptian cotton caterpillar)
Acokanthera oblongifolia
Ajuga remota
Allium sativum
Azadirachta indica
Catharanthus roseus
Dioscorea batatas
Disanthus cercidifolius
Physalis peruviana
Ruscus hypoglossum
Stachytarpheta mutabilis
Urginea maritima
Xylocarpus granatum
Zanthoxylum chalybea
Zanthoxylum holstii

Spodoptera litura
(Cotton leafworm)
Aconitum kusnezoffii
Acorus calamus
Alisma orientale
Allium sativum
Angelica japonica
Annona reticulata
Annona squamosa
Argemone mexicana
Aristolochia indica
Artemisia absinthium
Artemisia argyi
Artemisia capillaris
Azadirachta indica
Callicarpa japonica
Caryopterix divaricata
Cecropia mexicana
Cleome viscosa
Clerodendrum calamitosum
Clerodendrum crytophyllum
Clerodendrum phillipinum
Clerodendrum trichotomum
Cocculus trilobus
Colchicum autumnale
Croton tiglium
Datura metel
Datura stramonium

Derris elliptica
Dillenia indica
Entada polystachia
Enterolobium cyclocarpum
Eupatorium staechadosmum
Ficus carica
Humulus lupulus
Justicia gendarussa
Madhuca latifolia
Matricaria chamomila
Matricaria recutita
Melia azedarach
Millettia pachycarpa
Mundulea suberosa
Nepeta cataria
Ocimum sanctum
Origanum majorana
Orixa japonica
Ougeinia dalbergioides
Pachyrhizus erosus
Parabenzoin praecox
Parabenzoin trilobum
Parthenium hysterophorus
Piper kadzura
Pogostemon patchouli
Polygonum orientale
Pongamia pinnata
Ryania speciosa
Sapindus trifoliatus
Solanum xanthocarpum
Tephrosia purpurea
Tephrosia villosa
Tribulus terrestris
Tripterygium forrestii
Vitex negundo

Stegobium paniceum
(Drugstore beetle)
Annona squamosa
Azadirachta indica
Maranta arundinaceae
Nigella sativa
Ocimum basilicum
Pogostemon heyneanus
Pogostemon patchouli

Stored grain pests
(General category)
Acorus calamus
Afrormosia laxiflora
Agave americana
Annona senegalensis
Artemisia vulgaris
Atractylis ovata
Azadirachta indica

Butryospermum parkii
Cannabis sativa
Capsicum frutescens
Clematis vitalba
Cyperus rotundus
Datura stramonium
Erythrophleum suaveolans
Hyptis spicigera
Justicia adhatoda
Lantana rugosa
Lipia geminata
Luffa aegyptiaca
Lysimachia nummularia
Melia azedarach
Nicotiana sp.
Piper longum
Polygonum hydropiperoides
Punica granatum
Sapindus marginatus
Swartzia madagascariensis
Tacca pinnatifida
Trigonella foenum-graecum
Vitex agnus-castus
Vitex negundo
Xeromphis spinosa

Stored rice pests
(General category)
Azadirachta indica
Clerodendrum infortunatum
Mangifera indica
Melia azedarach

Sylepta derogata
(Cotton leafroller)
Sanquisorba officinalis

Taeniothrips gladioli
(Gladiolus thrips)
Derris sp.
Veratrum album

Tanymecus dilaticollis
(Weevil)
Artemisia vulgaris
Brassica napus var. arvensis
Clematis vitalba
Convolvulus arvensis
Datura stramonium
Helianthus annuus
Lepidium draba
Plantago major
Solanum tuberosum
Solidago virgaurea

Tenebrio molitor
(Yellow mealworm)
Anacyclus pyrethrum
Chamaecyparis lawsoniana
Echinacea angustifolia

Termites
(General category)
Acacia nilotica
Afrormosia laxiflora
Agave americana
Albizia odoratissima
Anacardium occidentale
Argemone mexicana
Boswellia dalzielii
Calotropis gigantea
Camellia sinensis
Capparis aphylla
Cedrus deodora
Chamaecyparis formosensis
Cinnamomum cecicodaphne
Cleistanthus collinus
Commiphora africana
Consolida regalis
Cordia dichotoma
Dalbergia latifolia
Daniellia oliveri
Detarium senegalense
Diospyros ebenum
Dodonaea viscosa
Erythrophleum suaveolens
Hardwickia mannii
Hyptis spicigera
Juniperus virginiana
Khaya nyasica
Melia azedarach
Mesua ferrea
Prosopis africana
Quassia indica
Samadera indica
Samanea saman
Santalum album
Semecarpus anacardium
Strychnos nux-vomica
Swartzia madagascariensis
Tectona grandis
Zanthoxylum xanthoxyloides

Tetroda histeroides
(Rice stink bug)
Millettia pachycarpa

Therioaphis maculata
(Spotted alfalfa aphid)
Nicotiana spp.

Thermobia domestica
(Firebrat)
Acorus calamus
Chrysanthemum cinerariifolium
'Rotenone'

Thrips
(General category)
Allium sativum
Chrysanthemum cinerariifolium
Consolida regalis
Gossypium herbaceum
Ryania speciosa
Tephrosia vogelii

Thrips nigropilosus
(Pyrethrum thrip)
Derris sp.

Thrips tabaci
(Tobacco thrip)
Derris sp.
Nicotiana sp.
'Rotenone'
Schoenocaulon officinale
Veratrum album

Tinea granella
Clothes moth)
Artemisia absinthium

Tineola bisselliella
(Webbing clothes moth)
Allionia incarnata
Annona cherimola
Annona glabra
Annona squamosa
Ardisia picardae
Baccharis floribunda
Bersama paulinioides
Dichapetalum toxicaria
Gynocardia odorata
Heliopsis longipes
Juniperus virginiana
Leucothoe axillaris
Luffa acutangula
Maesa indica
Maesa rufescens
Malouetia obtusiloba
Mammea americana
Mandevilla foliosa
Melia azedarach
Ochrocarpus africanus
Paeonia brownii
Peltandra virginica

Peltophorum suringari
Piscidia piscipula
Psorospermum febrifugum
Ryania angustifolia
Ryania speciosa
Schoenocaulon officinale
Spigelia humboltiana
Spigelia marilandica
Syngonium auritum
Tephrosia toxicaria
Wikstroema sandwicensis
Willardia mexicana

Toxoptera aurantii
(Citrus aphid)
Barringtonia racemosa
Cassia didymobotiya
Chrysanthemum cinerariifolium
Euphorbia tirucalli
Hydnocarpus wightiana
Mundulea suberosa
Nicotiana sp.
Tephrosia candida
Tephrosia nyikensis
Tephrosia toxicaria
Tephrosia vogelii

Trialeurodes packardii
(Strawberry white fly)
Nicotiana sp.

Trialeurodes vaporariorum
(Greenhouse white fly)
Derris sp.
Nicandra physalodes

Tribolium castaneum
(Red flour beetle)
Acacia concinna
Acorus calamus
Ageratum conyzoides
Ailanthus excelsa
Allium cepa
Anethum sowa
Angelica glauca
Annona reticulata
Annona squamosa
Aristolochia bracteata
Artemisia maritima
Artemisia roxburghiana
Asphodelus teniifolius
Atropa accuminata
Azadirachta indica
Berberis aristata

Cassia absus
Coriandrum sativum
Curcuma domestica
Dolichos buchani
Dolichos kilimandshauricus
Elephantopus scaber
Helleborus niger
Impatiens longipes
Ipomoea hederacea
Justicia adhatoda
Kaempferia galanga
Launaea nudicaulis
Maranta arundinaceae
Melia azedarach
Mentha arvensis
Mentha longifolia
Miliusa velutina
Nigella sativa
Ocimum basilicum
Olea cuspidata
Pavetta indica
Peganum harmala
Picrorrhiza kurroa
Piper nigrum
Pogostemon patchouli
Saccharum spontaneum
Sapindus trifoliatus
Skimmia laureala
Spigella marilandica
Tagetes minuta
Tithonia diversifolia
Tridax procumbens
Valeriana wallichii
Zanthoxylum armatum
Zingiber officinale

Tribolium confusum
(Confused flour beetle)
Amorpha fruticosa
Angelica sylvestris
Azadirachta indica
Brassica rapa var. rapifera
Curcuma domestica
Dalbergia retusa
Rauvolfia serpentina

Trichodectus canis
(Dog lice)
Tephrosia virginiana

Trichodetes sp.
Stemona japonica
Stemona sessilifolia

Trichoplusia ni
(Cabbage looper)
Delphinium sp.
Gliricidia sepium
Pteridium aquilinum
Schoenocaulon officinale
Tripterygium forrestii

Trimeresia miranda
Cocculus tribolus
Dioscorea batatas
Disanthus cercidifolius
Parabenzoin trilobum

Tristoma infestans
Ageratum houstonianum

Trogoderma granarium
(Khapra beetle)
Acorus calamus
Allium salivum
Angelica sylvestris
Azadirachta indica
Maranta arundinaceae
Melia azedarach
Saussurea lappa
Valeriana officinalis

Tryporyza incertulas
(Yellow rice stem borer)
Azadirachta indica
Catharanthus roseus
Celastrus orbiculatus
Melia azedarach
Melia toosendan
Rhododendron molle
Tripterygium wilfordii
Vitex negundo

Typhlocyba comes
(Grape leafhopper)
'Rotenone'

Udea rubigalis
(Celery leaftiller)
Celastrus angulatus
Celastrus orbiculatus
Erigeron flagellaris
Haplophyton cimicidum
Tripterygium wilfordii
Yucca schidigera

Urbanus proteus
(Bean leafroller)
Delphinium sp.

Yucca schidigera
Zanthoxylum clava-herculis

Urentius echinus
(Lacewing)
Annona squamosa
Arachis hypogea
Azadirachta indica

Urentius hystricellus
(Bringle lacewing)
Azadirachta indica

Utetheisa pulchella
(Caterpillar)
Azadirachta indica

Vermin
(General category)
Annona squamosa
Asclepias curassavica
Balanites aegyptica
Barringtonia asiatica
Bersama paulinioides
Calotropis gigantea
Carica papaya
Cassytha filiformis
Citrullus colocynthis
Citrus aurantium
Cleome pentaphylla
Datura metel
Delphinium staphisagria
Indigofera tinctoria
Khaya nyasica
Leonotis nepetifolia
Pachygone ovata
Pachyrhizus tuberosus
Pistia stratiotes
Polygonum flaccidum
Portulaca oleracea
Psidium guajava
Pteridium aquilinum
Salvia plebeia
Schoenocaulon officinale
Stachytarpheta jamaicensis
Strychnos ignatii
Trichilia trifoliata
Tylophora fasciculata
Wedelia natalensis
Zanthoxylum xanthoxyloides

Weevils
(General category)
Cannabis sativa
Carica papaya
Delonix regia
Madhuca latifolia
Momordica foetida
Oxalis corniculata
Sapindus marginatus

Wireworms
(General category)
Chrysanthemum cinerariifolium

Woolly aphids
(General category)
Solanum nigrum

Xylorycetes jamaicensis
Coconut rhinoceros beetle)
Hydnocarpus wightiana

Zabrotes subfasciatus
(Mexican bean weevil)
Cocus nucifera
Glycine max
Gossypium hirsutum
Zea mays

Zonocerus variegatus
(Edible grasshopper)
Dennettia tripetala

LEECHES AND MOLLUSKS (including snails)

LEECHES
(General category)
Anagallis arvensis
Careya arborea
Nicotiana rustica
Pogostemon patchouli

MOLLUSKS
Lymnaea auricularia rubiginosa
Croton tiglium

Mollusks
(General category)
Maesa lanceolata
Oxalis anthelmintica
Punica granatum

Snails
(General category)
Balanites aegyptica
Blumea balsamifera
Citrofortunella mitis
Entada gigas
Euphorbia tirucalli
Jatropha curcas
Menispermum cocculus
Nicotiana tabacum
Prangos pabularia

MITES AND TICKS

A. MITES
Chiggers
(General category)
Datura metel
Dioscorea latifolia
Melinis minutiflora

Mites
(General category)
Acorus calamus
Anabasis aphylla
Calophyllum inophyllum
Cardaria draba
Cocos nucifera
Croton klotzschianus
Curcuma domestica
Cyclamen elegans
Cynodon dactylon
Delphinium staphisagria
Jatropha curcas
Jatropha multifida
Juniperus virginiana
Lepidium ruderale
Leucas cephalotes
Liquidambar orientali
Mangifera indica
Nicotiana tabacum
Piper nigrum
Plumbago zeylanica
Prangos pabularia
Ricinus communis
Terminalia catappa
Trichilia cuneata

Panonychus citri
(Citrus red mite)
Acer saccharium
Aleurites fordii
Allamanda neriifolia
Annona cherimola
Aristolochia elegans
Azadirachta indica
Caesalpinia pulcherrima
Chrysanthemum coccineum
Conium maculatum
Cycas revoluta
Eucalyptus cinerea
Euphorbia pulcherrima
Fagus grandifolia
Hydrocotyle americana
Lantana camara
Malus sylvestris
Melia azedarach

Melia toosendan
Pinus strobus
Rhus radicans
Trifolium repens
Tsuga canadensis
Ulmus americana
Viola septentrionalis
Zea mays

Paratetranychus pilosus
(European red mite)
Consolida regalis
Delphinium staphisagria

Spider mites
(General category)
Allium tuberosum
Anemone hupehensis
Arisaema serratum
Arisaema thunbergii
Artemisia annua
Artemisia argyi
Cibotium barometz
Clematis aethusifolis
Clerodendrum bungei
Coriaria sinica
Dichroa febrifuga
Ehretia dicksonii
Euphorbia fischeriana
Euphorbia helioscopia
Gossypium herbaceum
Houttuynia cordata
Lobelia chinensis
Luffa aegyptiaca
Momordica cochinchinensis
Myrica rubra
Pinellia ternata
Platycarya strobilacea
Polygonum nodosum
Pseudolarix kaempferi
Pueraria lobata
Rumex dentatus
Sanquisorba officinalis
Sapium sebiferum
Semiaquilegia adoxoides
Swertia chinensis
Veronicastrum axillare
Vitex cannabifolia
Woodwardia japonica
Xanthium sibiricum

Tetranychus atlanticus
(Spider mite)
Brassica rapa var. rapifera
Pastinaca sativa

Tetranychus cinnabarinus
(Carmine spider mite)
Aloysia triphylla
Camellia sinensis
Chrysanthemum cinerariifolium
Consolida regalis
Coriandrum sativum
Cymbopogon citratus
Delphinium staphisagria
Derris sp.
Dracocephalum moldavica
Lavandula angustifolia
Lycopersicum hirsutum
Nicotiana sp.
Ocimum basilicum
Pyrrosia drakeana
'Rotenone'
Solanum pennellii

Tetranychus urticae
(Two-spotted spider mite)
Cucumis sativus
Lycopersicon hirsutum
Solanum pennellii

B. TICKS

Boophilus microplus
(Cattle tick)
Stylosanthes scabra
Stylosanthes viscosa

Dermacentor marginatus
Achillea millefolium
Allium cepa
Allium sativum
Anethum graveolens
Armoracia rusticana
Artemisia absinthium
Artemisia campestris
Cannabis sativa
Cichorium intybus
Conium maculatum
Filipendula vulgaris
Hypericum perforatum
Juglans regia
Levisticum officinale
Matricaria recutita
Mentha arvensis
Petroselinum crispum
Pinus sylvestris
Ranunculus illyricus
Salvia verticillata
Thymus pannonicus

Haemaphysalis punctata
Achillea millefolium
Allium cepa
Allium sativum
Anethum graveolens
Armoracia rusticana
Artemisia absinthium
Artemisia campestris
Cannabis sativa
Cichorium intybus
Conium maculatum
Filipendula vulgaris
Hypericum perforatum
Juglans regia
Levisticum officinale
Matricaria recutita
Mentha arvensis
Petroselinum crispum
Pinus sylvestris
Ranunculus illyricus
Salvia verticillata
Thymus pannonicus

Ixodes redikorzevi
Achillea millefolium

Allium cepa
Allium sativum
Anethum graveolens
Armoracia rusticana
Artemisia absinthium
Artemisia campestris
Cannabis sativa
Cichorium intybus
Conium maculatum
Filipendula vulgaris
Hypericum perforatum
Juglans regia
Levisticum officinale
Matricaria recutita
Mentha arvensis
Petroselinum crispum
Pinus sylvestris
Ranunculus illyricus
Salvia verticillata
Thymus pannonicus

Ixodes ricinus
(Castor bean tick)
Prunus laurocerasus
Prunus racemosus

Ixodes sp.
Citrus limon

Rhipicephalus rossicus
Achillea millefolium
Allium cepa
Allium sativum
Anethum graveolens
Armoracia rusticana
Artemisia absinthium
Artemisia campestris
Cannabis sativa
Cichorium intybus
Conium maculatum
Filipendula vulgaris
Hypericum perforatum
Juglans regia
Levisticum officinale
Matricaria recutita
Mentha arvensis
Petroselinum crispum
Pinus sylvestris
Ranunculus illyricus
Salvia verticillata
Thymus pannonicus

Rhipicephalus sanguineus
(Dog brown tick)
Mammea americana

NEMATODES

Anguina tritici
(Wheat ear-cockle nematode)
Anagallis arvensis
Tagetes nana

Aphelenchus avenae
(Oat nematode)
Arachis hypogaea
Azadirachta indica
Brassica nigra
Madhuca indica
Ricinus communis

Aphelencoides besseyi
(Rice white tip nematode)
Angelica pubescens
Carthamus tinctorius
Daphne odora
Iris japonica

Belonolaimus longicaudatus
(Sting nematode)
Crotalaria spectabilis
Indigofera hirsuta
Tagetes minuta

Bursaphelenchus xylophilus
(Pinewood nematode)
Cirsium japonicum

Criconemoides ornatum
(Root lesion nematode)
Crotalaria spectabilis
Desmodium tortuosum
Indigofera hirsuta
Tagetes minuta

Ditylenchus cypei
(Stem and bulb nematode)
Arachis hypogaea
Azadirachta indica
Brassica nigra
Madhuca indica
Ricinus communis

Ditylenchus dipsaci
(Stem and bulb nematode)
Tagetes nana

Helicotylenchus erythrinae
(Spiral nematode)
Arachis hypogaea
Azadirachta indica

Madhuca indica
Ricinus communis

Helicotylenchus indicus
(Spiral nematode)
Artocarpus heterophyllus
Azadirachta indica
Carica papaya
Cymbopogon flexuosus
Ficus carica
Ficus elastica
Ficus racemosa
Ipomoea fistulosa
Melia azedarach
Nerium oleander
Tabernaemontana divaricata
Tagetes erecta

Helicotylenchus nannus
(Spiral nematode)
Asparagus officinalis

Helicotylenchus sp.
(Spiral nematode)
Tagetes erecta

Heterodera glycines
(Cyst nematode of soybean)
Asparagus officinalis

Heterodera rostochiensis
(Golden cyst nematode of potato)
Asparagus officinalis
Brassica hirta
Brassica nigra
Nasturtium officinale
Ricinus communis
Solanum ballsii
Solanum sucrense
Tagetes nana

Heterodera schachtii
(Cyst nematode of sugar beetle)
Ricinus communis

Hirschmanniella oryzae
(Rice knot nematode)
Azadirachta indica

Hoplolaimus indicus
(Lance nematode)
Arachis hypogaea
Artocarpus heterophyllus
Azadirachta indica
Brassica nigra

Brassica rapa
Carica papaya
Carthamus tinctorius
Cymbopogon flexuosus
Eruca vesicaria
Ficus carica
Ficus elastica
Ficus racemosa
Ipomoea fistulosa
Madhuca indica
Melia azedarach
Nerium oleander
Ricinus communis
Sesamum indicum
Tabernaemontana divaricata
Tagetes erecta

Hoplolaimus sp.
(Lance nematode)
Nicotiana tabacum
Phaseolus vulgaris
Tagetes erecta

Meloidogyne arenaria
(Root knot nematode)
Asparagus racemosus
Azadirachta indica
Crotalaria spectabilis
Desmodium tortuosum
Tagetes erecta
Tagetes minuta

Meloidogyne hapla
(Root knot nematode)
Asparagus officinalis

Meloidogyne incognita
(Root knot nematode)
Ageratum conyzoides
Ailanthus excelsa
Allium sativum
Aloe barbadensis
Anacardium occidentale
Annona squamosa
Arachis hypogaea
Argemone mexicana
Artemisia absinthium
Artemisia cina
Artemisia dracunculus
Asparagus officinalis
Avena sativa
Azadirachta indica
Brassica nigra
Calophyllum inophyllum
Carica papaya

Catharanthus roseus
Centrosema pubescens
Chenopodium ambrosioides
Chromolaena odorata
Citrus reticulata
Crotalaria juncea
Crotalaria spectabilis
Curcuma domestica
Cuscuta reflexa
Cymbopogon citratus
Cymbopogon flexuosus
Cynodon dactylon
Cyperus rotundus
Datura stramonium
Derris elliptica
Desmodium tortuosum
Eragrostis amabilis
Helianthus annuus
Holarrhena antidysenterica
Hydnocarpus laurifolia
Imperata cylindrica
Jasminium arborescence
Justicia adhatoda
Lepidium sativum
Leucaena leucocephala
Madhuca indica
Matricaria recutita
Mimosa pudica
Momordica charantia
Moringa pterygosperma
Nigella sativa
Ocimum sanctum
Oryza sativa
Phaseolus lunatus
Portulaca oleracea
Raphanus sativus
Ricinus communis
Secale cereale
Tagetes erecta
Tagetes minuta
Tagetes patula
Tamarindus indica
Trichosanthes anguina
Vernonia anthelmintica

Meloidogyne indica
(Root knot nematode)
Calotropis gigantea
Chrysanthemum morifolium

Meloidogyne javanica
(Javanese root knot nematode)
Ageratum conyzoides
Ailanthus excelsa
Allium sativum

Anacardium occidentale
Arachis hypogaea
Argemone mexicana
Asparagus racemosus
Azadirachta indica
Brassica juncea
Calophyllum inophyllum
Calotropis gigantea
Cassia fistula
Cassia occidentalis
Catharanthus roseus
Citrus reticulata
Crotalaria spectabilis
Curcuma domestica
Datura stramonium
Desmodium tortuosum
Holarrhena antidysenterica
Hydnocarpus laurifolia
Justicia adhatoda
Linum usitatissimum
Madhuca indica
Melia azedarach
Momordica charantia
Ocimum sanctum
Pongamia pinnata
Ricinus communis
Sesbania aculeata
Tagetes erecta
Tagetes minuta
Vernonia anthelmintica

Meloidogyne sp.
(Root knot nematode)
Asparagus officinalis
Avena sativa
Crotalaria breviflora
Crotalaria mucronata
Crotalaria pumila
Croton sparsiflorum
Eclipta alba
Fagopyrum esculentum
Gossypium hirsutum
Lycopersicon lycopersicum
Oryza sativa
Raphanus sativus
Ricinus communis
Solanum hyporhodium
Trifolium pratense
Triticum aestivum
Zea mays

Pratylenchus alleni
(Lesion nematode)
Chrysanthemum morifolium
Tagetes patula

Pratylenchus brachyurus
(Lesion nematode)
Azadirachta indica
Frageria chiloensis
Ipomoea batatas
Lactuca sativa
Setaria italica

Pratylenchus curvitatus
(Lesion nematode)
Asparagus officinalis

Pratylenchus delattrei
(Root lesion nematode)
Azadirachta indica
Ricinus communis

Pratylenchus penetrans
(Lesion nematode)
Arachis hypogaea
Asparagus officinalis
Foeniculum vulgare
Glycine max
Gossypium hirsutum
Secale cereale
Tagetes erecta
Tagetes nana
Tagetes patula
Zea mays

Pratylenchus sp.
(Lesion nematode)
Azadirachta indica
Tagetes erecta

Pratylenchus zea
(Lesion nematode)
Citrullus lanatus
Crotalaria mucronata
Festuca elatior
Gossypium hirsutum
Lespedeza cuneata
Lycopersicon lycopersicum
Medicago sativa
Phaseolus vulgaris
Tagetes erecta
Trifolium repens
Vicia villosa

Radopholus similis
(Burrowing nematode of grapefruit)
Crotalaria mucronata
Crotalaria spectabilis
Tagetes erecta

Rotylenchulus reniformis
(Reniform nematode)
Arachis hypogaea
Artocarpus heterophyllus
Asparagus officinalis
Azadirachta indica
Carica papaya
Cymbopogon flexuosus
Ficus carica
Ficus elastica
Ficus racemosa
Ipomoea fistulosa
Madhuca indica
Nerium oleander
Pongamia pinnata
Ricinus communis
Tabernaemontana divaricata
Tagetes erecta

Rotylenchulus sp.
(Reniform nematode)
Asparagus officinalis
Croton sparsiflorum
Eclipta alba
Oryza sativa
Raphanus sativus
Tagetes erecta
Tagetes patula
Zea mays

Trichodorus christiei
(Stubby-root nematode)
Asparagus officinalis
Crotalaria spectabilis
Indigofera hirsuta
Tagetes minuta

Tylenchorhynchus brassicae
(Stunt nematode)
Arachis hypogaea
Artocarpus heterophyllus
Azadirachta indica
Brassica nigra
Carica papaya
Cymbopogon flexuosus
Ficus carica
Ficus elastica
Ficus racemosa
Ipomoea fistulosa
Madhuca indica
Nerium oleander
Ricinus communis
Tabernaemontana divaricata
Tagetes erecta

Tylenchorhynchus claytoni
(Tesselate stylet nematode)
Tagetes patula

Tylenchorhynchus dubius
(Stunt nematode)
Nicotiana tabacum
Phaseolus vulgaris

Tylenchorhynchus elegans
(Stunt nematode)
Azadirachta indica

Tylenchorhynchus sp.
(Stunt nematode)
Tagetes erecta

Tylenchulus semipenetrans
(Citrus nematode)
Brassica kaber
Cephalaria syriaca
Citrullus colocynthis
Consolida ambigua
Eminium intortum
Lepidium draba
Medicago sativa
Papaver rhoeas
Ricinus communis
Urtica urens
Xanthium strumarium

Tylenchus filiformis
Artocarpus heterophyllus
Azadiracta indica
Carica papaya
Ficus carica
Ficus elastica
Ficus racemosa
Ipomoea fistulosa
Melia azedarach
Nerium oleander
Tabernaemontana divaricata
Tagetes erecta

Xiphinema americanum
(Dagger nematode)
Asparagus officinalis
Crotalaria spectabilis
Desmodium tortuosum
Indigofera hirsuta
Tagetes minuta

Nematodes

(General category)

Andrographis paniculata
Asparagus officinalis
Azadirachta indica
Carthamus oxycantha
Carthamus tinctorius
Cirsium arvense
Cirsium lipskyi

Daphne odora
Fleurya interrupta
Glycine max
Peristrophe bicalyculate
Tagetes lucida
Tagetes minuta
Tagetes patula
Tagetes tenuifolia

RODENTS

Mice
(General category)
Anthemis arvensis
Antiaris toxicaria
Aristolochia indica
Solidago canadensis
Solidago flexicaulis
Solidago hispida
Solidago juncea
Solidago rugosa
Solidago serotina
Solidago squarrosa
Solidago uliginosa
Verbascum lychnitis
Verbascum phlomoides

Rats
(General category)
Acorus calamus
Acrostichum aureum
Aeschynomene indica
Albizia odoratissima
Antiaris toxicaria
Ardisia neriifolia
Butea monosperma
Caesalpinia coriaria
Caloncoba glauca
Caltha palustris
Capsella bursa-pastoris
Cassia javanica
Chamaecyparis funebris
Clerodendrum serratum
Curcuma domestica
Cynanchum auriculatum
Datura metel
Daucus carota
Dianella ensifolia
Dichapetalum toxicaria
Diospyros insignis
Dipcadi cowanii
Drimia cowanni
Embelia ribes
Embelia viridiflora
Eriolaena quinquelocularis
Erythrophleum suaveolens
Eucalyptus cloeziana
Euphorbia esula
Euphorbia fischeriana
Euphorbia primulaefolia
Garcinia indica
Gliricidia sepium
Grewia tiliifolia
Holigarna grahamii

Humboldtia brunonis
Hydrocotyle podantha
Hyoscyamus niger
Ixora coccinea
Leea aequata
Madhuca latifolia
Mentha arvensis
Momordica charantia
Nerium oleander
Perilla frutescens
Polygonum hydropiperoides
Samanea saman
Sapindus trifoliatus
Schefflera capitata
Spondianthus preussi
Spondianthus ungandensis
Strophanthus divaricatus
Symplocos gardneriana
Syzygium montanum
Tectaria cicutaria
Tephrosia vogelii
Terminalia arjuna
Thespesia populnea
Trigonella foenum-graecum
Tylophora fasciculata
Tylophora ovata
Urginea maritima
Vaccinium leschenaultii
Vateria chinensis
Vateria indica
Veratrum album
Verbena bonariensis
Wendlandia wallichii
Zamia loddigesii

Rodents
(General category)
Acer platanoides
Aesculus hippocastanum
Agrostemma githago
Aster lateriflorus
Aster sagittifolius
Citrullus colocynthis
Collinsonia canadensis
Comptonia aspleniifolia
Dryopteris austriaca
Euphorbia esula
Euphorbia maculata
Gliricidia sepium
Hypoxis latifolia
Ipomoea purpurea
Lasiosiphon krausii
Lespedeza intermedia
Lophira alata
Lysimachia ciliata

Millettia auriculata
Podophyllum peltatum
Polygala sanguinea
Rhus glabra
Rhus typhina
Saponaria officinalis
Solidago caesia
Strychnos nux-vomica
Urginea maritima
Vernonia colorata

VIRUSES

Apple chlorotic leaf spot virus
Chenopodium quinoa

Cucumber mosaic virus
Capsicum frutescens
Cucumis sativus
Datura stramonium
Dianthus barbatus
Spinacia oleracea

Cucumis virus
Datura metel

Gomphrena mosaic virus
Boerhavia diffusa
Datura metel

Latent potato ring spot virus
Spinacia oleracea

Physalis shoestring mosaic virus
Bougainvillea spectabilis

Potato virus X
Capsicum frutescens
Chenopodium album
Chenopodium amaranticolor
Datura metel
Datura stramonium
Lycopersicon lycopersicum
Nicandra physalodes
Pelargonium hortorum
Phaseolus vulgaris
Solanum integrifolium
Solanum tuberosum
Spinacia oleracea
Trifolium pratense
Vicia faba

Potato virus Y
Aeonium arboreum
Aeonium haworthi

Agave americana
Mesembryanthemum caprohetum

Red clover vein mosaic virus
Trifolium pratense

Southern mosaic bean virus
Ginkgo biloba
Phytolacca americana

Sunnhemp rosette virus
Boerhavia diffusa
Bougainvillea spectabilis
Chenopodium ambrosioides
Datura metel

Tobacco etch virus
Capsicum frutescens
Dianthus barbatus

Tobacco mosaic virus
Abutilon striatum
Aeonium arboreum
Aeonium balsamiferum
Aeonium haworthii
Agapanthus africanus
Agave americana
Agrostemma githago
Bergenia cordifolia
Boerhavia diffusa
Bougainvillea spectabilis
Brassica oleracea
Camptotheca acuminata
Capsicum frutescens
Cedrus deodora
Chamaecyparis lawsoniana
Cheiridopsis aspera
Chenopodium album
Chenopodium amaranticolor
Chenopodium ambrosioides
Crassula arborescens
Crassula argentea
Crassula falcata
Crassula multicava
Crassula prealtum
Crassula rupestris
Cuscuta reflexa
Cyrtomium flacerum
Datura metel
Datura stramonium
Dianthus barbatus
Dianthus caryophyllus
Duboisia myoporoides
Echinops setifer

Fragaria vesca
Ipomoea nil
Juniperus communis
Kalanchoe beharensis
Kalanchoe marmorata
Lonicera caprifolium
Lycium chinense
Mesembryanthemum caprohetum
Nicotiana glutinosa
Nicotiana tabacum
Opuntia robusta
Orthosiphum stamineus
Oryza sativa
Pelargonium hortorum
Phytolacca americana
Phytolacca esculenta
Phytolacca rigida
Platycladus orientalis
Psidium guajava
Rauvolfia serpentina
Sansevieria anthispica
Schinus molle
Schinus terebinthifolius
Scopolia japonica
Sedum dendroideum
Sedum nussbaumerianum
Sequoia sempervirens
Simmondsia chinensis
Solanum melongena
Spinacia oleracea
Syzygium paniculatum
Thevetia peruviana

Tobacco necrotic virus
Beta nana
Celosia plumosa
Chenopodium album
Chenopodium amaranticolor
Chenopodium ficifolium
Chenopodium opulifolium
Chenopodium rubrum
Chenopodium urbicum
Halimione portulacoides
Spinacia oleracea

Tobacco ring spot virus
Boerhavia diffusa
Capsicum frutescens
Cuscuta reflexa
Datura metel
Dianthus barbatus
Solanum melongena
Spinacia oleracea
Tetragonia tetragonioides

Tomato aucuba mosaic virus
Cucumis sativus
Dahlia pinnata
Phytolacca americana

Tomato bushy stunt virus
Cucumis sativus
Dahlia pinnata
Datura stramonium
Phytolacca americana

Tomato yellow mottle mosaic virus
Bougainvillea spectabilis

Tungro virus of rice
Azadirachta indica

WEEDS

Agropyron spicatum
(Bluebunch)
Artemisia tridentata
Chrysothamnus viscidiflorus
Tortula ruralis

Amaranthus retroflexus
(Redroot)
Digitaria sanguinalis
Helianthus annuus
Sorghum halepense

Amaranthus spinosus
(Spiny amaranth)
Cassia occidentalis
Cestrum diurnum
Cestrum occidentalis
Coffea arabica
Nicotiana tabacum
Ocimum sanctum

Ambrosia artemisiifolia
(Common ragweed)
Secale cereale
Sorghum bicolor
Sorghum sudanense
Triticum aestivum

Ambrosia elatior
(Ragweed)
Digitaria sanguinalis

Ambrosia psilostachya
(Western ragweed)
Platanus occidentalis

Andropogon gerardi
(Big bluestem)
Celtis laevigata

Andropogon virginicus
(Virginia beard grass)
Platanus occidentalis

Argemone mexicana
(Mexican prickly poppy)
Echinops echinatus
Solanum surattense

Aristida oligantha
(Prairie three-awn)
Digitaria sanguinalis
Helianthus annuus

Sorghum halpense

Avena fatua
(Wild oat)
Artemisia californica
Aretmisia fasciculata
Cofea arabicaa
Eucalyptus camaldulensis
Salvia apiana
Salvia leucophylla

Bromus japonicus
(Japanese brome grass)
Digitaria sanguinalis
Helianthus annuus
Sorghum halepense

Bromus mollis
(Soft chess)
Artemisia californica
Artemisia fasciculata
Eucalyptus camaldulensis
Salvia leucophylla

Bromus rigidus
(Brome grass)
Adenostoma fasciculatum
Artemisia californica
Artemisia fasciculata
Eucalyptus camaldulensis
Salvia leucophylla

Bromus rubens
(Foxtail chess)
Artemisia californica
Artemisia fasciculata
Salvia leucophylla

Bromus tectorum
(Downy brome)
Sorghum halepense

Cerastium vulgatum
(Mouse ear chickweed)
Secale cereale
Sorghum bicolor
Sorghum sudanense
Triticum aestivum

Chenopodium album
(Pigweed)
Secale cereale
Sorghum bicolor
Sorghum sudanense
Triticum aestivum

Coronilla varia
(Crown vetch)
Lepidium virginicum

Croton glandulosus
(Tropic croton)
Helianthus annuus

Cynodon dactylon
(Bermuda grass)
Platanus occidentalis

Cyperus rotundus
(Purple nutsedge)
Croton bonplandianum
Ocimum canum

Digitaria sanguinalis
(Hairy crabgrass)
Secale cereale
Sorghum bicolor
Sorghum halepense
Sorghum sudanense
Triticum aestivum

Echinochloa colonum
(Barnyard grass)
Coffea arabica

Echinochloa crus-galli
(Barnyard grass)
Coffea arabica

Erigeron canadensis
(Canadian fleabane)
Helianthus annuus

Erodium cicutarium
(Stork's bill)
Artemisia californica
Artemisia fasciculata
Salvia leucophylla

Erodium medicago
Eucalyptus camaldulensis

Festuca megalura
(Foxtail fescue)
Artemisia californica
Artemisia fasciculata
Salvia leucophylla

Haplopappus ciliatus
(Goldenrod genus)
Helianthus annuus

Hordeum distichon
(Grass family)
Cirsium arvense

Hordeum leporinum
(Grass family)
Eucalyptus camaldulensis

Hordeum stebbinsii
(Grass family)
Eucalyptus camaldulensis

Hypochoeris glabra
(Smooth cat's ear)
Eucalyptus camaldulensis

Imperata cylindrica
(Cogon grass)
Leucaena leucocephala

Lathyrus aphaca
(Sweetpea genus)
Coffea arabica

Lepideum sativum
(Garden cress)
Sorghum bicolor

Linum usitatissimum
(Flax)
Camelina sativa
Datura stramonium

Lolium multiflorum
(Italian ryegrass)
Platanus occidentalis

Lolium perenne
(English ryegrass)
Cirsium arvense

Lolium sp.
(Ryegrass family)
Eucalyptus camaldulensis

Monarda fistulosa
(Wild bergamot)
Helianthus rigidus

Panicum miliaceum
(Guinea grass)
Cucumis sativus

Panicum scribnerianum
(Grass family)
Platanus occidentalis

Panicum virgatum
(Switchgrass)
Celtis laevigata
Platanus occidentalis

Parabenzoin trilobum
Cocculus trilobus

Poa pratensis
(Kentucky bluegrass)
Helianthus rigidus
Platanus occidentalis

Polygonum persicariap
(Lady's thumb)
Secale cereale
Sorghum bicolor
Sorghum sudanense
Triticum aestivum

Rumex crispulus
(Yellow dock)
Cupressus lusitania

Schizachyrium scoparium
(Bunchgrass)
Celtis laevigata

Setaria viridis
(Green foxtail)
Platanus occidentalis
Secale cereale
Sorghum bicolor
Sorghum halepense
Sorghum sudanense
Triticum aestivum

Sitanion hystrix
(Squirrel tail)
Artemisia tridentata
Chrysothamnus viscidiflorus
Tortula ruralis

Sorghastrum nutans
(Woodgrass)
Celtis laevigata

Spergula arvensis
(Toad flax)
Eucalyptus camaldulensis

Stipa pulchra
(Grass family)
Salvia leucophylla

Stipa thurberiana
(Grass family)
Artemisia tridentata
Chrysothamnus viscidiflorus
Tortula ruralis

Tridens flavus
(Tall red-top)
Platanus occidentalis

Trifolium hirtum
(Clover genus)
Eucalyptus camaldulensis

Trifolium subterraneum
(Subterranean clover)
Cirsium arvense

Vicia sativa
(Spring vetch)
Coffea arabica

SECTION III: POISONOUS PLANTS & PLANTS WHICH CONTROL NON-INSECT ANIMAL PARASITES AND DISEASES

Abies lasiocarpa (Pinaceae) H-20, I-4,14 (504)

Abutilon indicum (Malvaceae) H-20, I-2,4,7 (504)

Acacia koa (Mimosaceae) H-19,20, I-6 (172) Alk=I-6,7,9 (1363)

A. nilotica subsp. adonsonii (Mimosaceae) H-19, I-? (443,444)

A. pruinescens (Mimosaceae) H-32, I-? (105)

A. salicina (Mimosaceae) H-32, I-? (105) Alk=I-7 (1392)

A. seyal (Mimosaceae) H-19, I-7 (443,444) Tan=I-4 (1353)

Acalypha ganitrus (Euphorbiaceae) H-19, I-7 (1076)

A. hispida (Euphorbiaceae) H-19, I-7 (982)

A. wilkesiana (Euphorbiaceae) H-19, I-7 (1076)

Acanthophora spicifera (Rhodomelaceae) H-19,22, I-1 (1114)

Acer pensylvanicum (Aceraceae) H-19, I-19 (851)

A. spicatum (Aceraceae) H-19, I-? (594)

Acokanthera abyssinica (Apocynaceae) H-33, I-5 (504)

A. longiflora (Apocynaceae) H-33, I-6,7 (105,507)

Aconitum balfourii (Ranunculaceae) H-33,34, I-? (220) Alk=I-2 (1392)

A. chamessonianum (Ranunculaceae) H-33, I-2 (620)

A. delphinifolium (Ranunculaceae) H-33, I-2 (620)

A. fischeri (Ranunculaceae) H-33, I-2 (839) Alk=I-2 (1392)

A. maximum (Ranunculaceae) H-33, I-2 (839) Alk=I-? (1392)

A. sachalinense (Ranunculaceae) H-33, I-2 (839) Alk=I-2 (1392)

A. yezoene (Ranunculaceae) H-33, I-2 (839) Alk=I-? (1392)

Acronychia acidula (Rutaceae) H-19, I-? (776) Alk=I-4,7,9 (1392)

A. baueri (Rutaceae) H-19, I-? (776) Alk=I-4,7 (1392)

A. laevis (Rutaceae) H-19, I-12 (775,776) Alk=I-6,7 (1392)

A. pedunculata (Rutaceae) H-32, I-2 (73,220,504)

A. resinosa (Rutaceae) H-32, I-2 (504)

Adenia palmata (Passifloraceae) H-33, I-2,9 (220) Sap=I-? (220)

A. wightiana (Passifloraceae) H-33, I-2,9 (220) Sap=I-? (220)

Adenium honghel (Apocynaceae) H-32,33 I-2,6,8,14 (504)

A. multiflorum (Apocynaceae) H-32, I-1 (504)

A. obesum (Apocynaceae) H-32,33, I-2,6 (504)

A. somalense (Apocynaceae) H-32,33, I-2,6 (504)

A. speciosum (Apocynaceae) H-33, I-? (504)

Adesmia retusa (Fabaceae) H-19, I-6,7 (940)

Adonis aestivalis (Ranunculaceae) H-33, I-? (220)

Aegiceras corniculatum (Myrsinaceae) H-32, I-4 (983)

A. majus (Myrsinaceae) H-32, I-4 (504)

Aegle marmelos (Rutaceae) H-22, I-2 (881) H-32, I-4 (983) Alk=I-4,6,7,9 (1361) Alk=I-10 (1374) Alk=I-4,5,7 (1392)

Aeschrion excelsa (Simaroubaceae) H-17, I-5 (504)

Aesculus glabra (Hippocastanaceae) H-32, I-9 (504)

A. chinensis (Hippocastanaceae) H-32, I-4 (504)

Agave bovicornuta (Agavaceae) H-32, I-2,7 (401)

A. lecheguilla (Agavaceae) H-32, I-2,7 (401)

A. schottii (Agavaceae) H-32, I-2,7 (401) Alk=I-10 (1391) Sap=I-2,6 (1367)

Aglaia odoratisima (Meliaceae) H-19, I-? (411)

Agonis linearifolia (Myrtaceae) H-19, I-8 (777)

Agrimonia aeupatoria (Rosaceae) H-22, I-16 (977)

A. striata (Rosaceae) H-19, I-? (1171)
Albizia acle (Mimosaceae) H-19,32,33, I-4 (968,982,983)
Alchornea parviflora (Euphorbiaceae) H-32, I-1 (983)
A. sicca (Euphorbiaceae) H-32,33, I-7,9 (968,983,1349) Sap=I-7,9 (968)
Aletris farinosa (Liliaceae) H-19, I-1 (1262)
Aleurites moluccana (Euphorbiaceae) H-19, I-10 (1320) H-19, I-? (172)
  Alk=I-4,6,7,9 (1363) Alk=I-10 (1392) Sfr, Tan=I-7 (1320)
Alliaria officinalieueus (Brassicaceae) H-19, I-14 (1080) Alk=I-2,6,7,9
  (1375) Alk=I-2,6,7 (1370)
Allium dilutum (Amaryllidaceae) H-19, I-16 (844)
A. helleri (Amaryllidaceae) H-22, I-? (909,975)
Allophylus serratus (Sapindaceae) H-22, I-16 (882) Str=I-6 (1384)
Alnus glutinosa (Betulaceae) H-19, I-6,7,8,9 (1101)
A. nitida (Betulaceae) H-22, I-16 (882)
Alocasia denudata (Araceae) H-33, I-3 (507)
A. longiloba (Araceae) H-33, I-3 (103)
Aloe principis (Liliaceae) H-19, I-7 (1262)
Alpinia purpurata (Zingiberaceae) H-19, I-7 (172)
Alstonia constricta (Apocynaceae) H-19, I-4 (776) Alk=I-2,4,7 (1392)
A. scholaris (Apocynaceae) H-33, I-14 (968) Alk=I-6,7 (1318) Alk=I-4
  (1392) Sap=I-6 (1318) Tri=I-4 (1381)
A. macrophylla (Apocynaceae) H-19, I-7 (982) Alk=I-2 (1372) Alk=I-4 (1392)
  Alk, Sap=I-6,7 (1318) Alk, Tri=I-7 (1390)
Alstroemeria ligtu (Alstroemeriaceae) H-19, I-6,7 (940)
Amansia multifida (Rhodomenaceae) H-19, I-1 (994)
Amantia musicaria (Amanitaceae) H-22, I-? (969)
Amaryllis belladonna (Amaryllidaceae) H-19, I-7 (982) Alk=I-3 (1392)
Ambrosia apters (Asteraceae) H-22, I-? (909,975)
A. artemisiifolia (Asteraceae) H-19, I-? (592) Alk=I-6,7,8 (1372)
A. chamissonis (Asteraceae) H-19, I-6,7,9 (940)
A. elatir (Asteraceae) H-19, I-1,14 (704,1022)
Amomum aromaticum (Zingiberaceae) H-17, I-2 (881)
Amoora wallichii (Meliaceae) H-17,21, I-6 (881)
Amorphophallus minor (Araceae) H-33, I-1 (103)
Amphipteryngium abstringens (Julianiaceae) H-22, I-? (975)
Anaphalis cuneifolia (Asteraceae) H-19, I-1 (882)
A. margaritacea (Asteraceae) H-19, I-? (1171) Fla=I-8 (1378)
Andropogon citratus (Poaceae) H-19, I-7 (982)
Anemone alpina (Ranunculaceae) H-33, I-2,7 (620) Alk=I-2,7 (620)
A. apennina (Ranunculaceae) H-19, I-? (186)
A. canadensis (Ranunculaceae) H-19, I-10 (580)
A. decapetala (Ranunculaceae) H-22, I-? (975)
A. narcissiflora (Ranunculaceae) H-33, I-2,7 (620) Alk=I-2,7 (620)
A. obtusiloba (Ranunculaceae) H-33, I-? (220)
A. occidentalis (Ranunculaceae) H-19, I-12 (625)
A. parviflora (Ranunculaceae) H-33, I-2,7 (620) Alk=I-2,7 (620)
A. patens (Ranunculaceae) H-33, I-1 (620) Alk=I-1 (620)
A. pulsatilla (Ranunculaceae) H-19, I-1 (186,628)
A. richardsoni (Ranunculaceae) H-33, I-2,7 (620) Alk=I-2,7 (620)
A. rupicola (Ranunculaceae) H-19, I-? (186)
Angelica arguta (Apiaceae) H-19, I-? (1171)
Angophora intermedia (Myrtaceae) H-19, I-7,8 (777)
Anisochilus carnosus (Lamiaceae) H-19, I-1 (882)

Anomodon rostratrus (Brachytheciaceae) H-19, I-7 (730)

Anomospermum grandifolium (Menispermaceae) H-15,19,33, I-2 (620) Alk=I-6
  (1392)

A. lucidum (Menispermaceae) H-15,19,33, I-2 (620)

A. ovatum (Menispermaceae) H-15,19,33, I-2 (620)

A. reticulatum (Menispermaceae) H-15,19,33, I-2 (620)

A. schomburgkii (Menispermaceae) H-15,19,33, I-2 (620)

Anthocephalus indicus (Rubiaceae) H-17, I-4 (881)

Antiaris africana (Moraceae) H-33, I-14 (102,507)

A. sp. (Moraceae) H-33, I-7,14 (139)

Antidesma stipulare (Euphorbiaceae) H-33, I-2 (103)

Apama tomentosa (Aristolochiaceae) H-32, I-? (73)

Aphania senegalensis (Sapindaceae) H-20, I-10 (504)

Aphelandra squarrosa (Acanthaceae) H-19, I-1 (413)

Aquilegia vulgaris (Ranunculaceae) H-33, I-? (220)

Aralidium pinnatifidum (Araliaceae) H-33, I-2 (103)

Arcangelisa flava (Menispermaceae) H-17, I-2 (504) H-20, I-4,5 (504)
  Alk=I-6 (1392) Alk, Gly, Tan=I-6,7 (1318)

Arenga obtusifolia (Arecaceae) H-32, I-? (73) H-32, I-9,14 (220)

A. pinnata (Arecaceae) H-33, I-9 (968)

Argemone subfusiformis (Papaveraceae) H-19, I-6,7,8,9 (940)

Aristotelia chilensis (Elaeocarpaceae) H-19, I-6,7 (940)

Arnebia nobilis (Boraginaceae) H-15,19, I-2 (1395)

Artemisia parviflora (Asteraceae) H-22, I-1 (882)

A. vestita (Asteraceae) H-15,19, I-7,12 (898)

Artocarpus camansi (Moraceae) H-19, I-4 (982)

Asarum canadense (Aristolochiaceae) H-19, I-6,7 (771,980)

Asclepias decumbens (Asclepidiaceae) H-22, I-? (975)

Asparagus densiflorus cv. sprengeri (Liliaceae) H-15, I-6 (651)

A. setaceus (Liliaceae) H-19, I-9 (776) Alk=I-2 (1392)

Astartea fasciculata (Myrtaceae) H-19, I-7,8 (777)

Aster divaricatus (Asteraceae) H-22, I-1 (909, 944)

Astragalus emoryanus (Fabaceae) H-33, I-? (1304)

A. lentiginosus (Fabaceae) H-33, I-7 (1304)

A. miser var. oblongifolius (Fabaceae) H-33, I-7 (1304)

A. mollissimus (Fabaceae) H-33, I-7 (1304) Alk=I-10 (1391)

A. praelongus (Fabaceae) H-33, I-7 (1304)

A. wootonii (Fabaceae) H-33, I-7 (1304) Alk=I-? (1392)

Atalantia kwangtungensis (Rutaceae) H-33, I-2 (103)

A. racemosus (Rutaceae) H-22, I-16 (790)

Atherosperma moschatum (Monimiaceae) H-19, I-12 (775,821) Alk=I-4,7
  (1392)

Athyrium macrocarpum (Polypodiaceae) H-19, I-2 (419)

Atropa wallichi (Solanaceae) H-22, I-? (729)

Atylosia trinerva (Fabaceae) H-19, I-16 (1394)

Austromyrtus acmeneoides (Myrtaceae) H-19, I-12 (775)

Baccharis pedicellata (Asteraceae) H-19, I-6,7,8 (940)

B. sagitalis (Asteraceae) H-19, I-6,7,8 (940)

Backhousia angustifolia (Myrtaceae) H-19, I-12 (775)

B. citriodora (Myrtaceae) H-19, I-12 (775,822) Alk=I-7 (1392)

Balbisia peduncularis (Geraniaceae) H-19, I-6,7,8 (940)

Baliospermum montanum (Euphorbiaceae) H-33, I-10 (504) Alk=I-2,6,7,9
  (1360) Sap=I-9 (1360)

Baloghia lucida (Euphorbiaceae) H-19, I-4 (776) Alk=I-4,7 (1392)
Baptisia australis (Fabaceae) H-19, I-7 (1103) Alk=I-2,6,7,10 (1392) Alk,
  Tan=I-? (1391)
Barleria cuspidata (Acanthaceae) H-22, I-1 (1394)
Barringtonia acutangula (Barringtoniaceae) H-32, I-2,4,6,10 (73,504,789,983)
  Fla=I-4 (1378)
B. calyptrata (Barringtoniaceae) H-32, I-4 (504)
B. careya (Barringtoniaceae) H-32, I-4 (105,504)
B. insignis (Barringtoniaceae) H-32, I-4 (504)
B. luzoniensis (Barringtoniaceae) H-32, I-4 (864)
Bauhinia thonningii (Caesalpiniaceae) H-19, I-4 (836)
Beaumontia grandiflora (Apocynaceae) H-33, I-10 (102)
B. fragrans (Apocynaceae) H-33, I-14 (102)
Begonia acuminatissima (Begoniaceae) H-19, I-7 (982)
B. rex (Begoniaceae) H-28, I-14 (220)
Berberis lycium (Berberidaceae) H-22, I-2 (881)
Berlinia globiflora (Caesalpiniaceae) H-33, I-2,4,7 (504)
Bidens frondosa (Asteraceae) H-19, I-? (732,1171) Alk=I-7 (1360)
Blepharodon mucronatum (Asclepiadaceae) H-20, I-14 (504)
Boisduvalia subulata (Onagraceae) H-19, I-6,7,8 (940)
Boletus edulis (Polyporaceae) H-22, I-9 (969) Alk=I-9 (1392)
Bombax bounopozense (Bombacaceae) H-19, I-4 (836)
B. ceiba (Bombacaceae) H-19, I-4 (982) H-22, I-? (881) Alk=I-7 (1360)
Boophone disticha (Amaryllidaceae) H-33,34, I-3,7 (504)
Boronia citriodora (Rutaceae) H-19, I-12 (821)
Brachythecium kamouense (Brachytheciaceae) H-19, I-? (570)
Brasenia schreberi (Nymphaeaceae) H-15, I-7 (652) Alk=I-2,7 (1392)
Bridelia retusa (Euphorbiaceae) H-22, I-4 (790)
B. squamosa (Euphorbiaceae) H-22, I-16 (882)
Bridilin ferruginea (Euphorbiaceae) H-19, I-19 (836)
Brucea sumatrana (Simaroubaceae) H-33, I-? (220) Alk=I-10 (1369)
Brunfelsia uniflora (Solanaceae) H-33, I-2 (504)
Buddleia brasiliensis (Loganiaceae) H-32, I-? (504)
B. japonica (Loganiaceae) H-32, I-7 (5020, I-7 (504)
Bulvine asphodeloides (Liliaceae) H-20,22, I-3,6,7 (504)
Bupleurum falcatum (Apiaceae) H-22, I-16 (977)
Buxus microphylla var. japonica (Buxaceae) H-15, I-7,10 (489)
B. rolfei (Buxaceae) H-32, I-9 (983,1349)
Cacalia decomposita (Asteraceae) H-32, I-1 (401)
C. suaveolons (Asteraceae) H-19, I-1,14 (704)
Caesalpina angulicaulis (Caesalpiniaceae) H-19, I-6,7 (940)
C. bonducella (Caesalpiniaceae) H-22, I-2,6 (881,1040) Alk=I-7,10 (1392)
  Fla=I-2 (1378)
C. crista (Caesalpiniaceae) H-19, I-7 (982) H-33, I-10 (968) Alk,
  Tan=I-9 (1360)
C. eriostachys (Caesalpiniaceae) H-32, I-4 (504)
C. nuga (Caesalpiniaceae) H-32, I-6,9 (220,459,504) Fla, Tan=I-9 (1377)
C. sappan (Caesalpiniaceae) H-19, I-4 (982) Sap, Tan=I-6,7 (1318)
C. sepiaria (Caesalpiniaceae) H-22, I-2 (1395)
Calamintha umbrosa (Lamiaceae) H-17, I-1 (881)
Calamus javensis (Arecaceae) H-33, I-14 (103)
Calceolaria mimuloides (Scrophulariaceae) H-19, I-6,7,8 (940)
C. purpurea (Scrophulariaceae) H-19, I-6,7 (940)

Callicarpa bicolor (Verbenaceae) H-19, I-7 (982)

C. erioclana (Verbenaceae) H-32, I-7 (528 983)

C. formosona (Verbenaceae) H-32,33, I-1,7 (528,968,983)

C. longifolia var. lanceolaria (Verbenaceae) H-32, I-? (73)

Callistemon linearis (Myrtaceae) H-19, I-9 (1262)

C. pallidus (Myrtaceae) H-19, I-7, 8 (777)

C. paludosus (Myrtaceae) H-19, I-8 (777)

C. phoeniceus (Myrtaceae) H-19, I-8 (777)

C. salignus (Myrtaceae) H-19, I-8 (777)

C. viminalis (Myrtaceae) H-19, I-7,8 (777)

C. violaceus (Myrtaceae) H-19, I-7,8,9 (777)

Callitriche turfosa (Callitrichaceae) H-19, I-6,7 (940)

Calophyllum amoenum (Clusiaceae) H-32, I-14 (504)

C. muscigerum (Clusiaceae) H-32, I-14 (504)

C. spectabile (Clusiaceae) H-32, I-? (105)

Camellia assamica (Theaceae) H-19, I-7 (1439)

C. kissi (Theaceae) H-32, I-18 (504)

Canavalia polystachya (Fabaceae) H-33, I-10 (504)

C. virosa (Fabaceae) H-33, I-9 (220)

Canscora diffusa (Gentianaceae) H-22, I-1 (1395) Alk=I-7 (1360)

Capparis longispina (Capparaceae) H-22, I-16 (881)

C. tomentosa (Capparaceae) H-33, I-2,9 (1353) Alk=I-9 (1392)

Cardiospermum halicacabum (Sapindaceae) H-33, I-7 (220) Alk=I-1,6,7,10
  (1360,1392)

Carex cernua (Cyperaceae) H-33, I-16 (220)

Carya illinoinensis (Juglandaceae) H-32, I-4,7 (401)

Caryocar toxiferum (Ternstroemiaceae) H-32, I-4,7 (1276)

Casearia graveolens (Flacourtiaceae) H-32, I-9 (73,220)

C. tomentosa (Flacourtiaceae) H-22, I-16 (1395) H-32, I-9 (73,220,504)

Casimiroa edulis (Rutaceae) H-32, I-1 (401) H-33, I-10 (1351) Alk=I-2,4,
  6,7,9,10 (1368,1371,1374,1392)

C. sapota (Rutaceae) H-32, I-1 (401)

Cassia arereh (Fabaceae) H-19, I-4 (836)

C. pistula (Caesalpiniaceae) H-19, I-4 (1052,1287)

C. reticulata (Caesalpiniaceae) H-19, I-7 (579,629)

C. sieberana (Caesalpiniaceae) H-19, I-2 (836)

C. sophera (Caesalpiniaceae) H-19, I-7 (982) Alk=I-6,7,9 (1360,1392)

Cassine glauca (Celastraceae) H-33, I-2 (220,504) Tri=I-4 (1383)

Castanopis indica (Fagaceae) H-22, I-4,6 (1394)

Casuarina decaisneana (Casuarinaceae) H-19, I-7 (777)

C. equisetifolia (Casuarinaceae) H-19, I-4 (982)

Catharanthus pusilla (Apocynaceae) H-33, I-16 (220)

C. rosea (Apocynaceae) H-33, I-16 (220) Alk=I-2,7,10 (1388) Tri=I-7 (1390)

Cautleya spicata (Zingberaceae) H-22, I-2 (881)

Ceanothus velutinus (Rhamnaceae) H-19, I-? (732,1171) Alk=I-4 (1392)

Cecropia guarumo (Moraceae) H-19, I-1 (1262)

Cedrela deodora (Meliaceae) H-22, I-6 (881)

C. febrifuga (Meliaceae) H-22, I-4,6 (881)

Celastrus dispermum (Celastraceae) H-19, I-4 (776) Alk=I-? (1392)

C. paniculatus (Celastraceae) H-22, I-16 (1395) Alk=I-10 (1392)

Celsia coromandeliana (Scrophulariaceae) H-22, I-1 (1395)

Centaurea chilensis (Asteraceae) H-19, I-6,7 (940)

Centella erecta (Apiaceae) H-15, I-6,7 (652)

Centrosema plumier (Fabaceae) H-19, I-7 (982)
Ceodes brunoniana (Nyctaginaceae) H-33, I-9 (504)
Cephalanthus occidentalis (Rubiaceae) H-19, I-1,14 (704)
Ceramium byssoideum (Ceramiaceae) H-19, I-1 (994)
Ceratiola ericoides (Empetraceae) H-32, I-7 (211,507)
Cerbera odollan (Apocynaceae) H-32, I-2,6,10 (220,504,983,1350) H-33,
   I-4,7,9 (220,504,983,1276) Str=I-2 (1385)
Cestrum nocturnum (Solanaceae) H-19, I-7 (982) Alk=I-6,7 (1392)
Chamaecyparis pisifera (Cupressaceae) H-19, I-6,7 (1262)
Chamelaucium axillare (Myrtaceae) H-19, I-8 (777)
C. megalopetalum (Myrtaceae) H-19, I-8 (777)
C. uncinatum (Myrtaceae) H-19, I-8,12 (744,775,777)
Cheilanthes glauca (Polypodiaceae) H-19, I-6,7 (940)
Chenopodium bonus-henricus (Chenopodiaceae) H-22, I-? (1061)
Chiliotrichium diffusum (Asteraceae) H-19, I-6,7 (940)
Chlorella pyrenoidosa (Chlorophyta-Green Algae) H-19, I-1 (1044)
C. stigmaphora (Chlorophyta- Green Algae) H-19, I-1 (987)
Chondodendron candicans (Menispermaceae) H-33, I-2 (504)
C. limacifolium (Menispermaceae) H-33, I-2 (504)
C. polyathemum (Menispermaceae) H-33, I-2 (504)
C. tomentosum (Menispermaceae) H-33, I-2 (504)
Chondria littoralis (Cryptogamae-Algae) H-19, I-1 (994)
Chorizanthe glabrescens (Polygonaceae) H-19, I-6,7,8 (940)
Chrysanthemum anethifolium (Asteraceae) H-20, I-2 (832)
C. leucanthemum (Asteraceae) H-22, I-1 (881)
C. maximum (Asteraceae) H-22, I-8 (969)
C. sinense (Asteraceae) H-19, I-7 (982) Alk=I-7,8 (1392)
Chrysophyllum cainito (Sapotaceae) H-17, I-14 (1318) Alk, Tan=I-6,7 (1318)
   Sap=I-6,10 (1318)
Chrysopsis mariana (Asteraceae) H-19, I-? (732,1171)
Cicuta virosa (Apiaceae) H-33, I-? (220)
Cinnamomum iners (Lauraceae) H-17, I-16 (882)
C. oliveri (Lauraceae) H-19, I-12 (775) Alk=I-4,7 (1392)
Cissampelos pareira (Menispermaceae) H-32, I-2 (983) Alk=I-2 (1392)
   Alk=I-6,7 (1360)
Citrus bergami (Rutaceae) H-15,19, I-12 (908)
C. microcarpus (Rutaceae) H-19, I-? (865)
C. nobilis (Rutaceae) H-19, I-7 (982) Alk=I-7 (1392)
Cladophoropsis gracillima (Siphonocladaceae-Algae) H-19, I-? (1023)
Clausena brevistyla (Rutaceae) H-19, I-4 (776) Alk=I-6 (1392)
Cleidon spiciflorum (Euphorbiaceae) H-33, I-7 (968)
Clematis buchananiana (Ranunculaceae) H-22, I-16 (1395)
C. crispa (Ranunculaceae) H-15,19, I-7 (651)
C. dioscoreifolia (Ranunculaceae) H-15,19, I-6,7 (651)
C. fremontii (Ranunculaceae) H-19, I-? (186)
C. glycinoides (Ranunculaceae) H-19, I-1 (776) Alk=I-7 (1392)
C. heracleifolia (Ranunculaceae) H-19, I-? (186)
C. hirsutissima (Ranunculaceae) H-20, I-14 (1353) Alk=I-2,6 (1371)
C. recta (Ranunculaceae) H-19, I-10 (186)
C. stans (Ranunculaceae) H-19, I-10 (186)
C. texensis (Ranunculaceae) H-19, I-10 (186)
Cleome acutifolia (Capparaceae) H-19, I-1 (982)
Clerodendrum villosum (Verbenaceae) H-33, I-7 (983)

Clitoria guianensis (Fabaceae) H-32, I-? (1276)
Clivia cyrtanthiflora (Amaryllidaceae) H-22, I-7 (847)
C. miniata (Amaryllidaceae) H-22, I-7 (847) Alk=I-2,3 (1392)
Clytostoma callistegioides (Bignoniaceae) H-15, I-7 (651)
Cnestis ferruginea (Connaraceae) H-19, I-2 (836)
Cnicus wallichi (Asteraceae) H-22, I-1 (881)
Cocculus amazonum (Menispermaceae) H-15,19,33, I-4 (620)
C. ferrandianus (Menispermaceae) H-32, I-9 (504)
C. filipendula (Menispermaceae) H-33, I-? (504)
C. laurifolius (Menispermaceae) H-33, I-2,7 (983) Alk=I-1,4,5,7 (1392)
C. macrophyllus (Menispermaceae) H-15,19,33, I-16 (620)
C. pendulus (Menispermaceae) H-22, I-1, (1395)
C. sarmentosus (Menispermaceae) H-33, I-2 (504)
C. suberosus (Menispermaceae) H-32, I-14 (605)
C. thunbergii (Menispermaceae) H-22, I-16 (977)
Cochlospermum tinctorium (Cochlospermaceae) H-19, I-4 (836)
Codiaeum variegatum (Euphorbiaceae) H-19, I-7 (982)
Coelospermum reticulatum (Rubiaceae) H-19, I-2 (776) Alk=I-4,7 (1392)
Colchicum luteum (Liliaceae) H-33, I-? (220) Alk=I-3 (1392)
Coleus hybridus (Lamiaceae) H-19, I-7 (1076)
Colliguaja integerrima (Euphorbiaceae) H-19, I-6,7,8 (940)
C. odorifera (Euphorbiaceae) H-19, I-6,7 (940) H-33, I-14 (620)
C. salicifolia (Euphorbiaceae) H-19, I-6,7 (940)
Colocasia gigantea (Araceae) H-33, I-14 (504)
C. esculenta (Araceae) H-33, I-? (220)
Combretum comosum (Combretaceae) H-19, I-2 (836)
C. confertum (Combretaceae) H-33, I-4 (504)
C. micranthum (Combretaceae) H-19, I-2 (836) H-17, I-2 (1353) Alk=I-7
   (1392) Tan=I-7 (1353)
Connarus wightii (Connaraceae) H-22, I-16 (882)
Convolvulus hermanniae (Convolvulaceae) H-19, I-6,7,8 (940)
Conyza viscidula (Asteraceae) H-22, I-1 (790)
Coptis chinensis (Ranunculaceae) H-19, I-2,6 (655,735)
Coreopsis cardaminifolia (Asteraceae) H-32, I-? (211)
C. maritima (Asteraceae) H-19, I-2,8,10 (748)
Coriaria nepalensis (Coriariaceae) H-33, I-? (220)
C. sarmentosa (Coriariaceae) H-33, I-10,15 (504)
Cornus florida (Cornaceae) H-15, I-7,8 (651,652) Alk=I-6,7 (1392)
C. nuttallii (Cornaceae) H-19, I-6,7,10 (732,1171)
C. officinalis (Cornaceae) H-19, I-9 (735)
Corypha umbraculifera (Arecaceae) H-32, I-9 (73,220,504)
Coscinium blumeanum (Menispermaceae) H-32, I-? (105)
C. fenestratum (Menispermaceae) H-15, I-? (411) Alk=I-? (1392)
Costus speciosus (Zingiberaceae) H-17, I-2 (1320) H-22, I-1 (882)
   Sap=I-2 (790) Tan=I-2 (1320)
Cotinus coggygria (Anacardiaceae) H-22, I-16 (881)
Cotoneaster affinis var. bacillaris (Rosaceae) H-22, I-16 (882)
Crepis japonica (Asteraceae) H-19, I-7, (982)
Crinum latifolium (Amaryllidaceae) H-19, I-7 (982) Alk=I-3,10 (1392)
Crocus sativus (Iridaceae) H-32, I-3 (220) Sap=I-12 (220)
Crotalaria retzii (Fabaceae) H-19, I-9,10 (651)
Croton setigerus (Euphorbiaceae) H-32, I-4,7 (504)
Cryptocarya angulata (Lauraceae) H-19, I-4 (776) Alk=I-4 (1392)

C. hyspodia (Lauraceae) H-19, I-4,7 (776) Alk=I-4 (1392)
C. obovata (Lauraceae) H-19, I-4 (776) Alk=I-4,7 (1392)
C. pleurosperma (Lauraceae) H-19, I-4 (776) Alk=I-4,7 (1392)
Cryptomeria japonica (Taxodioicaceae) H-19, I-6,7 (1262) H-22, I-16 (1395)
Ctenitis spectabilis (Polypodiaceae) H-19, I-2,6,7 (940)
Cucumis naudinianus (Cucurbitaceae) H-33, I-2 (504)
Cucurbita maxima (Cucurbitaceae) H-17, I-10 (910,1353) Alk=I-10 (1391)
Culcasia scandens (Araceae) H-32, I-14 (1353) Alk=I-6,7 (1372)
Cunninghamia lanceolata (Taxodiaceae) H-19, I-7 (652)
Cupania pseudorhus (Sapindaceae) H-32, I-4 (504)
Cupressus torulosa (Cupressaceae) H-22, I-16 (1394)
Cyathocline lyrata (Asteraceae) H-17, I-12,16 (872)
Cycas media (Cycadaceae) H-19, I-10 (776)
Cyclamen persicum (Primulaceae) H-32, I-2 (220,459)
Cydista aequinoctialis (Bignoniaceae) H-19, I-7 (982)
Cynanchum caudatum (Asclepiadaceae) H-33, I-2 (839)
C. macrophyllum (Asclepiadaceae) H-33, I-6,10 (620)
C. vincetoxicum (Asclepiadaceae) H-33, I-? (220)
Cynara cardunculus (Asteraceae) H-19, I-1 (581)
C. scolymus (Asteraceae) H-19, I-1 (581)
Cynometra ramiflora (Caesalpiniaceae) H-12, I-12 (105)
Cyperus longus (Cyperaceae) H-33, I-? (220)
C. niveus (Cyperaceae) H-22, I-1 (881)
Dalbergia stipulacea (Fabaceae) H-32, I-2,4 (73,220,504)
Daphne bholua (Thymelaeaceae) H-33, I-? (220)
D. gnidium (Thymelaeaceae) H-32, I-2 (504)
D. kamtstratica (Thymelaeaceae) H-33, I-4,9 (839)
D. oleiodes (Thymelaeaceae) H-33, I-? (220)
Darwinia citriodora (Myrtaceae) H-19, I-7,8,12 (744,777)
D. grandiflora (Myrtaceae) H-19, I-? (819)
Datura metel var. alba (Solanaceae) H-33, I-1 (968) Alk=I-6,7,9,10 (1392)
D. quercifolia Solanaceae) H-17, I-? (881) Alk=I-7,10 (1392)
Delphinium barbeyi (Ranunculaceae) H-33, I-16 (800)
D. bicolor (Ranunculaceae) H-33, I-2 (800) Alk=I-2 (800)
D. menziesii (Ranunculaceae) H-33, I-2 (800) Alk=I-2 (800)
D. nelsonii (Ranunculaceae) H-33, I-2 (800) Alk=I-2 (800)
D. scopulorum (Ranunculaceae) H-33, I-2 (800) Alk=I-2 (800)
Dendrophthoe falcata (Loranthaceae) H-22, I-16 (790 1395) Alk=I-7 (1360)
  Fla=I-7 (1377) Sap=I-6 (1360) Tan=I-6,7,8 (1360,1377)
Dendropogon usneoides (Bromeliaceae) H-19, I-7 (982)
Derbesia prolifica (Derbesiaceae-Green Algae) H-19, I-? (1023)
Deringa canadensis (Apiaceae) H-19, I-? (1171)
Derris ferruginea (Fabaceae) H-32, I-2 (220,459)
D. grandifolia (Fabaceae) H-32, I-16 (459)
D. koolgibberah (Fabaceae) H-32, I-2 (504)
D. pterocarpus (Fabaceae) H-32, I-2 (504)
D. scandens (Fabaceae) H-19, I-16 (1393) H-32, I-2 (73,459,983)
Desmarestia herbacea (Desmarestiaceae) H-19, I-1 (940)
Desmodium triquetrum (Fabaceae) H-22, I-16 (882)
Dichapetalum graffithii (Dichapetalaceae) H-33, I-2 (103)
Dicranopteris linearis (Gleicheniaceae) H-19, I-6,7 (172) Fla=I-2 (1379)
Dictyopteris justii (Dictyotaceae-Algae) H-19, I-1 (994)
Dictyota acutiloba (Dictyotaceae-Algae) H-19,22, I-1 (1114)

D. divaricata (Dictyotaceae-Algae) H-19, I-1 (1023)
D. flabellata (Dictyotaceae-Algae) H-19,22, I-1 (1114)
Dieffenbachia maculata (Araceae) H-19, I-? (413) Alk=I-2,6,7 (1392)
Diervilla lonicera (Capritolliaceae) H-19, I-? (592)
Dioscorea alata (Dioscoreaceae) H-19, I-7 (982) H-33, I-2 (620) Alk=I-2
  (620)
D. cayenensis (Dioscoreaceae) H-33, I-2,6 (620)
D. composita (Dioscoreaceae) H-32, I-2 (1115)
D. hirsuta (Dioscoreaceae) H-33, I-2,6 (620)
D. macabiha (Dioscoreaceae) H-33, I-2 (504)
D. macrostachya (Dioscoreaceae) H-32, I-6,7 (1276)
D. montana (Dioscoreaceae) H-32, I-9 (1276)
D. poilanei (Dioscoreaceae) H-32,33, I-2 (504)
D. sansibariensis (Dioscoreaceae) H-32,33, I-2 (504)
D. tokora (Dioscoreaceae) H-32, I-2 (504)
D. trifida (Dioscoreaceae) H-33, I-2,6 (620)
D. triphylla (Dioscoreaceae) H-33,34, I-2,14 (103,504)
Diospyros canomoi (Ebenaceae) H-32, I-9 (864) H-33, I-? (863)
D. chloroxylon (Ebenaceae) H-22, I-16 (882)
D. ebenaster (Ebenaceae) H-32, I-9 (983)
D. maritima (Ebenaceae) H-32, I-4,9 (192,504,983)
D. multiflora (Ebenaceae) H-32, I-4,9 (504, 528, 983) H-33, I-4,9 (504,
  968,983)
D. paniculata (Ebenaceae) H-32, I-7 (73,220)
D. peregina (Ebenaceae) H-22, I-4,6 (881)
D. wallichi (Ebenaceae) H-32, I-9 (105,504)
Dolichandrone falcata (Bignoniaceae) H-32, I-4 (73,220,504)
Dolichos longeracemosa (Fabaceae) H-33,35, I-10 (504)
D. lupiniflorus (Fabaceae) H-32, I-? (105)
D. phaseoloides (Fabaceae) H-33, I-10 (504)
Dorema ammoniacum (Apiaceae) H-20, I-14 (504)
Dracaena mannii (Agavaceae) H-19, I-4 (836)
Drimys winteri var. chilensis (Winteraceae) H-19, I-6,7 (940)
Drosera intermedia (Droseraceae) H-19, I-? (594)
D. whittakeri (Droseraceae) H-19, I-6,7 (744)
Dryopteris odontoloma (Polypodiaceae) H-19, I-2 (419)
D. paleacea (Polypodiaceae) H-19, I-7 (419)
Duranta plumieri (Verbenaceae) H-33, I-16 (220) Alk=I-9 (1392)
Dysoxylum decandrum (Meliaceae) H-32, I-4 (968,983) H-33, I-7 (968)
  Alk=I-7 (1392)
Dyssodia anomala (Asteraceae) H-32, I-2,7 (401)
Echium vulgare (Boraginaceae) H-19, I-6,7,8 (940) Alk=I-1 (1392)
Edgeworthia gardneri (Thymelaeaceae) H-32, I-? (73,220)
Ekebergia senegalensis (Meliaceae) H-19, I-4 (836)
Elaeocarpus grandiflorous (Elaeocarpaceae) H-19, I-7 (1076)
E. virosa (Elaeocarpaceae) H-19, I-7 (1076)
Elaeophorbia drupifera (Euphorbiaceae) H-32,33, I-1 (504)
Elasteriospermum tapos (Euphorbiaceae) H-33, I-10 (504)
Eleocharis dulcis (Cyperaceae) H-22, I-3 (969)
E. smalli (Cyperaceae) H-19, I-1 (837)
E. tuberosa (Cyperaceae) H-19, I-10 (740,980)
Elephantopus mollis (Asteraceae) H-19, I-7 (982) Tan=I-7 (1321)
Embelia schimperi (Myrsinaceae) H-17, I-10 (1143)

Engelhardtia colebrookiana (Juglandaceae) H-22, I-4,6 (790)
Entada phaseoloides (Mimosaceae) H-32,33, I-4,7 (220,968,983,1349)
  Alk=I-4 (1392) Sap=I-? (1349)
E. pursaetha (Mimosaceae) H-19, I-2,4 (836)
Enteromorpha compressa (Ulvaceae-Algae) H-19, I-1 (1023)
E. kylinii (Ulvaceae-Algae) H-19, I-1 (1023)
E. prolifera (Ulvaceae-Algae) H-19, I-1 (1023)
Epilobium coloratum (Onagraceae) H-19, I-7 (1101)
Equisetum diffusum (Equisetaceae) H-19, I-7 (419)
Eremocitrus glauca (Rutaceae) H-19, I-? (776) Alk=I-7 (1392)
Eremostachys acanthocalyx (Lamiaceae) H-33, I-? (220)
E. superba (Lamiaceae) H-32, I-? (73)
E. vicaryi (Lamiaceae) H-32, I-1 (459) H-33, I-? (220)
Erigeron vernus (Asteraceae) H-19, I-6,7 (652)
Eriocaulon sieboldianum (Eriocaulaceae) H-22, I-16 (977)
Eriodictyon colifornicum (Hydrophyllaceae) H-19, I-6,7 (565)
Eriostemon myoporoides (Rutaceae) H-19, I-? (821) Alk=I-7 (1392)
Erythrophleum le-tustui (Caesalpiniaceae) H-33, I-? (835)
Erythroxylum cuneatum (Erythroxylaceae) H-32,33, I-? (968,983)
Eucalyptus alba (Myrtaceae) H-19, I-7,9 (1100)
E. australiana (Myrtaceae) H-19, I-7,12 (775,928,1041)
E. citriodora (Myrtaceae) H-15, I-? (912) H-19, I-12 (647,775,912,928)
E. cneorifolia (Myrtaceae) H-19 I-12 (775,928)
E. consideniana (Myrtaceae) H-19, I-? (1041)
E. dives (Myrtaceae) H-19, I-12 (647,775,928)
E. fasciculata (Myrtaceae) H-19, I-8 (777)
E. fruticetorum (Myrtaceae) H-19, I-12 (647)
E. laceta (Myrtaceae) H-19, I-12, (823)
E. lehmannii (Myrtaceae) H-19, I-7,8,9 (777)
E. leucoxylon (Myrtaceae) H-19, I-? (777)
E. macarthurii (Myrtaceae) H-19, I-12 (823)
E. megacarpa (Myrtaceae) H-19, I-8 (777)
E. moluccana var. albens (Myrtaceae) H-19, I-12 (823)
E. moorei (Myrtaceae) H-19, I-12, (823)
E. nova-anglica (Myrtaceae) H-19, I-12 (823)
E. phellandra (Myrtaceae) H-19, I-12 (775,823)
E. phlebophylla (Myrtaceae) H-19, I-12 (823)
E. polybractea (Myrtaceae) H-19, I-12 (775,823,1041)
E. populnea (Myrtaceae) H-19, I-12 (775)
E. pulchella (Myrtaceae) H-19, I-? (1041)
E. radiata (Myrtaceae) H-19, I-12 (823)
E. robertsonii (Myrtaceae) H-19, I-? (775)
E. sepulcralis (Myrtaceae) H-19, I-8 (777)
E. straigeriana (Myrtaceae) H-19, I-12 (823)
E. tereticornis (Myrtaceae) H-19, I-12 (982)
Eucryphia glutinosa (Eucryphiaceae) H-19, I-6,7 (940)
Euonymus javanicus (Celastraceae) H-19, I-? (982)
Eupatorium adenophorum (Asteraceae) H-19, I-6,7,12 (775,776)
E. altissimum (Asteraceae) H-19, I-? (1171)
E. riparium (Asteraceae) H-19, I-12 (775) Alk=I-6,7,9 (1392)
E. rugosum (Asteraceae) H-19, I-? (1171)
E. salvia (Asteraceae) H-19, I-? (940)
Euphorbia calycina (Euphorbiaceae) H-32, I-? (754)

E. candelabrum (Euphorbiaceae) H-33, I-14 (504)

E. characias (Euphorbiaceae) H-32, I-? (504)

E. corollata (Euphorbiaceae) H-19, I-6,7 (1171)

E. cotinifolia (Euphorbiaceae) H-32,33, I-? (504)

E. cotinoides (Euphorbiaceae) H-33, I-6,7 (105) H-33, I-14 (620)
  Alk=I-14 (620)

E. fulgens (Euphorbiaceae) H-32, I-2 (504)

E. hyberna (Euphorbiaceae) H-32, I-? (105)

E. kamerunica (Euphorbiaceae) H-33, I-14 (504)

E. laro (Euphorbiaceae) H-32, I-? (1276)

E. milii (Euphorbiaceae) H-19, I-? (172)

E. pilulifera (Euphorbiaceae) H-19, I-7 (982,1076) Alk=I-7 (1392)

E. plorifera (Euphorbiaceae) H-22, I-? (881)

E. plumerioides (Euphorbiaceae) H-19, I-6,7 (776)

E. serpens (Euphorbiaceae) H-19, I-6,7,8 (940)

E. supina (Euphorbiaceae) H-19, I-1,14 (704)

E. trigona (Euphorbiaceae) H-33, I-14 (504) H-32, I-14 (983)

E. venefica (Euphorbiaceae) H-33, I-14 (504)

Eurycles amboinensis (Amaryllidaceae) H-19, I-? (982) Alk=I-2,3 (1392)

Eurycoma apiculata (Simaroubaceae) H-33, I-1 (103)

Evodia lunu-ankenda (Rutaceae) H-19,22, I-4,6 (1394)

E. xanthoxyloides (Rutaceae) H-19, I-7 (776)

Fabiana imbricata (Solanaceae) H-19, I-6 (940) Alk=I-? (1392)

Fagonia cretica (Zygophyllaceae) H-22, I-16 (881)

Fagus sylvatica (Fagaceae) H-22, I-16 (790)

Falkenbergia hillebrandi (Bonnemaisoniaceae-Red Algae) H-19, I-1 (994)

Feijoa sellowiana (Myrtaceae) H-15, I-7 (652)

Ficus atroy (Moraceae) H-33, I-2,4 (620)

F. erecta (Moraceae) H-15, I-15 (917)

F. hauli (Moraceae) H-33, I-14 (968)

F. reticulata (Moraceae) H-22, I-? (882)

F. stipulosa (Moraceae) H-19, I-7 (982)

Flagraea racemosa (Loganiaceae) H-33, I-2,14 (103)

Flindersia dissosperma (Rutaceae) H-19, I-12 (775) Alk=I-4,5,6,7 (1392)

F. maculosa (Rutaceae) H-19, I-4 (776) Alk=I-4,5,7 (1392)

F. oxyleyana (Rutaceae) H-19, I-6,7 (776) Alk=I-4,5,7 (1392)

F. xanthoxyla (Rutaceae) H-19, I-6,7 (776) Alk=I-7 (1392)

Fouquieria diguetii (Fouquieriaceae) H-19, I-4, (1262)

Francoa appendiculata (Saxifragaceae) H-19, I-2,7,8 (940)

Frankenia ericifolia (Frankeniaceae) H-32, I-? (504)

Freycinetia arborea (Pandanaceae) H-19, I-? (172) Alk=I-6,7 (1363)

Fritillaria imperialis (Liliaceae) H-33, I-3 (220) Alk=I-3 (1392)

Fuchsia magellanica (Onagraceae) H-19, I-6,7,8 (940)

Furcraea hexapetala (Agavaceae) H-32, I-7 (1276)

Fusanus spicatus (Santalaceae) H-19, I-12 (820)

Gaillardia lanceolata (Asteraceae) H-15, I-6,7,8 (652)

Galphimia gracilis (Malpighiaceae) H-19, I-7 (1076)

Ganoderma applanatum (Polypodiaceae) H-19, I-6 (940)

Ganophyllum falcatum (Sapindaceae) H-19, I-7 (982) H-32, I-4 (968,983,1349)

G. obliquum (Sapindaceae) H-32, I-4 (864)

Garcinia morrella (Clusiaceae) H-19, I-9,10,13 (220,1082)

G. talbotii (Clusiaceae) H-22, I-16 (790)

Gardenia curranii (Rubiaceae) H-32, I-9 (528,983)

G. jovis tonantis (Rubiaceae) H-32, I-9 (504)

G. turgida (Rubiaceae) H-22, I-9 (881)

Gaura biennis (Onagraceae) H-19, I-? (1171)

Geijera paviflora (Rutaceae) H-19, I-4 (776) Alk=I-4,7 (1392)

G. salicifolia (Rutaceae) H-19, I-12 (775) Alk=I-4,7 (1392)

Gelidium cartilagenium (Gelidiaceae-Red Algae) H-22, I-? (978)

G. corneum (Gelidiaceae-Algae) H-22, I-1 (969)

Genista roetam (Fabaceae) H-33, I-7 (504)

G. tinctoria (Fabaceae) H-19, I-7,8 (1262) Alk=I-6,7,9 (1392)

Geranium carolinianum (Geraniaceae) H-15, I-6,7 (652)

G. phaeum (Geraniaceae) H-15, I-? (1062)

G. platypetalum (Geraniaceae) H-15, I-? (1062)

G. pratense (Geraniaceae) H-15, I-? (1062)

Geum macrophyllum (Rosaceae) H-19, I-? (594)

Gilia macombii (Polemoniaceae) H-32, I-6,7 (401)

Gisekia parnaceoides (Ficoidaceae) H-17, I-? (1353) Fla=I-1 (1377)

Glochidion hohenackerii (Euphorbiaceae) H-22, I-16 (881)

Glycosmis chlorosperma (Rutaceae) H-33, I-1 (102)

Gnaphalium decurrens (Asteraceae) H-19, I-7,8 (1262)

G. luteo-album (Asteraceae) H-33, I-16 (220) Alk=I-16 (1392)

Gnetum gnemon (Gnetaceae) H-19, I-7 (982) H-33, I-1 (103)

G. scandens (Gnetaceae) H-32, I-7 (73,220,459)

Gossypium arboreum (Malvaceae) H-32, I-8,9 (968)

Gracilariopsis sjoestedtii (Graciliariaceae-Red Algae) H-19, I-1 (1023)

Grevillea bipinnatifida (Proteaceae) H-19, I-7 (777)

G. dallaceana (Proteaceae) H-19, I-8 (777)

Grewia hirsuta (Tiliaceae) H-22, I-1 (790)

G. latifolia (Tiliaceae) H-22, I-16 (1395)

Guatheria venificiorum (Annonaceae) H-33, I-9 (620)

Gutierrezia texana (Asteraceae) H-22, I-? (975)

Gymnophyton isatidicarpum (Misodendraceae) H-19, I-6 (940)

Gynochthodes sublanceolata (Rubiaceae) H-33, I-2 (103)

Habenaria psycodes (Orchidaceae) H-19, I-? (592)

Haemodorum corymbosum (Haemodoraceae) H-19, I-2 (776)

Hagenia abyssinica (Rosaceae) H-17, I-7,8 (1143)

Hakea saligna (Proteaceae) H-22, I-16 (1394)

Halfordia kendach (Rutaceae) H-19, I-4 (776)

H. scleroxyla (Rutaceae) H-19, I-4 (776)

Halimeda opuntia (Udoteaceae) H-19, I-? (1138)

Haplopappus cuneifolius (Asteraceae) H-19, I-6,7 (940)

H. gracilis (Asteraceae) H-19, I-19 (851)

H. multifolius (Asteraceae) H-19, I-6,7,8 (940)

Haplophragma adenophyllum (Bignoniaceae H-22, I-16 (790)

Haplophyton apocynacia (Apocynaceae) H-32, I-? (1276)

Hardwickia binata (Caesalpiniaceae) H-22, I-16 (790)

Harpullia cupanioides (Sapindaceae) H-32, I-? (73,220)

Harungana madagascariensis (Gutifferae) H-17,19, I-4 (836,1353)

Hedera colchica (Araliaceae) H-22, I-1 (790)

Helenium amarum (Asteraceae) H-33, I-16 (929) Alk=I-2,6,7,8,9 (1369)

Helianthus helianthoides (Asteraceae) H-19, I-? (1171)

H. microcephalus (Asteraceae) H-19, I-? (1171)

Heliotropium amplexi (Boraginaceae) H-19, I-1,6 (776)

H. eichwaldii (Boraginaceae) H-33, I-? (220) Alk=I-2,8 (1379)

Helleborus viridis (Ranunculaceae) H-19, I-? (186) Alk=I-2 (1392)
Hemerocallis liloasphodelus (Liliaceae) H-19, I-8 (413)
Hemichroa diandra (Amaranthaceae) H-19, I-8 (777)
Hemidesmus indicus (Asclepiadaceae) H-22, I-1 (881,1040) Alk=I-2,6 (1360)
    Tan=I-7 (1360)
Heuchera micrantha (Saxifragaceae) H-22, I-? (1061)
Hibiscus cannabinus (Malvaceae) H-17, I-7 (1353) Alk=I-10 (1391)
H. mutabilis (Malvaceae) H-15,19, I-7,9,10 (651) Alk=I-7 (1390,1392)
Hippobroma longiflora (Campanulaceae H-19, I-7 (982) Alk=I-2,16 (1392)
Hippomane mancinella (Euphorbiaceae) H-32, I-? (1276) H-33, I-9,14
    (504,620) Alk=I-9 (1392)
Hippophae rhamnoides (Elaeagnaceae) H-19, I-7 (1262) Alk=I-6,7,9 (1392)
    Fla=I-4 (1379)
Holigarna arnottiana (Anacardiaceae) H-33, I-14 (220)
H. longifolia (Anacardiaceae) H-33, I-14 (220)
Homalanthus fastuosus (Euphorbiaceae) H-32, I-7 (968,983,1349)
H. populens (Euphorbiaceae) H-19, I-7 (982)
Homalium circumbinnatum (Flacourtiaceae) H-19, I-? (776)
Homalomena rubescens (Araceae) H-33, I-7 (220,504)
Homaranthus flavescens (Euphorbiaceae) H-19, I-12 (822)
H. virgatus (Euphorbiaceae) H-19, I-12 (822)
Hunnemannia fumariifolia (Papaveraceae) H-19, I-2 (846) H-16, I-16 (832)
    Alk=I-? (1392)
Hydnocarpus iliciflolius (Flacourtiaceae) H-19, I-? (186)
Hydrocotyle umbellata (Apiaceae) H-15, I-7 (652)
Hydrophylum capitatum (Hydrophyllaceae) H-19, I-16 (625)
Hymenocardia acida (Hymenocardiaceae) H-20, I-4 (504)
Hyoscyamus muticus (Solanaceae) H-33, I-? (220) Alk=I-? (1392)
H. pusillus (Solanaceae) H-33, I-? (220)
H. reticulatus (Solanaceae) H-33, I-? (220) Alk=I-5,10 (1392)
Hypericum androsaemum (Hypericaceae) H-19, I-6,7,8 (940)
H. moseranum (Hypericaceae) H-19, I-8 (1262)
H. mutilum (Hypericaceae) H-19, I-? (1171)
H. mysorense (Hypericaceae) H-15, I-? (1394)
Hypochoeris radicata (Asteraceae) H-15, I-1 (946)
Hyptis brevipes (Lamiaceae) H-19, I-7 (982) Alk, Tri=I-7 (1390)
Iberis sempervirens (Brassicaceae) H-19, I-10 (186)
Ichnocarpus frutescens (Apocynaceae) H-22, I-1 (881) Alk, Fla=I-2 (1361)
Ilex coriacea (Aquifoliaceae) H-15, I-7 (652)
I. cymosa (Aquifoliaceae) H-33, I-2 (103)
Illicium anisatum (Illiciaceae) H-32, I-10 (504)
I. griffithii (Illiciaceae) H-33, I-12 (220)
Impatiens irvingii (Balsaminaceae) H-20, I-14 (504)
I. noli-tangere (Balsaminaceae) H-20, I-? (504)
Indigofera pulchella (Fabaceae) H-22, I-2 (881,1040)
I. simplifolia (Fabaceae) H-19, I-? ((982) H-33, I-2 (504)
Intsia bijuga (Fabaceae) H-19, I-? (982)
Inula graveolens (Asteraceae) H-33, I-16 (220)
Ipomoea altissima (Convulvulaceae) H-33, I-2 (504)
I. violacea (Convolvulaceae) H-15, I-2 (945) H-35, I-10 (504) Alk=I-10
    (1370)
Ipomopsis thurberi (Polemoniaceae) H-32, I-2,7 (401)
Isanthus brochiatus (Lamiaceae) H-19, I-? (1171)

Ixora arborea (Rubiaceae) H-22, I-16 (790)
I. nigricans (Rubiaceae) H-22, I-16 (790)
Jacquinia arboreo (Theophrastaceae) H-33, I-? (620)
J. armillaris (Theophrastaceae) H-32, I-2 (1276) H-33, I-9 (620)
J. axillaris (Theophrastaceae) H-32, I-2,6,7 (1276)
J. caracasana (Theophrastaceae) H-33, I-? (620)
J. donnell-smithii (Theophrastaceae) H-32, I-? (1276)
J. sprucei (Theophrastaceae) H-32, I-9 (105)
Jasminum grandiflorum (Oleaceae) H-19, I-7 (982)
Juglans ailantifolia (Juglandaceae) H-33, I-4 (839)
J. microcarpa (Juglandaceae) H-32, I-4,7 (401)
Juncus effusus (Juncaceae) H-33, I-? (220)
Juniperus conferta (Cupressaceae) H-19, I-9 (972)
J. occidentalis (Cupressaceae) H-19, I-9 (625)
J. rigida (Cupressaceae) H-19, I-9 (972)
Kalmia angustifolia (Ericaceae) H-19, I-? (592)
Karwinskia humboldtiana (Rhamnaceae) H-33, I-10 (504)
Khaya senegalensis (Meliaceae) H-19, I-4 (836)
Kibatalia blancoi (Apocynaceae) H-32, I-4,7 (504,983)
Kirganelia reticulata (Euphorbiaceae) H-22, I-16 (881)
Krameria cistoidea (Krameriaceae) H-19, I-6,7 (940)
Kreysigia multiflora (Liliaceae) H-19, I-6,7 (776) Alk=I-2,6,7 (1392)
Laburnum anagyroides (Fabaceae) H-33, I-1,10 (504,507)
Lagascea mollis (Asteraceae) H-22, I-1 (882)
Lagerstroemia speciosa (Lythraceae) H-22, I-16 (1395) Alk=I-6 (1392)
  Tan=I-6,7 (1318)
Laggera pterodonta (Asteraceae) H-22, I-1 (790)
Laminaria angustifolia (Laminariales-Algae) H-19, I-1 (1084)
L. digitata (Laminariales-Algae) H-19, I-1 (1083,1084)
Lapageria rosea (Liliaceae) H-19, I-6,7 (940)
Laportea meyeniana (Urticaceae) H-32, I-17 (968)
Larrea tridentata (Zygophyllaceae) H-20, I-6,7 (504)
Lasianthus maingay (Rubiaceae) H-33, I-? (103)
Laurentia obtusa (Lobeliaceae) H-19, I-1 (994,995)
Lechea intermedia (Cistaceae) H-19, I-? (594)
Leea indica (Vitaceae) H-22, I-7 (881)
Lepidium hyssopifolium (Brassicaceae) H-19, I-2,6,7,10 (776)
Lepisanthes kunstleri (Sapindaceae) H-33, I-2,6,8,9 (103)
Leptochillus decurrens (Polypodiaceae) H-19, I-9 (419)
Leptospermum flavescens (Myrtaceae) H-19, I-12 (822) Alk=I-7 (1392)
L. laevigatum (Myrtaceae) H-19, I-7 (777) Alk=I-10 (1373)
L. liversidgei (Myrtaceae) H-19, I-12 (775)
L. petersonii (Myrtaceae) H-19, I-12 (647,775,822)
Leuceria senecioides (Asteraceae) H-19, I-6,7,8 (940)
Libertia chilensis (Iridaceae) H-19, I-2,6,7 (940)
Ligusticum ponteri (Apiaceae) H-32, I-2 (401)
Linaria kurdica (Scrophulariaceae) H-19, I-10,16 (844)
L. linaria (Scrophulariaceae) H-19, I-? (1171)
Linnaea borealis var. americana (Scrophulariaceae) H-19, I-? (592)
Linostoma decandrum (Thymelaeaceae) H-32, I-4 (73,459)
Lithrae caustica (Anacardiaceae) H-19, I-6 (940)
Lobelia cardinalis (Lobeliaceae) H-22, I-1,8 (945) H-33, I-7 (504)
  Alk=I-2,16 (1392)

Lolium temulentum (Poaceae) H-33, I-10 (220) Alk=I-7 (1392)
Lomatia ferruginea (Proteaceae) H-19, I-6,7 (940)
Lonchocarpus velutinus (Fabaceae) H-32, I-2 (192)
Lonicera involucrata (Caprifoliaceae) H-22, I-? (667)
Lophira pallidum (Celastraceae) H-33, I-4 (103)
Lophopetalum javanicum (Celastraceae) H-33, I-? (102)
Lophophora williamsii (Cactaceae) H-19, I-1,7 (402,1155) Alk=I-16 (1392)
Lophosoria quadripinnata (Cyatheaceae) H-19, I-7 (940)
Lotus gebelia (Fabaceae) H-19, I-16 (844)
L. uliginosus (Fabaceae) H-19, I-? (940)
Ludwigia alternifolia (Onagraceae) H-15, I-7,8 (652)
L. breots (Onagraceae) H-15, I-6,7,8 (652)
L. octovalvis (Onagraceae) H-22, I-16 (1394)
L. perensis (Onagraceae) H-22, I-1 (1395)
L. peruviana (Onagraceae) H-15,19, I-6,7 (652)
Lunasia amara (Rutaceae) H-19,32 33, I-? (776,968,983) Alk=I-4,7 (1392)
Lupinus polyphyllus (Fabaceae) H-19, I-2,6,7 (580) Alk=I-10 (1373,1392)
Luzuriaga erecta (Liliaceae) H-19, I-6,7,8 (940)
Lycium barbaratum (Solanaceae) H-33, I-16 (220)
Maba nigrescens (Ebenaceae) H-22, I-16 (882) Tri=I-2 (1384)
Macaranga tanarius (Euphorbiaceae) H-19, I-7 (982) Alk=I-9 (1392)
Macrozamia pauli-guilielmi (Zamiaceae) H-19, I-10 (776)
Madia elegans (Asteraceae) H-19, I-? (732,1022)
Maesa chisia (Myrsinaceae) H-22, I-16 (1394)
M. cumingii (Myrsinaceae) H-32, I-4 (983)
M. denticulata (Myrsinaceae) H-32, I-1,4 (192,864,983)
M. indica var. angustifolia (Myrsinaceae) H-22, I-16 (1395)
M. laxa (Myrsinaceae) H-32, I-9 (983)
Magnolia kobus (Magnoliaceae) H-22, I-? (1015) Alk=I-4,6,7,9 (1368,1392)
M. quinquepeta (Magnoliaceae) H-15, I-7,8 (652) Alk=I-2,4,5,7 (1392)
Mahinot ultilissima (Euphorbiaceae) H-32, I-2 (220,968)
Mahonia aquifolium (Berberidaceae) H-19, I-9 (414) Alk=I-4,5 (1392)
Mallotus apelta (Euphorbiaceae) H-32, I-7 (1276)
M. ricinoides (Euphorbiaceae) H-19, I-7 (982)
Malouetia nitida (Apocynaceae) H-33, I-14 (504)
Malus prunifolia (Rosaceae) H-19, I-? (594)
M. toringoides (Rosaceae) H-19, I-? (594)
Malva parviflora (Malvaceae) H-33,35, I-? (220)
M. rotundifolia (Malvaceae) H-19, I-6,7 (732)
Malvaviscus arboreus var. mexicanus (Malvaceae) H-19, I-7 (651)
Mammillaria lepidotus (Cactaceae) H-19, I-6 (982)
M. triangularis (Cactaceae) H-19, I-7 (982)
Mandragora caulescens (Solanaceae) H-33, I-? (220)
Marsdenia rostrata (Asclepiadaceae) H-19, I-4,5 (776) Alk=I-? (1392)
Martynia louisianica (Martyniaceae) H-19, I-1,14 (704) Alk=I-10 (1391)
Maytenus magellanica (Celastraceae) H-19, I-6,7 (940)
Meconopsis aculeata (Papaveraceae) H-33,35, I-2 (220) Alk=I-2 (1378)
Meibomia canadensis (Fabaceae) H-19, I-? (1171)
M. rigida (Fabaceae) H-19, I-? (1171)
Melaleuca alternifolia (Myrtaceae) H-19,20, I-12 (504,647,775,821,1161)
M. decora (Myrtaceae) H-19, I-12 (775)
M. ericifolia (Myrtaceae) H-19, I-12 (775)

M. hypercifolia (Myrtaceae) H-19, I-8 (777)
M. linariifolia (Myrtaceae) H-19, I-12 (821) H-20, I-? (504)
M. platycalyx (Myrtaceae) H-19, I-7 (777)
M. squarrosa (Myrtaceae) H-19, I-8 (777)
M. uncinata (Myrtaceae) H-19, I-12 (775,821) Alk=I-7 (1392)
M. violacea (Myrtaceae) H-19, I-7 (777)
M. viridifolia (Myrtaceae) H-19, I-12 (775)
M. wilsonii (Myrtaceae) H-19, I-8 (777)
Melanolepsis multiglandulosa (Euphorbiaceae) H-19, I-7 (982)
Melanorrhoea inappendiculata (Anacardiaceae) H-33, I-7 (504)
Melanthera hastata (Asteraceae) H-15, I-6,7,8 (652)
Melastoma normale (Melastomaceae) H-22, I-1 (1395)
Melianthus major (Sapindaceae) H-33, I-8 (220)
Melissa cinalis (Lamiaceae) H-22, I-? (1119) Tan=I-7 (1119)
Melodinus monogynus (Apocynaceae) H-32, I-? (73,220)
Melothria pendula (Cucurbitaceae) H-15, I-6,7,8 (652)
Memecylon umbellatum (Melastomataceae) H-22, I-7 (881) Fla, Tan=I-7 (1377)
Menyanthes trifoliata var. minor (Gentianaceae) H-19, I-? (592) Alk=I-2,7
   (1392)
Meryta capitata (Araliaceae) H-32, I-9 (416)
Mesua ferrea (Clusiaceae) H-15,19, I-? (411)
Metopium toxiferum (Anacardiaceae) H-33, I-14 (504)
Metrosideros macropus (Myrtaceae) H-19, I-6 (172) Alk=I-6,7,8 (1363)
Metroxylon sagu (Arecaceae) H-33, I-9,14 (504)
Mictomelum pubescens (Rutaceae) H-19, I-7,9 (776) Alk=I-4,6,7 (1389,
   1392)
Mikania scandens (Asteraceae) H-19, I-7 (982)
Millettia barteri (Fabaceae) H-32, I-4 (504)
M. ferruginea (Fabaceae) H-32, I-? (1143)
M. ichtyochtona (Fabaceae) H-32, I-9 (504)
M. merillii (Fabaceae) H-32, I-2,4,7 (983)
M. servicea (Fabaceae) H-32, I-2 (504)
M. splendens (Fabaceae) H-32, I-2,7 (968) Sap=I-2,7 (968)
Mimosa dysocarpa (Mimosaceae) H-32, I-2 (401)
M. hamata (Mimosaceae) H-19, I-16 (845)
Mimusops djave (Sapotaceae) H-33, I-9 (504)
Mitchella repens (Rubiaceae) H-19, I-? (1171)
Mnium cuspidatum (Bryophyta-Moss) H-19, I-7 (730)
M. longirostrum (Bryophyta-Moss) H-19, I-7 (570)
Momordica balsamina (Cucurbitaceae) H-33, I-9 (220) Alk=I-10 (1391)
Morinda citrifolia (Rubiaceae) H-19, I-? (172) Alk=I-6,7,9 (1362,1363,
   1374,1392)
Mubomia canadense (Fabaceae) H-19, I-? (732)
Mucuna nigricans (Fabaceae) H-33, I-? (968)
M. poggei (Fabaceae) H-32, I-6 (504)
M. pruriens (Fabaceae) H-33, I-17 (983) Alk=I-10,16 (1392)
M. prurita (Fabaceae) H-17, I-11 (1135)
Muntingia calabura (Tiliaceae) H-19, I-7,9 (982) Sap, Tan=I-6,7 (1319)
Murraya koenigii (Rutaceae) H-15,19, I-7,12 (902)
Muscadinia rotundifolia (Vitaceae) H-15,19, I-7 (488)
Mussaenda philippica (Rubiaceae) H-19, I-7 (982)
Mutisia spinosa (Asteraceae) H-19, I-6,7,8 (940)
Myoporum acuminatum (Myoporaceae) H-19, I-12 (775) Alk=I-7 (1392)

Myrceugenia leptospermoides (Myrtaceae) H-19, I-6,7 (940)

Myrica javanica (Myricaceae) H-19, I-7 (1076)

M. nagi (Myricaceae) H-22,32, I-4 (220,459,881,1040) Fla=I-4,6,7 (1379)
   Str, Tri=I-2 (1382)

Myroxlyon balsamum (Fabaceae) H-20, I-13 (504,1262)

Myzodendron punctulatum (Myzodendraceae) H-19, I-6,7 (940)

Narcissus pseudonarcissus (Amaryllidaceae) H-22, I-7 (847)

Nardostachys jatamansi (Valerianaceae) H-19, I-2,12 (450)

Naucleopsis mello-barretoi (Moraceae) H-33, I-14 (415)

Nelumbo nucifera (Nymphaeaceae) H-19, I-? (1171)

Nertera granadensis (Rubiaceae) H-19, I-2,6,7,9 (940)

Newbouldia laevis (Bignoniaceae) H-19, I-4 (836) Alk=I-4,7 (1372)

Nicotiana acutifolia (Solanaceae) H-33, I-10 (620)

N. glabra (Solanaceae) H-19, I-? (940)

N. miersii (Solanaceae) H-33, I-7 (620)

N. obtusifolia (Solanaceae) H-33, I-7 (620)

N. plumbaginifolia (Solanaceae) H-22, I-1 (1395) Alk=I-7 (1392)

Nothofagus alpina (Fagaceae) H-19, I-6,7 (940)

N. obligua (Fabaceae) H-19, I-6,7 (940)

N. pumilio (Fagaceae) H-19, I-6,7 (940)

Nymphaea stellata (Nymphaeaceae) H-19, I-7 (982)

Oenothera parviflora (Onagraceae) H-15, I-7 (652)

Oplopanax elatum (Araliaceae) H-15, I-2,12 (1157)

Opuntia cholla (Cactaceae) H-19, I-6 (1155)

O. ficus-indica (Cactaceae) H-19, I-7 (1155)

O. lindheimeri (Cactaceae) H-19, I-6 (1155)

O. littoralis (Cactaceae) H-19, I-6 (1155)

O. littoralis var. vaseyi (Cactaceae) H-19, I-6 (1155)

O. phaecantha (Cactaceae) H-19, I-6 (1155)

O. stricta (Cactaceae) H-19, I-6 (1155)

O. violacea var. macrocentra (Cactaceae) H-19, I-6 (1155)

Oreobolus obtusangulus (Cyperaceae) H-19, I-2,6,7 (940)

Orites myrtoidea (Proteaceae) H-19, I-6,7 (940)

Osyris arborea (Santalaceae) H-22, I-6,7 (881,1040)

Ougeinia oojeinensis (Fabaceae) H-22, I-4 (881)

Ourisia macrophylla (Scrophulariaceae) H-19, I-6,7 (940)

Oxalis europaea (Oxalidaceae) H-19, I-1,14 (592,704)

O. grandia (Oxalidaceae) H-19, I-1,14 (704)

Oxytropis lambertii (Fabaceae) H-33, I-16 (1304) Alk=I-2,6,7,8,9 (1369,
   1392)

Pachyelasma tessmannii (Caesalpiniaceae) H-32, I-9,10 (504)

Paeonia emodi (Paeoniaceae) H-33, I-? (220) Alk=I-10 (1392)

P. lactiflora (Paeoniaceae) H-19, I-? (413)

P. officinalis (Paeoniaceae) H-19, I-? (729)

Pangium edule (Flacourtiaceae) H-32, I-4,10 (504,968) H-33, I-6,10 (983)

Papaver dubium (Papaveraceae) H-35, I-10 (220) Alk=I-7,9,10 (1392)

P. nudicaule (Papaveraceae) H-33, I-1 (620)

Paramignya scandens (Rutaceae) H-19, I-16 (882)

Parinari curatellifolia (Chryobalanaceae) H-19, I-4 (836)

Parkia bussei (Mimosaceae) H-33, I-10 (504)

Parkinsonia aculeata (Caesalpiniaceae) H-19, I-7 (940) Alk=I-6,7,8,10,11
   (1360, 1392) Sap=I-6,11 (1360) Tan=I-11 (1360)

Paspalum scrobiculatum (Poaceae) H-33, I-? (220) Alk=I-10 (1369)

Passiflora edulis f. flavicarpa (Passifloraceae) H-19, I-9 (172) I-6,7 (1392)
P. foetida (Passifloraceae) H-19, I-9 (172) I-6,7 (1379,1390) Alk=I-7
   (1392) Alk, Fla=I-6,7 1379)
Paullinia fuscensens (Sapindaceae) H-32,33, I-6,10 (1276)
Pedilanthus tithymaloides (Euphorbiaceae) H-19, I-7 (982) H-33, I-7 (620)
Pelargonium demosticum (Geraniaceae) H-19, I-8 (580)
Pelvetia canaliculata (Fucales-Brown Algae) H-19, I-1 (1083)
Pentanisia variabilis (Rubiaceae) H-19, I-? (746)
Pentchlethra filimentosa (Mimosaceae) H-33, I-10 (504)
Persea americana (Lauraceae) H-19, I-2,4 (982) H-22, I-10 (969) Tan=I-6,7
   (1319)
P. gamblei (Lauraceae) H-22, I-7 (881,1040)
Persicaria hydropiper (Polygonaceae) H-19, I-? (1171)
Phaseolus trilobus (Fabaceae) H-22, I-1 (790)
Photinia integrifolia (Rosaceae) H-22, I-16 (790)
Phyla lanceolata (Verbenaceae) H-19, I-? (1171)
P. nodiflora (Verbenaceae) H-22, I-1 (1395)
Phyllanthus brasiliensis (Euphorbiaceae) H-32, I-? (1276)
P. conami (Euphorbiaceae) H-32, I-? (105)
P. engleri (Euphorbiaceae) H-33, I-2 (504)
P. indicus (Euphorbiaceae) H-33, I-10 (504)
P. niruri (Euphorbiaceae) H-19, I-4,7 (982,1076) H-32, I-6 (983,1349)
   Sap=I-2, Tan=I-6 (1318)
P. urinaria (Euphorbiaceae) H-32, I-? (220) Alk=I-? (1392)
Physalis heterophyla (Solanaceae) H-19, I-? (1171)
Physochlaina praealta (Solanaceae) H-33, I-? (220) Alk=I-2,7 (1392)
Phytolacca dioica (Phytolaccaceae) H-19, I-9 (776)
Pimenta acris (Myrtaceae) H-32, I-? (1276)
Pimpinella diversifolia (Apiaceae) H-22, I-1 (881)
Pinus caribeae (Pinaceae) H-19, I-12 (775) Alk=I-10 (1373)
P. mugo (Pinaceae) H-19, I-6 (1262) H-20, I-7,12 (504)
P. resinosa (Pinaceae) H-19, I-9 (1262) Alk=I-7 (1392)
P. roxburghii (Pinaceae) H-15,19, I-6 (411)
P. wallichiana (Pinaceae) H-19,22 I-7 (1052)
Piper angustifolium (Piperaceae) H-20, I-7 (504)
P. geniculatum (Piperaceae) H-33, I-2,9 (620)
P. retrofractum (Piperaceae) H-19, I-7 (982)
p. unguiculatum (Piperaceae) H-33, I-7 (968)
Pistacia integrrima (Anacardiaceae) H-19, I-6 (450) H-22, I-4,6 (881)
Pithecellobium acle (Mimosaceae) H-32, I-4 (864)
P. bigeminum (Mimosaceae) H-32, I-4,6,7 (73,220,459,504) Alk=I-4,10 (1392)
P. dulce (Mimosaceae) H-19, I-2,4 (982) Alk=I-7 (1390,1392)
P. ellipticum (Mimosaceae) H-32, I-2,4,5,7,14 (105,983,1389) Alk=I-10
   (1389)
P. jupunba (Mimosaceae) H-32, I-4 (504)
Pittosporum pentandrum (Pittosporaceae) H-19, I-7 (982)
P. rhytidocarpum (Pittosporaceae) H-32, I-9 (1276)
P. tobira (Pittosporaceae) H-22, I-? (847)
Plagiobothrys tinctorium (Boraginaceae) H-19, I-6,7,8 (940)
Planchonia spectabilis (Lecythidaceae) H-19, I-7 (982)
Plantago truncata subsp. firma (Plantaginacae) H-19, I-6,7,8,9 (940)
Plesmonium margaritiferum (Araceae) H-33, I-10 (220)
Pluchea dioscorides (Rhizophoraceae) H-19, I-12 (775)

Plumeria acutifolia (Apocynaceae) H-33, I-? (220) H-22, I-? (847) Alk=I-6,7
    (1318,1370,1372,1392)
Podanthus ovatifolius (Asteraceae) H-19, I-6,7 (940)
Podophyllum hexandrum (Berberidaceae) H-33, I-2,9,13 (220,504)
Polanisia icosandra (Capparaceae) H-33, I-2 (968)
Polyachus fuscus (Asteraceae) H-19, I-6,7,9 (940)
Polygonatum commutatum (Liliaceae) H-19, I-9 (1171)
Polygonum baratum (Polygonaceae) H-32, I-2,6,7 (968,983)
P. pensylvanicum (Polygonaceae) H-19, I-1 (944) H-32, I-1 (401)
Polypodium alternifolium (Polypodiaceae) H-19, I-2 (419)
P.ebenipes (Polypodiaceae) H-19, I-9 (410)
Polyscias nodosa (Araliaceae) H-32, I-7 (504,968,983) H-33, I-? (968
    Sap=I-7 (968)
Polysiphonia fastigiata (Ceramiales-Red Algae) H-19, I-1 (1083)
Polystichum squarrosum (Polypodiaceae) H-19, I-9 (419)
Pomaderris elliptica (Rhamnaceae) H-19, I-8 (777)
Pongamia glabra (Fabaceae) H-32, I-? (459)
Popowia tomentosa (Annonaceae) H-33, I-2 (504)
Populus alba (Salicaceae) H-19, I-19 (851)
P. davidiana (Salicaceae) H-19, I-19 (851)
P. grandidentata (Salicaceae) H-19, I-19 (851)
P. tremuloides (Salicaceae) H-19, I-4,19 (851)
P. trichocarpa (Salicaceae) H-15, I-? (752)
Porlieria chilensis (Zygophyllaceae) H-19, I-6 (940)
Primula malacoides (Primulaceae) H-19, I-1,6,7,8 (446,1101)
P. reticulata (Primulaceae) H-33, I-16 (220)
Prismatomeris malayana (Rubiaceae) H-33, I-2 (103)
Prostanthera cineoliefa (Lamiaceae) H-19, I-? (1041)
P. ovalifolia (Lamiaceae) H-19, I-12 (775)
Proustia cuneifolia (Asteraceae) H-19, I-6,7 (940)
Prunus avium (Rosaceae) H-33, I-10 (220)
P. capuli (Rosaceae) H-32, I-4,7 (401)
P. cerasifera divariata (Rosaceae) H-19, I-? (1262)
P. cerasoides (Rosaceae) H-19, I-6,7 (1262) H-34, I-10 (220)
P. cerasus (Rosaceae) H-33, I-10 (220)
P. cornuta (Rosaceae) H-22, I-16 (882)
P. domestica (Rosaceae) H-19, I-? (594)
P. mahaleb (Rosaceae) H-33, I-10 (220) Alk=I-10 (1392)
P. serotina (Rosaceae) H-19, I-7 (652) Alk=I-10 (1374)
P. umbellata (Rosaceae) H-19, I-7 (652)
P. undulata (Rosaceae) H-33, I-10 (220)
Pseuderanthemum reticulatum (Acanthaceae) H-19, I-7 (982)
Psidium guineense (Myrtaceae) H-19, I-7 (1262)
Psilotum nudum (Lycopodiaceae) H-19, I-? (172) Alk=I-6,7,10 (1363)
Psychotria luconensis (Rubiaceae) H-19, I-7 (982) Tan=I-6,7 (1318)
P. sarmentosa (Rubiaceae) H-33, I-2 (103)
P. truncata (Rubiaceae) H-22, I-16 (882)
Ptelea trifoliata (Rutaceae) H-15,19, I-6,7 (842) H-20, I-6,7 (832) Alk=I-6
    7,8 (1369,1371)
Pterospermum diversifolium (Byttneriaceae) H-32, I-2 (983)
P. obliquum (Byttneriaceae) H-19, I-4 (982)
Pygeum gardneri (Rosaceae) H-32, I-9 (73,220)
Pyrrosia manii (Polypodiaceae) H-19, I-9 (419)

Pyrus pashia (Rosaceae) H-32, I-9 (993)
Quercus lamellosa (Fagaceae) H-22, I-4,6 (881)
Q. lanceaefolia (Fagaceae) H-22, I-4,6 (881)
Q. pachyphylla (Fagaceae) H-22, I-4,6 (881,1040)
Quinchamalium chilense (Santalaceae) H-19, I-6,7,8 (940)
Randia uliginosa (Rubiaceae) H-32, I-? (73)
R. walkeri (Rubiaceae) H-32, I-? (504)
Ranunculus arvensis (Ranunculaceae) H-33, I-16 (220)
R. bulbosus (Ranunculaceae) H-19, I-? (186)
R. flammula (Ranunculaceae) H-19, I-? (186)
R. hyperboreus (Ranunculaceae) H-33, I-? (620)
R. lapponicus (Ranunculaceae) H-33, I-? (620)
R. lingua (Ranunculaceae) H-19, I-? (186)
R. nivalis (Ranunculaceae) H-33, I-2,7 (620)
R. occidentalis (Ranunculaceae) H-19, I-6,7,8, 12,14 (625,727)
R. pollasii (Ranunculaceae) H-33, I-2,7 (620)
R. palmatus (Ranunculaceae) H-15, I-7 (652)
R. pensylvanicus (Ranunculaceae) H-19, I-7,9 (735)
R. purshii (Ranunculaceae) H-33, I-2,7 (620)
R. pygmaeus (Ranunculaceae) H-33, I-2,7 (620)
R. septentrionalis (Ranunculaceae) H-19, I-? (1171)
R. verticillatus (Ranunculaceae) H-33, I-2,7 (620)
Raphia sassandrensis (Arecaceae) H-32, I-9 (504)
Raphidophora korthalsii (Araceae) H-33, I-14 (504)
R. merrillii (Araceae) H-33, I-14 (968)
Raputia alba (Rutaceae) H-32, I-4 (504)
Ratibida pinnata (Asteraceae) H-32, I-? (1171)
Regelia ciliata (Myrtaceae) H-19, I-7,8,9 (777)
R. grandiflora (Myrtaceae) H-19, I-8 (777)
Rheum palmatum (Polygonaceae) H-19, I-2 (735)
R. rhabarbarum (Polygonaceae) H-22, I-14 (975)
Rhizophora mucronata (Rhizophoraceae) H-19, I-2 (776) Fla, Tan=I-6,7 (1377)
Rhododendron anthopogon (Ericaceae) H-33, I-16 (220) Fla=I-6,7 (1378)
R. arboreum (Ericaceae) H-33, I-16 (220) Fla=I-4,6,7 (1361)
R. barbatum (Ericaceae) H-32, I-? (73) H-33, I-16 (220)
R. campanulatum (Ericaceae) H-33, I-16 (220) Fla=I-6,7 (1378)
R. cinnabarinum (Ericaceae) H-33, I-16 (220)
R. falconeri (Ericaceae) H-32, I-15 (459) H-33, I-16 (220)
R. setosum (Ericaceae) H-33, I-16 (220)
Rhodomyrtus macrocarpa (Myrtaceae) H-19, I-? (776)
Rhus copallena (Anacardiaceae) H-15,19, I-7 (651)
R. insignis (Anacardiaceae) H-33, I-14 (220)
R. parviflora (Anacardiaceae) H-22, I-16 (790)
R. punjabensis (Anacardiaceae) H-33, I-14 (220)
R. succedanea (Anacardiaceae) H-22, I-7 (881) H-33, I-14 (220)
R. wallichii (Anacardiaceae) H-33, I-14 (220)
Rhynchosia suaveolens (Fabaceae) H-19, I-1 (882)
Ribes bracteosum (Saxifragaceae) H-19, I-6 (1171)
R. vulgare (Saxifragaceae) H-19, I-14 (497)
Rosa conina (Rosaceae) H-19, I-8 (580)
R. dilecta (Rosaceae) H-22, I-8 (969)
R. indica (Rosaceae) H-15, I-8 (412)
R. moschata (Rosaceae) H-19, I-6,7 (940) Fla=I-2,6,7 (1361) Sap=I-8

(1361)
Rourea erecta (Connaraceae) H-33, I-5 (968)
R. glabia (Connaraceae) H-33, I-10 (504)
R. volubilis (Connaraceae) H-33, I-9 (504)
Rubus moluccanus (Rosaceae) H-33, I-? (220)
Rudbeckia laciniata (Asteraceae) H-19, I-6 (732,1171) Alk=I-2,6,7 (1369)
Rumex maritimus (Polygonaceae) H-15, I-1 (882)
Rumorha adiantiformis (Polypodiaceae) H-19, I-6,7 (940)
Ruta tuberculata (Rutaceae) H-33, I-? (220)
Saccopetalum tomentosum (Annonaceae) H-19, I-7,12 (897)
Sagittaria cuneata (Alismataceae) H-19,34, I-1 (837)
Salacia roxburghii (Hippocrateaceae) H-22, I-16 (882)
Salix alba (Salicaceae) H-22, I-4,6 (882)
Salmea eupatoria (Asteraceae) H-32,34, I-2 (504)
Salvia divinorum (Lamiaceae) H-35, I-7 (504)
S. farinacea (Lamiaceae) H-22, I-? (975) Alk=I-6,7,8 (1370)
S. greggii (Lamiaceae) H-22, I-? (975)
Sambucus caerulea (Caprifoliaceae) H-19, I-? (1171)
Sambucus ebulus (Caprifoliaceae) H-33, I-13 (220) Glu=I-? (220)
Sambucus sieboldiana (Caprifoliaceae) H-22, I-4 (1014)
Sandoricum koetjape (Meliaceae) H-15, I-4 (1318) H-19, I-4 (982)
  Sap, Tan=I-6,7 (1318)
Sanicula crassicaulis (Apiaceace) H-19, I-? (1171)
Santalum lanceolatum (Santalaceae) H-19, I-? (904)
Sapium biloculare (Euphorbiaceae) H-32, I-4,6,7 (397,401,504) H-32,
  I-7,14 (397,504)
S. discolor (Euphorbiaceae) H-32, I-? (1276)
S. insigne (Euphorbiaceae) H-32, I-10 (1276) H-33, I-4,7 (102) Fla=I-4
  (1379) Str=I-6 (1385)
S. madagascariensis (Euphorbiaceae) H-33, I-14 (504)
Saponaria vaccaria (Caryophyllaceae) H-33, I-? (220) Sap=I-? (220)
Saritaea magnifica (Bignoniaceae) H-19, I-7 (1076)
Sarcolobus narcoticus (Asclepiadaceae) H-33, I-2,4 (504)
S. spanoghei (Asclepiadaceae) H-33, I-2,4 (504)
Saurauja roxburghii (Dilleniaceae) H-22, I-16 (1394)
Scaevola gaudichaudiana (Goodeniaceae) H-19, I-? (172) Alk=I-6,7,8 (1363)
Scaphocalyx spathacea (Flacourtiaceae) H-33, I-4 (103)
Schoendum chilense (Liliaceae) H-33, I-10 (504)
Schumanniophyton arboreum (Rubiaceae) H-32, I-4 (504)
S. magnificum (Rubiaceae) H-32, I-4 (504)
Scilla ridigifolia (Liliaceae) H-33, I-3 (504)
Scoparia dulcis (Scrophulariaceae) H-19, I-? (982) Alk=I-12,6,7 (1392)
Scopolia anomala (Solanaceae) H-33, I-? (220)
Scrophularia marilandica (Scrophulariaceae) H-19, I-? (1171) Alk=I-6,7
  (1371)
Scutia myrtina (Rhamnaceae) H-22, I-1,16 (881,1040)
Sebastiana palmeri (Euphorbiaceae) H-33, I-14 (620)
S. plinglei (Euphorbiaceae) H-32, I-4 (401)
Selaginella involvens (Selaginellaceae) H-19, I-7 (419)
S. ludoviciana (Selaginellaceae) H-15, I-1 (651)
Semecarpus cuneiformis (Anacardiaceae) H-33, I-4,7 (968)
S. phillipinensis (Anacardiaceae) H-33, I-7,12 (968)
S. travancoricus (Anacardiaceae) H-33, I-? (220)

Senecio adenotrichius (Asteraceae) H-19, I-2,6,7 (940)
S. chinensis (Asteraceae) H-19, I-2,6,7 (940)
S. hartwegii (Asteraceae) H-32, I-2,7 (401)
S. otites (Asteraceae) H-19, I-6,7,8 (940)
S. sylvaticus (Asteraceae) H-19, I-8 (1171) Alk=I-7 (1392)
S. tenuifolius (Asteraceae) H-22, I-1 (1393)
Serjania atrolineata (Sapindaceae) H-32, I-2,6,7 (1276)
S. curassavica (Sapindaceae) H-33, I-2,9 (620)
S. cuspidata (Sapindaceae) H-32, I-? (1276)
S. inebrians (Sapindaceae) H-32, I-7,15 (1276)
S. mexicana (Sapindaceae) H-32, I-6 (504)
S. nodosa (Sapindaceae) H-33, I-2,9 (620) Alk=I-? (620)
Sida acuta (Malvaceae) H-22, I-1 (1040) Alk=I-2,6,7 (1392)
S. javensis (Malvaceae) H-19, I-7 (982)
Siegesbeckia orientalis (Asteraceae) H-22, I-1 (881) Alk=I-16 (1392)
Silene coniflora (Caryophyllaceae) H-19, I-1 (844)
Silphium asteriscus (Asteraceae) H-15, I-7 (652)
S. perfoliatum (Asteraceae) H-19, I-7,8 (732,1171)
S. terebinthinaceum (Asteraceae) H-19, I-? (1171)
Sisymbrium altissimum (Brassicaceae) H-19, I-? (592) Alk=I-2,6,7,8,9 (1370)
Sisyrinchium arizonicum (Brassicaceae) H-32, I-2 (401)
Solanum insanum (Solanaceae) H-33, I-? (220) Alk=I-9 (1374,1392)
S. malacoxylon (Solanaceae) H-33, I-6,7 (414)
S. nodiflorum (Solanaceae) H-19, I-9 (172) Alk=I-9,14 (1363,1392)
S. sodomeum (Solanaceae) H-19, I-? (172) Alk=I-? (1392)
S. spirale (Solanaceae) H-33, I-? (220)
Sophora cassiodes (Fabaceae) H-19, I-? (940)
S. glauca (Fabaceae) H-22, I-16 (1394)
Sorbus aucuparia (Rosaceae) H-19, I-? (592,728)
Sparganium fluctans (Sparganiaceae) H-19,34, I-1 (837)
S. lurycarpum (Sparganiaceae) H-19,34, I-1 (837)
Spartina pectinata (Poaceae) H-19, I-? (594)
Spigelia fruticulosa (Loganiaceae) H-33, I-? (620)
Spiraea aruncus (Rosaceae) H-19, I-7,8 (980)
S. tomentosa (Rosaceae) H-19, I-1,14 (704)
Spondias purpurea (Anacardiaceae) H-19, I-4,7,9 (982)
Stachys macraei (Lamiaceae) H-19, I-6,7 (940)
Stellaria cuspidata (Caryophyllaceae) H-19, I-6,7 (940)
Stephania aculeata (Menispermaceae) H-19, I-? (776) Alk=I-2 (1392)
S. hernandiifolia (Menispermaceae) H-33, I-? (73)
Steudnera virosa (Araceae) H-33, I-? (220)
Stevia salicifolia (Asteraceae) H-32, I-2,15 (401)
Strobilanthes wightianus (Acanthaceae) H-22, I-16 (1394)
Strombosia cumingi (Apocynaceae) H-33, I-10 (504)
S. eminii (Apocynaceae) H-33, I-10 (504)
S. gratus (Apocynaceae) H-33, I-10 (504)
S. hispidus (Apocynaceae) H-33, I-10 (504)
S. kombe (Apocynaceae) H-33, I-10 (504)
S. preussii (Apocynaceae) H-33, I-10 (504)
S. sarmentosus (Apocynaceae) H-33, I-10 (504)
S. thollonii (Apocynaceae) H-33, I-10 (504)
Strophanthus cumingii (Apocynaceae) H-33, I-4 (863,983)
Strychnos angulensis (Loganiaceae) H-33, I-14 (835) Alk=I-2,4 (1392)

S. atherstonei (Loganiaceae) H-33, I-? (504)
S. axillaris (Loganiaceae) H-32,33, I-2 (103)
S. balansae (Loganiaceae) H-33, I-14 (102) H-33, I-? (504)
S. beccarii (Loganiaceae) H-33, I-14 (102) H-33, I-? (504)
S. brasiliensis (Loganiaceae) H-33, I-14 (620)
S. castelnaei (Loganiaceae) H-33, I-4 (504) H-33, I-14 (620)
S. cogens (Loganiaceae) H-33, I-14 (620) Alk=I-? (1392)
S. colubria (Loganiaceae) H-32,33, I-2,4,10 (73,102,220) Alk=I-7 (1371)
   Str=I-1 (1383)
S. crevauxii (Loganiaceae) H-33, I-14 (620) Alk=I-? (620,1392)
S. curare (Loganiaceae) H-33, I-14 (620)
S. cuspidata (Loganiaceae) H-33, I-14 (102) H-33, I-? (504)
S. dekindtiana (Loganiaceae) H-33, I-? (504)
S. densiflora (Loganiaceae) H-33, I-? (835)
S. depavperata (Loganiaceae) H-33, I-14 (620)
S. gardneri (Loganiaceae) H-33, I-14 (620)
S. gaulthieriana (Loganiaceae) H-33, I-4 (504)
S. gubleri (Loganiaceae) H-33, I-14 (620) Alk=I-? (1392)
S. guianensis (Loganiaceae) H-33, I-14 (504,620) Alk=I-2,4 (1392)
S. henningsii (Loganiaceae) H-33, I-? (504)
S. hirsuta (Loganiaceae) H-33, I-14 (620) Alk=I-4 (1392)
S. icaja (Loganiaceae) H-33, I-? (504,835) Alk=I-2,4,7 (1392)
S. ignatii (Loganiaceae) H-33, I-2,4 (849) Alk=I-2,4,5,7 (1389,1392)
S. innocua (Loganiaceae) H-33, I-10 (504)
S. jobertiana (Loganiaceae) H-33, I-14 (620)
S. jollyana (Loganiaceae) H-33, I-? (504)
S. kipapa (Loganiaceae) H-33, I-2 (504)
S. krabiensis (Loganiaceae) H-33, I-14 (102) H-33, I-? (504)
S. lanceolaris (Loganiaceae) H-33, I-? (103,504)
S. ligustrina (Loganiaceae) H-33, I-4 (504)
S. malaccensis (Loganiaceae) H-33, I-4 (504)
S. melinoniana (Loganiaceae) H-33, I-? (504)
S. nigricans (Loganiaceae) H-33, I-14 (620)
S. nux-blanda (Loganiaceae) H-33, I-10 (504)
S. obertiana (Loganiaceae) H-33, I-? (620)
S. odorata (Loganiaceae) H-33, I-? (504)
S. ovalifolia (Loganiaceae) H-33, I-14 (102) H-33, I-? (504)
S. ovata (Loganiaceae) H-33, I-? (102)
S. pedunculata (Loganiaceae) H-33, I-14 (620)
S. potatorum (Loganiaceae) H-33, I-? (504)
S. pseudoquina (Loganiaceae) H-17, I-4 (504) H-33, I-? (504)
S. psilosperma (Loganiaceae) H-19, I-7 (776) Alk=I-4,7,10 (1392)
S. pubescens (Loganiaceae) H-33, I-? (504)
S. quadrangularis (Loganiaceae) H-31, I-10 (103) H-33, I-? (103,504)
S. rubiginosa (Loganiaceae) H-33, I-14 (620)
S. rufa (Loganiaceae) H-33, I-2 (103,504)
S. samba (Loganiaceae) H-33, I-? (835)
S. spinosa (Loganiaceae) H-33, I-10 (504)
S. subcordota (Loganiaceae) H-33, I-14 (620)
S. tieute (Loganiaceae) H-33, I-2,14 (102) H-33, I-1 (504)
S. toxifera (Loganiaceae) H-33, I-2,4 (504) H-33, I-14 (620)
S. triclisioides (Loganiaceae) H-33, I-? (504)
S. triplinervia (Loganiaceae) H-33, I-14 (620)

S. unguacha (Loganiaceae) H-33, I-? (504)
S. wallichiana (Loganiaceae) H-33, I-? (103,504)
S. yopurensis (Loganiaceae) H-33, I-14 (620)
Styrax tonkinense (Styracaceae) H-20, I-13 (504)
Suaeda fruticosa (Chenopodiaceae) H-33, I-? (220)
Suillus phallus-ravenelli (Boletaceae) H-22, I-? (968)
Symplocos stawelli (Symplocaceae) H-19, I-7 (776)
S. tinctoria (Symplocaceae) H-32, I-4 (192)
Syzygium jambos (Myrataceae) H-19, I-4 (982) Alk=I-2,4 (1392)
S. malaccense (Myrtaceae) H-19, I-4,7,10 (172)
S. samarangense (Myrtaceae) H-19, I-7 (982)
Tabernaemontana acuminata (Apocynaceae) H-33, I-4 (620)
T. citrifolia (Apocynaceae) H-33, I-14 (507) Alk=I-6,7 (1375)
T. malaccensis (Apocynaceae) H-33, I-2 (103) Alk=I-2,6,7 (1388,1389)
T. orientalis (Apocynaceae) H-19, I-4 (776) Alk=I-7,9 (1392)
T. solanifolia (Apocynaceae) H-33, I-4 (620)
T. speciosa (Apocynaceae) H-33, I-4 (620)
Tamarix dioica (Tamaricaceae) H-22, I-2 (1395)
Taxithelium nepalense (Bryophyte-Moss) H-19, I-? (570)
Taxus media (Taxaceae) H-19, I-6,7 (1101)
Telosma cordata (Asclepiadaceae) H-33, I-6 (968) Alk=I-6 (1392)
Tephrosia densiflora (Fabaceae) H-32, I-2 (504)
T. leiocarpa (Fabaceae) H-32, I-2, (401)
T. nitens (Fabaceae) H-32, I-7, (504)
T. periulosa (Fabaceae) H-32, I-2 (504)
T. piscatoria (Fabaceae) H-32, I-2 (416)
T. talpa (Fabaceae) H-32, I-2 (401)
Tepualia stipularis (Myrtaceae) H-19, I-6,7,8 (940)
Terminalia avicennoides (Combretaceae) H-15, I-4 (945) H-19, I-? (836)
T. laxifolia (Combretaceae) H-19, I-4 (836)
Ternstroemia rhoderiana (Theaceae) H-19, I-10 (776)
T. robinsonii) (Theaceae) H-32, I-4 (504)
T. toquians (Theaceae) H-32, I-4,9 (983,1349)
Tetraglochin alatum (Rosaceae) H-19, I-6,7 (940)
Tetragonia maritima (Tetragoniaceae) H-19, I-2,6,7 (940)
Tetrapleura thonningii (Mimocaceae) H-32, I-6 (1276)
Teucrium bicolor (Lamiaceae) H-19, I-6,7,8 (940)
T. polium (Lamiaceae) H-15, I-7 (504)
T. scordium (Lamiaceae) H-15, I-7 (504)
Thalictrum collinum (Ranunculaceae) H-20, I-2,8 (504)
T. rugosum (Ranunculaceae) H-15, 19, I-2 (833,838) H-20, I-16 (832)
Thevetia thevetioides (Apocynaceae) H-33, I-10 (504)
Tilia europaea (Tiliaceae) H-22, I-16 (790)
Tillandsia usneoides (Bromeliaceae) H-19,34, I-1 (657,980)
Tinomiscium philippinense (Menispermaceae) H-32, I-9 (504,528,983)
  Alk=I-? (1392)
Tithonia tagetiflora (Asteraceae) H-19, I-7 (1076)
Toddalia aculeata (Rutaceae) H-17, I-2 (504) H-20, I-9 (504)
Tovaria virginiana (Tovariaceae) H-19, I-? (732)
Trachelospermum asiaticum (Apocynaceae) H-22, I-? (847)
Trachypogon plumosus (Poaceae) H-19, I-2 (746)
Tradescantia foliosa (Commelinaceae) H-19, I-7 (652)
Tricholoma gambosum (Agaricaceae) H-20, I-9 (504)

Trichosanthes bracteata (Cucurbitaceae) H-33, I-9 (220)
Triclisia gelletii (Menispermaceae) H-33, I-7 (504)
Trigonostemon longifolius (Euphorbiaceae) H-33, I-4 (103)
Tulipa gesnerana (Liliaceae) H-19, I-8 (580) Alk=I-? (1392)
Turbina corymbosa (Convolvulaceae) H-20, I-19 (1209)
Tweedia hookeri (Asclepiadaceae) H-19, I-6,7 (940)
Tylophora indica (Asclepiadaceae) H-33, I-? (220) Alk=I-2,6,7 (1392)
Typha domingensis (Typhaceae) H-15, I-7 (652)
Ulmus wallichiana (Ulmaceae) H-22, I-4,6 (882)
Uncinia multifaria (Rubiaceae) H-19, I-7,8 (940)
Undaria pinnatifida (Laminariaceae) H-19, I-? (1084)
Unonopis veneficiorum (Annonaceae) H-33, I-? (504)
Uraria lagopoides (Fabaceae) H-22, I-1 (1395)
Urechites suberecta (Apocynaceae) H-32, I-7,16 (211,504)
Urena lobata (Malvaceae) H-17, I-10 (1319) Alk=I-1,10 (1391,1392)
   Tan, Sfr=I-6,7 (1319)
Urginea indica (Liliaceae) H-22, I-3 (1040)
Vaccaria pyramidata (Caryophyllaceae) H-22, I-16 (977)
Vaccinium angustifolium var. laeviforium (Ericaceae) H-19, I-? (592)
V. corymbosum (Ericaceae) H-19, I-? (594)
Vanda spathulata (Orchidaceae) H-19, I-16 (1394)
V. virgatum (Scrophulariaceae) H-19, I-4,6,7 (776) Alk=I-6,7 (1392)
Verbena angustifolia (Verbenaceae) H-19, I-2,6,10 (1171)
V. bipinnatifida (Verbenaceae) H-22, I-? (975)
Verbesina podocephala (Asteraceae) H-32, I-2,7 (401)
V. virginica (Asteraceae) H-15, I-? (652)
Vernonia cinerea (Asteraceae) H-19, I-? (776) H-22, I-1 (881) Alk=I-7
   (1392) Alk, Tri=I-7 (1390)
Verticordia brownii (Myrtaceae) H-19, I-7,8 (777)
V. monodelpha (Myrtaceae) H-19, I-8 (777)
V. plumosa (Myrtaceae) H-19, I-7,8 (777)
Viburnum erubscens (Caprifoliaceae) H-22, I-1 (881,1040)
V. plicatum (Caprifoliaceae) H-19, I-7 (1103)
Viguiera decurrens (Asteraceae) H-32, I-2 (401)
Viscum orientale (Loranthaceae) H-33, I-? (504)
Vitex parviflora (Verbenaceae) H-32, I-4,9 (968,983) Sap=I-4,9 (968)
Vitis vulpina (Vitaceae) H-19, I-? (1171) Alk=I-10 (1391)
Voacanga globosa (Apocynaceae) H-32, I-9 (983)
Vogelia indica (Plumbaginaceae) H-22, I-1 (1395) Alk, Tan=I-6,7 (1360)
Wendtia gracilis (Geraniaceae) H-19, I-6,7 (940)
Whitfordiodendron atropurpureum (Fabaceae) H-32, I-2 (504)
Woodfordia fruticosa (Lythraceae) H-22, I-1 (881) Alk=I-8 (1360)
   Sap, Tan=I-6,7,8 (1360) Str=I-6 (1386)
Xanthium amicanum (Asteraceae) H-19, I-? (1171)
Ximeria americana (Olacaceae) H-22, I-16 (882)
Xylopia aethiopica (Annonaceae) H-19, I-9 (836,848)
Yucca arkansana (Agavaceae) H-22, I-? (975)
Y. baccata (Agavaceae) H-19, I-7 (1262)
Y. decipiens (Agavaceae) H-32, I- 2,7 (401)
Y. glauca (Agavaceae) H-33, I-6 (620)
Y. gloriosa (Agavaceae) H-15, I-7 (882)
Zamia latifolia (Zamiaceae) H-33, I-2 (504)
Z. pumila (Zamiaceae) H-33, I-2 (504)

Zanthoxylum brachyacanthum (Rutaceae) H-19, I-12 (775) Alk=I-4,6,7,9 (1392)
Z. elephantiasis (Rutaceae) H-19, I-4 (834)
Z. piperitum (Rutaceae) H-15, I-? (971) Alk=I-2,10 (1392)
Z. torvum (Rutaceae) H-32, I-6,7 (983)
Zizyphus glaberina (Rhamnaceae) H-22, I-16 (882)
Z. jujuba (Rhamnaceae) H-32, I-? (983)

REFERENCES

1. Yule, W. N. (1964). *Ann. Appl. Biol.* 53:15-28.

2. Martin, J. T. (1960). *Ann. Appl. Biol.* 48(4):837-846.

3. Glynne-Jones, G. D. (1960). *Ann. Appl. Biol.* 48(2):352-362.

4. David, W. A. L., and B. O. C. Gardiner (1953). *Ann. Appl. Biol.* 40:91-112.

5. Ahmed, S., and M. Grainge (1986). *Econ. Bot.* 40:201-209.

6. Gnadinger, C. B., L. E. Evans, and C. S. Corl (1936). *Exp. Sta. Col. Bull.* #428. 29pp.

7. Schaffer, P. S., F. Acree, Jr., and H. L. Haller (1936). *J. Econ. Ento.* 29:601-604.

8. Hoyer, D. G., and M. D. Leonard (1936). *J. Econ. Ento.* 29:605-606.

9. Fulton, R. A., and N. F. Howard (1938). *J. Econ. Ento.* 31:405-410.

10. Ejercito, J. M. (1937). *Philip. J. Agr.* 8:89-96.

11. Ejercito, J. M. (1939). *Philip. J. Agr.* 10:187-191.

12. De Ong, E. R. (1937). *J. Econ. Ento.* 30:921-927.

13. De Ong, E. R. (1923). *J. Econ. Ento.* 16:486-493.

14. Davidson, W. M. (1930). *J. Econ. Ento.* 23:868-874.

15. Harper, S. H., C. Potter, and E. M. Gillham (1947). *Ann. Appl. Biol.* 34:104-112.

16. Gilliver, K. (1947). *Ann. Appl. Biol.* 34:136-143.

17. Owen, R. W., and N. Waloff (1946). *Ann. Appl. Biol.* 33:387-389.

18. Tattersfield, F., and C. Potter (1943). *Ann. Appl. Biol.* 30:259-279.

19. Parkin, E. A., and A. A. Green (1943). *Ann. Appl. Biol.* 30:279-292.

20. Martin, J. T. (1943). *Ann. Appl. Biol.* 30:293-300.

21. Martin, J. T. (1940). *Ann. Appl. Biol.* 27:274-294.

22. Tattersfield, F., and C. Potter (1940). *Ann. Appl. Biol.* 27:262-273.

23. Le, R. R., and G. Worsley (1937). *Ann. Appl. Biol.* 24:659-664.

24. Martin, J. T., H. H. Mann, and F. Tattersfield (1939). _Ann. Appl. Biol._ 26:14-24.

25. Le, R. R., and G. Worsley (1936). _Ann. Appl. Biol._ 23:311-328.

26. Higbee, E. C. (1947). _Econ. Bot._ 1:427-438.

27. Morallo-Rejesus, B., and L. C. Eroles (1978). _Philip. Ento._ 4: 87-98.

28. Marques, A. (1910). _Hawaiian Almanac and Annual_ 36:101-105.

29. Lee, C. S., and R. Hansberry (1943). _J. Econ. Ento._ 36: 915-921.

30. Busbey, R. L. (1950). _USDA Yearbook of Agriculture_ 1950-1951, pp. 765-772.

31. Nawrot, J., E. Bloszyk, J. Harmatha, L. Novotny, and B. Drozdz (1986). _Acta Ent. Bohemoslov._ 83:327-335.

32. Potter, C. (1935). _Ann. Appl. Biol._ 22:769-803.

33. Tattersfield, F., and J. T. Martin (1935). _Ann. Appl. Biol._ 22:578-605.

34. Tattersfield, F., and J. T. Martin (1935). _Ann. Appl. Biol._ 19:253-262.

35. Martin, J. T., and F. Tattersfield (1934). _Ann. Appl. Biol._ 21:682-689.

36. Martin, J. T., and F. Tattersfield (1934). _Ann. Appl. Biol._ 21:670-681.

37. Marten, G. G. (1986). _J. Trop. Med. Hygiene_ 89:213-222.

38. Tattersfield, F., C. T. Gimingham, and H. M. Morris (1926). _Ann. Appl. Biol._ 13:424.

39. Tattersfield, F., C. T. Gimingham, and H. M. Morris (1925). _Ann. Appl. Biol._ 12:61-65.

40. Fryer, J. C. F., and R. Stenson (1923). _Ann. Appl. Biol._ 10:18-34.

41. Tattersfield, F., and R. P. Hobson (1931). _Ann. Appl. Biol._ 18:203-243.

42. Tattersfield, F. (1931). _Ann. Appl. Biol._ 18:602-635.

43. Fryer, J. C. F., F. Tattersfield, and C. T. Gimingham (1928). _Ann. Appl. Biol._ 15:423-472.

44. Gimingham, C. T., and F. Tattersfield (1928). _Ann. Appl. Biol._ 15:649-658.

45. Lefroy, H. M. (1915). _Ann. Appl. Biol._ 1:280-298.

46. Swingle, W. T. (1941). _Science_ 93:60-61.

47. Jacobson, M. (1954). _Science_ 120:1028-1029.

48. Heal, R. E., E. F. Rogers, R. T. Wallace, and O. Starnes (1950). _Lloydia_ 13:89-162.

49. Sharma, H. L., O. P. Vimal, and D. Prasad (1985). _Int. J. Trop. Plt. Dis._ 3:51-55.

50. Saxena, R. C., N. J. Liquido, and H. D. Justo (1982). Ento. Dept., International Rice Research Institute, Philippines. Unpublished mss.

51. Saxena, R. C., G. P. Waldbauer, N. J. Liquido, and B. C. Puma (1980). Ento. Dept., International Rice Research Institute, Philippines. Unpublished mss.

52. Katsura, M. (May, 1980). Paper presented at the ICIPE (International Center Insect Physiol. and Ecology) conference, Nairobi, Kenya.

53. Gebreyesus, T. (1980). Bioassay Unit, ICIPE, Nairobi, Kenya. Unpublished paper.

54. Blau, P. A., P. Feeny, L. Contardo, and D. S. Robson (1978). _Science_ 200:1296-1298.

55. Amonkar, S. V., and A. Banerji (1971). _Science_ 174:1343.

56. Bowers, W. S., T. Ohta, J. S. Cleere, and P. A. Marsella (1976). _Science_ 193:542-547.

57. Little, V. A. (1931). _Science_ 73:315-316.

58. Fryer, J. C. F., and C. T. Gimingham (1931). _Nature_ 127:573-574.

59. Barton, R. (1957). _Nature_ 180:613-614.

60. Kircher, H. W., and F. V. Lieberman (1967). _Nature_ 215:97-98.

61. Crombie, L. (1954). _Nature_ 174:832.

62. Reyes, F. R., and A. C. Santos (1931). _Philip. J. Sci._ 44:409-410.

63. Jotwani, M. G., and K. P. Srivastava (1981). _Pesticides_ 15(11): 40-47.

64. Santos, A.C. (1930). _Philip. J. Sci._ 43:561-564.

65. Tattersfield, F. (1925). _Nature_ 116:243.

66. Parkin, E. A. (1944). _Nature_ 154:16-17.

67. Ripper, W. E. (1944). Nature 153:448-452.

68. Clark, E. P. (1933). Science 77:311-312.

69. Worsley, G., and R. R. Le (1937). Ann. Appl. Biol. 24:651-658.

70. Worsley, G., R. R. Le, and F. J. Nutman (1937). Ann. Appl. Biol. 24:698-702.

71. Blackith, R. E. (1953). Ann. Appl. Biol. 40:113-120.

72. Kubo, I., P. S. Tanis, Y. W. Lee, I. Miura, K. Nakanishi, and A. Chapya (1976). Heterocycles 5:485-498.

73. Chopra, R. N., R. L. Badhwar, and S. L. Nayar (1941). J. Bombay Natur. Hist. Soc. 42:856-902.

74. Matthews, D., A. D. Edwards, S. Ganser, and E. Weinsteiger (1980). Hort. and New Crops Depts., Organic Gardening and Farming Research Ctr., Rodale Press, Inc., Emmaus, PA.

75. Matthews, D. (1981). Hort. and New Crops Depts. Organic Gardening and Farming Research Ctr., Rodale Press, Inc., Emmaus, PA.

76. Miyazaki, S. (1982). Pesticide Research Center, Mich. State University. Personal comm.

77. Patterson, B. D., S. K. Wahba Khalil, L. J. Schermeister, and M. S. Quraishi (1975). Lloydia 38:391-403.

78. Khan, R. A. (1982). Pakistan Council Sci. Ind. Res., Karachi, Pakistan. Personal comm.

79. Tattersfield, F., and J. T. Martin (1938). Ann. Appl. Biol. 25:411-429.

80. Khan, Z. S. (1982). ESCAP, Bangkok, Thailand. Personal comm.

81. Poe, S. L. (1982). Dept. Ento., Virginia Polytechnical Institute, Blacksburg, VA. Personal comm.

82. Wai-Hong-Loke (1982). MARDI, Malaysia. Personal comm.

83. Omar, D. (1983). Dept. Plant Protection, Univ. Pertanian, Malaysia. Personal comm.

84. Yingchol, P. (1983). Dept. Agron., Kasetsart Univ., Thailand. Personal comm.

85. Saxena, R. C. (1983). Dept. Ento., IRRI, Philippines. Personal comm.

86. Gunasena, H. P. M. (1983). Dept. Crop Sci., Univ. Peradeniya, Sri Lanka. Personal comm.

87. Jacobson, M. (1975). _Insecticides from Plants - A Review of the Literature, 1953-1971._ USDA Agr. Hdbk # 461, ARS-USDA, Beltsville, MD.

88. Balasubramanian, M. (1982). Dept. Ento., Tamil Nadu Agr. Univ., India. Personal comm.

89. Hameed, S. F. (1982). Dept. Ento., Rajendra Agr. Univ., Pusa, India. Personal comm.

90. Morallo-Rejesus, B. (1982). Dept. Ento., Univ. Philippines, Los Banos. Personal comm.

91. Borle, M. N. (1965). Dept. of Ento., Punjabrao Krishi Vidyapeeth, Akda, Maharashtra, India. Unpublished Ph.D. thesis.

92. Nguyen, C. T. (1983). Plant Protection Bureau, Hanoi, Vietnam. Personal comm.

93. Krishnamurthy Rao, D. B. H. (1982). Andra Pradesh Agr. Univ., Hyderabad, India. Personal comm.

94. Snyder, J. K. (1983). Dept. Chem., Columbia Univ. Personal comm.

95. Satpathy, J. M. (1983). Dept. Ento., Orissa Agr. and Tech. Univ., India. Personal comm.

96. Feinstein, L. (1952). _USDA Yearbook_ 1952, pp. 222-229.

97. Roark, R. C. (1947). _Econ. Bot._ 1:437-445.

98. Hartzell, A., and F. Wilcoxon (1943). _Contr. Boyce Thompson Inst._ 12:127-141.

99. Hartzell, A. (1944). _Contr. Boyce Thompson Inst._ 13:243-252.

100. Mukerjee, S. K. (1983). Div. Agr. Res., Indian Agr. Res. Inst., New Delhi, India. Personal comm.

101. Jacobson, M. (1958). _Agr. Hdbk_ #154, ARS-USDA, Beltsville, MD.

102. Bisset, N. G. (1966). _Lloydia_ 29:1-18.

103. Bisset, N. G., and M. C. Woods (1966). _Lloydia_ 29:172-195.

104. McIndoo, N. E. (1924). _USDA Bull_ #1201.

105. McIndoo, N. E. (1945). Bureau of Entomology and Plant Quarantine Publication E-661 (USDA).

106. Gunter, F. A., and L. R. Jeppson (1960). Botanical Compounds in _Modern Insecticides and World Food Production_, Wiley and Sons, New York. 284pp.

107. Kubo, I. (1983). Div. Ento. and Parasitology, Univ. Calif., Berkeley. Personal comm.

108. Kubo, I., and J. A. Klocke (1982). Div. Ento. and Parasitology, Univ. of Calif., Berkeley. Personal comm.

109. Miyazaki, S. (1982). Pest Research Center, Mich. State Univ. Personal comm.

110. Elliot, M. (1983). Dept. Insecticides and Fungicides, Rothampstead Exp. Sta., England. Personal comm.

111. Wai-Hong-Loke (1983). MARDI, Malaysia. Personal comm.

112. Jilani, G. (1980). Univ. Agr., Faisalabad, Pakistan. Ph.D. thesis.

113. Lapis, D. B., and E. E. Dumancas (1978). _Philip. Phytopath._ 14: 23-37.

114. Lapis, D. B., and E. E. Dumancas (1979). _Philip. Phytopath._ 15: 23-34.

115. Le, Thi-Hoan H., and R. G. Davide (1979). _Philip. Agr._ 62(4): 285-295.

116. Hameed, S. F. (1983). Dept. Ento., Rajendra, Agr. Univ., India. Personal comm.

117. Nagarajah, T. M. (1983). MARGA Inst., Colombo, Sri Lanka. Personal comm.

118. Ines Frias D. (1983). INIREB, Xalapa, Mexico. Personal comm.

119. Abraham, C. C., and B. Ambika (1979). _Current Sci._ 48(12):554-556.

120. Khanvilkar, V. G. (1983). Dept. Ento., Konkan Krishi Vidyapeeth Univ., Dapoli, India. Personal comm.

121. Abraham, C. C., B. Thomas, K. Karunakaran, and R. Gopalakrishnan (1972). _Agr. Res. J. Kerala_ 10(1):59-60.

122. Abraham, C. C., and B. Ambika (1979). _Current Sci._ 48(12):554-556.

123. Patel, R. C. (1983). Gujarat Agr. Univ., India. Personal comm.

124. Balasubramanian, M. (1982). Dept. Ento., Tamil Nadu Agr. Univ., India. Personal comm.

125. Carino, F. A. (1983). Dept. Plant Pathology, Visayas State College, Baybay Leyte, Philippines. Personal comm.

126. Krishnamuthy Rao, D. B. H. (1983). Andhra Pradesh Agr. Univ. Personal comm.

127. Khan, A. A. (1983). Dept. Plant Pathology, Bangladesh Agr. Univ., Mymensingh. Personal comm.

128. Morgan, E. D. (1983). Dept. Chem., Univ. Keele, England. Personal comm.

129. Ladd, T. L. (1983). USDA Hort. Insects Res. Lab., Wooster, Ohio. Personal comm.

130. Redfern, R. E. (1983). Livestock Insects Lab., USDA, Beltsville, MD. Personal comm.

131. Ascher, K. R. S. (1981). Phytoparasitica 9(3):197-205.

132. Fagoonee, I. (1983). Univ. Mauritus, Reduit. Personal comm.

133. Wen, T. (1983). Dept. Plant Protection, Jiangsu Academy Sci., Nanjing, China. Personal comm.

134. Padolina, W. G. (1983). Dept. Chem., Univ. Philippines, Los Banos. Personal comm.

135. Subasinghe, T. B. (1983). Agrarian Research and Training Inst., Colombo, Sri Lanka. Personal comm.

136. Radwanski, S. A. (1983). Paris, France. Personal comm.

137. Litsinger, J. A., E. C. Price, and R. T. Herrera (1978). Inter. Rice Res. Newsletter (Oct.). 3(5):15-16.

138. Anonymous (1982). New Scientist (April), p. 19.

139. Bengwayan, M. A. (1982). Sugar News (Jan.), p. 35.

140. Mariappan, V., R. C. Saxena, and K. C. Ling (May, 1982). Paper presented Philip. Assoc. Ento. convention, Bagio, Philippines.

141. Epino, P. B. and R. C. Saxena (May, 1982). Paper presented, Philip. Assoc. of Ento. Convention, Bagio, Philippines.

142. Jacobson, M. (1983). USDA, Beltsville, MD. Personal comm.

143. Waespe, H. R. (1983). Ciba-Geigy, Basel, Switzerland. Personal comm.

144. Bowers, W. S. (1983). Dept. Ento., State Agr. Exp. Sta., Cornell Univ., New York. Personal comm.

145. Kuan C. (1983). Dept. Plant Pathology, Beijing, Agr. Univ., China. Personal comm.

146. Mukherjee, N. (1983). BCKVV Univ. Kalyani, West Bengal, India. Personal comm.

147. Dar, D. W., Mt. State Agr. College, Benquet, Philippines. Personal comm.

148. Choubey, S. D., and J. P. Tiwari. (1983). JNKVV Univ., Dept. Agronomy, Jabalpur, India. Personal comm.

149. Singh, G. (1983). Dept. Chem., Ghorakpur Univ., India. Personal comm.

150. Mitchell, W. (1983). Dept. Ento., Univ., Hawaii. Personal comm.

151. Butterworth, J. H., and E. D. Morgan (1971). J. Insect Physiol. 17:969-977.

152. Putman, A. R. (1983). Pest Res. Ctr., Mich. State Univ. Personal comm.

153. Topham, M., and J. W. Beardsley (1975). Proc. Hawaii Ento. Soc. 22(1):145-154.

154. Jiron, L. F. (1983). Ento. Dept., Universidad de Costa Rica. Personal comm.

155. Zaman, S. M. H. (1983). Bangladesh Rice Res. Inst. Personal comm.

156. Schermeister, L. J. (1983). Dept. Pharmacognosy, N. Dakota State Univ., Fargo. Personal comm.

157. Crooker, P. (1983). Rural Dev. and Small Business Trng. School, Nadi, Fiji. Personal comm.

158. Agustin, J. (1983). Mindanao State Univ., Philippines. Personal comm.

159. Rizvi, S. J. H., D. Mukerjii, and S. N. Mathur (1980). Ind. J. Exp. Biol. 18:777-778.

160. Murashige, T. (1982). Science (Washington) 3(1):74.

161. Islam, B. N. (1983). Dept. Ento. Bangladesh Agr. Univ. Mymensingh, India. Personal comm.

162. Chiu, S. (1982). Laboratory of Insect Toxicology, Dept. Plant Protection, South-China Agr. College, Guangzhou, China. Unpublished mss.

163. Campos, F. F. (1983). Central Luzon State Univ., Philippines. Personal comm.

164. Fagoonee, I. (1982). Univ. Mauritus, Reduit. Personal comm.

165. Schermeister, L. J., Dept. of Pharmacognosy, N. Dakota State Univ. Unpublished mss.

166. Herrera, L. M. (1983). Pampanga Agr. College, Philippines. Personal comm.

167. Elamin, E. T. M., M. A. Ahmed, and A. A. Mirghani (1987). Unpublished mss.

168. Warthen, J. D. (April, 1976). USDA Pub. ARM-NE-4, (April).

169. Nakanishi, K. (1977). Insect growth regulators from plants, in Natural Products and the Protection of Plants, (G. B. Marini-Bettolo, editor), Elsevier Scientific Pub. Co.,New York. 846 pp.

170. Marini-Bettolo, G.B. (1983). The role of natural products in plant-insect and plant-fungi interaction, in Natural Products for Innovative Pest Management, (D. L. Whitehead and W. S. Bowers, editors), Pergamon Press, New York. 586 pp.

171. Gebreyesus, T., and A. Chapya (1983). Antifeedants from "Clausenia anisata", in Natural Products for Innovative Pest Management, (D. L. Whitehead and W.S. Bowers editors), Pergamon Press, New York. 586 pp.

172. Bushnell, D. A., M. Fukuda, and T. Makinodan (1950). Pacific Sci. 4:167-183.

173. Rodriquez, E. (1981). Tech. Rev. (April), p. 90.

174. Dath, A. P., and S. Devadath (1979). Central Rice Res. Inst. Ann. Rept. 1979.

175. Anonymous (1982). New Scientist 94:19.

176. Hernandez, R. C., A. L. Tejeda, and R. D. Rivero (1982). Chapingo Ano 7(37):35-39.

177. Sutherst, R. W., R. J. Jones, and H. J. Schnitzerling (1982). Nature 295:320-321.

178. Lovett, J. V., and A. M. Duffield (1981). J. Appl. Ecol. 18: 280-290.

179. Jilani, G. (1984). Pak. Agr. Res. Council, Islamabad, Pakistan. Personal comm.

180. Blair, J. G. (1982). Science 82, (Jan.-Feb.), p. 82.

181. Misra, S. B., and S. N. Dixit (1979). Acta Bot. Ind. 7:147-150.

182. Misra, S. B., and S. N. Dixit (1979). Ind. J. Mycol. Plant Path. 9(2):250-251.

183. Zhang, X., and S. Chiu (1983). J. South China Agr. College 4(3): 1-7.

184. Simons, J. N., R. Swidler, and L. M. Moss (1963). _Phytopath._ 53:677-683.

185. Maruzzella, J. C., and J. Balter (1959). _Plant Dis. Rept._ 43(11):1143-1147.

186. Osborn, E. M. (1942). _Brit. J. Exp. Path._ 24(6):227-231.

187. Murakata, K. (1980). Paper presented at Inter. Ctr. Insect Phys. and Ecol. (ICIPE), Nairobi, Kenya, May 12-15.

188. Rahman, M. S., and G. N. Bhattacharya (1982). _Current Sci._ 51(8):434-435.

189. Agarwal, D. C., R. S. Deshpande, and H. P. Tipnis (1973). _Pesticides_ 7(4):21.

190. Sievers, A. F., W. A. Archer, R. H. Moore, and E. R. McGovran (1949). _J. Econ. Ento._ 42:549-551.

191. Rosenthal, G. A. (1983). _Scientific Amer._ (Nov., 1983) pp. 164-171.

192. Spies, J. R. (1933). _J. Econ. Ento._ 26:285-288.

193. Forbes, S. A. (1908). _J. Econ. Ento._ 1:81-83.

194. Taylor, E. P. (1908). _J. Econ. Ento._ 1:83-91.

195. Cory, E. N. (1921). _J. Econ. Ento._ 14:345-347.

196. Strickland, E. H. (1922). _J. Econ. Ento._ 15:214-220.

197. Parrott, P. J., and H. Glasgow (1923). _J. Econ. Ento._ 16:90-95.

198. Parrott, P. J., and G. F. MacLeod (1923). _J. Econ. Ento._ 16: 424-430.

199. Bishopp, F. C., F. C. Cook, D. C. Parman, and E. W. Laake (1923). _J. Econ. Ento._ 16:222-224.

200. De Ong, E. R. (1923). _J. Econ. Ento._ 16:486-493.

201. De Ong, E. R., and L.T.W. White (1924). _J. Econ. Ento._ 17:499-501.

202. Ballou, C. H. (1929). _J. Econ. Ento._ 22:289-293.

203. Davidson, W. M. (1929). _J. Econ. Ento._ 22:226-235.

204. Davidson, W. M. (1930). _J. Econ. Ento._ 23:877-879.

205. Headlee, T. J. (1930). _J. Econ. Ento._ 23:251-259.

206. Darley, M. M. (1931). _J. Econ. Ento._ 24:111-115.

207. De Ong, E. R. (1931). J. Econ. Ento. 24:736-743.

208. Little, V. A. (1931). J. Econ. Ento. 24:743-754.

209. Driggers, B. F. (1931). J. Econ. Ento. 24:319-325.

210. Simanton, F. L., F. F. Dicke, and G. T. Bottger (1931). J. Econ. Ento. 24:395-404.

211. Drake, N. L., and T. R. Spies (1932). J. Econ. Ento. 25:129-133.

212. Carter, W. (1932). J. Econ. Ento. 25:1031-1035.

213. Ginsburg, J. M. (1932). J. Econ. Ento. 25:918-922.

214. Haegele, R. W. (1932). J. Econ. Ento. 25:1073-1077.

215. Richardson, H. H. (1932). J. Econ. Ento. 25:592-607.

216. Turner, N. (1932). J. Econ. Ento. 25:1228-1237.

217. Campbell, F. L., and W. N. Sullivan (1933). J. Econ. Ento. 26:500-509.

218. Metzger, F. W. (1933). J. Econ. Ento. 26:299-300.

219. Vansell, G. H. (1933). J. Econ. Ento. 26:168-170.

220. Chopra, R. N., and R. L. Badhwar (1940). Poisonous Plants of India, School of Tropical Medicine, Calcutta.

221. Hamilton, C. C., and L. G. Gemmell (1934). J. Econ. Ento. 27:446-453.

222. Huckett, H. C. (1934). J. Econ. Ento. 27:440-445.

223. Smith R. H., H. U. Meyer, and C. O. Persing (1934). J. Econ. Ento. 27:1192-1195.

224. Walker, H. G. (1934). J. Econ. Ento. 27:388-392.

225. Osburn, M. R. (1934). J. Econ. Ento. 27:293.

226. Headlee, T. J. (1935). J. Econ. Ento. 28:605-607.

227. Howard, N. F., L. W. Brannon, and H. C. Mason (1935). J. Econ. Ento. 28:444-448.

228. Thomas, C. A. (1936). J. Econ. Ento. 29:313-317.

229. Dudley, J. E., T. E. Bronson, and F. E. Carroll (1936). J. Econ. Ento. 29:501-508.

230. Bronson, T. E. (1936). J. Econ. Ento. 29:1170-1172.

231. Richardson, H. C., L. C. Craig, and T. R. Hansberry (1936). J. Econ. Ento. 29:850-855.

232. Huckett, H. C. (1936). J. Econ. Ento. 29:575-580.

233. LePelley, R. H., and W. N. Sullivan (1935). J. Econ. Ento. 29: 791-797.

234. Wisecup, C. B. (1936). J. Econ. Ento. 29:1000-1003.

235. Ritcher, P. O., and R. K. Calfee (1937). J. Econ. Ento. 30: 166-174.

236. De Ong, E. R. (1937). J. Econ. Ento. 30:921-927.

237. Huckett, H. C. (1937). J. Econ. Ento. 30:323-328.

238. Jones, H. A., and W. N. Sullivan (1937). J. Econ. Ento. 30: 679-680.

239. Fulton, R. A., and N. F. Howard (1938). J. Econ. Ento. 31(3): 405-410.

240. Wisecup, C. B., and L. B. Reed (1938). J. Econ. Ento. 31(6) 690-695.

241. Pepper, B. B., and C. M. Haenseler (1939). J. Econ. Ento. 32: 291-296.

242. Driggers, B. F., and W. J. O'Neill (1939). J. Econ. Ento. 32: 286-290.

243. Hough, W. S. (1938). J. Econ. Ento. 31:216-221.

244. Bushland, R. C. (1939). J. Econ. Ento. 32:430-431.

245. Brooks, J. W., and T. C. Allen (1940). J. Econ. Ento. 33:416-417.

246. Fisher, R. A. (1940). J. Econ. Ento. 33:728-734.

247. Bronson, T. E., and J. E. Dudley, Jr. (1940). J. Econ. Ento. 33:736-738.

248. Goodhue, L. D., and W. N. Sullivan (1940). J. Econ. Ento. 33: 329-332.

249. Coon, B. F., and C. Wakeland (1940). J. Econ. Ento. 33:389-393.

250. Gnadinger, C. B., J. B. Moore, and R. W. Coultier (1940). J. Econ. Ento. 33:143-153.

251. Brindley, T. A., F. G. Hinman, and R. A. Fisher (1940). J. Econ. Ento. 33:881-886.

252. Smith, F. F., and W. N. Sullivan (1940). J. Econ. Ento. 33: 807-810.

253. Yothers, M. A. (1940). J. Econ. Ento. 33:800-803.

254. Haller, H. L. (1940). J. Econ. Ento. 33:941.

255. Brooks, J. W., and T. C. Allen (1941). J. Econ. Ento. 34:295-297.

256. Roark, R. C. (1941). J. Econ. Ento. 34:684-692.

257. Norton, L. B., and O. B. Billings (1941). J. Econ. Ento. 34: 630-635.

258. Huckett, H. C. (1941). J. Econ. Ento. 34:566-571.

259. Pyenson, L. (1941). J. Econ. Ento. 34:473-474.

260. Michelbacher, A. E., G. F. MacLeod, and R. F. Smith (1941). J. Econ. Ento. 34:709-716.

261. Goodhue, L. D., and W. N. Sullivan (1941). J. Econ. Ento. 34: 77-78.

262. Arant, F. S. (1942). J. Econ. Ento. 35:873-878.

263. Ginsburg, J. M., Schmitt, and T. S. Reid (1942). J. Econ. Ento. 35:276-280.

264. Little, V. A. (1942). J. Econ. Ento. 35:54-57.

265. Yothers, M. A., F. W. Carlson, and C. C. Cassil (1942). J. Econ. Ento. 35:450-452.

266. Haller, H. L., F. B. LaForge, and W. N. Sullivan (1942). J. Econ. Ento. 35:247-248.

267. Richardson, H. H., and A. H. Casanges (1942). J. Econ. Ento. 35:242.

268. Sorenson, C. J. (1942). J. Econ. Ento. 35:884-886.

269. Seiferle, E. J., I. B. Johns, and C. H. Richardson (1942). J. Econ. Ento. 35:35-45.

270. Chiu, S. F., S. Lin, and Y. S. Chiu (1942). J. Econ. Ento. 35: 80-82.

271. Turner, N. (1943). J. Econ. Ento. 36:266-272.

272. Hansberry, R., and C. Lee (1943). J. Econ. Ento. 36:351-352.

273. Smith, F. F., and L. D. Goodhue (1943). J. Econ. Ento. 36:911-914.

274. Hartzell, A. (1943). J. Econ. Ento. 36:320-325.

275. Haller, H. L., and N. E. McIndoo (1943). J. Econ. Ento. 36(4):638.

276. Pepper, B. B., and R. S. Filmer (1944). J. Econ. Ento. 37:248-252.

277. McGregor, E. A. (1944). J. Econ. Ento. 37:78-80.

278. Plank, H. K. (1944). J. Econ. Ento. 37:737-739.

279. Bowen, C. V. (1944). J. Econ. Ento. 37:293.

280. Allen, T. C., R. J. Dicke, and H. H. Harris (1944). J. Econ. Ento. 37:400.

281. Sweetman, H. L., and G. G. Gyrisko (1944). J. Econ. Ento. 37: 746-749.

282. Brunn, L. K., and T. C. Allen (1945). J. Econ. Ento. 38:392.

283. Pepper, B. P., and L. A. Carruth (1945). J. Econ. Ento. 38:59-60.

284. Hansberry, R., and R. T. Clausen (1945). J. Econ. Ento. 38: 305-307.

285. Dicke, R. J., F. J. Dexheimer, and T. C. Allen (1945). J. Econ. Ento. 38:389.

286. Goodhue, L. D., and H. L. Haller (1939). J. Econ. Ento. 32: 877-879.

287. Allen, T. C., F. J. Dexheimer, and E. Cole (1945). J. Econ. Ento. 38:389-390.

288. Tate, H. D., and D. B. Gates (1945). J. Econ. Ento. 38:391.

289. Cheng, T. H. (1945). J. Econ. Ento. 38:491-492.

290. Metcalf, R. L., and C. E. Wilson (1945). J. Econ. Ento. 38:499.

291. Allen, T. C., K. P. Link, M. Ikawa, and L. K. Brunn (1945). J. Econ. Ento. 38:293-296.

292. Fisher, E. H., and W. W. Stanley (1945). J. Econ. Ento. 38: 125-126.

293. Frazler, N. W. (1945). J. Econ. Ento. 38:720.

294. Anderson, R. F. (1945). J. Econ. Ento. 38:564-566.

295. Lindgren, D. L., J. P. LaDue, and R. C. Dickson (1945). J. Econ. Ento. 38:567-572.

296. Lathrop, F. H., and L. G. Keirstead (1946). J. Econ. Ento. 39:534.

297. Siegler, E. H., and C. V. Bowen (1946). J. Econ. Ento. 39:673-674.

298. Jones, M. A., W. A. Gersdorff, and E. R. McGovran (1946). J. Econ. Ento. 39:281-283.

299. Brett, C. H. (1945). J. Econ. Ento. 39:810.

300. Scholl, J. M., and J. T. Medler (1947). J. Econ. Ento. 40:446-448.

301. Freeborn, S. B., and F. H. Wymore (1929). Ecoc. Bot. 22:666-671

302. Walton, R. R. (1947). J. Econ. Ento. 40:389-395.

303. Hamilton, D. W. (1947). J. Econ. Ento. 40:234-236.

304. Sievers, A. F., W. A. Archer, R. H. Moore, and E. R. McGovran (1949). J. Econ. Ento. 42:549-551.

305. Berry, R. C. (1950). J. Econ. Ento. 43:112.

306. Brett, C. H., and R. W. Brubaker (1955). J. Econ. Ento. 48:343.

307. Pagan, C., and M. P. Morris (1953). J. Econ. Ento. 46:1092-1093.

308. Huddle, H. B., and A. P. Mills (1952). J. Econ. Ento. 45:40-42.

309. Beroza, M., and G. T. Bottger (1954). J. Econ. Ento. 47:188-189.

310. Allen, T. C., and L. K. Brunn (1945). J. Econ. Ento. 38:291-293.

311. Dicke, R. J., F. J. Dexheimer, and T. C. Allen, Jr. (1945). J. Econ. Ento. 38:389.

312. Van Kammen, A., D. Noordam, and T. H. Thung (1961). Virology 14:100-108.

313. Fleming, W. E., and R. D. Chisholm (1944). J. Econ. Ento. 37:116.

314. Van Leeuwen, E. R. (1948). J. Econ. Ento. 41:345-351.

315. Bowen, C. V., and C. A. Weigel (1948). J. Econ. Ento. 41:117.

316. Hansberry, R. (1940). J. Econ. Ento. 33:734-735.

317. Haley, D. E., O. Olson, and F. L. Follweiler (1925). J. Econ. Ento. 18:807-817.

318. Smith, R. H. (1961). J. Econ. Ento. 54:365-369.

319. Hardee, D. D., and T. B. Davich (1966). J. Econ. Ento. 59:1267-1270.

320. Smith, C. R. (1931). J. Econ. Ento. 24:1108.

321. Maxwell, F. G., W. L. Parrott, J. N. Jenkins, and H. N. Lafever (1965). J. Econ. Ento. 58:985-988.

322. Beroza, M., and G. C. LaBrecque (1967). J. Econ. Ento. 60:196-199.

323. Branson, T. F., P. L. Guss, and E. E. Ortman (1969). J. Econ. Ento. 62:1375-1378.

324. McMillian, W. W., M. C. Bowman, R. L. Burton, K. J. Starks, and B. R. Wiseman (1969). J. Econ. Ento. 62:708-710.

325. Amonkar, S. V., and E. L. Reeves (1970). J. Econ. Ento. 63: 1172-1175.

326. Su, H. C. F. (1977). J. Econ. Ento. 70:18-21.

327. Thurston, R. (1970). J. Econ. Ento. 63:272-274.

328. Parr, J. C., and R. Thurston (1968). J. Econ. Ento. 61:1525-1531.

329. Kawano, Y., W. C. Mitchell, and H. Matsumoto (1968). J. Econ. Ento. 61:986-988.

330. Su, H. C. F., R. D. Speirs, and P. G. Mahany (1972). J. Econ. Ento. 65:1438-1444.

331. Su, H. C. F., R. D. Speirs, and P. G. Mahany (1972). J. Econ. Ento. 65:1433-1436.

332. Ladd, T. L., M. Jacobson, and C. R. Buriff (1978). J. Econ. Ento. 71:810-813.

333. Scott, W. P., and G. H. McKibben (1978). J. Econ. Ento. 71: 343-344.

334. Schoonhoven, A. V. (1978). J. Econ. Ento. 71:254-256.

335. Fletcher, B. S., M. A. Bateman, N. K. Hart, and J. A. Lamberton (1975). J. Econ. Ento. 68:815-816.

336. Shepard, H. H. (1931). J. Econ. Ento. 24:725-731.

337. Iwuala, M. O. E., I. U. W. Osisiogu, and E. O. P. Agbakwuru (1981). J. Econ. Ento. 74:249-252.

338. Khanna, S. G. S., Y. L. Nene, C. K. Banerjee, and P. N. Thapliyal (1967). Ind. Phytopath. 20:64-69.

339. Ragetli, H. W. J., and M. Weintraub (1962). Virology 18:241-248.

340. Doskotch, R. W., T. M. Odell, and P. A. Godwin (1977). Envir. Ento. 6:563-566.

341. Cantelo, W. W., and M. Jacobson (1979). Envir. Ento. 8:444-447.

342. Kareem, A. A., S. Sadakathulla, M. S. Venugopal, and T. R. Subramanium (1974). Phytoparasitica 2(2):127-129.

343. Joshi, B. G., and G. Ramaprasad (1975). Phytoparasitica 3(1):
     59-61.

344. Hassid, E., S. W. Applebaum, and Y. Birk (1976). Phytoparasitica
     4(3):173-183.

345. Meisner, J., M. Wysoki, and K. R. S. Ascher (1976). Phytoparasitica
     4(3):185-192.

346. Meisner, J., M. Kehat, M. Zur, and C. Eizick (1978).
     Phytoparasitica 6(2):85-88.

347. Joshi, B. G., and S. Sitaramaiah (1979). Phytoparasitica 7(3):
     199-202.

348. Meisner, J., K. R. S. Ascher, R. Aly, and J. D. Warthen, Jr. (1981).
     Phytoparasitica 9: 27-32.

349. Ascher, K. R. S., H. Schmutterer, E. Glotter, and I. Kirson (1981).
     Phytoparasitica 9: 197-205.

350. Fagoonee, I., and G. Lauge (1981). Phytoparasitica 9:111-118.

351. Davide, R. G. (1979). Phil. Phytopath. 15(2):141-144.

352. Barnes, G. L. (1963). Plant Dis. Rept. 47:114-117.

353. Davey, C. B., and G. C. Papavizas (1960). Phytopath. 50:516-522.

354. Johnson, L. F. (1962). Phytopath. 52(5):410-413.

355. Wilhelm, S. (1951). Phytopath. 41:684-690.

356. Patrick, Z. A., R. M. Sayre, and H. J. Thorpe (1965). Phytopath.
     55:702-704.

357. Zentmyer, G. A. (1963). Phytopath. 53:1385-1387.

358. Sayre, R. M., Z. A. Patrick, and H. Thorpe (1964). Phytopath.
     54:905.

359. Oswald, J. W., and O. A. Lorenz (1956). Phytopath. 46:22.

360. Adams, P. B., J. A. Lewis, and G. C. Papavizas (1968). Phytopath.
     58:1603-1608.

361. Huber, D. M., and R. D. Watson (1970). Phytopath. 60:22-26.

362. Adams, P. B., G. C. Papavizas, and J. A. Lewis (1968). Phytopath.
     58:373-377.

363. Bell, D. T., and D. E. Koeppe (1972). Agron. J. 64:321-325.

364. Bieber, G. L., and C. S. Hoveland (1968). Agron. J. 60:185-188.

365. Babu, T. H., and Y. P. Beri (1969). <u>Andhra</u> <u>Agr.</u> <u>J.</u> 16:107-111.

366. Weinhold, A. R., J. W. Oswald, T. Bowman, J. Bishop, and D. Wright (1964). <u>Amer.</u> <u>Pot.</u> <u>J.</u> 41:265-273.

367. Skaptason, J. B. (1938). <u>Amer.</u> <u>Pot.</u> <u>J.</u> 15:271-277.

368. Mukerjee, K., L. P. Awasthi, and H. N. Verma (1981). <u>Z.</u> <u>Pflanzenkrankheiten</u> 88: 228-234.

369. Kubo, I., J. A. Klocke, and S. Asano (1981). <u>Agr.</u> <u>Biol.</u> <u>Chem.</u> 45:1925-1927.

370. Nakajima, S., and K. Kawazu (1980). <u>Agr.</u> <u>Biol.</u> <u>Chem.</u> 44:2893-2899.

371. Bhan, P., R. Soman, and S. Dev (1980). <u>Agr.</u> <u>Biol.</u> <u>Chem.</u> 44: 1483-1487.

372. Nakajima, S., and K. Kawazu (1980). <u>Agr.</u> <u>Biol.</u> <u>Chem.</u> 44:1529-1533.

373. Nakajima, S., and K. Kawazu (1977). <u>Agr.</u> <u>Biol.</u> <u>Chem.</u> 41:1801-1802.

374. Kawazu, K., M. Ariwa, and Y. Ku (1977). <u>Agr.</u> <u>Biol.</u> <u>Chem.</u> 41: 223-224.

375. Tomar, S. S., M. L. Maheshwari, and S. K. Mukerjee (1979). <u>Agr.</u> <u>Biol.</u> <u>Chem.</u> 43:1479-1483.

376. Kogiso, S., K. Wada, and K. Munakata (1976). <u>Agr.</u> <u>Biol.</u> <u>Chem.</u> 40:2119-2120.

377. Kogiso, S., K. Wada, and K. Munakata (1976). <u>Agr.</u> <u>Biol.</u> <u>Chem.</u> 40:2085-2089.

378. Hosozawa, S., N. Kato, K. Munakata, and Y. L. Chen (1974). <u>Agr.</u> <u>Biol.</u> <u>Chem.</u> 38:1045-1048.

379. Kato, N., M. Takahashi, M. Shibayama, and K. Munakata (1972). <u>Agr.</u> <u>Biol.</u> <u>Chem.</u> 36:2579-2582.

380. Isogai, A., S. Murakoshi, A. Suzuki, and S. Tamura (1973). <u>Agr.</u> <u>Biol.</u> <u>Chem.</u> 37:889-895.

381. Wada, K., and K. Munakata (1967). <u>Agr.</u> <u>Biol.</u> <u>Chem.</u> 31:336-339.

382. Wada, K., K. Matsui, Y. Enomoto, O. Ogiso, and K. Munakata (1970). <u>Agr.</u> <u>Biol.</u> <u>Chem.</u> 34:941-945.

383. Singh, D. B., S. P. Singh, and R. C. Gupta (1979). <u>Trans</u> <u>Brit.</u> <u>Mycol.</u> <u>Soc.</u> 73:349-350.

384. Miller, P. M., and J. F. Ahrens (1969). <u>Bull.</u> <u>Con.</u> <u>Exp.</u> <u>Sta.</u> #701, New Haven, CT. 10 pp.

385. Agrawal, P. (1978). <u>Trans.</u> <u>Brit.</u> <u>Mycol.</u> <u>Soc.</u> 70:439-441.

386. Tansey, M. R. (1975). _Mycologia_ 67:409-413.

387. Lichtenstein, E. P., F. M. Strong, and D. G. Morgan (1962). J. Agr. Food Chem. 10:30-33.

388. Lichtenstein, E. P., and J. E. Casida (1963). J. Agr. Food Chem. 11:410-415.

389. Nalbandov, O., R. T. Yamamoto, and G. S. Fraenkel (1964). J. Agr. Food Chem. 12: 55-59.

390. Lichtenstein, E. P., D. G. Morgan, and Mueller (1964). J. Agr. Food Chem. 12:158-161.

391. Wada, K., and K. Munakata (1968). J. Agr. Food Chem. 16:471-474.

392. Jurd, L., and G. D. Manners (1980). J. Agr. Food Chem. 28:183-188.

393. Su, H. C. F., and R. Horvat (1981). J. Agr. Food Chem. 29:115-118.

394. Jacobson, M., M. M. Crystal, and D. J. Warthen (1981). J. Agr. Food Chem. 29:591-593.

395. Singh, G., and R. M. Pandey (1982). J. Agr. Food Chem. 30(3): 604-608.

396. Blackmon, G. H. (1947). Econ. Bot. 1:161-175.

397. Bradley, C. E. (1956). Econ. Bot. 10:362-366.

398. Tinker, R. B., and W. M. Lauter (1956). Econ. Bot. 10:254-257.

399. Boyle, W. (1956). Econ. Bot. 10:257.

400. Rao, D. S. (1957). Econ. Bot. 11:274-276.

401. Pennington, C.W. (1958). Econ. Bot. 12:95-102.

402. McCleary, J. A., P. S. Sypherd, and D. L. Walkington (1960). Econ. Bot. 14:47-249.

403. Fly, L. B., and I. Kiem (1963). Econ. Bot. 17:46-49.

404. Barnes, D. K., and R. H. Freyre (1966). Econ. Bot. 20:279-284.

405. Barnes, D. K., and R. H. Freyre (1966). Econ. Bot. 20:368-371.

406. Morton, J. F. (1967). Econ. Bot. 21:57-68.

407. Worthley, E. G., C. D. Schott, and G. A. Hauptmann (1967). Econ. Bot. 21:238-242.

408. Conklin, H. C. (1967). Econ. Bot. 21:243-272.

409. Margetts, E. L. (1967). Econ. Bot. 21:358-362.

410. Gupta, S. K., and A. B. Banerjee (1972). Econ. Bot. 26:255-259.

411. Ray, P. G., and Majumdar (1976). Econ. Bot. 30:317-320.

412. Dixit, S. N., S. C. Tripathi, and R. R. Upadhyay (1976). Econ. Bot. 30:371-74.

413. Roia, F. C., and R. A. Smith (1977). Econ. Bot. 31:28-37.

414. Okada, K. A., B. J. Carrillo, and M. Tilley (1977). Econ. Bot. 31:225-236.

415. Bisset, N. G., and P. J. Hylands (1977). Econ. Bot. 31:307-311.

416. Cox, P. A. (1979). Econ. Bot. 33:397-399.

417. Singh, A. K., A. Dikshit, M. L. Sharma, and S. N. Dixit (1980). Econ. Bot. 34:186-190.

418. Misra, S. B., and S. N. Dixit (1980). Econ. Bot. 34:362-367.

419. Banerjee, R. D., and S. P. Sens (1980). Econ. Bot. 34:284-290.

420. Lewis, W. H., and M. P. F. Elvin-Lewis (1983). Econ. Bot. 37:69-70.

421. Lear, B. (1959). Plant Dis. Rept. 43:459-460.

422. Maier, C. R. (1959). Plant Dis. Rept. 43:1027-1030.

423. Mankau, R., and R. J. Minteer (1962). Plant Dis. Rept. 46:375-378.

424. Johnson, L. F. (1959). Plant Dis. Rept. 43:1059-1062.

425. Ark, P. A., and J. P. Thompson (1959). Plant Dis. Rept. 43:276-282.

426. Maruzzella, J. C., and J. Balter (1959). Plant Dis. Rept. 43: 1143-1147.

427. Gupta, K. C., and R. Viswanathan (1955). Antibiotics Chemother. 5:22-23.

428. Maruzzella, J. C., D. A. Scrandis, J. B. Scrandis, and G. Grabon (1960). Plant Dis. Rept. 44:789-792.

429. Chopra, I. C., K. S. Jamwal, and B. N. Khajuria (1954). Ind. J. Med. Res. 42:381-384.

430. Payawal P. C. (1985). Personal comm.

431. Taylor, D. A. H. (1953). Brit. J. Phar. 8:237.

432. Dumancas, E. E. (March, 1976). Univ. Philippines-Los Banos. M.S. thesis.

433. Joshi, B. G., and R. S. N. Rao (1968). Ind. Farming 18(7):33.

434. Pradhan, S., M. G. Jotwani, and B. K. Rai (1962). Ind. Farming 12(8):7-8.

435. Rajendran, B., and M. Gopalan (1979). Ind. J. Agr. Sci. 49:295-297.

436. Deb-Kirtaniya, S., M. R. Ghosh, N. Adityachaudhury, and A. Chatterjee (1980). Ind. J. Agr. Sci. 50:507-509.

437. Deb-Kirtaniya, S., M. R. Ghosh, S. R. Mitra, N. Adityachaudhury, and A. Chatterjee (1980). Ind. J. Agr. Sci. 50:510-512.

438. Sudhakar, T. R., N. D. Pandey, and G. C. Tewari (1978). Ind. J. Agr. Sci. 48:16-18.

439. Biswas, P., A. Bhattacharyya, A. Bose, N. Mukherjee, and N. Adityachaudhury (1981). Experientia 37:397-398.

440. Joshi, B. G., G. Ramaprasad, and S. V. V. Satyanarayana (1978). Ind. J. Agr. Sci. 48:19-22.

441. Rajendran, B., and M. Gopalan (1978). Ind. J. Agr. Sci. 48:306-308.

442. Akeson, W. R., F. A. Haskins, H. J. Gorz, and Manglitz (1968). Crop Sci. 8:574-576.

443. Johnstone, D. B., M. W. Foote, W. I. Rogers, and J. E. Little (1953). Antibiotics Chemother. 3:203-207.

444. Little, J. E., M. W. Foote, W. I. Rogers, and D. B. Johnstone (1953). Antibiotics Chemother. 3:183-191.

445. Marcus, S., & D. W. Esplin (1953). Antiobiotics Chemother. 3:393-98.

446. Weller, L. E., C. T. Redemann, R. Y. Gottshall, J. M. Roberts, E. H. Lucas, and H. M. Sell (1953). Antibiotics Chemother. 3:603-606.

447. Featherly, H. I., and K. S. Harmon (1944). Amer. Vet. Med. Assoc. J. 105:291-292.

448. Gupta, K. C., and R. Viswanathan (1955). Antibiotics Chemother. 5:18-21.

449. Gupta, K. C., and R. Viswanathan (1956). Antibiotics Chemother. 6:194-195.

450. Chopra, I. C., K. S. Jamwal, and B. N. Khajuria (1954). Ind. J. Med. Res. 42:385-388.

451. Kurup, P. A., and N. P. L. Rao (1954). Ind. J. Med. Res. 42:85-95.

452. Gopalakrishna, K. S., P. A. Kurup, and N. P. C. Rao (1954). Ind. J. Med. Res. 42:97-100.

453. Robinson, J. A., and H. W. Ling (1953). Brit. J. Phar. 8:79-82.

454. Girish, G. K., and S. K. Jain (1974). <u>Bull. Grain Tech.</u> 12(3): 226-228.

455. Kashyap, N. P., V. K. Gupta, and A. N. Kaushal (1974). <u>Bull. Grain Tech.</u> 12(1):41-44.

456. Biswal, L. D., and G. Rout (1973). <u>Bull. Grain Tech.</u> 11(3-4): 211-213.

457. Tiwari, G. C., and T. P. S. Teotia (1971). <u>Bull. Grain Tech.</u> 9(1):7-12.

458. Saramma, P. U., and A. N. Verma (1971). <u>Bull. Grain Tech.</u> 9(3): 207-210.

459. Lamba, S. S. (1970). <u>Econ. Bot.</u> 24:134-136.

460. Pandey, G. P., and B. K. Varma (1977). <u>Bull. Grain Tech.</u> 15: 100-104.

461. Singh, K. N., and P.K. Srivastava (1980). <u>Bull. Grain Tech.</u> 18:127.

462. Tikku, K., O. Koul, and B. P. Saxena (1978). <u>Bull. Grain Tech.</u> 16:3-9.

463. Hartzell, A. (1947). <u>Cont. Boyce Thom. Inst.</u> 15:21-34.

464. Hartzell, A., and F. Wilcoxon (1932). <u>Cont. Boyce Thom. Inst.</u> 4:107-117.

465. Synerholm, M. E., A. Hartzell, and V. Cullmann (1947). <u>Cont. Boyce Thom. Inst.</u> 15:35-45.

466. Synerholm, M. E., A. Hartzell, and J. M. Arthur (1945). <u>Cont. Boyce Thom. Inst.</u> 13:433-442.

467. Wilcoxon, F., and A. Hartzell (1939). <u>Cont. Boyce Thom. Inst.</u> 11: 1-4.

468. Chadha, S. S. (1977). <u>East Africa Agr. For. J.</u> 42(3):257-262.

469. Jacobson, M., D. K. Reed, M. M. Crystal, D. S. Moreno, and E. L. Soderstrom (1978). <u>Ento. Exp. Appl.</u> 24:448-457.

470. Cherrett, J. M. (1969). <u>Trop. Agr.</u> 46:81-90.

471. Prabhu, V. K. K., and M. John (1975). <u>Ento. Exp. Appl.</u> 18:87-95.

472. Smith, J. B. (1910). <u>Ento. Newsletter</u> 21:437-441.

473. Metcalf, R. L. (1979). <u>Ento. Soc. Amer. Bull.</u> 25:30-35.

474. Latheef, M. A., and R. D. Irwin (1980). <u>Envir. Ento.</u> 9:195-198.

475. Rout, G., and B. Senapati (1967). <u>Farm. J.</u> 8(4):12-13.

476. Deshpande, R. S., P. R. Adhikary, and H. P. Tipnis (1974). _Bull. Grain Tech_. 12:232.

477. Ebeling, W., F. A. Gunther, J. P. Ladue, and J. J. Ortega (1944). _Hilgardia_ 15: 675-701.

478. Morton, J. F. (1962). _FL State Hort. Soc_. 75:400-407.

479. Ochse, J. J., and W. S. Brewton (1954). _FL State Hort. Soc_. 67: 218-219.

480. Shekhawat, P. S., and R. Prasada (1971). _Ind. Phyto_. 24:800-802.

481. McIndoo, N. E. (1917). _J. Agr. Res_. 10:497-531.

482. Clayton, E. E., T. E. Smith, K. J. Shaw, J. G. Gaines, T. W. Graham, and C. C. Yeager (1943). _J. Agr. Res_. 66:261-276.

483. McIndoo, N. E. (1916). _J. Agr. Res_. 7:89-122.

484. Brett, C. H. (1946). _J. Agr. Res_. 73:81-96.

485. McIndoo, N. E., A. F. Sievers, and W. S. Abbott (1917). _J. Agr. Res_. 17:177-200.

486. Harries, F. H., J. D. Decoursey, and R. N. Hofmaster (1945). _J. Agr. Res_. 71:553-565.

487. Hansberry, R., R. T. Clausen, and L. B. Norton (1947). _J. Agr. Res_. 74:55-64.

488. Fulton, R. A., and H. C. Mason (1937). _J. Agr. Res_. 55:903-907.

489. Ginsburg, J. M., J. B. Schmitt, and P. Granett (1935). _J. Agr. Res_. 51:349-354.

490. Richardson, H. H. (1934). _J. Agr. Res_. 49:359-373.

491. Walker, J. C. (1923). _J. Agr. Res_. 24:1019-1039.

492. Martin, J. T., and F. Tattersfield (1931). _J. Agr. Sci_. 21:115-135.

493. Martin, H., and E. S. Salmon (1931). _J. Agr. Sci_. 21:638-658.

494. Tattersfield, F., and J. T. Martin (1934). _J. Agr. Sci_. 24:598-626.

495. Tattersfield, F. (1932). _J. Agr. Sci_. 22:396-417.

496. Martin, H. and E. S. Salmon (1933). _J. Agr. Sci_. 23:228-251.

497. Huddleson, I. F., J. Dufrain, K. C. Barrons, and M. Giefel (1944). _Amer. Vet. Med. Assoc. J._ 105:394-397.

498. Ryan, J., M. F. Ryan and F. McNaeidhe (1980). _J. Appl. Ecol_. 17: 31-40.

499. Webb, L. J., J. G. Tracey, and K. P. Haydock (1967). _J. Appl. Ecol._ 4:13-25.

500. Radwanski, S. A. (1969). _J. Appl. Ecol._ 6:507-511.

501. Pitts, O. M., H. S. Thompson, and J. H. Hoch (1969). _J. Phar. Sci._ 58:379-380.

502. Srivastava, B. K. (1956). _Madras Agr. J._ 43:518-519.

503. Subramanian, T. R. (1958). _Madras Agr. J._ 45:122-123.

504. Usher, G. (1973). _Dictionary of Plants Used by Man_, Hasner Press, NY.

505. Patel, H. K., V. C. Patel, M. S. Chari, J. C. Patel, and J. R. Patel (1968). _Madras Agr. J._ 55:509-510.

506. Graf, A. B. (1978). _Tropica._, Roehrs Co., East Rutherford, NJ.

507. Bailey, L. H., and E. Z. Bailey (1941). _Hortus Third_, MacMillan Co., NY.

508. Bailey, L. H., and E. Z. Bailey (1941). _Hortus Second_, MacMillan Co., NY.

509. Henderson, M. R. (1934). _Malay Agr. J._ 22:125-130.

510. Georgi, C. D. V., J. L. Greig, and G. L. Teik (1936). _Malay Agr. J._ 24:268-281.

511. Grist, D. H. (1935). _Malay Agr. J._ 23:477-482.

512. Milsum, J. N., and C. D. V. Georgi (1937). _Malay Agr. J._ 25:239-245.

513. Georgi, C. D. V. (1937). _Malay Agr. J._ 25:300-301.

514. Georgi, C. D. V. (1937). _Malay Agr. J._ 25:334-337.

515. Milsum, J. N. (1938). _Malay Agr. J._ 26:18-19.

516. Sangappa, H. K. (1977). _Mysore J. Agr. Sci._ 11:391-397.

517. Qadri, S. S. H. (1973). _Pesticides_ 7(12):18-19.

518. Qadri, S. S. H., and S. K. Majumder (1968). _Pesticides_ 2(1):25-30.

519. Deshpande, R. S., and H. P. Tipnis (1977). _Pesticides_ 11(5):11-12.

520. Qadri, S. S. H., and B. B. Rao (1977). _Pesticides_ 11(12):21-23.

521. Pillai, S. N., and M. V. Desai (1975). _Pesticides_ 9(4):37-39.

522. Mukherjee, N. (1974). _Pesticides_ 8(11):38-39.

523. Shekhawat, P.S., and R. Prasada (1971). *Pesticides* 5(8):27-28.

524. Viado, G. B., A. F. Banaag, and R. A. Luis (1957). *Phil. Agr.* 41:402-411.

525. Castillo, M. B., M. S. Alejar, and J. A. Litsinger (1977). *Phil. Agr.* 60:285-292.

526. Castillo, N. (1926). *Phil. Agr.* 15:257-275.

527. de Peralta, F., and R. P. Estioko (1923). *Phil. Agr.* 11:205-216.

528. Villadolid, D. V., and M. D. Sulit (1932). *Phil. Agr.* 21:25-35.

529. Ruelo, J. S., and R. G. Davide (1979). *Phil. Agr.* 62:159-165.

530. Timonin, M. I., and R. H. Thexton (1950). *Soil Sci. Soc. Amer.* 15:186-189.

531. Jermy, T. (1966). *Ento. Exp. Appl.* 9:1-12.

532. Plank, H. K. (1950). *Trop. Agr.* 27:38-41.

533. Giles, P. H. (1964). *Trop. Agr.* 41:197-212.

534. Bendall, G. M. (1975). *Weed Res.* 15:77-81.

535. Von Langenbuch, R. *Z. Pflanzenkrankheiten* 59:179-189.

536. Fischer, G. (1952). *Acta Path. Micro. Immuno. Scand.* 31:433-447.

537. Fischer, G. (1954). *Acta Path. Micro. Immuno. Scand.* 34:482-492.

538. Mashkoor, A. M., S. K. Saxena, and A. M. Khan (1977). *Acta Bot. Ind.* 5:33-39.

539. Khan, M. W., M. M. Alam, A. M. Khan, and S. K. Saxena (1974). *Acta Bot. Ind.* 2:120-128.

540. Mohammad, H. Y., S. I. Husain, and J. Al-Zarari (1981). *Acta Bot. Ind.* 9:198-200.

541. Khan, M. W., A. M. Khan, and S. K. Saxena (1973). *Acta Bot. Ind.* 1:49-54.

542. Husain, S. I., and A. Masood (1975). *Acta Bot. Ind.* 3:142-146.

543. Saxena, B. P., K. Tikku, and O. Koul (1977). *Acta Ento. Bohemoslov* 74:381-387.

544. Slama, K. (1978). *Acta Ento. Bohemoslov* 75:65-82.

545. Nemec, V., T. T. Chen, and G. R. Wyatt (1978). *Acta Ento. Bohemoslov* 75:285-286.

546. Mulkern, G. B., and D.R. Toczek (1970). Ann. Ento. Soc. Amer. 63:272-284.

547. All, J. N., D. M. Benjamin, and F. Matsumura (1975). Ann. Ento. Soc. Amer. 68:1095-1101.

548. Hsiao, T. H., and G. Fraenkel (1968). Ann. Ento. Soc. Amer. 61: 485-493.

549. Krukoff, B. A., and A. C. Smith (1937). Amer. J. Bot. 24:573-587.

550. Walker, J. C., S. Morell, and H. H. Foster (1937). Amer. J. Bot. 24:536-541.

551. Davis, E. F. (1928). Amer. J. Bot. 15:620.

552. Greathouse, G. A., and G. M. Watkins (1938). Amer. J. Bot. 25: 743-748.

553. Gray, R., and J. Bonner (1948). Amer. J. Bot. 35:52-57.

554. Bennett, E. L., and J. Bonner (1953). Amer. J. Bot. 40:29-33.

555. Muller, C. H. (1953). Amer. J. Bot. 40:53-60.

556. Muller, W. H., and C.H. Muller (1956). Amer. J. Bot. 43:354-361.

557. Verma, H. N., L. P. Awasthi, and K. Mukerjee (1979). Z. Pflanzenkrankheiten 86: 735-740.

558. Bowers, W. S. (1981). Amer. Zool. 21:737-742.

559. Garb, S. (1961). Bot. Rev. 27:422-443.

560. Kim, M., H. Koh, T. Ichikawa, H. Fukami, and S. Ishu (1975). Appl. Ento. Zool. 10:116-122.

561. Kim, M., H. Koh, T. Obata, H. Fukami, and S. Ishu. Appl. Ento. Zool. 11:53-57.

562. Little, J. E., and D. B. Johnstone (1951). Arch. Biochem. Biophy. 30:445-452.

563. Fontaine, T. D., G. W. Irving, and S. P. Doolittle (1947). Arch. Biochem. Biophy. 12: 395-404.

564. Fontaine, T. D., G. W. Irving, R. Ma, J. B. Poole, and S. P. Doolittle (1948). Arch. Biochem. Biophy. 18:467-475.

565. Salle, A. J., G. J. Jann, and L. G. Wayne (1951). Arch. Biochem. Biophy. 32:121-123.

566. Michener, N. D., N. Snell, and E. F. Jansen (1948). Arch. Biochem. Biophy. 19:199-208.

567. Ma, R., and T. D. Fontaine (1948). Arch. Biochem. Biophy. 16: 399-402.

568. Little, J. E., M. W. Foote, and D. B. Johnstone (1950). Arch. Biochem. Biophy. 27: 247-254.

569. Evenari, M. (1949). Bot. Rev. 15:153-194.

570. Banerjee, R. D., and S. P. Sen (1979). Bryologist 82:141-153.

571. LePelley, R. H. (1933). Bull. Ento. Res. 24:1-32.

572. Charles, L. J. (1954). Bull. Ento. Res. 45:403-410.

573. Crowe, T. J. (1961). Bull. Ento. Res. 52:31-41.

574. Wilson, R. E., and E. L. Rice (1968). Bull. Torrey Bot. Club 95:432-448.

575. Al-Naib, F. A., and E.L. Rice (1971). Bull. Torrey Bot. Club 98: 75-82.

576. Lodhi, M. A. K., and E. L. Rice (1971). Bull. Torrey Bot. Club 98:83-89.

577. Abdul-Wahab, A. S., and E. L. Rice (1967). Bull. Torrey Bot. Club 94:486-497.

578. Muller, C. H. (1966). Bull. Torrey Bot. Club 93:332-351.

579. Robbins, W. J., F. Kavanagh, and J. D. Thayer (1947). Bull. Torrey Bot. Club 74:287-292.

580. Lucas, E. H., A. Lickfeldt, R. Y. Gottshall, and J. C. Jennings (1951). Bull. Torrey Bot. Club 78:310-321.

581. Harris, H. A. (1949). Bull. Torrey Bot. Club 76:244-254.

582. Parenti, R. L., and E. L. Rice (1969). Bull. Torrey Bot. Club 96:70-78.

583. Muller, C. H., and R. del Moral (1966). Bull. Torrey Bot. Club. 93:130-137.

584. Curtis, J. T., and G. Cottam (1950). Bull. Torrey Bot. Club. 77:187-191.

585. Hayes, L. E. (1946). Bot. Gaz. 108:408-414.

586. Muller, W. H. (1965). Bot. Gaz. 126:195-200.

587. Mergen, F. (1959). Bot. Gaz. 121:32-36.

588. Trial, H., and J. B. Dimond (1979). Can. Ento. 111:207-212.

589. Srivastava, K. M., S. Chandra, B. P. Singh, and S. M. H. Abidi (1976). Ind. J. Exp. Biol. 14:377-378.

590. Verma, H. N., and L. P. Awasthi (1980). Can. J. Bot. 58:2141-2144.

591. MacDonald, R. E., and C. J. Bishop (1952). Can. J. Bot. 30:486-489.

592. Bishop, C. J., and R. E. MacDonald (1951). Can. J. Bot. 29:260-269.

593. Chinn, S. H. F., and R. J. Ledingham (1957). Can. J. Bot. 35: 697-701.

594. MacDonald, R. E., and C.J. Bishop (1953). Can. J. Bot. 31:123-131.

595. Erickson, J. M., and P. Feeny (1974). Ecology 55:103-111.

596. Bernays, E. A., and R. F. Chapman (1977). Ecol. Ento. 2:1-18.

597. McPherson, J. K., and C. H. Muller (1969). Ecol. Monog. 39:177-198.

598. Rao, D. S. (1955). Ind. J. Ento. 17:121-127.

599. Pandey, N. D., S. R. Singh, and G. C. Tewari (1976). Ind. J. Ento. 38:110-113.

600. Atwal, A. S., and H. R. Pajni (1964). Ind. J. Ento. 26:221-227.

601. Jotwani, M. G., and P. Sircar (1965). Ind. J. Ento. 27:160-164.

602. Paul, C. F., P. N. Agarwal, and A. Ausat (1965). Ind. J. Ento. 27:114-117.

603. Murkerjea, T. D., and R. Govind (1959). Ind. J. Ento. 21:194-205.

604. Jotwani, M. G., and P. Sircar (1967). Ind. J. Ento. 29:21-24.

605. Puttarudriah, M., and K. L. Bhatta (1955). Ind. J. Ento. 17: 165-174.

606. Gupta, K. M. (1973). Ind. J. Ento. 35:276.

607. Pandey, N. D., T. R. Sudhakar, G. C. Tewari, and U. K. Pandey (1979). Ind. J. Ento. 41:107-109.

608. Pandey, U. K., M. Pandey, and S. P. S. Chuahan (1981). Ind. J. Ento. 43:404-407.

609. Subrahmanyam, T. V. (1942). Ind. J. Ento. 4:238.

610. Kumar, A., G. D. Tewari, and N. D. Pandey (1979). Ind. J. Ento. 41:103-106.

611. Bhatia, D. R., and H. L. Sikka (1956). Ind. J. Ento. 18:205-210.

612. Deshmukh, M. G., and S. K. Prasad (1969). Ind. J. Ento. 31:273-276.

613. Mishra, S. D., and S. K. Prasad (1977). _Ind. J. Ento._ 39:228-231.

614. Bhatia, D. R. (1940). _Ind. J. Ento._ 2:187-192.

615. Husain, M. A., and C. B. Mathur (1946). _Ind. J. Ento._ 8:141-163.

616. Desai, M. V., H. M. Shah, and S. N. Pillai (1973). _Ind. J. Nema._ 3:77-78.

617. Alam, M. M., and A. M. Khan (1974). _Ind. J. Nema._ 4:239-240.

618. Fagoonee, I., and G. Umrit (1981). _Insect. Sci. Appl._ 1(4):373-376.

619. Rehr, S. S., P. P. Feeny, and D. H. Janzen (1973). _J. Animal Ecol._ 42:405-416.

620. Cheney, R. H. (1931). _Amer. J. Bot._ 18:136-145.

621. Little, J. E., and K. K. Grubaugh (1946). _J. Bact._ 52:587-591.

622. Cavallito, C. J., and J. H. Bailey (1949). _J. Bact._ 57:207-212.

623. Carlson, H. J., H. G. Douglas, and H. D. Bissell (1948). _J. Bact._ 55:607-614.

624. Carlson, H. J., and H. G. Douglas (1948). _J. Bact._ 55:615-621.

625. Carlson, H. J., H. D. Bissell, and M. G. Mueller (1946). _J. Bact._ 52:155-168.

626. Irving, G. W., T. D. Fontaine, and S. P. Doolittle (1946). _J. Bact._ 52:601-607.

627. Ikawa, M., R. J. Dicke, T. C. Allen, and K. P. Link (1945). _J. Biol. Chem._ 159:517-524.

628. Baer, W., M. Holden, and B. C. Seegal (1946). _J. Biol. Chem._ 162:65-68.

629. Anchel, M. (1949). _J. Biol. Chem._ 177:169-177.

630. Little, J. E., T. J. Sproston, and M. W. Foote (1948). _J. Biol. Chem._ 174:335-342.

631. Russell, G. B., O. R. W. Sutherland, R. F. N. Hutchins, and P. E. Christmas (1978). _J. Chem. Ecol._ 4:571-579.

632. Capinera, J. L., and F. R. Stermitz (1979). _J. Chem. Ecol._ 5:767-771.

633. Sumimoto, M., M. Shiraga, and T. Kondo (1975). _J. Insect. Physiol._ 21:713-722.

634. Butterworth, J. H., and E. D. Morgan (1971). _J. Insect. Physiol._ 17:969-977.

635. Bergamann, E. D., Z. H. Levinson, and R. Mechoulam (1958). _J. Insect. Physiol._ 2:162-177.

636. Feeny, P. P. (1968). _J. Insect. Physiol._ 14:805-817.

637. Seligmann, C. G. (1903). _J. Insect. Physiol._ 29:39-57.

638. McGray, R. J., and E. S. McDonough (1954). _Mycologia_ 46:463-469.

639. Sutherland, O. R. W., N. D. Hood, and J. R. Hillier (1975). _N.Z. J. Zool._ 2:93-100.

640. Sutherland, O. R. W., J. Mann, and J. R. Hillier (1975). _N.Z. J. Zool._ 2:509-512.

641. Sutherland, O. R. W., R. F. N. Hutchins, and W. J. Greenfield (1982). _N.Z. J. Zool._ 9:511-514.

642. Sutherland, O. R. W., and W. J. Greenfield (1978). _N.Z. J. Zool._ 5:173-175.

643. Anderson, J. M., and K. Fisher (1956). _Physiol. Zool._ 29:314-323.

644. Janzen, D. H., H. B. Juster, and E. A. Bell (1977). _Phytochem._ 16:223-227.

645. Morgan, E. D., and M. D. Thornton (1973). _Phytochem._ 12:391-392.

646. Rose, A. F., K. C. Jones, W. F. Haddon, and D. L. Dreyer (1981). _Phytochem._ 20: 224-229.

647. Low, D., B. D. Rawal, and W. J. Griffin (1974). _Planta Medica_ 26:184-189.

648. Bisset, N. G. (1962). _Planta Medica_ 10:143-151.

649. Jain, S. R., and M. R. Jain (1972). _Planta Medica_ 22:136-139.

650. Jain, S. R., and A. Kar (1971). _Planta Medica_ 20:118-123.

651. Madsen, G. C., and A. L. Pates (1952). _Bot. Gaz._ 113:293-300.

652. Pates, A. L., and G. C. Madsen (1954). _Bot. Gaz._ 116:250-261.

653. Salle, A. J., G. J. Jann, and M. Ordanik (1949). _Proc. Soc. Exp. Biol. Med._ 70:409-411.

654. Jensen, K. K. (1948). _Proc. Soc. Exp. Biol. Med._ 66:625-630.

655. Chang, N. C. (1948). _Proc. Soc. Exp. Biol. Med._ 69:141-143.

656. Shimkin, M. B., and H. H. Anderson (1936). *Proc. Soc. Exp. Biol. Med.* 34:135-139.

657. Weld, J. T. (1945). *Proc. Soc. Exp. Biol. Med.* 59:40-41.

658. Southam, C. M. (1946). *Proc. Soc. Exp. Biol. Med.* 61:391-396.

659. McKnight, R. S., and C. C. Lindegren (1936). *Proc. Soc. Exp. Biol. Med.* 35:477-479.

660. Vollrath, R. E., L. Walton, and C. C. Lindegren (1937). *Proc. Soc. Exp. Biol. Med* 36:55-61.

661. Howlett, F. M. (1912). *Roy. Ento. Soc. Trans. (Lond.)* 60:412-418.

662. Belen, E. H. (1982). *Monitor* 10(1):6-7.

663. Evans, P. H., W. S. Bowers, and E. J. Funk (1984). *J. Agr. Food Chem.* 32:1254-1256.

664. Valle, E. (1957). *Acta Chem. Scand.* 11:395.

665. Don, D. (1948). *Acta Chem. Scand.* 2:644.

666. Virtanen, A. I, and P. K. Hietala (1958). *Acta Chem Scand.* 12:579-580.

667. Anonymous, (1966). *Natural pest control agents.* Amer. Chem. Ser. #53, Amer. Chem. Soc. Wash., D.C.

668. del Moral, R., and C. H. Muller (1970). *Amer. Midland Nat.* 83:254-282.

669. Dethier, V. G. (1980). *Amer. Nat.* 115:45-66.

670. Thompson, L. (1880). *Amer. Nat.* 14:48-49.

671. Cook, A. J. (1881). *Amer. Nat.* 15:145-147.

672. Smith, J. R. C., D. J. Dickinson, J. E. King, and R. W. Holt (1968). *Ann. Appl. Biol.* 62:103-111.

673. Martin, J. T. (1940). *Ann. Appl. Biol.* 27:274-275.

674. Tattersfield, F., and O. C. Potter (1940). *Ann. Appl. Biol.* 27:262-273.

675. Ellenby, C. (1951). *Ann. Appl. Biol.* 38:859-875.

676. Tattersfield, F., and C. T. Gimingham (1932). *Ann. Appl. Biol.* 19:253-262.

677. Ellenby, C. (1945). *Ann. Appl. Biol.* 32:67-70.

678. Worsley, G., and R. R. Le (1936). *Ann. Appl. Biol.* 23:311-328.

679. Worsley, G., and R. R. Le (1937). *Ann. Appl. Biol.* 24:659-664.

680. Martin, J. I., E. A. Baker, and R. J. W. Byrde (1966). *Ann. Appl. Biol.* 57:491-500.

681. Worsley, G., and R. R. Le (1934). *Ann. Appl. Biol.* 21:649-669.

682. Ellenby, C. (1945). *Ann. Appl. Biol.* 32:237-239.

683. Russell, G. B., P. G. Fenemore, and P. Singh (1972). *Aust. J. Biol. Sci.* 25:1025-1029.

684. Slama, K., and C. M. Williams (1966). *Biol. Bull.* 130:235-246.

685. Takasugi, M., Y. Yachida, M. Anetai, T. Masamune, and K. Kegasawa (1975). *Chem. Let.* pp. 43-44.

686. Chauvin, R. (1946). *Compt. Rend. Acad. Sci.* 222:412-414.

687. Gupta, S. C., U. M. Khanolkar, O. Koul, and B. P. Saxena (1977). *Curr. Sci.* 46:304-305.

688. Koul, O., K. Tikku, and B. P. Saxena (1977). *Curr. Sci.* 46:724-725.

689. Narasimhan, T. R., M. Ananth, M. N. Swamy, M. R. Babu, A. Mangala, and P. V. S. Rao (1977). *Curr. Sci.* 46:15-16.

690. Prabhu, V. K. K., M. John, and B. Ambikamma (1973). *Curr. Sci.* 42:725-726.

691. Lehle, F. R., and A. R. Putnam (1982). *Plant Physiol.* 69:1212-1216.

692. Abraham, C. C., and B. Ambika (1979). *Curr. Sci.* 48:554-556.

693. Mathur, A. C., J. B. Srivastava, and I. C. Chopra (1961). *Curr. Sci.* 30:223-224.

694. Alam, S. M. (1952). *Curr. Sci.* 21:344.

695. Majumder, S. K., and H. R. Gundurao (1962). *Curr. Sci.* 31:238-239.

696. Shah, A. H., and J. V. Vora (1976). *Curr. Sci.* 45:313-314.

697. Murthy, R. S. R., and D. K. Basu (1981). *Curr. Sci.* 50:64-66.

698. Saxena, B. P., and J. B. Srivastava (1972). *Experientia* 28:112-113.

699. Ascher, K. R. S., N. E. Nemmy, M. Eliyahu, I. Kirson, A. Abraham, and E. Glotter (1980). *Experientia* 36:998-999.

700. Chenevert, R., J. M. Perron, R. Paquin, M. Robitaille, and Y. K. Wang (1980). *Experientia* 36:379-380.

701. Unnithan, G. C., K. K. Nair, and A. Syed (1980). *Experientia* 36:135-136.

702. Bowers, W. S., and J. R. Aldrich (1980). *Experientia* 36:362-364.

703. Landers, M. H., and G. M. Happ (1980). *Experientia* 36:619-620.

704. Sanders, D. W., P. Weatherwax, and L. S. McClung (1942). *J. Bact.* 49:611-615.

705. Unnithan, G. C., K. K. Nair, and C. J. Kooman (1978). *Experientia* 34:411-412.

706. Bernays, E., and C. de Luca (1981). *Experientia* 37:1289-1290.

707. Saxena, B. P., and A. C. Mathur (1976). *Experientia* 32:315-316.

708. Pakrashi, A., B. Chakrabarty, and A. Dasgupta (1976). *Experientia* 32:394-395.

709. Tripathi, R. D., H. S. Srivastava, and S. N. Dixit (1978). *Experientia* 34:51-52.

710. Koul, O., K. Tikku, and B. P. Saxena (1977). *Experientia* 33:29-34.

711. Dosa, A. (1950). *Experientia* 6:18-19.

712. Prabhu, V. K. K., and M. John (1975). *Experientia* 31:913.

713. Kwazu, K., S. Nakajima, and M. Ariwa (1979). *Experientia* 35:1294-1295.

714. Saxena, B. P., and E. B. Rohdendorf (1974). *Experientia* 30:1298-1300.

715. Akeson, W. R., F. A. Haskins, and H. J. Gorz (1969). *Science* 163:293-294.

716. Williams, W. G., G. G. Kennedy, R. T. Yamomoto, J. D. Thacker, and J. Bordner (1980). *Science* 207:888.

717. Casida, J. E. (1964). *Science* 146:1011-1017.

718. Mai, W. F., and L. C. Peterson (1952). *Science* 116:224-225.

719. Smissman, E. E., S. D. Beck, and M. R. Boots (1961). *Science* 133:462.

720. Major, R. T. (1967). *Science* 157:1270-1273.

721. Bowers, W. S., H. M. Fales, M. J. Thompson, and E. C. Uebel (1966). *Science* 154:1020-1021.

722. Eisner, T. (1964). *Science* 146:1318-1320.

723. Bowers, W. S., and R. Nishida (1980). *Science* 209:1030-1032.

724. Rosenthal, G. A., D. L. Dahlman, and D. H. Janzen (1978). Science 202:528-529.

725. Meinwald, J., G. D. Prestwich, K. Nakanishi, and I. Kubo (1978). Science 199:1167-1173.

726. Anonymous (1981). Science 212: 430.

727. Seegal, B. C., and M. Holden (1945). Science 101:413-414.

728. Goth, A. (1945). Science 101:383.

729. Lucas, E. H., and R. W. Lewis (1944). Science 100:597-599.

730. McCleary, J. A., P. S. Sypherd, and D. L. Walkington (1960). Science 131:108.

731. Herz, W., A. L. Pates, and G. C. Madsen (1951). Science 114:206.

732. Carlson, H. J., and H. G. Douglas (1948). J. Bact. 55:235-240.

733. Muller, C. H., W. H. Muller, and B. L. Haines (1964). Science 143:471-473.

734. Irving, G. W., T. D. Fontaine, and S. P. Doolittle (1945). Science 102:9-11.

735. Gaw, H. Z. and H. P. Wang (1949). Science 110:11-12.

736. Gerretsen, F. C., and N. Haagsma (1951). Nature 168:659.

737. Erdtman, H., and J. Gripenberg (1948). Nature 161:719.

738. Spencer, D. M., J. H. Topps, and R. L. Wain (1957). Nature 179: 651-652.

739. Topps, J. H., and R. L. Wain (1957). Nature 179:652-653.

740. Chen, S. L., B. L. Cheng, W. K. Cheng, and P. S. Tang (1945). Nature 156:234.

741. Rudman, P., E. W. B. da Costa, F. J. Gay, and A. H. Wetherly (1958). Nature 181: 721-722.

742. Bowden, K., A. C. Drysdale, and G. A. Mogey (1965). Nature 206:1359-1361.

743. Brooks, G. T., G. E. Pratt, and R. C. Jennings (1979). Nature 281:570-577.

744. Atkinson, N. (1946). Nature 158:876-877.

745. Rao, R. R., M. George, and K. M. Pandalai (1946). Nature 158: 745-746.

746. Stiven, G. (1952). Nature 170:712-713.

747. McDonough, E. S., L. Bell, and G. Arnold (1950). Nature 166:1034.

748. Osborn, E. M., and J. L. Harper (1951). Nature 167:685-686.

749. Klopping, H. L., and G. J. M. Vander Kerk (1951). Nature 167: 996-997.

750. Ivanovics, G., and S. Horvath (1947). Nature 160:297-298.

751. Gupta, K. C., and I. C. Chopra (1954). Nature 173:1194.

752. Grosjean, J. (1950). Nature 165:853-854.

753. Cruickshank, I. A. M., and D. R. Perrin (1960). Nature 187:799-800.

754. Vanderplank, F.L. (1945). Nature 156:782.

755. Raghunandana Rao, R., S. Srinivasa Rao, S. Natarajan, and P. R. Venkataraman (1946). Nature 157:441.

756. Gill, J. S., and C. T. Lewis (1971). Nature 232:402-403.

757. Ingham, J. L., and R. L. Millar (1973). Nature 242:125-126.

758. Mukherjee, N. (1975). J. Plant Dis. Prot. 83:305-308.

759. Saxena, B. P., O. Koul, K. Tikku, and C. K. Atal (1977). Nature 270:512-513.

760. Martin, H., and E. S. Salmon (1930). Nature 126:58.

761. Cavallito, C. J., and J. H. Bailey (1944). J. Amer. Chem. Soc. 66:1950-1951.

762. Jacobson, M. (1956). J. Amer. Chem. Soc. 78:5084-5087.

763. Kubo, I., I. Miura, and K. Nakanishi (1976). J. Amer. Chem. Soc. 98:6704-6705.

764. Gnadinger, C. B., and C. S. Corl (1930). J. Amer. Chem. Soc. 52:680-684.

765. Cavallito, C. J., J. H. Bailey, and F. K. Kirchner (1945). J. Amer. Chem. Soc. 67:948-950.

766. Jacobson, M. (1951). J. Amer. Chem. Soc. 73:100-103.

767. Jones, H. A. (1953). J. Amer. Chem. Soc. 55:1737-1738.

768. Jones, M. A., and H. K. Plank (1945). J. Amer. Chem. Soc. 67: 2266-2267.

769. Smith, C. R. (1935). J. Amer. Chem. Soc. 57:959-960.

770. Anderson, A. B., and E. C. Sherrard (1933). _J. Amer. Chem. Soc._ 55:3813-3818.

771. Cavallito, C. J., and J. H. Bailey (1946). _J. Amer. Chem. Soc._ 68:489-492.

772. Jacobson, M., and H. L. Haller (1947). _J. Amer. Chem. Soc._ 69: 709-710.

773. Jacobson, M. (1948). _J. Amer. Chem. Soc._ 70:4234-4237.

774. Misra, S. B., and S. N. Dixit (1977). _Geobios_ 4:29-30.

775. Atkinson, N., and H. E. Brice (1955). _Aust. J. Expt. Biol. Med. Sci._ 33:547-554.

776. Atkinson, N. (1956). _Aust. J. Expt. Biol. Med. Sci._ 34:17-26.

777. Atkinson, N., and K. M. Rainsford (1946). _Aust. J. Expt. Biol. Med. Sci._ 24:49-51.

778. Misra, S. B., and S. N. Dixit (1977). _Geobios_ 4:176.

779. Nakajima, S., and K. Kawazu (1978). _Heterocycles_ 10:117-121.

780. Adityachaudhury, N., and D. Ghosh (1969). _Ind. Chem. Soc. J._ 46:95.

781. Tiwari, B. K., V. N. Bajpai, and P. N. Agarwal (1965). _Ind. J. Exp. Biol._ 4:128-129.

782. Desai, V. B., and M. Sirsi (1966). _Ind. J. Exp. Biol._ 4:164-166.

783. Murthy, N. B. K., and S. V. Amonkar (1974). _Ind. J. Exp. Biol._ 12:208-209.

784. Visweswariah, K., and M. Jayaram (1971). _Ind. J. Exp. Biol._ 9: 519-521.

785. Atal, C. K., J. B. Srivastava, B. K. Wali, R. B. Chakravarty, B. N. Dhawan, and R.P. Rastogi (1978). _Ind. J. Exp. Biol._ 16:330-349.

786. Gupta, S., and A. B. Banerjee (1970). _Ind. J. Exp. Biol._ 8:148-149.

787. Saxena, B. P., and J. B. Srivastava (1972). _Ind. J. Exp. Biol._ 11:56-58.

788. Saxena, B. P., and J. B. Srivastava (1971). _Ind. J. Exp. Biol._ 10:391-393.

789. Chakraborty, D. P., A. C. Nandy, and M. T. Philipose (1971). _Ind. J. Exp. Biol._ 10:78-80.

790. Bhakuni, D. S., M. L. Dhar, M. M. Dhar, B. N. Dhawan, B. Gupta, and R. C. Srima (1971). _Ind. J. Exp. Biol._ 9:91-102.

791. Saxena, B. P., O. Koul, K. Tikku, C. K. Atal, O. P. Suri, and K. A. Suri (1978). Ind. J. Exp. Biol. 17:354-360.

792. Srivastava, J. B. (1970). Ind. J. Exp. Biol. 8:224-225.

793. Mashkoor, A., A. Masood, and I. Husain (1975). Ind. J. Exp. Biol. 13:412-414.

794. Verma, H. N., and K. Mukerjee (1975). Ind. J. Exp. Biol. 13: 416-417.

795. Deshpande, R. S., P. R. Adhikary, and H. P. Tipnis (1974). Ind. J. Exp. Biol. 12:574-575.

796. Swarup, G., and R. D. Sharma (1967). Ind. J. Exp. Biol. 5:59.

797. Ghosh, S. B., S. Gupta, and A. K. Chandra (1979). Ind. J. Exp. Biol. 18:174-176.

798. Agarwal, R., M. D. Kharya, and R. Shrivastava (1979). Ind. J. Exp. Biol. 17:1264-1265.

799. Garg, S. K., V. S. Mathur, and R. R. Chaudhury (1978). Ind. J. Exp. Biol. 16:1077-1079.

800. Kuder, R. C. (1947). J. Chem. Educ. 24:418-422.

801. Bhuyan, B. R. (1968). J. Bombay Nat. Hist. 65:236-239.

802. Sukramaniam, T. V. (1949). J. Bom. Nat. Hist. 48:338-341.

803. Gullard, J. M., and G. U. Hopton (1930). J. Chem. Soc. (Lond). pp. 6-14.

804. Bowden, K., and W. J. Ross (1963). J. Chem. Soc. (Lond). pp. 3503-3505.

805. Gokhale, V. G., and B. V. Bhide (1945). J. Ind. Chem. Soc. 22: 250-252.

806. Mookerjee, A. (1940). J. Ind. Chem. Soc. 17:593-600.

807. Khuda, M. Q., A. Mukherjee, and S. K. Ghosh (1939). J. Ind. Chem. Soc. 16:583-588.

808. Siddiqui, S., and R. H. Siddiqui (1931). J. Ind. Chem. Soc. 8: 667-680.

809. Ghosh, S., N. R. Chatterjee, and A. Dutta (1929). J. Ind. Chem. Soc. 6:517-522.

810. Acree, F., M. Jacobson, and H. L. Haller (1945). J. Org. Chem. 10:236-242.

811. Jacobson, M., F. Acree, Jr., and H. L. Haller (1945). J. Org. Chem. 10:449-451.

812. Haller, H. L., E. R. McGovran, L. D. Goodhue, and W. N. Sullivan (1941). J. Org. Chem. 7:183-184.

813. Laforge, F. B., and W. F. Barthel (1944). J. Org. Chem. 9:250-253.

814. Jacobson, M., F. Acree, Jr., and H. L. Haller (1947). J. Org. Chem. 12:731-732.

815. Jacobson, M., F. Acree, Jr., and H. L. Haller (1943). J. Org. Chem. 8:572-574.

816. Doskotch, R. W., E. H. Fairchild, C. T. Huang, J. H. Wilton, M. A. Beno, and G. G. Christoph (1980). J. Org. Chem. 45:1441-1446.

817. Jones, W. A., M. Beroza, and E. D. Becker (1962). J. Org. Chem. 27:3232-3235.

818. Jacobson, M. (1966). J. Org. Chem. 32:1646-1647.

819. Grant, R. (1924). J. Proc. Roy. Soc. N.S. Wales 58:117-127.

820. Penfold, A. R., and R. Grant (1926). J. Proc. Roy. Soc. N.S. Wales 60:167-171.

821. Penfold, A. R., and R. Grant (1925). J. Proc. Roy. Soc. N.S. Wales 59:346-350.

822. Penfold, A. R., and R. Grant (1923). J. Proc. Roy. Soc. N.S. Wales 57:211-215.

823. Penfold, A. R. (1923). J. Proc. Roy. Soc. N.S. Wales 57:80-89.

824. Morrison, F. R. (1931). J. Proc. Roy. Soc. N.S. Wales 65:153-177.

825. Dixit, R. S., S. L. Perti, and P. N. Agarwal (1965). Labdev 3: 273-274.

826. Srivastava, A. S., H. P. Saxena, and D. R. Singh (1965). Labdev 3:138.

827. Nene, Y. L., P. N. Thapliyal, and K. Kumar (1968). Labdev 6: 226-228.

828. Zalkow, L. H., J. T. Baxter, R. J. McClure, Jr., and M. M. Gordon (1980). Lloydia 43: 598-608.

829. El-Naggar, S. F., and R. W. Doskotch (1980). Lloydia 43:524-526.

830. El-Naggar, S. F., and R. W. Doskotch (1980). Lloydia 43:617-631.

831. Chen, C. R., J. L. Beal, R. W. Doskotch, L. A. Mitscher, and G. H. Svoboda (1974). Lloydia 37:493-500.

832. Mitscher, L. A., R. P. Leu, M. S. Bathala, W. Wu, and J. L. Beal (1972). Lloydia 35: 157-166.

833. Mitscher, L. A., W. Wu, R. W. Doskotch, and J. L. Beal (1972). Lloydia 35:167-176.

834. Mitscher, L. A., H. D. H. Showalter, M. T. Shipchandler, R. P. Leu, and J. L. Beal (1972). Lloydia 35:177-180.

835. Bisset, N. G., and A. J. M. Leeuwenberg (1968). Lloydia 31:208-222.

836. Malcolm, S. A., and E. A. Sofowora (1969). Lloydia 32:512-517.

837. Su, K. L., Y. Abul-hajj, and E. J. Staba (1973). Lloydia 36:80-87.

838. Wu, W., J. L. Beal, G. W. Clark, and L. A. Mitscher (1976). Lloydia 39:65-75.

839. Bisset, N. G. (1976). Lloydia 39:87-124.

840. Jacobson, M. (1976). Lloydia 39:412-419.

841. Verma, H. N., and L. P. Awasthi (1978). Can. J. Bot. 57:926-932.

842. Mitscher, L. A., M. S. Bathala, G. W. Clark, and J. L. Beal (1975). Lloydia 38:109-124.

843. Patterson, B. D., S. K. W. Khalil, L. J. Schermeister, and M. S. Quraishi (1975). Lloydia 38:391-403.

844. Al-Shamma, A., and L.A. Mitscher (1979). Lloydia 42:633-647.

845. Hussain, N., M. H. Modan, S. G. Shabbir, and S. A. H. Zaidi (1979). Lloydia 42:525-527.

846. Mitscher, L. A., Y. H. Park, D. Clark, G. W. Clark III, P. D. Hammesfahr, W. Wu, and J. L. Beal (1978). Lloydia 41:145-150.

847. Van Den Berghe, D. A., M. Leven, F. Mertens, and A. J. Vlietinck (1978). Lloydia 41:463-471.

848. Bookye-Yiadom, K., N. I. Y. Fiagbe, and J. S. K. Ayim (1977). Lloydia 40:543-545.

849. Bisset, N. G., and K. H. C. Baser (1977). Lloydia 40:546-560.

850. Maradufu, A. (1978). Lloydia 41:181-183.

851. Mathes, M. C. (1967). Lloydia 30:177-181.

852. Leuschner, K. (1972). Naturwissenschaften 59:217-218.

853. Hikino, H., and T. Takemoto (1972). Naturwissenschaften 59:91-98.

854. Maruzzella, J. C. (1961). Nature 191:518.

855. Nene, Y. L., and P. N. Thapliyal (1965). <u>Naturwissenschaften</u> 52: 89-90.

856. Mathur, A. C., and B. P. Saxena (1975). <u>Naturwissenschaften</u> 62: 576-577.

857. Rizvi, S. J. H., V. Jaiswal, D. Mukerjii, and S. N. Mathur (1980). <u>Naturwissenschaften</u> 67:459-460.

858. Renu, K. M., R. D. Tripathi, and S. N. Dixit (1980). <u>Naturwissenschaften</u> 67:150-151.

859. Nene, Y. L., and K. Kumar (1966). <u>Naturwissenschaften</u> 53:363-364.

860. Lowe, M. D., R. F. Henzell, and H. J. Taylor (1971). <u>N.Z. J. Sci.</u> 14:322-326.

861. Ehsan, A. (1968). <u>Pak. J. Sci. Ind. Res.</u> 11:218.

862. Jilani, G., and M. M. Malik (1973). <u>Pak. J. Sci. Ind. Res.</u> 16: 251-254.

863. Bacon, R. F. (1908). <u>Phil. J. Sci.</u> 3A:41-44.

864. Bacon, R. F. (1906). <u>Phil. J. Sci.</u> 1:1007-1036.

865. Schobl, O., and H. Kusama (1924). <u>Phil. J. Sci.</u> 24:443-445.

866. Sinha, N. P., and K. C. Gulati (1965). <u>Proc. Nat. Acad. Sci. Ind.</u> 35B:338-342.

867. Munakata, K. (1975). <u>Pure Appl. Chem.</u> 42:57-66.

868. Uhlenbroek, J. H., and J. D. Bijloo (1958). <u>Recueil Travaux Chim. Pays Bas</u> 77:1004-1009.

869. Meijer, T. M. (1947). <u>Recueil Travaux Chim. Pays Bas</u> 66:395-400.

870. Meijer, T. M. (1946). <u>Recueil Travaux Chim. Pays Bas</u> 65:835-842.

871. Srivastava, P. D. (1953). <u>Sci. Cult.</u> 19:139-140.

872. Shrivastava, R. (1979). <u>Sci. Cult.</u> 47:39-40.

873. Bhuyan, B. R. (1967). <u>Sci. Cult.</u> 33:82-83.

874. Babu, N. (1965). <u>Sci. Cult.</u> 31:308-310.

875. Singh, R. S., and K. R. Pande (1965). <u>Sci. Cult.</u> 31:534-535.

876. Gundurao, H. R., and S. K. Majumder (1966). <u>Sci. Cult.</u> 32:461-462.

877. Pajni, H. R. (1964). <u>Punjab. Univ. Res. Bull.</u> 15:345-346.

878. Pradhan, S., and M. G. Jotwani (1968). <u>Chem. Age Ind.</u> 19:756-760.

879. Lovell, T. H. (1937). _Food Res. J. Food Sci._ 2:435-438.

880. Fuller, J. E., and E. R. Higgins (1940). _Food Res. J. Food Sci._ 5:503-507.

881. Dhar, M. L., M. M. Dhar, B. N. Dhawan, B. N. Mehrotra, and C. Ray (1968). _Ind. J. Exp. Biol._ 6:232-247.

882. Dhar, M. L., M. M. Dhar, B. N. Dhawan, B. N. Mehrotra, R. C. Srimal, and J. S. Tandon (1973). _Ind. J. Exp. Biol._ 11:43-54.

883. Labaw, G. D., and N. W. Desrosier (1953). _Food Res. J. Food Sci._ 18:186-190.

884. Buckingham, D. E. (1930). _Indust. Eng. Chem._ 22:1133-1134.

885. Huddle, H. B. (1936). _Indust. Eng. Chem._ 28:18-21.

886. Howe, H. E. (1939). _Indust. Eng. Chem._ 32:135-136.

887. Sowder, A. M. (1929). _Indust. Eng. Chem._ 21:981-984.

888. Qadri, S. S. H., and S. B. Hasan (1978). _J. Food Sci. Tech._ 15: 121-123.

889. Bhatnagar-Thomas, P. L., and A. K. Pal (1973). _J. Food Sci. Tech._ 11:153-158.

890. Bhatnagar-Thomas, P. L., and A. K. Pal (1973). _J. Food Sci. Tech._ 11:110-113.

891. Broadbent, J. H., and G. Shone (1963). _J. Sci. Food Agr._ 14: 524-527.

892. Mackie, A., and A. L. Misra (1956). _J. Sci. Food Agr._ 7:203-209.

893. Blackith, R. E. (1952). _J. Sci. Food Agr._ 3:219-224.

894. Coombs, C. W., C. J. Billings, and J. E. Porter (1977). _J. Stored Prod. Res._ 13:53-58.

895. Anonymous (1938). _Sci. Amer._ (July), p. 33.

896. Roark, R. C. (1951). _Soap Sanit. Chem._ 27:125 and 137.

897. Bhargava, A. K., and C. S. Chauhan (1968). _Ind. J. Phar._ 30: 150-151.

898. Kaul, V. K., S. S. Nigam, and K. L. Dhar (1976). _Ind. J. Phar._ 38:21-22.

899. Rao, J. T. (1975). _Ind. J. Phar._ 38:53-54.

900. Goutam, M. P., and R. M. Purohit (1973). _Ind. J. Phar._ 35:118-119.

901. Garg, S. C. (1974). Ind. J. Phar. 36:46-47.

902. Goutam, M. P., and R. M. Purohit (1973). Ind. J. Phar. 36:11-12.

903. Banerjee, A., and S. S. Nigam (1977). Ind. J. Phar. 39:143-145.

904. Okazaki, K., and S. Oshima (1952). Yakugaki Zasshi 73:344-347.

905. Okazaki, K., and S. Oshima (1952). Yakugaki Zasshi 72:588-560.

906. Okazaki, K., and T. Kawaguchi (1952). Yakugaki Zasshi 72:561-564.

907. Okazaki, K., and S. Oshima (1952). Yakugaki Zasshi 72:564-567.

908. Okazaki, K., and S. Oshima (1952). Yakugaki Zasshi 72:131-1135.

909. Farnsworth, N. R. (1966). J. Phar. Sci. 55:225-276.

910. Srivastava, M. C., S. W. Singh, J. P. Tewari, and V. Kant (1967). J. Res. India Med. 2(1):11-15.

911. Chaurasia, S. C., and K. M. Vyas (1976). J. Res. Ind. Med. 12(3):139-142.

912. Kaul, V. K, and S. S. Nigam (1976). J. Res. Ind. Med. 12(3): 132-135.

913. Tarrant, C. A., and E. W. Cupp (1978). Roy. Soc. Trop. Med. Hyg. Trans. 72:666-668.

914. Feldlaufer, M. F., and M. W. Eberle (1980). Roy. Soc. Trop. Med. Hyg. Trans. 74:398-399.

915. Filho, A. M. O., R. Pinchin, C. E. Santos, and W. S. Bowers (1980). Roy. Soc. Trop. Med. Hyg. Trans. 74:545-547.

916. Reeves, E. L., and C. Garcia (1969). Mosq. New. 29:601-607.

917. Shirata, A., and K. Takahashi (1982). Ann. Phytopath. Jap. 48: 141-146.

918. Shirata, A. (1982). Ann. Phytopath. Jap. 48:147-152.

919. Fletcher, R. A., and A. J. Renney (1963). Can. J. Plant Sci. 43:475-481.

920. Nene, Y. L., and P. N. Thapliyal (1966). Ind. Phytopath. 19:26-29.

921. Kelling, C. L., I. A. Schipper, L. J. Schermeister, and J. P. Vacik (1976). Amer. J. Vet. Res. 37:215-218.

922. Singh, R. S., and K. Sitaramaiah (1968). Ind. Phytopath. 20: 349-355.

923. Khan, M. W., A. M. Khan, and S. K. Saxena (1974). <u>Ind. Phyto.</u> 27:480-484.

924. Khan, A. M., S. K. Saxena, and Z. A. Siddiqi (1970). <u>Ind. Phyto.</u> 24:166-169.

925. Goswami, B. K., and G. Swarup (1971). <u>Ind. Phyto.</u> 24:491-494.

926. Misra, S. B., and S. N. Dixit (1976). <u>Ind. Phyto.</u> 29:448.

927. Misra, S. B., and S. N. Dixit (1976). <u>Ind. Phyto.</u> 29:448-449.

928. Marcus, C., and E. P. Lichtenstein (1979). <u>J. Agr. Food Chem.</u> 27:1217-1223.

929. Ivie, G. W., D. A. Witzel, and D. D. Rushing (1975). <u>J. Agr. Food Chem.</u> 23:845-849.

930. Lichtenstein, E. P., T. T. Liang, K. R. Schulz, H. K. Schnoes, and G. T. Carter (1974). <u>J. Agr. Food Chem.</u> 22:658-664.

931. Su, H. C. F., and R. Horvat (1981). <u>J. Agr. Food Chem.</u> 29:115-118.

932. Beroza, M., T. McGovern, L. F. Steiner, and D. H. Miyashita (1964). <u>J. Agr. Food. Chem.</u> 12:158-159.

933. Hooker, W. J., J. C. Walker, and K. P. Link (1945). <u>J. Agr. Res.</u> 70:63-78.

934. Fleming, W. E., and F. E. Baker (1936). <u>J. Agr. Res.</u> 53:197-205.

935. Walker, J. C., C. C. Lindegren, and F. M. Bachmann (1925). <u>J. Agr. Res.</u> 30:175-187.

936. Asenjo, C. F., L. M. Marin, W. Torres, and A. del Campillo (1958). <u>J. Agr. Univ. Puerto Rico</u> 42:185-195.

937. Zalkow, L. H., M. M. Gordon, and N. Lanir (1979). <u>J. Econ. Ento.</u> 72:812-815.

938. Meisner, J., M. Weissenberg, D. Palevitch, and N. Aharonson (1981). <u>J. Econ. Ento.</u> 74:131-135.

939. Gentile, A. G., R. E. Webb, and A. K. Stoner (1969). <u>J. Econ. Ento.</u> 62:834-840.

940. Bhakuni, D. S., M. Bittner, C. Marticorena, M. Silva, and E. Weldt (1974). <u>Lloydia</u> 37: 621-632.

941. Russell, A., and E. A. Kaczka (1944). <u>J. Amer. Chem. Soc.</u> 66: 548-550.

942. Mariappan, V., and R. C. Saxena (1983). <u>J. Econ. Ento.</u> 76:573-576.

943. Dale, D., and K. Saradamma (1981). <u>Pesticides</u> 15(7):21-22.

944. Farnsworth, N. R., L. K. Henry, G. H. Svoboda, R. N. Blomster, M. J. Yates, and K. L. Euler (1966). Lloydia 29:101-122.

945. Fong, H. H. S., N. R. Farnsworth, L. K. Henry, G. H. Svoboda, and M. J. Yates (1972). Lloydia 35:35-48.

946. Farnsworth, N. R., L. K. Henry, G. H. Svoboda, R. N. Blomster, H. H. S. Fong, M. W. Quimby, and M. J. Yates (1968). Lloydia 31:237-248.

947. Nienstaedt, H. (1953). Phytopath. 43:32-38.

948. Gilbert, R. G., J. D. Menzies, and G. E. Griebel (1968). Phytopath. 58:1051.

949. Taijan, A. C. (1960). Phytopath. 50:577.

950. Massey, A. B. (1925). Phytopath. 15:773-784.

951. Patrick, Z. A., T. A. Toussoun, and W. C. Snyder (1963). Phytopath. 53:152-161.

952. Sequeira, L. (1962). Phytopath. 52:976-982.

953. Mankau, R. (1963). Phytopath. 53:881.

954. Southam, C. M., and J. Ehrlich (1943). Phytopath. 33:517-524.

955. Good, J. M., N. A. Minton, and C. A. Jaworski (1965). Phytopath. 55:1026-1030.

956. Keiser, I., E. J. Harris, and D. H. Miyashita (1975). Lloydia 38:141-153.

957. Smale, B. C., R. A. Wilson, and H. L. Keif (1964). Phytopath. 54:748.

958. Jacobson, M. (1975). Lloydia 38:455-472.

959. Singh, R. S., and N. Singh (1970). Phytopath. Z. 69:160-167.

960. Snyder, W. C., M. N. Schroth, and T. Christou (1959). Phytopath. 49:755-756.

961. Miller, P. M., and L. V. Edgington (1962). Plant Dis. Rept. 46:745-747.

962. Singh, R. S., and K. Sitaramaiah (1965). Plant Dis. Rept. 50:668-672.

963. Miller, P. M., and J. F. Ahrens (1969). Plant Dis. Rept. 53:642-646.

964. Abbassy, M. A., A. El-Shazli, and F. El-Gayar (1977). Z. Angew. Ento. 83:317-322.

965. Yadava, R. L. (1971). Z. Angew. Ento. 68:289-294.

966. Rajendran, R., and A. A. Kareem (1977). S. India Hort. 25:14-19.

967. Manresa, M. C. (1924). Phil. Agr. 13:213-214.

968. Kalaw, M. M., and F. M. Sacay (1925). Phil. Agr. 14:421-427.

969. Cochran, K. W., T. Nishikawa, and E. S. Beneke (1967). Antimicro. Agents Chemoth. 1966, pp. 515-520.

970. Imai, K. (1956). Yakugaku Zasshi 76:405-408.

971. Aihara, T., and T. Suzuki (1951). Yakugaku Zasshi 71:1323-1324.

972. Ukita, T., and R. Matsuda (1951). Yakugaku Zasshi 71:1050-1052.

973. Garg, L. C., and C. K. Atal (1963). Ind. J. Phar. 25:422.

974. Burlage, H. M., S. M. Gibson, G. F. McKenna, and A. Taylor (1954). Tex. Rep. Biol. Med. 12:229-235.

975. Taylor, A., G. F. McKenna, H. M. Burlage, and D. M. Stokes (1954). Tex. Rep. Biol. Med. 12:551-557.

976. Verma, H. N., L. P. Awasthi, and K. C. Saxena (1979). Can. J. Bot. 57:1214-1217.

977. Cutting, W., E. Furusawa, S. Furusawa, and Y. K. Woo (1965). Soc. Exp. Biol. Med. Proc. 120:330-333.

978. Gerber, P., J. D. Dutcher, E. V. Adams, and J. H. Sherman (1958). Soc. Exp. Biol. Med. Proc. 99:590-593.

979. Cohen, R. A., L. S. Kucera, and J. Herrmann (1964). Soc. Exp. Biol. Med. Proc. 117:431-434.

980. Kavanagh, F. (1947). Adv. Enzyme 7:461-511.

981. Wolfenbarger, D. O. (1947). Fla. Ento. 29:37-44.

982. Masilungan, V. A., J. Maronon, V. V. Valencia, N. C. Diokno, and P. de Leon (1955). Phil. J. Sci. 84:275-299.

983. Quisumbing, E. (1947). Phil. J. Sci. 77:127-177.

984. Bendz, G. (1956). Physiol. Plant. 9:243-246.

985. Miyamoto, T. (1962). Physiol. Plant. 15:409-412.

986. Rennerfelt, E. (1948). Physiol. Plant. 1:245-254.

987. Accorinti, J. (1964). Phyton. 21(1):95.

988. Muller, C. H., R. B. Hanawalk, and J. K. McPherson (1968). Bull. Torrey Bot. Club 95: 225-231.

989. Muller, C. H. (1965). Bull. Torrey Bot. Club 92:38-45.

990. Asakawa, Y., M. Toyota, T. Takemoto, I. Kubo, and K. Nakanishi (1980). Phytochem. 19:2147-2154.

991. Perrin, D. R., and I. A. M. Cruickshank (1969). Phytochem. 8: 971-978.

992. Singh, R. S., and K. Sitaramaiah (1971). Ind. J. Mycol. Plant Path. 1:20-29.

993. Nadal, N. G. M., C. M. C. Chapel, L. V. Rodriguez, J. R. R. Perazza, and L. T. Vera (1966). Bot. Marina 9:21-26.

994. Burkholder, P. R., L. M. Burkholder, and L. R. Almodovar (1960). Bot. Marina 2:149-155.

995. Almodovar, L. R. (1963). Bot. Marina 6:143-146.

996. Bate-Smith, E. C. (1973). Phytochem. 12:1809-1812.

997. Rose, A. F., B. A. Butt, and T. Jermy (1980). Phytochem. 19: 563-566.

998. Evans, C. S., and E. A. Bell (1979). Phytochem. 18:1807-1810.

999. Hackney, R. W., and O. J. Dickerson (1975). J. Nematol. 7:84-90.

1000. Rohde, R. A. (1960). Helm. Soc. Wash. Proc. 27:121-123.

1001. Roonwal, M. L. (1953). Zool. Soc. Sci. Ind. 5:44-58.

1002. Lundgren, L. (1975). Zool. Scripta 4:253-258.

1003. Sato, Y. (1968). Appl. Ento. Zool. 3:155-162.

1004. Buranday, R. P., and R. S. Raros (1975). Phil. Ento. 2:369-374.

1005. Morallo-Rejesus, B. (1983). Dept. Ento. Univ. Philipp. Personal comm.

1006. Gilbert, B. L., J. E. Baker, and D. M. Norris (1967). J. Insect Physiol. 13:1453.

1007. Chan, B. G., A. C. Waiss, and M. Lukefahr (1978). J. Insect Physiol. 24:113-118.

1008. Saxena, V. S. (1979). Ind. J. Ento. 42:780-782.

1009. Singh, R. P., and N. C. Pant (1980). Ind. J. Ento. 42:460-464.

1010. Gour, A. C., and S. K. Prasad (1970). Ind. J. Ento. 32:186-188.

1011. Pandey, N. D., M. Singh, and G. C. Tewari (1977). Ind. J. Ento. 39:60-64.

1012. Chiu, S. (1950). J. Sci. Food Agr. 1:276-286.

1013. Furusawa, E., S. Ramanathan, S. Furusawa, and W. Cutting (1971). Proc. Soc. Exp. Biol. Med. 138:790-795.

1014. Furusawa, E., S. Furusawa, M. Kroposki, and W. Cutting (1968). Proc. Soc. Exp. Biol. Med. 128:1196-1199.

1015. Furusawa, E., and W. Cutting (1966). Proc. Soc. Exp. Biol. Med. 122:280-282.

1016. Goulet, N. R., K. W. Cochran, and G. C. Brown (1960). Proc. Soc. Exp. Biol. Med. 103:96-100.

1017. Furusawa, E., S. Ramanathan, S. Furusawa, Y. K. Woo, and W. Cutting (1967). Proc. Soc. Exp. Biol. Med. 125:234-240.

1018. Doskotch, R. W., H. Y. Cheng, T. M. Odell, and L. Girard (1980). J. Chem. Ecol. 6: 845-851.

1019. Dreyer, D. L., J. C. Reese, and K. C. Jones (1980). J. Chem. Ecol. 7:273-284.

1020. Alfaro, R. I., H. D. Pierce, Jr., J. H. Borden, and A. C. Oehlschlager (1981). J. Chem. Ecol. 7:39-48.

1021. Maizel, J. V., H. J. Burkhardt, and H. K. Mitchell (1963). Biochem. 3:424-426.

1022. Sanders, D., P. W. Weatherwax, and L. S. McClung (1945). J. Bact. 49:206.

1023. Allen, M. B., and E. Dawson (1960). J. Bact. 79:459-460.

1024. Welch, A. M. (1962). J. Bact. 83:97-99.

1025. Gomilevsky, V. (1915). Rev. Appl. Ento. 4:58-59.

1026. Schreiber, A. F. (1915). Rev. Appl. Ento. 4:59.

1027. Vostrikov, P. (1915). Rev. Appl. Ento. 3:340.

1028. Medynsky, V. E. (1915). Rev. Appl. Ento. 3:611-612.

1029. Bhatta, K. L., and B. T. Narayanan (1938). Rev. Appl. Ento. 26: 360-361.

1030. Kayumov, S. (1938). Rev. Appl. Ento. 26:249-251.

1031. Nanta (1938). Rev. Appl. Ento. 26:206.

1032. Thiem, H. (1939). Rev. Appl. Ento. 27:297-298.

1033. Goryainovui, A. A., and F. V. Koblova (1936). <u>Rev. Appl. Ento</u>. 24:351-352.

1034. Hurd-Karrer, A. M., and F. W. Poos (1936). <u>Rev. Appl. Ento</u>. 24: 768-769.

1035. Goriainov, A. (1916). <u>Rev. Appl. Ento</u>. 5:24-26.

1036. Nakatani, M., J. C. James, and K. Nakanishi (1981). <u>J. Amer. Chem. Soc</u>. 103:1228-1229.

1037. Cooper, R., P. H. Solomon, I. Kubo, K. Nakanishi, J. N. Shoolery, and J. L. Occolowit (1980). <u>J. Amer. Chem. Soc</u>. 102:7953-7955.

1038. Brotherson, J. D., L. A. Szyska, and W. E. Evenson (1980). <u>Great Basin Nat</u>. 40:229-253.

1039. Rao, P. J., and K. N. Mehrotra (1977). <u>Ind. J. Exp. Biol</u>. 15: 148-150.

1040. Babbar, O. P., B. L. Chowdhury, M. P. Singh, S. K. Khan, and S. Bajpai (1970). <u>Ind. J. Exp. Biol</u>. 8:304-312.

1041. Greig-Smith, R. (1919). <u>Linn. Soc. N.S. Wales</u> 44:72-92.

1042. Melin, E., and T. Wiken (1946). <u>Nature</u> 158:200-201.

1043. Fantes, K. H., and C. F. O'Neill (1964). <u>Nature</u> 203:1048-1050.

1044. Nielsen, E. S. (1955). <u>Nature</u> 176:553.

1045. Stiven, G. (1952). <u>Nature</u> 170:712.

1046. Banerji, A. (1972). <u>Sci. Today</u> (March), p. 21.

1047. Janardhanan, K. K., D. Ganguly, J. N. Baruah, and P. R. Rao (1963). <u>Curr. Sci</u>. 32:226-227.

1048. Dixit, S. N., and S. C. Tripathi (1975). <u>Curr. Sci</u>. 44:279.

1049. Joseph, A. (1967). <u>Curr. Sci</u>. 36:433.

1050. Tripathi, S. C., and S. N. Dixit (1977). <u>Experientia</u> 33:207-209.

1051. Singh, R. P., and N. C. Pant (1980). <u>Experientia</u> 36:552-553.

1052. Fischer, G., S. Gardell, and E. Jorpes (1954). <u>Experientia</u> 10: 329-330.

1053. Kucera, L. S., R. A. Cohen, and E. C. Herrmann (1965). <u>Ann. N.Y. Acad. Sci</u>. 130:474-482.

1054. Walker, J. T., C. H. Specht, and S. Mavrodineau (1967). <u>Plant Dis. Rept</u>. 51:1021-1024.

1055. Takemoto, T., S. Ogawa, and N. Nishimoto (1967). _Yakugaku Zasshi_ 87:325-327.

1056. Green, R. H. (1949). _Soc. Exp. Biol. Med. (Proc.)_ 71:84-85.

1057. Cheo, P. C., and R. C. Lindner (1964). _Virology_ 24:414-425.

1058. Ragetli, H. W. J., and M. Weintraub (1962). _Virology_ 18:232-240.

1059. Thresh, J. M. (1956). _Ann. Appl. Biol._ 44:608-618.

1060. Weintraub, M., and J. D. Gilpatrick (1952). _Can. J. Bot._ 30: 549-557.

1061. Manil, P. (1949). _Compt. Rend. Soc. Biol._ 143:101-105.

1062. Galanti, M., and P. Manil (1954). _Compt. Rend. Soc. Biol._ 148: 1892-1894.

1063. Isogai, A., S. Murakoshi, A. Suzuki, and S. Tamura (1977). _Agr. Biol. Chem._ 41:1779-1784.

1064. Yajima, T., N. Kato, and K. Munakata (1977). _Agr. Biol. Chem._ 41:1263-1268.

1065. Hosozawa, S., N. Kato, and K. Munakata (1974). _Agr. Biol. Chem._ 38:823-826.

1066. Oda, J., N. Ando, Y. Nakajima, and Y. Inouye (1977). _Agr. Biol. Chem._ 41:201-204.

1067. Yajima, T., and K. Munakata (1979). _Agr. Biol. Chem._ 43:1701-1706.

1068. Matsui, K., K. Wada, and K. Munakata (1976). _Agr. Biol. Chem._ 40:1045-1046.

1069. Duggar, B. M. (1925). _Ann. Missouri Bot. Gard._ 12:359-366.

1070. McKeen, C. D. (1956). _Can. J. Bot._ 34:891-903.

1071. Benda, G. T. A. (1956). _Virology_ 2:438-454.

1072. Walker, J. C., and W. H. Sill, Jr. (1952). _Phytopath._ 42:349-352.

1073. Kuntz, J. E., and J. C. Walker (1947). _Phytopath._ 37:561-579.

1074. Johnson, J. (1941). _Phytopath._ 31:679-701.

1075. McKeen, C. D. (1954). _Science_ 120:229.

1076. Collier, W. A., and L. van de Pijl (1950). _Biol. Abstr._ 24:6653.

1077. Pollacci, G., and M. Gallotti (1943). _Chem. Abstr._ 37:2875.

1078. Sokolov, A. G., and F. V. Koblova (1940). _Chem. Abstr._ 34:6001.

1079. Chen, C. (1936). _Chem. Abstr._ 30:225.

1080. Rudat, K. D. (1960). _Chem. Abstr._ 54:19868.

1081. Galachyan, R. M. (1959). _Chem. Abstr._ 53:8229.

1082. Rajagopala, R., T. R. Gupta, V. S. Gupta, K. V. Nageswara Rao, and P. L. Narasimha Rao (1963). _Chem. Abstr._ 59:14293.

1083. Chesters, C. G. C., and J. A. Stott (1958). _Chem. Abstr._ 52:505.

1084. Saito, K., and M. Sameshima (1958). _Chem. Abstr._ 52:9299.

1085. Munch, J. C., J. Silver, and E. E. Horn (1929). _USDA Tech. Bull._ #134. 36pp.

1086. Cressman, A. W. (1942). _USDA Tech. Bull._ #801. 15pp.

1087. Sievers, A. F., G. A. Russell, M. S. Locoman, E. D. Flower, C. O. Erlanson, and V. A. Little (1938). _USDA Tech. Bull._ #595. 40pp.

1088. Higbee, E. C. (1948). _USDA Misc. Pub._ #650. 36 pp.

1089. Back, E. A. (1922). _USDA Bull._ #1051. 14pp.

1090. Leach, B. R., and J. P. Johnson (1925). _USDA Bull._ #1332. 17pp.

1091. Sievers, A. F. (1941). _USDA Circ._ #581. 28pp.

1092. Plummer, C. C. (1938). _USDA Circ._ #455. 10pp.

1093. Warthen, J. D. (1979). _USDA ARM-NE-4_ (April).

1094. Warthen, J. D., R. E. Redfern, E. C. Uebel, and G. D. Mills, Jr. (1978). _USDA ARR-NE-1_, (Nov.). 47pp.

1095. Warthen, J. D., E. C. Uebel, S. R. Dutky, W. R. Lusby, and H. Finegold (1978). _USDA ARR-NE-2_, (Nov.). 11pp.

1096. Anonymous. _USDA AR_ 25(12):15.

1097. Schaffer, P. S., W. E. Scott, and T. D. Fontaine. _USDA Yearbook_, 1950-1951. 727pp.

1098. Sievers, A. F., and E. C. Higbee (1943). _USDA: For. Agr. Rept._ #8.

1099. Gnadinger, C. B., L. E. Evans, and C. S. Corl (1933). _Exp. Sta. Col. Bull._ #401. 19 pp.

1100. Rohde, R. A., and W. R. Jenkins (1958). _Exp. Sta. Maryland Bull. A-97._

1101. Frisbey, A., J. M. Roberts, J. C. Jennings, R. Y. Gottshall, and E. H. Lucas (1953). _Exp. Sta. Mich. Quart. Bull._ 35:392-404.

1102. Lucas, E. H., A. Frisbey, R. Y. Gottshall, and J. C. Jennings (1955). Exp. Sta. Mich. Quart. Bull. 37:425-426.

1103. Gottshall, R. Y., E. H. Lucas, A. Frisbey, and S. Geis (1955). Exp. Sta. Mich. Quart. Bull. 38:52-59.

1104. McCrory, S. A. (1942-1943). Exp. Sta. S. Dak. Ann. Rept. (1943-1944), p. 23.

1105. Sproston, T., and J. E. Little (1948). Exp. Sta. Ver. Bull. #543, p. 7.

1106. Brooks, M. G. (1951). Exp. Sta. W.Virg. Bull. #347. 31pp.

1107. Singh, R. S. (1965). FAO Plant Prot. Bull. 13:35-37.

1108. Plank, H. K. (1950). Exp. Sta. Puerto Rico (Mayaquez) Bull. #49. 17pp.

1109. Moore, R. H. (1936). Exp. Sta. Puerto Rico (Mayaquez) Ann. Rept., pp. 72-74.

1110. Moore, R. H. (1938). Exp. Sta. Puerto Rico (Mayaquez) Ann. Rept., pp. 55-59.

1111. Moore, R. H. (1939). Exp. Sta. Puerto Rico (Mayaquez) Ann. Rept., pp. 71-93.

1112. Plank, H. K. (1944). Exp. Sta. Puerto Rico (Mayaquez) Ann. Rept., pp. 15-16.

1113. Funke, G. L. (1943). Blumea 5:281-293.

1114. Starr, T. J., E. F. Deig, K. K. Church, and M. B. Allen (1962). Tex. Rept. Biol. Med. 20:271-278.

1115. Kemp, M. S. (1977). Phytochem. 17:1002.

1116. Virtanen, A. I. (1965). Phytochem. 4:207-208.

1117. Yardeni, D., and M. Evenari (1952). Phyton. 2:11-16.

1118. Jermy, T., B. A. Butt, and L. McDonough (1980). Insect. Sci. Appl. 1:237.

1119. Kucera, L. S., and E. C. Herrmann (1967). Proc. Soc. Exp. Biol. Med. 124:865.

1120. Herrmann, E. C., and L. S. Kucera (1967). Proc. Soc. Exp. Biol. Med. 124:869-874.

1121. Herrmann, E. C., and L. S. Kucera (1967). Proc. Soc. Exp. Biol. Med. 124:874-878.

1122. Maroon, C. J. M., O. S. Opina, and A. B. Molina (1984). _Phil._
Phytopath. 20:27-38.

1123. Siddiqui, M. A., and M. M. Alam (1985). _Neem Newsletter_ 2(1):1-4.

1124. Golob, P., and D. J. Webley (1980). _The Use of Plants and Minerals_
as Traditional Protectants of Stored Products. Tropical Products
Institute, London. 32pp.

1125. Takayama, S., M. Misawa, K. Ko, and T. Misato (1977). _Physiol._
Plant. 41:313-320.

1126. Misawa, M., M. Hayashi, T. Hozumi, K. Kudo, and T. Misata (1975).
Biotech. Bioeng. 17:1335-1347.

1127. Wyatt, S. D., and R. J. Shepherd (1969). _Phytopath._ 59:1787-1794.

1128. El-Kandelgy, S. M., and R. D. Wilcoxon (1966). _Phytopath._ 56:
832-837.

1129. Francki, R. I. B. (1964). _Virology_ 24:193-199.

1130. Van Kammen, A., D. Noordam, and T. H. Thung (1961). _Virology_
14:100-108.

1131. Apablaza, G. E., and C. C. Bernier (1972). _Can. J. Bot._ 50:
1473-1478.

1132. Gendron, Y., and B. Kassanis (1954). _Ann. Appl. Biol._ 41:183-188.

1133. Reznik, P. A., and Y. G. Imbs (1965). _Zool. Zhur._ 44:1861-1864.

1134. Nielsen, J. K. (1978). _Ento. Exp. Appl._ 24:41-54.

1135. Neogi, N. C., P. A. Baliga, and R. K. Srivastava (1963). _Ind. Med._
Assoc. J. 41:435-439.

1136. Blaszczak, W., A. F. Ross, and R. H. Larson (1959). _Phytopath._
49:784-791.

1137. Thomas, C. A., and E. H. Allen (1970). _Phytopath._ 60:261.

1138. Abivardi, C. (1971). _Phytopath. Z._ 71:300-308.

1139. Jacobson, M., D. K. Reed, M. M. Crystal, D. S. Moreno, and E. L.
Soderstrom (1978). _Ento. Exp. Appl._ 24:448-457.

1140. Tumbleson, M. E. (1960). _N. Central Weed Contr. Conf. 1960._ 106pp.

1141. Yokum, H. C., M. W. Jutras, and R. A. Peters (1961). _N. East Weed_
Contr. Conf. Proc. 15:341-349.

1142. Peters, R. A., and H. C. Yokum (1961). _N. East Weed Contr. Conf._
Proc. 15:350-354.

1143. Wilson, R. T., and W. G. Mariam (1979). Econ. Bot. 33:29-34.

1144. Fery, R. L., and F. P. Cuthbert, Jr. (1975). Amer. Soc. Hort. Sci. Proc. 100:276-279.

1145. Chou, F. Y., K. Hostettman, I. Kubo, and K. Nakanishi (1977). Heterocycles 7:969-975.

1146. Ohigashi, H., H. Katsumata, K. Kawazu, K. Koshimizu, and T. Mitsui (1974). Agr. Biol. Chem. 38:1093-1095.

1147. Sakata, K., K. Kawazu, and T. Mitsui (1971). Agr. Biol. Chem. 35:2113-2126.

1148. Khanna, P., S. Mohan, and T. N. Nag (1971). Lloydia 34:168-169.

1149. Gupta, S. K., and A. B. Banerjee (1976). Lloydia 39:218-222.

1150. Rother, A., and A. E. Schwarting (1975). Lloydia 38:477-488.

1151. Mitscher, L. A., Y. H. Park, and D. Clark (1980). Lloydia 43:256-269.

1152. Elliger, C. A., D. F. Zinkel, B. G. Chan, and A. C. Waiss (1976). Experientia 32:1364-1366.

1153. da Costa, C. P., and C. M. Jones (1971). Science 172:1145-1146.

1154. Muller, W. H., and C. H. Muller (1964). Bull. Torrey Bot. Club 91:327-330.

1155. McCleary, J. A., and D. L. Walkington (1964). Bull. Torrey Bot. Club 91:361-369.

1156. Nielson, J. K., L. M. Larsen, and H. Sorensen (1977). Phytochem. 16:1519-1522.

1157. Vichkanova, S. A., V. U. Adgina, and S. B. Izosimova (1973). Chem. Abstr. 78:67456g.

1158. Hanriot, M. (1907). Acad. Sci. Comptes Rendus 144:498-500.

1159. Orekhoff, A. (1929). Acad. Sci. Comptes Rendus 189:945.

1160. Rao, R. R. (1949). Ind. J. Med. Res. 37:159-167.

1161. Turkheim, H. J. (1949). J. Dental Res. 28:677-678.

1162. Haag, H. B. (1931). J. Phar. Exp. Therap. 43:193-208.

1163. Larson, P. S., J. K. Finnegan, and H. B. Haag (1949). J. Phar. Exp. Therap. 95:506-508.

1164. Kuna, S., and R. E. Heal (1948). J. Phar. Exp. Therap. 93:407-413.

1165. Dreyer, D. L., J. C. Reese, and K. C. Jones (1981). J. Chem. Ecol. 7:273-284.

1166. Carpenter, C. W. (1945). Hawaiian Planters Record 49:41-67.

1167. Lehmann, F. A. (1930). Chem. Abstr. 24:5066.

1168. Walton, L., M. Herbold, and C. C. Lindegren (1936). Food Res. 1:163-169.

1169. Cooper-Driver, G. A., and T. Swain (1976). Nature 260:604.

1170. Hendrix, S. D. (1977). Oecologia 26:347-361.

1171. Carlson, H. J., H. G. Douglas, and J. Robertson (1948). J. Bact. 55:241-248.

1172. McCahon, C. B., R. G. Kelsey, R. P. Sheridan, and F. Shafizadeh (1973). Bull. Torrey Bot. Club 100:23-28.

1173. Muller, W. H., and R. Hauge (1967). Bull. Torrey Bot. Club 94: 182-191.

1174. Schlatterer, E. F., and E. W. Tisdale (1969). Ecology 50:869-873.

1175. Todd, G. W., A. Getahun, and D. C. Cress (1971). Ann. Ento. Soc. Amer. 64:718-722.

1176. Sequeira, L., R. J. Hemingway, and S. M. Kupchan (1968). Science 161:789-790.

1177. Hedin, P. A., editor (1977). Host Plant Resistance to Pests, ACS Symposium Series 62, Amer. Chem. Soc., Wash., D.C. 286pp.

1178. Parker, V. T., and C. H. Muller (1979). Oecologia 37:315-320.

1179. Ballester, A., J. M. Albo, and E. Vieitez (1977). Oecologia 30: 55-61.

1180. Moral, R. D., and R. G. Cates (1971). Ecology 52:1030-1037.

1181. Grant, R. E., and E. E. C. Clebsch (1975). Ecology 56:604-615.

1182. Vivrette, N. J., and C. H. Muller (1977). Ecol. Monog. 47:301-318.

1183. Gliessman, S. R. (1978). Trop. Ecol. 19:200-208.

1184. Rao, J. V. S., K. R. M. Rao, and S. S. Murthy (1979). Trop. Ecol. 20:5-8.

1185. Sarma, K. K. V. (1974). Trop. Ecol. 15:156-157.

1186. Sarma, K. K. V., G. S. Giri, and K. Subrahmanyan (1967). Trop. Ecol. 17:76-77.

1187. Colton, C. E., and F. A. Einhellig (1980). Amer. J. Bot. 67: 1407-1413.

1188. Rice, E. L. (1972). Amer. J. Bot. 59:752-755.

1189. Ballester, A., A. M. Vieitez, and E. Vieitez (1979). Bot. Gaz. 140:433-436.

1190. Groner, M. G. (1975). Bot. Gaz. 136:207-211.

1191. Friedman, T., and M. Horowitz (1970). Weed Res. 10: 382-385.

1192. Gressel, J. B., and L. G. Holm (1964). Weed Res. 4:44-53.

1193. Welbank, P. J. (1963). Weed Res. 3:205-214.

1194. Toai, T. V., and D. L. Linscott (1979). N. East Weed Sci. Soc. Proc. 33:331.

1195. Horowitz, M., and T. Friedman (1971). Weed Res. 11:88-93.

1196. Lovett, J. V., J. Levitt, A. M. Duffield, and N. G. Smith (1981). Weed Res. 21:165-170.

1197. Anonymous (1981). Pesticides 15(3):40-41.

1198. Palmiter, D. H. (1959). Phytopath. 49:228.

1199. Zaitlin, M., and A. Siegel (1963). Phytopath. 53:224-227.

1200. Schneiderhan, F. J. (1927). Phytopath. 17: 529-540.

1201. Norby, R. J., and T. Kozlowski (1980). Plant Soil 57:363-374.

1202. Young, C. C., and D. P. Bartholomew (1981). Crop Sci. 21:770-774.

1203. Akhtar, N., H. H. Naqvi, and F. Hussain (1976). Pak. J. For. 28:194-200.

1204. Larson, M. M., and E. L. Schwarz (1980). For. Sci. 26:511-520.

1205. Gabriel, W. J. (1975). J. For. 73:234-237.

1206. Ashraf, N., and D. N. Sen (1980). Ind. J. Weed Sci. 12:69-74.

1207. Tiedemann, C. (1978). Amer. Hort. 57(6):8-9.

1208. Sarma, K. K. V. (1974). Geobios 1:137.

1209. Staba, E. J., and P. Khanna (1968). Lloydia 31:180-189.

1210. Pandya, S. M. (1977). Sci. Cult. 43(8):343-344.

1211. Sugha, S. K. (1978). Sci. Cult. 44:461-462.

1212. Singh, S. C., and D. N. Sen (1982). Curr. Sci. 51:45-46.

1213. Tinnin, R. O., and C. H. Muller (1971). Bull. Torrey Bot. Club. 98:243-250.

1214. Lodhi, M. A. K., and G. L. Nickell (1973). Bull. Torrey Bot. Club. 100:159-165.

1215. Bonasera, J., J. Lynch, and M. A. Leck (1979). Bull. Torrey Bot. Club. 106:217-222.

1216. Jobidon, R., and J. R. Thibault (1981). Bull. Torrey Bot. Club. 108:413-418.

1217. Saiki, H., and K. Yoneda (1981). J. Chem. Ecol. 8:185-193.

1218. Lodhi, M. A. K. (1979). J. Chem. Ecol. 5:429-437.

1219. Lodhi, M. A. K. (1975). J. Chem. Ecol. 1:171-182.

1220. Stewart, R. E. (1974). J. Chem. Ecol. 1:161-169.

1221. Anaya, A. L., and S. del Amo (1978). J. Chem. Ecol. 4:289-304.

1222. Anderson, R. C., A. J. Katz and M. R. Anderson (1978). J. Chem. Ecol. 4:9-16.

1223. de la Parra, M. G., A. L. Anaya, F. Espinosa, M. Jimenez, and R. Castillo (1981). J. Chem. Ecol. 7:509-515.

1224. Alsaadawi, I. S., E. L. Rice, and K. B. Karns (1983). J. Chem. Ecol. 9:761-774.

1225. Alsaadawi, I. S., and E. L. Rice (1982). J. Chem. Ecol. 8:1011-1023.

1226. Alsaadawi, I. S., and E. L. Rice (1982). J. Chem. Ecol. 8:993-1009.

1227. Chou, C. H., and Z. A. Patrick (1976). J. Chem. Ecol. 2:369-387.

1228. Leather, G. R. (1983). J. Chem. Ecol. 9:983-989.

1229. Putnam, A. R., J. Defrank, and J. P. Barnes (1983). J. Chem. Ecol. 9:1001-1010.

1230. Liebl, R. A., and A. D. Worsham (1983). J. Chem. Ecol. 9:1027-1043.

1231. Barnes, J. P., and A. R. Putnam (1983). J. Chem. Ecol. 9:1045-1057.

1232. Miller, D. A. (1983). J. Chem. Ecol. 9:1059-1072.

1233. Dalrymple, R. L., and J. L. Rogers (1983). J. Chem. Ecol. 9:1073-1078.

1234. Friedman, J., and G. R. Waller (1983). J. Chem. Ecol. 9:1107-1117.

1235. Kil, B. S., and Y. J. Yim (1983). J. Chem. Ecol. 9:1135-1151.

1236. Rietveld, W. J., R. C. Schlesinger, and K. J. Kessler (1983). J. Chem. Ecol. 9:1119-1133.

1237. Lehle, F. R., and A. R. Putnam (1983). J. Chem. Ecol. 9:1223-1234.

1238. Schumacher, W. J., D. C. Thill, and G. A. Lee (1983). J. Chem. Ecol. 9:1235-1245.

1239. Granato, T. C., W. L. Banwart, P. M. Porter, and J. J. Hassett (1983). J. Chem. Ecol. 9:1281-1292.

1240. Tingle, F. C., and E. R. Mitchell (1984). J. Chem. Ecol. 10: 101-113.

1241. Tang, C., and C. Young (1982). Plant Physiol. 69:155-160.

1242. Khan, I. M. (1982). Physiol. Plant. 54:323-328.

1243. Metzger, F. W., and D. H. Grant (1932). USDA Tech. Bull. #299.

1244. Naqvi, H. H., and C. H. Muller (1975). Pak. J. Bot. 7:139-147.

1245. Junttila, O. (1975). Physiol. Plant. 33:22-27.

1246. McPherson, J. K. (1971). Phytochem. 10:2925.

1247. Chou, C. (1980). Comp. Physiol. Ecol. 5:222-234.

1248. Datta, S. C., and A. K. Chatterjee (1980). Comp. Physiol. Ecol. 5:54-59.

1249. Kawazu, K., Y. Nishii, and S. Nakajima (1980). Agr. Biol. Chem. 44:903-906.

1250. Tamura, S., C. F. Chang, A. Suzuki, and S. Kumai (1969). Agr. Biol. Chem. 33:391-397.

1251. Reid, C. P. P., and W. Hurtt (1970). Nature 225:291.

1252. Bottger, G. T., and M. Jacobson (1950). USDA E-796. 35pp.

1253. Tsuji, H., Y. Tani, and H. Ueda (1977). Agr. Chem. Soc. Jap. 51:609-615.

1254. LeTourneau, D., G. D. Failes, and H. G. Heggeness (1956). Weed Sci. 4:363-368.

1255. Gabor, W. E., and C. Veatch (1981). Weed Sci. 29:155-159.

1256. Drost, D. C., and J. D. Doll (1980). Weed Sci. 28:229-233.

1257. Frank, P. A., and N. Dechoretz (1980). Weed Sci. 28:499-505.

1258. Steenhagen, D. A., and R. L. Zimdahl (1979). Weed Sci. 27:1-3.

1259. Kommendahl, T., J. B. Kotheimer, and J. V. Bernardini (1959). Weed Sci. 7:1-12.

1260. Carnahan, G., and A. C. Hull, Jr. (1962). Weed Sci. 10:87-90.

1261. Ohman, J. H., and T. Kommedahl (1964). Weed Sci. 12:222-231.

1262. Frisbey, A., R. Y. Gottshall, J. C. Jennings, and E. H. Lucas (1954). Exp. Sta. Mich. Quart. Bull. 36:477-488.

1263. LeTourneau, D., and H. G. Heggenes (1957). Weed Sci. 5:12-19.

1264. Leather, G. R. (1983). Weed Sci. 31:37-42.

1265. Gupta, D. C. (1980). Ind. J. Nemat. 10:96-98.

1266. Attri, B. S. (1975). Ind. J. Ento. 37:417-418.

1267. Attri, B. S., and R. Prasad (1980). Ind. J. Ento. 42:371-374.

1268. Narayanan, C. R., R. P. Singh, and D. D. Sawaikar (1980). Ind. J. Ento. 42:469-472.

1269. Verma, S. K., and R. Prasad (1970). Ind. J. Ento. 32:68-73.

1270. Attri, B. S., and R. P. Singh (1976). Ind. J. Ento. 39:383-384.

1271. Miller, P. M., N. C. Turner, and H. Tomlinson (1973). J. Nemat. 5:173-177.

1272. Butterworth, J. H., and E. D. Morgan (1971). J. Insect Physiol. 17:969-977.

1273. Egunjobi, O. A., and S. O. Afolami (1976). Nematologica 22:125-132.

1274. Endo, B. Y. (1959). Phytopath. 49:417-421.

1275. Birchfield, W., and F. Bistline (1956). Plant Dis. Rept. 40: 398-399.

1276. Roark, R. C. (1931). USDA - Consular Corresp., Bur., Chem. Soils. (June).

1277. Narayanan, C. R., R. P. Singh, and D. D. Sawaikar (1978). Pesticides 12(11):31-32.

1278. Bodhade, S,. N. and M. N. Borle (1979). Pesticides 13(1):37.

1279. Attri, B. S., and R. Prasad (1980). Pestology 4:16-20.

1280. Radwanski, S. (1977). World Crops 29:62-66.

1281. Radwanski, S. (1977). World Crops 29:111-113.

1282. Radwanski, S. (1977). <u>World Crops</u> 29:167-168.

1283. Radwanski, S. (1977). <u>World Crops</u> 29:222-224.

1284. Srivastava, Y. N., and R. K. Bhanotar (1980). <u>Ind. J. For.</u> 3: 353-356.

1285. Noonan, J. C. (1953). <u>Fla. St. Hort. Soc.</u> 66:205-210.

1286. Kanchan, S. D., and Jayachandra (1979). <u>Plant Soil</u> 53: 27-35.

1287. Rai, A., and M. S. Sethi (1972). <u>Ind. J. Animal Sci.</u> 42:1066-1070.

1288. Joshi, B. G., G. Ramaprasad, and S. Sitaramaiah (1978). <u>Ind. Farming</u> 28(9):17-18.

1289. Rao, P. R. M., and P. S. Prakasa Rao (1979). <u>Ind. J. Agr. Sci.</u> 49(11):905-906.

1290. Foter, M. J., and A. M. Golick (1938). <u>J. Food Sci.</u> 3:609-613.

1291. Walton, L., M. Herbold, and C. C. Lindegren (1936). <u>J. Food Sci.</u> 1:163-169.

1292. Foter, M. J. (1940). <u>J. Food Sci.</u> 5:147-152.

1293. Ladd, T. L., M. Jacobson, and C. R. Buriff (1978). <u>J. Econ. Ento.</u> 71:810-813.

1294. Putnam, A. R., and W. B. Duke (1974). <u>Science</u> 185:370-372.

1295. Mesra, S. B., and S. N. Dixit (1977). <u>Geobios</u> 4:129-132.

1296. Khanna, K. K., and S. Chandra (1972). <u>Nat. Acad. Sci. Ind. Proc.</u> 42:300-302.

1297. Nandu, V. D., and V. T. John (1981). <u>IRR Newsletter</u> 6(5):12.

1298. Grainge, M., L. Berger, and S. Ahmed (1984). <u>Curr. Sci.</u> 54(2):90.

1299. Saxena, R. C., and Z. R. Khan (1985). <u>J. Econ. Ento.</u> 78:222-226.

1300. Mukherjee, N., and P. Biswas (1984). <u>Ind. Agr.</u> 28(3):145-151.

1301. Verma, H. N., and S. D. Dwivedi (1984). <u>Physiol. Plant Path.</u> 25: 93-101.

1302. Silva, A. B. (1983). <u>Plant Prot. News</u> 12(1):42-44.

1303. Silva, A. B. (1982). <u>Plant Prot. News</u> 11(2):41-44.

1304. Williams, M. C. (1982). <u>Weeds Today</u>, (Spring). 27pp.

1305. Anonymous (1979). <u>Central Rice Res. Inst. Ann. Rept. 1979.</u> 198pp.

1306. Kassanis, B., and A. Kleczkowski (1948). J. Gen. Microb. 2:143-153.

1307. Varma, J. P. (1973). Ind. Phytopath. 26:713-722.

1308. Verma, H. N., and K. Mukerjee (1979). Ind. Phytopath. 32:95-97.

1309. Saksena, K. N., and G. I. Mink (1969). Phytopath. 59:61-63.

1310. Moraes, W. B. C., E. M. F. Martins, E. de Conti, and A. R. Oliveira (1974). Phytopath. Z. 81:145-152.

1311. Alberghina, A. (1976). Phytopath. Z. 87:17-27.

1312. Taniguchi, T., and T. Goto (1976). Ann. Phytopath. Soc. Jap. 42: 42-45.

1313. Thomson, A. D., and B. A. Peddie (1965). N.Z. J. Agr. Res. 8: 825-831.

1314. Verma, H. N., and V. K. Baranwal (1983). Proc. Ind. Acad. Sci.. (Plant Science) 92: 461-465.

1315. Jones, W. A., and M. Jacobson (1959). Nature 184:1146.

1316. Smookler, M. M. (1971). Ann. Appl. Biol. 69:157-168.

1317. Kahn, R. P., and T. C. Allen, Jr. (1957). Phytopath. 47:515.

1318. de Padua, L. S., G. C. Lugod, and J. V. Pancho (1977). Hndbk. on Philip. Med. Plants, Vol. 1, Univ. Philip. Tech. Bull., Los Banos, Philippines. 64pp.

1319. de Padua, L. S., G. C. Lugod, and J. V. Pancho (1978). Hndbk. on Philip. Med. Plants, Vol. 2, Univ. Philip. Tech. Bull., Los Banos, Philippines. 63pp.

1320. de Padua, L. S., G. C. Lugod, and J. V. Pancho (1981). Hndbk. on Philip. Med. Plants, Vol. 3, Univ. Philip. Tech. Bull., Los Banos, Philippines. 66pp.

1321. de Padua, L. S., and J. V. Pancho (1983). Hndbk. on Philip. Med. Plants, Vol. 4, Univ. Philip. Tech. Bull., Los Banos, Philippines. 68pp.

1322. Heywood, V. H., editor (1978). Flowering Plants of the World, Oxford Univ. Press, London.

1323. Wilkinson, R. E., and H. E. Jaques (1979). How to Know the Weeds, Wm. C. Brown Co. Pub., Dubuque, Iowa. 235pp.

1324. Dittmer, H. J. (1964). Phylogeny and Form in the Plant Kingdom, D. Van Nostrand Co., Ltd., New York. 642pp.

1325. Barnes, C. S., and J. R. Price (1975). Lloydia 38:135-140.

1326. Mariappan, V. (1986). National symposium on insecticidal plants and control of environment pollution, Jan. 9-11. Bharathidasan Univ., Tiruchirapalli, India.

1327. Brahmananda Rao, A. S. (1986). National symposium on insecticidal plants and control of environment pollution, Jan. 9-11. Bharathidasan Univ., Tiruchirapalli, India.

1328. Kathiresan, K., and T.S. Thangam (1986). National symposium on insecticidal plants and control of environment pollution, Jan.9-11. Bharathidasan Univ., Tiruchirapalli, India.

1329. Sartaj, A. T., M. A. Siddiqui, and M. M. Alam (1986). National symposium on insecticidal plants and control of environment pollution, Jan. 9-11. Bharathidasan Univ., Tiruchirapalli, India.

1330. Siddiqui, M. A., and M. M. Alam (1986). National symposium on insecticidal plants and control of environment pollution, Jan. 9-11. Bharathidasan Univ., Tiruchirapalli, India.

1331. Kareem, A. A. (1986). National symposium on insecticidal plants and control of environment pollution, Jan. 9-11, 1986, Bharathidasan Univ., Tiruchirapalli, India.

1332. Tewari, S. N., and A. Prakas (1986). National symposium on insecticidal plants and control of environment pollution, Jan. 9-11. Bharathidasan Univ., Tiruchirapalli, India.

1333. Panda, N., N. C. Patnaik, and K. B. Bhuyan (1986). National symposium on insecticidal plants and control of environment pollution, Jan. 9-11. Bharathidasan Univ., Tiruchirapalli, India.

1334. Jayakumar, M., M. Eyini, and S. Pannirselvam (1986). National symposium on insecticidal plants and control of environment pollution, Jan. 9-11. Bharathidasan Univ., Tiruchirapalli, India.

1335. Vijayalakshmi, K., H. S. Gaur, and B. K. Goswami (1986). Neem Newsletter 2(4):35-42.

1336. Adhikary, S. (1985). Neem Newsletter 2(4):48-49.

1337. Webb, R. E., H. G. Larew, A. M. Wieber, P. W. Fard, and J. D. Warthen (1984). Neem Newsletter 1(4):48-49.

1338. Saxena, R. C., and Z. R. Khan (1984). Neem Newsletter 1(3):25-27.

1339. Saxena, R. C., and Z. R. Khan (1984). Neem Newsletter 1(3):28-29.

1340. Saxena, R. C., Z. R. Khan, and N. B. Bajet (1986). Science (in press).

1341. Schmutterer, H., R. C. Saxena, and J. Von der Heyde (1984). Neem Newsletter 1(3): 31-32.

1342. Amonkar, S. V., and K. Vijayalakshmi (1979). Trans. Brit. Mycol. Soc. 73:350.

1343. Larew, H. (1986). USDA Agr. Res. (April), pp. 13-14.

1344. Stein, U., and M. P. Parrella (1985). Cal. Agr. (July-Aug.), pp. 19-20.

1345. Secoy, D. M., and A. E. Smith (1983). Econ. Bot. 37:28-58.

1346. Ahmed, S., and M. Grainge (1986). Econ. Bot. 40(2):201-209.

1347. Chopra, I. C., B. N. Khajuria, and C. C. Chopra (1957). Antibiotics Chemother. 7: 378-383.

1348. Brown, W. H. (1954). Useful plants of the Philippines, Dept. Agr. and Nat. Resources, Vol. I, Manila.

1349. Brown, W. H. (1954). Useful plants of the Philippines, Dept. Agr. and Nat. Resources, Vol. II, Manila.

1350. Brown, W. H. (1954). Useful plants of the Philippines, Dept. Agr. and Nat. Resources, Vol. III, Manila.

1351. Hendrick, U. P., editor (1972). Sturtevant's Edible Plants of the World, Dover Pub. NY.

1352. Dastur, J. F. (1964). Useful plants of India and Pakistan, D. B. Taraporevala Sons and Co. Ltd., Bombay, India.

1353. Dalziel, J. M. (1937). The useful plants of West tropical Africa, Crown Agents for Overseas Governments and Administrations, London.

1354. Kubo, I., and T. Matsumoto (1985). Bioregulators for pest control, P. A. Hedin, editor. American Chemical Society, Wash. D.C.

1355. Rodriquez, E. (1985). Bioregulators for pest control, P. A. Hedin, editor. American Chemical Society, Wash. D.C.

1356. Miles, D. H., B. L. Hankinson, and S. A. Randle (1985). Bioregulators for pest control, P. A. Hedin, editor. American Chemical Society, Wash. D.C.

1357. Grainge, M., and A. Alvarez (1987). Dept. Plant Path., Univ. of Hawaii. Unpublished paper.

1358. Yang, J. Z., and C. S. Tang (1987). Dept. Phytochem., South China Inst. Botany, Guangzhou, China and Dept. Biochem., Univ. of Hawaii. Submitted for publication.

1359. Bhattacharjee, A. K., and A. K. Das. Econ. Bot. 23:274-276.

1360. Hungund, B., and C. H. Pathak (1971). USDA Forest Serv. Res. Paper, NE-201.

1361. Kapoor, L. D., A. Singh, S. L. Kapoor, and S. N. Srivastava (1969). Lloydia 32(3):297-304.

1362. Swanholm, C. E., H. St. John, and P. J. Scheuer (1960). Pac. Sci. 14:68-74.

1363. Swanholm, C. E., and H. St. John (1959). Pac. Sci. 13:295-305.

1364. Cambie, R. C., B. F. Cain, and S. LaRoche (1961). N.Z. J. Sci. 4:707-714.

1365. Cambie, R. C., and B. F. Cain (1961). N.Z. J. Sci. 4:604-663.

1366. Arthur, H. R., and H. T. Cheung (1960). J. Phar. Pharmacol. 12:567.

1367. Dominguez, X. A., and P. Rojas (1960). Econ. Bot. 14:157-159.

1368. Smolenski, S. J., H. Silinis, and N. R. Farnsworth (1972). Lloydia 35:1-34.

1369. Fong, H. H. S., M. Tronjankova, J. Trojanek, and N. R. Farnsworth. Lloydia 35: 117-149.

1370. Smolenski, S. J., H. Silinis, and N. R. Farnsworth. Lloydia 36: 359-389.

1371. Smolenski, S. J., H. Silinis, and N. R. Farnsworth (1974). Lloydia 37:31-61.

1372. Smolenski, S. J., H. Silinis, and N. R. Farnsworth (1974). Lloydia 37:506-536.

1373. Smolenski, S. J., H. Silinis, and N. R. Farnsworth (1975). Lloydia 38:225-255.

1374. Smolenski, S. J., H. Silinis, and N. R. Farnsworth (1975). Lloydia 38:411-441.

1375. Smolenski, S. J., H. Silinis, and N. R. Farnsworth (1975). Lloydia 38:497-528.

1376. Lynch, B. A., A. D. A. Fay, and C. E. Seafarth (1970). Lloydia 33:285-287.

1377. Saxena, H. O. (1975). Lloydia 38:346-351.

1378. Kapoor, L. D., S. L. Kapoor, S. N. Srivastava, A. Singh, and P. C. Sharma (1971). Lloydia 34:95-102.

1379. Kapoor, L. D., S. N. Srivastava, A. Singh, S. L. Kapoor, and N. C. Shah (1972). Lloydia 35:288-295.

1380. Kapoor, L. D., A. Singh, S. L. Kapoor, and S. N. Srivastava (1975). Lloydia 38:221-223.

1381. Anjaneyulu, B., V. Babu Rao, A. K. Ganguly, T. R. Govindachari, B. S. Joshi, V. N. Kamat, A. H. Madmade, P.A. Mohamed, A. D. Rahimtula, A. K. Saksena, D. S. Vande, and N. Viswanathan (1965). Ind. J. Chem. 3:237-238.

1382. Desai, P. D., A. K. Gangulay, T. R. Govindachari, B. S. Joshi, V. N. Kamat, A. H. Manmade, P. A. Mohamed, S. K. Nagle, R. H. Nayak, A. K. Saksena, S. S. Sathe, and N. Viswanathan (1966). Ind. J. Chem. 4:457-459.

1383. Desai, P. D., M. D. Dutia, A. K. Ganguly, T. R. Govindachari, B. S. Joshi, V. N. Kamat, D. Prakash, D. F. Rane, S. S. Sathe, and N. Viswanathan (1967). Ind. J. Chem. 5:523-524.

1384. Govindachari, T. R., S. J. Jadhav, B. S. Joshi, V. N. Kamat, P. A. Mohamed, P. C. Parthasarathy, S. I. Patankar, D. Prakash, D. F. Rane, and N. Viswanathan (1969). Ind. J. Chem. 7:308-310.

1385. Desai, H. K., D. H. Gawad, B. S. Govindachari, B. S. Joshi, V. N. Kamat, J. D. Modi, P. A. Mohamed, P. C. Parthasarathy, S. J. Patankar, A. R. Sidhaye, and N. Viswanathan (1970). Ind. J. Chem. 8:851-853.

1386. Desai, H. K., D. H. Gawad, T. R. Govindachari, B. S. Joshi, V. N. Kamat, J.D. Modi, P.C. Parthasarthy, S.J. Patankar, A. R. Sidhaye, and N. Viswanathan (1971). Ind. J. Chem. 9:611-613.

1387. Tallent, W. H., V. L. Stromberg, and E. C. Horning (1955). J. Chem. Soc., pp. 6361-6364.

1388. Kiang, A. K., B. Douglas, and F. Morsingh (1961). J. Phar. Pharmacol. 13:98-104.

1389. Amarasingham, R. D., N. G. Bisset, A. H. Millard, and M. C. Woods (1964). Econ. Bot. 18:270-278.

1390. Arthur, H. R. (1954). J. Phar. Pharmacol. 6:66-72.

1391. Earle, F. R., and Q. Jones (1962). Econ. Bot. 16:221-250.

1392. Anonymous (1960). Alkaloid bearing plants and their contained alkaloids, USDA Tech. Bull. #1234.

1393. Dhawan, B. N., G. K. Patnaik, R. P. Rastogi, K. K. Singh, and J. S. Tandon (1977). Ind. J. Exp. Biol. 15:208-219.

1394. Dhar, M. L., B. N. Dhawan, C. R. Prasad, R. P. Rastogi, K. K. Singh, and J. S. Tandon (1974). Ind. J. Exp. Biol. 12:512-523.

1395. Bhakuni, D. S., M. L. Dhar, M. M. Dhar, B. N. Dhawan, and B. N. Mehrotra (1969). Ind. J. Exp. Biol. 7:250-262.

1396. Stachon, W. J., and R. L. Zimdahl (1980). Weed Sci. 28:83-86.

1397. Lockerman, R. H., and A. R. Putnam (1979). Weed Sci. 27:54-57.

1398. Nawrot, J., K. E. Bloszy, J. Harmatha, L. Novotny, B. Drozdz (1986). ACTA Ent. Bohemostov. 83:327-335.